Bacilli in Climate Resilient Agriculture and Bioprospecting

Microorganisms play an important role due to their presence all over the universe, and regulating biogeochemical systems in virtually all of our planet's environments. Among all microbes especially in prokaryotic genera, Bacillus spp. represents one of the most important unmapped pools of biodiversity with immense potential of applications in agriculture, industry and medicine. These Gram-positive spore forming bacteria are some of the most dominant groups that exist in various ecological niches on the earth due to their survivability in adverse environmental conditions by producing endospores. Recognizing the enormous potential of the bacteria in this genus, scientists all over the world have directed significant research towards selection and commercialization of the best organisms that may provide protection of plants from harmful microbes and/or enhance plant growth, produce industrially important enzymes, antibiotics, probiotics and other biochemicals. Innovative approaches are also being explored utilizing *Bacillus* mediated bioremediation of environmental pollutants such as pesticides, explosive wastes, dyes and polycyclic aromatic hydrocarbons. Discovery of insecticidal toxin from *Bacillus thuringiensis* (Bt) revolutionized insect pest management in many economically important crops by developing resistant crop varieties. Agricultural biotechnology that involves a wide range of insecticidal toxin producing genes from Bt has dominated the pesticide management research for last few decades. The emergence of resistant insect pests to Bt-based bioinsecticides and Bt-crops has created new challenges invoking more research on stacking resistant genes and further modifications of Bt toxin chemistries. While Bt transgenic crops are being released, questions on ecological aspects are getting louder provoking more research in this area. Genetics and genomics research on allied species of Bacillus has grown at a bewildering pace in the last decade.

More information about this series at http://www.springer.com/series/16030

Md Tofazzal Islam • M. Mahfuz Rahman
Piyush Pandey • Michael Henry Boehme
Geert Haesaert
Editors

Bacilli and Agrobiotechnology: Phytostimulation and Biocontrol

Volume 2

Springer

Editors
Md Tofazzal Islam
Institute of Biotechnology and Genetic Engineering (IBGE)
Bangabandhu Sheikh Mujibur Rahman Agricultural University
Gazipur, Bangladesh

Piyush Pandey
Department of Microbiology
Assam University
Silchar, Assam, India

Geert Haesaert
Department of Applied Biosciences
Ghent University
Ghent, Belgium

M. Mahfuz Rahman
West Virginia University
Morgantown, WV, USA

Michael Henry Boehme
Faculty of Life Sciences
Humboldt University of Berlin
Berlin, Germany

ISSN 2524-5120 ISSN 2524-5139 (electronic)
Bacilli in Climate Resilient Agriculture and Bioprospecting
ISBN 978-3-030-15177-5 ISBN 978-3-030-15175-1 (eBook)
https://doi.org/10.1007/978-3-030-15175-1

This Springer imprint is published by the registered company Springer Nature Switzerland AG.
The registered company address is: Gewerbestrasse 11, 6330 Cham, Switzerland

Contents

Chapter 1
Management of Fungal Diseases on Cucumber (*Cucumis sativus* L.) and Tomato (*Solanum lycopersicum* L.) Crops in Greenhouses Using *Bacillus subtilis*

Li Ni and Zamir K. Punja

1.1 Introduction

The greenhouse vegetable industry has expanded rapidly in recent decades, and cultivated areas of greenhouse vegetable production currently include over 1.6 million hectares worldwide (Peet and Welles 2005). The most extensive greenhouse vegetable production occurs in Asia, followed by Europe, with the Netherlands having the largest acreage of vegetable production under non-plastic glasshouses. In Canada, the greenhouse vegetable industry has been the largest and fastest-growing sector of Canadian horticulture. The high-technology greenhouses are equipped with computerized environmental control systems, automatic irrigation, heating and ventilation systems, and movable screens for shading or conservation of energy, which can provide optimal temperatures, humidity, nutrient regimes, lighting, and moisture conditions to produce high-quality fresh vegetables (Dickerson 1996). This type of greenhouse production is more cost-intensive compared to crops grown under field conditions because it requires a relatively higher investment in infrastructure for growers. However, greenhouse commodities provide more profits to greenhouse operators because they have always been considered a premium product by consumers. Canada's greenhouses produce many high-value vegetables, such as tomato, cucumber, lettuce, pepper, eggplant, greenhouse beans, and various herbs. Tomato (*Solanum lycopersicum* L.) and cucumber (*Cucumis sativus* L.) are the two most popular crops grown by Canadian greenhouse vegetable growers. In 2016, the total farm gate value of greenhouse vegetables was more than $1.3 billion in Canada. Of these commodities, tomatoes accounted for 41% ($544 million), and cucumbers accounted for 25% ($334 million) (Agriculture and Agri-Food Canada 2017).

L. Ni · Z. K. Punja (✉)
Department of Biological Sciences, Simon Fraser University, Burnaby, BC, Canada
e-mail: punja@sfu.ca

M. T. Islam et al. (eds.), *Bacilli and Agrobiotechnology: Phytostimulation and Biocontrol*, Bacilli in Climate Resilient Agriculture and Bioprospecting,
https://doi.org/10.1007/978-3-030-15175-1_1

Fungal diseases are the primary limiting factor during commercial production of tomatoes and cucumbers in greenhouses. Infection by these pathogens can result in plant death and reduced yields. Many management approaches have been taken to prevent the establishment of diseases and to minimize the development of pathogens in tomato and cucumber crops. These include the use of fungicides, the development of resistant or tolerant cultivars, and use of biological control agents. In recent years, the primary management approaches for diseases are growing resistant cultivars and use of newer fungicide chemistries. However, there are only a few resistant cultivars developed and available on the market, and although applying fungicides can be the most effective control method where available, pathogens can develop fungicide resistance (Brown 2002). More importantly, potential environmental contamination and residues of fungicides on fresh produce are of public concern. Consequently, the controlled environment of greenhouses, the high value of the crops, the limited number of resistant varieties, and the restrictions to fungicide use offer a unique opportunity for the development and application of biological control agents to reduce crop losses due to pathogen infection. A majority of biological control products available on the market have in fact been specifically developed for greenhouse crops (Paulitz and Belanger 2001). Currently, there are several registered commercial biological control products available for use in Canadian greenhouses, including *Bacillus subtilis* QST 713 (Rhapsody® and Serenade®), *Gliocladium catenulatum* strain J1446 (Prestop®), *Streptomyces lydicus* WYEC 108 (Actinovate®), *Streptomyces griseoviridis* K61 (Mycostop®), and *Trichoderma harzianum* strain T-22 (Rootshield®). Among these products, *B. subtilis* has great potential to be a broad and effective biological control agent for management of a number of important foliar, root-infecting, or postharvest fungal diseases since *B. subtilis* has the capacity to produce a broad range of antibiotics and antifungal enzymes (Cawoy et al. 2011). In this chapter, the efficacy of *B. subtilis* against specific diseases of greenhouse-grown cucumber and tomato crops will be discussed.

1.2 *Bacillus subtilis*

1.2.1 Biology and Ecology of B. subtilis

Bacillus subtilis is an extremely diverse bacterial species that is broadly adaptable to many environments. The bacterial cells are able to withstand extremes of environmental conditions due to the transient synthesis of heat shock or cold shock proteins at different temperatures – the bacterium has the capacity for completing its life cycle at temperatures up to 54–55 °C (Danchin 2001). The bacterium grows best at a pH range of 5.5 to 6.5. Temperatures required for optimal antifungal activities of *B. subtilis* against fungal pathogens can differ. For example, antifungal activities of *B. subtilis* were highest at 17 °C for *Fusarium oxysporum*, and higher temperatures impaired the ability of the bacterium to inhibit growth of *F. oxysporum* during in vitro tests (Krebs et al. 1998). In contrast, higher temperatures increased the inhibition zones of *B. subtilis* against *Sclerotinia sclerotiorum* and *Rhizoctonia solani* (Krebs et al. 1998). A reduced suppression of *F. oxysporum* was reported at higher pH value (pH 8.3) of the medium, while there was no relationship between pH value of the media from pH 5.7 to pH 8.3 and the antifungal activities for *R. solani* and *S. sclerotiorum* (Krebs et al. 1998).

B. subtilis has historically been thought of as strict aerobe, but many studies have reported that this Gram-positive bacterium can grow under anaerobic conditions (Nakano and Zuber 1998). It forms endospores when levels of nitrogen, carbon, and phosphorus are below the optimal threshold for growth (Earl et al. 2008). These environmentally resistant and metabolically dormant endospores allow the bacterium to survive under nutrient-deficient and extremes of environmental conditions, including high temperatures, ionizing radiation, unfavorable pH, and soil desiccation (Higgins and Dworkin 2011). The endospores are airborne and can be disseminated by wind (Merrill et al. 2006). They do not germinate until the environmental conditions become favorable for growth and development.

The availability of effective biocontrol formulations, and survival in storage and rapid multiplication and postinoculation colonization, determines the success of biological control agents for plant diseases (Becker and Schwinn 1993). *B. subtilis* has been successfully formulated, and many stable and effective formulations have been developed for application in biological control of plant diseases. Formulations are generally divided into two types: dry products such as dusts, granules, and wettable powders and liquid products such as oils, cell suspensions, and emulsions (Schisler et al. 2004). It has been reported that *B. subtilis* is able to form packed dense surface- or interface-associated multicellular aggregates, generally referred to as biofilms, under unfavorable environmental factors, including oxygen and nutrient deprivation (Morikawa 2006). The bacterium attaches to plant roots through production of an extracellular matrix to form biofilms (Beauregard et al. 2013). Previous research has shown the formation of a biofilm of *B. subtilis* required only a few hours after the settlement of cells on the plant root (Allard-Massicotte et al. 2016). One of the most important benefits of biofilms is to protect cells from environmental damage (Davey and O'Toole 2000). It has been suggested that biofilm formation by *B. subtilis* enhances its ability to act as a biological control agent for the suppression of pathogens infecting plants (Morikawa 2006). Previous research showed that the major components of cell walls (polysaccharides) serve as environmental triggers to the formation of biofilms by *B. subtilis* (Beauregard et al. 2013).

Bacillus subtilis is often considered as "soil dweller" because it is most commonly found in soil environments and can be isolated from the rhizospheres of various plants (Earl et al. 2008). The bacterium has also been isolated from other plant parts, including leaves, stems, and fruits. Bacterial cells can multiply rapidly and colonize root systems, leaf surfaces, and the surrounding soil horizons (Cawoy et al. 2011). It is beneficial for *B. subtilis* colonization of plant roots and the mutualistic symbiosis that develops between the bacterium and the host when there is abundant carbon available for bacterial growth (Allard-Massicotte et al. 2016). Plants support *B. subtilis* populations through secretion of approximately 30% of fixed carbon from root exudates (Allard-Massicotte et al. 2016). It also has been indicated that intact chemotaxis machinery is needed for root colonization by this bacterium at early stages (Allard-Massicotte et al. 2016). Many *B. subtilis* chemoreceptors are involved in root colonization, which plays an important role to colonize the root efficiently in soil (Allard-Massicotte et al. 2016). The presence of bacteria on the roots can enhance plant growth and disease resistance through several mechanisms: competition with other microbes that could cause adverse effects on the plant, induction of host defense responses against pathogens, and increased uptake of certain nutrients, such as nitrogen and phosphorus (Nagorska et al. 2007). These will be described in more detail below.

1.2.2 Competition with Other Microbes

At the root surface, competition with other root-inhabiting microorganisms for space and nutrients released from plant leaves or roots is an important aspect of the mechanism of action of *B. subtilis* against plant pathogens (Cawoy et al. 2011). Carbon plays an important role in fungistasis which is characterized by the suppression of germination of fungal spores in soil (Alabouvette et al. 2006). Competition for carbon by the bacterium is a critical factor for biological control by *B. subtilis* (Cawoy et al. 2011). *B. subtilis* also competes for several trace elements with other soilborne pathogens, including copper, manganese, iron, and zinc. Soil-inhabiting organisms have to compete for iron in particular due to its limited availability in soil (Loper and Henkels 1997). *B. subtilis* produces siderophores that have a high affinity for iron so that they can solubilize and obtain ferric ion efficiently; consequently, other soil microorganisms cannot grow and develop in the absence of iron (Haas and Defago 2005; Loper and Henkels 1997; Cawoy et al. 2011). In addition, it has been reported that *B. subtilis* can dissolve inorganic phosphate during phosphorylation and increase the use rate of phosphate in the soil (Wang et al. 2018).

1.2.3 Production of Inhibitory Chemicals

B. subtilis can produce antibiotic molecules at high levels during sporulation. Approximately 4–5% of the genome is utilized for the synthesis of antibiotic molecules (Stein 2005). More than two dozen antimicrobial compounds with diverse structures are produced by *B. subtilis* (Stein 2005), including polymyxin, subtilin, difficidin, and mycobacillin. The main class of antibiotics produced by *B. subtilis* is the peptides which usually consist of amino acids and other residues (Cawoy et al. 2011). Cyclic lipopeptides are a primary class of peptide antibiotics of *Bacillus*. There are three major families of cyclic lipopeptides (LPs) of *Bacillus* species:

iturins, surfactins, and fengycins. *B. subtilis* cells can produce these cyclic LPs at high rates in vitro and secrete them under environmental conditions in the rhizosphere (Cawoy et al. 2014). Each family of cyclic LPs exhibits particular antibiotic activities. They play different roles in the confrontation between *Bacillus* species and plant pathogens. The families of fengycin and iturin are capable of destroying surface tension of cell membranes of fungi so that micropores occur and essential ions are lost, which consequently triggers fungal cell death (Ongena and Jacques 2008; Mihalache et al. 2017). It appears that iturin is the major component effective against plant pathogens based on the size of inhibition zones on agar media (Cawoy et al. 2014). Experiments have shown that *B. subtilis* strain RB14 produces iturin A that positively influences the suppression of damping-off on tomato plants, caused by *Rhizoctonia solani* (Asaka and Shoda 1996). Not only are soilborne diseases inhibited but foliar diseases are also reduced by the production of cyclic lipopeptides synthesized by *B. subtilis*. For example, it has been demonstrated that the production of both iturins and fengycins in *B. subtilis* contributes to the inhibition of powdery mildew caused by *Podosphaera fusca* on melon leaves (Romero et al. 2007a). These cyclic lipopeptides are capable of inhibiting the germination of conidia of *P. fusca* (Cawoy et al. 2011). As for postharvest diseases, a strain of *B. subtilis* GA1 reduced gray mold caused by *Botrytis cinerea* which results in visible decay on apple fruits through effectively producing three families of cyclic lipopeptides – iturins, surfactins, and fengycins (Cawoy et al. 2011). It has been found that fengycin produced by *B. subtilis* 9407 showed strong antifungal activity and was important to the biocontrol of apple ring rot caused by *Botryosphaeria dothidea* (Fan et al. 2017). Touré et al. (2004) found that fengycins produced by *B. subtilis* GA1 show strong inhibitory effects on gray mold diseases after treating apple fruits with cyclic lipopeptide-enriched extracts and detecting the inhibitory amounts of fengycins. There is a clear connection between the antagonism activities against fungal diseases and the permeabilization of conidia for suppressing spore germination or perturbing hyphal cells, and these two phenomena are most likely caused by membrane injury resulting from the production of cyclic lipopeptides (Cawoy et al. 2011). Surfactin is probably not involved against fungal growth, but it assists the bacterium in colonizing root tissues and establishing in the rhizosphere of plants through the formation of a biofilm layer (Cawoy et al. 2014; Mihalache et al. 2017). Additionally, some biological control organisms may take advantage of other important inhibitory mechanisms such as predation or parasitism by the enzymatic damage of cell walls of the fungal pathogen (Cawoy et al. 2011). Whipps (2001) suggested that biological control agents can produce a variety of enzymes that degrade the cell wall to result in the parasitism and degradation of hyphae or spores of fungal pathogens. *B. subtilis* strain AF1 secretes *N*-acetyl glucosaminidase and glucanase to inhibit the growth of fungi (Manjula and Podile 2005). Chitinase enzyme produced by *B. subtilis* ATCC 11774 was able to inhibit growth of *R. solani* by 43.3% on agar medium (Saber et al. 2015).

1.2.4 Induction of Plant Resistance

The modes of action of some biocontrol bacteria strains are distinct from the mechanisms described above. These strains are able to activate defense systems when host plants are attacked by pathogens, resulting in an increased level of resistance (Conrath et al. 2006). Induced systemic resistance is observed as sequential processes. Firstly, elicitors in plant cells perceive the penetration by fungal pathogens, thereby initiating the induced systemic resistance responses of the host plant through cell signaling throughout the plant (Van Loon 2007). Invasion of pathogens is subsequently reduced by the expression of defense mechanisms (Van Loon 2007). There are several molecules produced for the defense response against pathogens, including pathogenesis-related proteins (chitinases, proteinase inhibitors, β-1,3-glucanases, etc.), phytoalexins, and lignin (Van Loon 2007). The thickened cell wall and dead tissues create a barrier to block off the transport of nutrients to pathogens, preventing the extensive penetration caused by fungal pathogens (Lugtenberg et al. 2002). *Bacillus* spp. can produce volatile compounds and lipopeptides. These compounds are considered as elicitors of induced systemic resistance. Ongena et al. (2007) reported that systemic resistance was induced by surfactin produced by strain *B. subtilis* S499. According to experimental results from tomato and bean, the overexpression of genes of surfactin and fengycin in *B. subtilis* strain 168 was related to the potential increased derivatives that stimulate resistance of the host plant (Cawoy et al. 2011). Siderophores produced by *B. subtilis* adhere to the biofilm surface so that *B. subtilis* can secrete broad-spectrum resistance compounds and improve the spectrum of disease resistance of plants (Hofte and Bakker 2007).

1.3 Control of Cucumber Diseases Using *B. subtilis*

Cucumber (*Cucumis sativus L.*), a member of the Cucurbitaceae family, is a crop that grows best under conditions of warm temperatures (20–25 °C), abundant sunlight, high humidity, and adequate fertilizers. It is sensitive to cold temperatures, which can significantly reduce yields as well as plant growth. Cucumber plants are grown in either greenhouses or under field conditions. Greenhouse cucumber production is an important segment of the Canadian agriculture food industry and is also very popular in many areas of the world, such as China, India, Russia, and the United States. In 2015, greenhouse cucumber was one of the primary greenhouse vegetable crops grown annually in Canada with about 180 million kg of production (Dey et al. 2017). The major greenhouse cucumber production areas are mainly in Ontario (144 million kg), British Columbia (23 million kg), and Alberta (9 million kg) (Dey et al. 2017). All greenhouse cucumbers are produced for fresh market consumption. It has been reported that the farm gate value of greenhouse cucumbers in Canada was $308 million in 2015 (Wang 2016). The most important diseases on cucumber are Pythium crown and root rot (*Pythium aphanidermatum*, *P. irregulare*, *P. sylvaticum*, and *P.*

ultimum), Fusarium root and stem rot (*Fusarium oxysporum* f. sp. *radicis-cucumerinum*), gummy stem blight (*Didymella bryoniae*, anamorph *Phoma cucurbitacearum*), powdery mildew (*Podosphaera xanthii* synonym *Sphaerotheca fusca* or *Podosphaera fusca*), and botrytis gray mold (*Botrytis cinerea*).

1.3.1 Powdery Mildew Control on Cucumber

Powdery mildew on cucurbits is a widespread fungal disease that is caused by *Podosphaera xanthii* and less commonly by *Golovinomyces* (*Erysiphe*) *cichoracearum* (Lebeda 1983; Jahn et al. 2002; Křıstková et al. 2003). Powdery mildew is easy to recognize because it forms colonies of whitish talcum-like powdery growth on the upper and occasionally lower leaf surfaces, petioles, and stems (Fig. 1.1). This disease can impair photosynthesis, resulting in a reduction in plant growth, premature senescence of infected foliage, and sometimes death of leaves (Nunez-Palenius et al. 2006). Powdery mildew is one of the most important limiting factors for cucurbit production in many countries around the world under a wide range of environmental conditions. Under controlled environmental conditions with optimal temperature and humidity conditions, greenhouses provide an ideal place for the development and spread of powdery mildews. Roberts and Kucharek (2005) reported that the incidence of outbreaks of powdery mildew has increased in recent years. Powdery mildews develop rapidly under favorable conditions because a large number of conidia can be produced within a short time and symptoms can appear on leaves after only 3–7 days of infection (McGrath 2017).

In Canada, powdery mildew is the most important foliar disease on cucumber plants grown in greenhouses (Fig. 1.1). Therefore, it is critical to control the disease

Fig. 1.1 Early stages of powdery mildew development on a greenhouse cucumber leaf (**a**) compared to severe infection later in the season (**b**)

in a timely manner to avoid extensive premature defoliation on the lower leaves of plants. Powdery mildew is managed through many different strategies. These include selecting resistant cultivars and applying fungicides, biological control agents, and chemical compounds (Nunez-Palenius et al. 2006). Currently, due to more frequently reported fungicide resistance development in cucurbit powdery mildew fungi (McGrath 2001) and the rising public concerns toward the potential side effects of pesticides on the environment and nontarget species, it is necessary to develop multiple ecologically safer chemicals or other complementary procedures for managing cucurbit powdery mildews (Perez-Garcia et al. 2009).

The use of biocontrol agents is one approach to manage powdery mildew on cucurbits. There are a number of biological control agents that have been shown to suppress powdery mildew development on cucumber, including *Bacillus subtilis* (Serenade®), *Ampelomyces quisqualis* Ces. (AQ10®), *Lecanicillium lecanii* (Mycotal®), and *Sporothrix flocculosa* (Sporodex®). *B. subtilis* strain QST713 (Serenade®) has been previously shown to be an effective biological control product against powdery mildew on squash and pumpkin. As mentioned previously, several lipopeptide antibiotics, such as iturin, fengycin, surfactins, and agrastatin/plipastatins, were reported to prevent infection by powdery mildew fungi on leaves (Romero et al. 2004, 2007c; Nunez-Palenius et al. 2006; Gilardi et al. 2008). Combination of surfactins at 25 ppm or less with the group of agrastatin/plipastatin or iturins significantly suppressed the development of spores and germ tubes (Manker 2005). Moreover, iturins and agrastatins closely cooperate to inhibit the growth of pathogen spores (Gilardi et al. 2008). Matheron and Porchas (2000) indicated that the alternation of *B. subtilis* QST713 with myclobutanil, sulfur, and trifloxystrobin resulted in effective control of powdery mildew on lettuce. In the case study of zucchini, it was shown that *P. xanthii* was better controlled by *B. subtilis* QST713 when applied with azoxystrobin than azoxystrobin alone (Gilardi et al. 2008). *B. subtilis* strain UMAF6639 is another antagonistic strain which protects melon plants against powdery mildew caused by *P. xanthii* by means of induced systemic resistance (ISR) in the host plant (Garcia-Gutierrez et al. 2013). The disease severity was reduced by approximately 50% on melon seedlings treated with *B. subtilis* UMAF6639 after 18-day inoculation of powdery mildew fungi (Garcia-Gutierrez et al. 2013). Experimental results suggested that the ISR response induced by *B. subtilis* UMAF6639 in melon to suppress powdery mildews is dependent on both jasmonate (JA) and salicylic acid (SA) signaling (Garcia-Gutierrez et al. 2013). Cucurbit powdery mildew caused by *P. xanthii* is sensitive to SA-regulated defense responses because the fungi are biotrophic (Glazebrook 2005; Pieterse et al. 2009). Systemic acquired resistance (SAR) inducers, including salicylic acid (SA), benzothiadiazole (BTH), and acibenzolar-S-methyl (ASM), activate SA-dependent signaling pathways to suppress the severity of cucurbit powdery mildew (Salmeron et al. 2002; Lin et al. 2009). Garcia-Gutierrez et al. (2013) indicated that resistance against powdery mildew was improved via the activation of JA signaling pathways. It appears that UMAF6639-bacterized melon plants demonstrated increased cooperative activity of SA and JA signaling pathways which may be associated with the ISR response, allowing the plant to produce more reactive oxygen species and reinforce cell walls which provide efficient defenses against *P. xanthii* (Garcia-Gutierrez et al. 2013). Besides, surfactin is essential to trigger ISR by *B. subtilis* UMAF6639 (Garcia-Gutierrez et al. 2013). Romero et al. (2007b) used butanolic extracts obtained from supernatants of *B. subtilis* (UMAF6614 and UMAF6639) to test the antifungal activity against *P. xanthii* on the cotyledons of zucchini. They observed conidia of powdery mildew failed to germinate due to these lipopeptides having the ability to attack biological membranes of target cells and induce changes in ultrastructure and morphology (Romero et al. 2007b). It also has been indicated that *Bacillus* species can reduce the infection of powdery mildew on foliage by limiting the germination of spores when seedlings, detached foliage, and plants are treated with the bacteria ahead of the time of inoculation of the pathogen, which possibly is due to the production of antifungal compounds by *Bacillus* species (Baruzzi et al. 2011; Cawoy et al. 2011). Viewed under the electron microscope, cells of *Bacillus* species were stuck firmly to the conidia and hyphae of powdery mildew, and also colonization by bacteria on the surface of foliage was observed (Romero et al. 2004). Kim et al. (2013) treated greenhouse-grown cucumber plants with a culture filtrate of *Bacillus* spp. strain BS061, which resulted in a 62.4% reduction in severity of powdery mildew on greenhouse cucumbers. Although *B. subtilis* successfully controlled powdery mildews in many previous studies, it has shortcomings as well. For example, high relative humidity is required for the bacteria to produce and secrete more antifungal compounds for the suppression of powdery mildew (Romero et al. 2007a). Previous studies demonstrated that the effects of the same formulation of *B. subtilis* on inhibition of powdery mildew on cucurbits are not consistent if *B. subtilis* was applied alone (Gilardi et al. 2008). It has been shown that *B. subtilis* was apparently more effective when alternated with the fungicide QoIs (Keinath and Dubose 2004).

In our research, *B. subtilis* QST713 formulated as Rhapsody applied weekly at a rate of 1.0 L product /100 L water, or in some experiments at 1.5 L product/100 L water, to the foliage of cucumber plants under greenhouse conditions, showed both preventative and eradicative effects on the development of powdery mildew (Fig. 1.2). Natural infection by powdery mildew was observed on cucumber plants in the greenhouse. In one trial, plants were sprayed with Rhapsody at the time when initial infection was observed and continued weekly for 3 weeks (preventative). In a second trial, after powdery mildew had established, Rhapsody was applied at weekly intervals for 3 weeks (eradicant). The number of mildew colonies that developed, and the leaf area covered with mildew, was determined for three representative leaves (third, fourth, and fifth from the top down) on each plant after 3 weeks, from control and treated plants. There were 24 plants per treatment in each trial, in 6 groups of 4 replicates. The results show that preventative applications were effective in reducing the development of powdery mildew by 70% following three applications (Figs. 1.2a and 1.3a). When Rhapsody was applied after powdery mildew had been established, there was an eradicative effect (Figs. 1.2b and 1.3b). The percent leaf area infected was reduced significantly after two or three applications (Fig. 1.4) compared to a nontreated or water control. These results demonstrate that *B. subtilis* is an effective biological control agent against powdery mildew on greenhouse cucumbers.

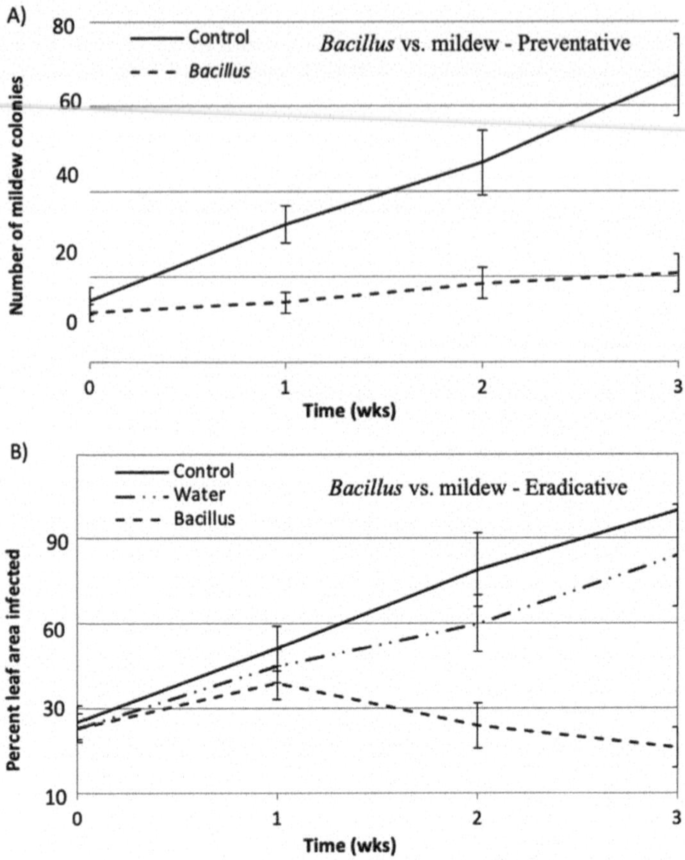

Fig. 1.2 Development of mildew colonies on control and *Bacillus*-treated leaves over a 3-week period. Applications were made weekly and the number of colonies was assessed visually. (**a**) Applications made prior to disease onset. (**b**) Applications made after disease symptoms were apparent

1.3.2 Gummy Stem Blight (Black Rot) Control on Cucumber

Gummy stem blight is a widespread fungal disease of greenhouse cucumber worldwide. It is caused by the fungus *Didymella bryoniae* (teleomorph) with *Phoma cucurbitacearum* as the asexual stage (Zitter 1992) (Fig. 1.5a). The pathogen can infect all parts of the plant, except for roots, and plants are susceptible to gummy stem blight at all stage of plant development under favorable environmental conditions (Sabaratnam 2018). Symptoms begin through wound sites, causing pale brown lesions to appear at the bottom of the main stem (Ferguson 2009). Stem cracks and amber-colored gummy sap are likely to appear on cucumber stems. This pathogen affects fruits internally and externally (Fig. 1.5b). Internal fruit rot is not visible from the outside at harvest, but often is distinguished by narrowing of the blossom

Fig. 1.3 (**a**) Effect of *B. subtilis* as a preventative application against powdery mildew development. Leaf on the left received two applications of Rhapsody before mildew infection appeared, and leaf on the right is the control. (**b**) Comparison of powdery development on a *B. subtilis*-treated leaf (left) with a control leaf (right). Three applications of Rhapsody were made to each leaf at 1-week intervals after the onset of mildew infection began. Mildew colony development was significantly suppressed but not eradicated

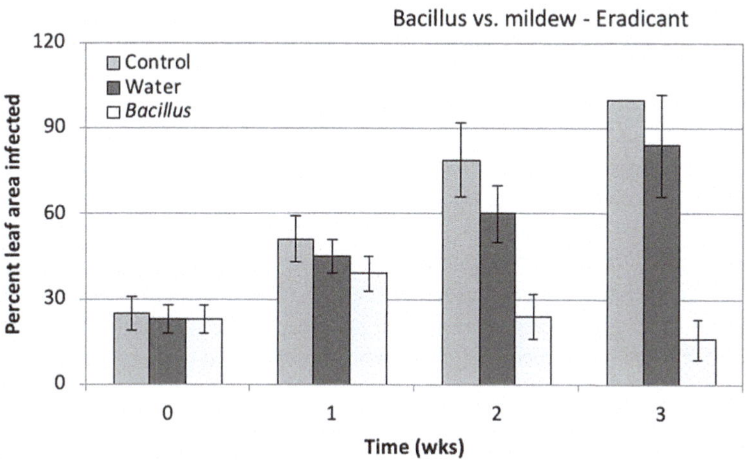

Fig. 1.4 Comparison of mildew development on plants treated with *B. subtilis* or water vs. untreated plants. Each time period (weeks) represents one, two, or three applications made after the onset of powdery mildew infection

end of fruit and central discoloration (Ferguson 2009) (Fig. 1.5c). External symptoms consist of spots of gummy and brown fruit lesions. Symptoms of gummy stem blight leaf infection first occur at the tip of the leaf as a pale brown discoloration, which expands into a V-shape lesion on mature leaves (Ferguson 2009). The development of gummy stem blight is favored by high humidity in the growing environment and free water on leaf surfaces.

Fig. 1.5 Cucumber fruit development and gummy stem blight infection. (**a**) Healthy cucumber fruit. (**b**) Fruit infected by mycelium of *D. bryoniae* at the blossom end. (**c**) Colonies of *D. bryoniae* on potato dextrose agar

The effect of *B. subtilis* on growth of *D. bryoniae* has been evaluated. In vitro experiments showed 11 common strains of *B. subtilis* produced antagonistic zones against the pathogen on potato dextrose agar (Utkhede and Koch, 2002). Mycelial growth of *D. bryoniae* was significantly inhibited by *B. subtilis* QST713 from in vitro dual culture experiments (Kaewkham et al. 2016). There was a significant reduction in lesion length on cucumber plants inoculated with *D. bryoniae* when a strain of *B. subtilis* (AGS-4) was applied as a postinoculation spray to plants under greenhouse conditions (Utkhede and Koch 2002). Kaewkham et al. (2016) reported that cucumber seeds treated with *B. subtilis* QST713 at 1×10^5 cfu (colony-forming units)/seed significantly inhibited the severity of gummy stem blight on cucumber plants when compared to nontreated seeds. Cucumber plants inoculated with spores of *D. bryoniae* followed by a foliar spray of methylcellulose-formulated *B. subtilis* QST713 at 1×10^5 cfu/ml and 1×10^7 cfu/ml significantly suppressed gummy stem blight disease severity under greenhouse conditions (Kaewkham et al. 2016). In our research, when Rhapsody was applied at a rate of 1 L product/100L water to cucumber fruit as a spray application made 24 h prior to pathogen inoculation, the extent of fruit rot was visibly reduced (Fig. 1.6) when compared to fruit sprayed with water. Mycelial growth was visibly reduced on the fruit surface by the *Bacillus* treatment.

1.3.3 Fusarium Root and Stem Rot Control on Cucumber

Fusarium root and stem rot caused by *Fusarium oxysporum* f. sp. *radicis-cucumerinum* (*F.o.r.c.*) (DJ Vakalounakis) is one of the most destructive diseases of greenhouse cucumbers. Symptoms of the disease include wilting of plants at the fruit-bearing stage, especially during hot weather (Punja and Parker 2000). There are masses of pink-orange or light-salmon spores formed and white mycelium grows on the outside of the stem on cucumber plants severely affected by *F.o.r.c.* (Punja and Parker 2000) (Fig. 1.7a), and the fungus also produces abundant spores in culture (Fig. 1.7b). The fungus can survive as chlamydospores in soil and plant debris for many years. It infects plants through root tips and wound sites during

Fig. 1.6 Comparison of development of *D. bryoniae* on cucumber fruits pretreated with *B. subtilis* QST713 (left) and control fruit sprayed with water (right). Mycelial development was reduced on the surface of the treated fruits (right)

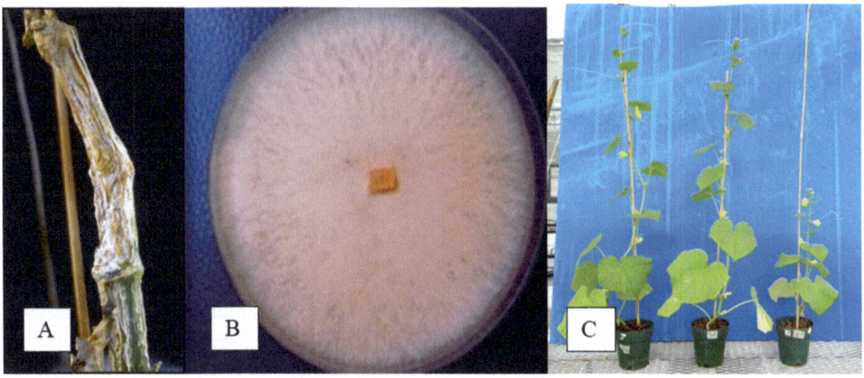

Fig. 1.7 (**a**) Stem lesion on cucumber caused by *F. oxysporum* f. sp. *radicis-cucumerinum* (*F.o.r.c.*). (**b**) Colony of *F.o.r.c.* on potato dextrose agar. (**c**) Effect of three drench applications of *B. subtilis* strain QST713 prior to pathogen inoculation on Fusarium root and stem rot development. Left plant, noninculated control; middle plant, *B. subtilis* applied thrice at 72, 48, and 24 h prior to inoculation by *F.o.r.c.*; right plant, *F.o.r.c.* only without *B. subtilis*. Photo was taken 3 weeks after inoculation

transplanting. The recirculating systems in the greenhouse offer opportunities for *F.o.r.c.* to spread through the irrigation system, resulting in contamination of the growth substrate (Cerkauskas 2012). The optimal temperature for development of *F.o.r.c.* on young plants is about 20 °C, and disease will not develop at 32 °C (Cerkauskas 2012; Punja and Parker 2000). Fusarium root and stem rot is generally controlled by soil fumigation with methyl bromide, greenhouse disinfestation with formaldehyde, and cucumber grafting on rootstocks with disease resistance (Pavlou et al. 2002). Resistant cultivars and chemical controls are not available. Fortunately, several registered biocontrol agents can be employed to diminish the impact of Fusarium root and stem rot, including *Bacillus subtilis*, *Streptomyces griseoviridis*, *Gliocladium catenulatum*, and *Trichoderma harzianum*. Of these four biocontrol agents, *B. subtilis* has been considered as a potential biocontrol agent for the control of root and stem rot on cucumber. From in vitro tests, *B. subtilis* strain MB10 inhibited the growth of *F.o.r.c.* by about 52% (Al-Tuwaigri 2008). The antifungal activity of *B. subtilis* on *Fusarium* spp. also has been demonstrated by Zhang et al. (1996). From in vivo tests, cucumber plants bacterized with *B. subtilis* MB10 increased the number of healthy plants by approximately 36% compared with control plants after infestation of *F.o.r.c.* in the soil (Al-Tuwaigri 2008). Also, *B. subtilis* strain TB10 significantly increased shoot length, fresh weight, and dry weight of cucumber plants grown under greenhouse conditions as compared to untreated healthy control plants (Al-Tuwaigri 2008). Seed treatment with *B. subtilis* TB10 reduced cucumber infection by *F.o.r.c.* (Al-Tuwaigri 2008). Certain metabolites may be secreted by *B. subtilis* to suppress the germination of conidia of *F.o.r.c.* (Al-Tuwaigri 2008). Also, our results have shown that three preventative applications of Rhapsody containing 1×10^9 CFU/g *B. subtilis* strain QST 713 made prior to *Fusarium* inoculation resulted in a significant increase in plant growth after 3 weeks (Fig. 1.7c), while Rhapsody was much less effective in the control of Fusarium root and stem rot when applied to plants 1 and 4 days after *Fusarium* inoculation.

1.3.4 Pythium Root Rot Control on Cucumber

Pythium root rot, caused by several *Pythium* spp., including *P. aphanidermatum*, *P. ultimum*, and *P. irregulare*, is a predominant problem for growers as there are no resistant varieties available and limited use of fungicides in Canada (Punja and Yip 2003). *P. aphanidermatum* is one of the most common and destructive root-infecting pathogens of greenhouse-grown cucumbers (Utkhede et al. 2000). *P. aphanidermatum* results in browning of roots within days under high disease pressures, while it may not immediately kill plants at low inoculum levels (Menzies and Stan 1996). Experimental results indicated that even a low inoculum density of 22 zoospores of the root rot pathogen per 100 L nutrient solution can lead to significant yield losses in cucumbers (Menzies and Stan 1996). The pathogen can survive in water, soil, and root debris and spread via the production of zoospores that infect and colonize wounds or root tips on plant roots, leading to a dramatic decrease in yield and plant quality (Utkhede et al. 2000). *P. aphanidermatum* adversely affected plant roots leading to a decrease in shoot growth and water uptake (Grosch et al. 1999). Symptoms are frequently difficult to notice until plants suddenly wilt, particularly in sunny and hot (32–37 °C) weather. Consequently, it is almost impossible for growers to determine if cucumber crops are suffering from Pythium root rot. It has been found that *B. subtilis* strain AG-1 inhibited the growth of *P. aphanidermatum* on potato dextrose agar (Utkhede et al. 1999). *B. subtilis* strain 8B-1 also showed antagonistic activity against *P. ultimum* from in vitro tests (Khabbaz and Abbasi 2013). *B. subtilis* strain FZB37 effectively inhibited *P. aphanidermatum* through the production of antifungal metabolites from in vitro tests (Krebs et al. 1998), but the bacterization of this strain on the plant increased the number of diseased cucumber plants compared with the infected control plants (Grosch et al. 1999). The findings suggested that effects of the bacterium on the inhibition of *P. aphanidermatum* in vitro had no association with effects of disease control obtained from in vivo experiments (Grosch et al. 1999). There was a significant increase in plant fresh biomass and healthy seedlings, and a 27–50% reduction in severity of Pythium root rot caused by *P. ultimum* when *B. subtilis* 8B-1 was applied to peat mix with the pathogen (Khabbaz and Abbasi 2013). Strains of *B. subtilis* FZB37, FZB38, and FZB44 significantly reduced yield losses on cucumber plants grown in rockwool bags following inoculation with *P. aphanidermatum* compared with the pathogen-treated control (Grosch et al. 1999). Of these three *B. subtilis* strains, FZB37 and FZB44 reduced yield losses caused by *P. aphanidermatum* primarily through increased fresh weights of cucumber fruits (Grosch et al. 1999). Although *B. subtilis* strain BACT-0 did not show antagonism against *P. aphanidermatum* from in vitro experiments, it enhanced plant growth and fruit yield of cucumber plants inoculated with *P. aphanidermatum* (Utkhede et al. 1999). The colonization of *B. subtilis* can promote the lignification on plant roots to increase resistance to the pathogen (Grosch et al. 1999). Studies suggested that *B. subtilis* BACT-0 increased growth and yield in cucumber plants mainly via the induction of resistance to root rot and the production of hormones by the plant (Utkhede et al. 1999). The strain *B. subtilis* BACT-0 is likely to be a plant growth-promoting rhizobacteria (PGPR) based on the results that it improved growth and yield of cucumber plants without the presence of *P. aphanidermatum* (Utkhede et al. 1999). The PGPR has potential to adversely affect pathogen establishment in roots or the formation of secondary inoculum (Utkhede et al. 1999). Similarly, results showed the applications of *B. subtilis* strain BACT-0 at a concentration of 1×10^9 CFU/L in nutrient solution significantly improved greenhouse lettuce production in the presence of *P. aphanidermatum* compared to the control (Utkhede et al. 2000). Corrêa et al. (2010) showed that applications of *B. subtilis* GB03 in the nutrient solution enhanced growth of lettuce plants inoculated with *P. aphanidermatum*. In our research, three applications of *B. subtilis* strain QST713 made to cucumber seedlings 72, 48, and 24 h prior to pathogen inoculation significantly reduced the development of Pythium root rot caused by *P. aphanidermatum* (Fig. 1.8a) compared to the control plants without *B. subtilis* (Fig. 1.8b).

Fig. 1.8 Effect of three drench applications of *B. subtilis* prior to pathogen inoculation on Pythium root development. (**a**) *B. subtilis* was applied at 72, 48, and 24 h prior to inoculation with *P. aphanidermatum*. (**b**) Control plants receiving pathogen inoculum only. Photo was taken 2 weeks after inoculation

1.4 Control of Tomato Diseases Using *B. subtilis*

Tomato (*Solanum lycopersicum* L.) is a member of the Solanaceae (nightshade) family. It is generally believed to have originated from the tropical and subtropical areas of South and Central America, and different varieties of tomatoes are cultivated in many temperate climates around the world (Knapp and Peralta 2016). The tomato plant is a short-lived perennial in natural environments, but it is grown and cultivated as an annual for its edible fruits. Tomato fruits provide health-protective effects to humans due to the production of lycopene, a strong antioxidant (Agriculture and Agri-Food 2006). Most tomato varieties are sensitive to chilling temperatures (0–15 °C) in the period of seed germination, vegetative growth, and reproduction (Foolad and Lin 2000). In many countries of northern Europe, tomatoes are grown and produced in greenhouses to better control environmental conditions. Canada is the primary producer of greenhouse tomatoes in North America. There are more than 280,000 t of tomatoes produced annually in Canadian greenhouses, valued at about $500 million (Agriculture and Agri-Food Canada 2016). Tomatoes can be grown in double-layered polyethylene and glass greenhouses, primarily in hydroponic systems using mineral nutrient solutions to enhance plant growth. The production greatly depends on integrated pest management practices, including a series of management approaches, such as crop monitoring, sanitation, cultural control, and biological control. These strategies are beneficial to reduce the use of chemical fungicides for some diseases. The most important diseases on greenhouse tomatoes are gray mold (*Botrytis cinerea*), Pythium damping-off (*Pythium aphanidermatum*), Fusarium wilt (*Fusarium oxysporum*), and powdery mildew (*Oidium neolycopersici*).

1.4.1 Gray Mold Control on Tomato

Tomato gray mold, caused by *Botrytis cinerea*, is a common and serious fungal disease of tomato plants in greenhouses. The pathogen can infect all aboveground parts, including leaves, petioles, stems, flowers, and fruits. It often first occurs at exposed sites, such as pruning wounds on stems and fallen petals on leaves (Harrison 1996). Tomato plants are more susceptible to infection during the period of flower and fruit enlargement (Hong 2012). The most obvious symptom of gray mold is the production of gray furry mold consisting of a large number of fungal spores that covers the infected tissue (Harrison 1996). The most destructive symptom related to gray mold is the stem canker, which can kill plants when stems are infected severely. Currently, suitable gray mold disease resistance sources have not been identified for greenhouse tomato plants due to the absence of resistant germplasm and resistant varieties (Wang et al. 2018). Growers rely on fungicides to control tomato gray mold (Wahab 2009; Corrêa and Soria 2010). Therefore, biological control approaches are gaining attention because alternative disease control methods are needed (Wang et al. 2018). As mentioned above, *B. subtilis* can secrete the antibiotics iturin A and surfactin against pathogen invasion (Asaka and Shoda 1996). Siripornvisal (2010) reported that *B. subtilis* strain BCB3-19 isolated from tomato rhizosphere has antifungal properties against *B. cinerea*, inhibiting fungal growth by approximately 53–56% from in vitro assays (Siripornvisal 2010). Furthermore, in vivo assays on tomato fruits showed the bacterium effectively reduced the development of gray mold at 4 °C or 23 °C, and there was no damage caused to fruits (Siripornvisal 2010). Experimental results showed that *B. subtilis* BCB3-19 was more effective with higher temperatures and concentrations of cells (Siripornvisal 2010). Dilutions of the culture filtrate of *B. subtilis* BS061 (1×10^8 CFU/ml) resulted in reduced gray mold on tomato seedlings and decreased incidence of disease on whole tomato plants at 20 °C (Kim et al. 2013). Wang et al. (2018) also found that *B. subtilis* strain WXCDD105 from the rhizosphere of tomato plants also controlled gray mold. Results showed this strain significantly suppressed mycelial growth of *B. cinerea* by over 90% and effectively reduced gray mold by around 75% in pot experiments (Wang et al. 2018). Additionally, *B. subtilis* strain WXCDD105 not only stimulated seed germination and seedling growth but also enhanced the firmness of tomato fruit by about 20% compared to the control (Wang et al. 2018).

1.4.2 Pythium Damping Off Control on Tomato

Damping-off caused by *Pythium aphanidermatum* is a major disease of greenhouse tomato crops. The pathogen affects both seedlings and mature plants. Seedling damping-off can result in significant yield losses (Kipngeno et al. 2015). Preemergence damping-off is characterized by brown water-soaked lesions on seedlings, while symptoms of postemergence damping-off first appear on roots. Infection is most prevalent in cool and wet soils. Seedlings affected by damping-off can fail to germinate or die soon after emergence. Under hot and sunny conditions, *P. aphanidermatum* can destroy small roots of mature plants, resulting in a sudden wilt (Agriculture and Agri-Food Canada 2006). Biological control agents have been evaluated to manage Pythium damping-off owing to the lack of resistant varieties and lack of effective control options (Kipngeno et al. 2015). *B. subtilis* strain MBI 600 is the active ingredient of the bio-fungicide that was registered in Canada for disease management of damping-off and root diseases on greenhouse tomato (Agriculture and Agri-Food Canada 2016). Several studies have been conducted to determine the efficacy of various strains of *B. subtilis* against *P. aphanidermatum*. In vitro experiments showed that *B. subtilis* CBE 4 and FZB37 effectively inhibited *P. aphanidermatum* growth (Shankar 2016; Krebs et al. 1998). However, *B. subtilis* strain FZB37 did not reduce the percentage of infected roots on tomato plants compared to control plants infected with *P. aphanidermatum* (Grosch et al. 1999). While strain FZB37 was able to produce antifungal metabolites in vitro, it failed to produce these antifungal metabolites in tomato plants grown under soilless culture (Krebs et al. 1998; Grosch et al. 1999). Soilless culture favors the development of *P. aphanidermatum* because the pathogen prefers high temperature and moisture levels (Favrin et al. 1988). The activity and population dynamics of *B. subtilis* greatly depended on environmental conditions (Burr and Caesar 1984; Weller 1988). Kipngeno et al. (2015) indicated that coating tomato seeds with *B. subtilis* BS01 at a concentration of 10^6 CFU/ml not only protected seedlings against damping-off caused by *P. aphanidermatum* but also promoted plant growth. Experimental results showed that tomato seeds treated with *B. subtilis* had increased germination, shoot length, vigor index, and root length compared to controls inoculated with *P. aphanidermatum* only (Zalte et al. 2013). It has been reported that postemergence damping-off on tomato seedlings was reduced by about 30% compared to controls after seed treatment with *B. subtilis* at 10 g/kg seed (Zalte et al. 2013). *B. subtilis* can inhibit *P. aphanidermatum* establishment in seedlings by competing with germinated oospores for nitrogen and soluble carbon from root exudates (Weller, 1988). Grosch et al. (1999) observed that the colonization by *B. subtilis* strains FZB13, FZB24, and FZB44 on tomato plants caused a reduction in the severity of disease caused by *P. aphanidermatum* at the beginning of the experiment, but no differences were observed by the end of the experiment. As mentioned previously, *B. subtilis* is able to produce several antibiotics and metabolites to protect against invasion of *P. aphanidermatum* in tomato seedlings (Kipngeno et al. 2015). Strain ATCC 6633 of *B. subtilis* inhibited the pathogenic infection caused by *P. aphanidermatum* in seedlings through the overexpression of mycosubtilin (Leclère et al. 2005).

1.4.3 Fusarium Wilt Control on Tomato

Fusarium wilt caused by *Fusarium oxysporum* f. sp. *lycopersici* (Jarvis and Shoemaker 1978) is a common soilborne fungal disease of greenhouse tomatoes. Symptoms of wilt begin with the yellowing of lower foliage, followed by irreversible wilting as the disease progresses. Brown discoloration may appear in the vascular system and is very obvious at leaf nodes (Agriculture and Agri-Food Canada 2006). The disease not only affects susceptible plants during the reproductive stage but also has an adverse effect on the development of seedlings. Fusarium wilt tends to occur in conditions of extremely high temperatures, particularly in drought-prone areas (Allen et al. 1996). The pathogen spreads mainly by water, soil, infected transplants, and contaminated equipment. This pathogen is capable of surviving as chlamydospores in root residues and soil (Agriculture and Agri-Food Canada 2006). *F. oxysporum* has the capability to synthesize many mycotoxins, which is the one of the primary determinants of Fusarium wilt (Khan et al. 2017). Toxins include beauvericin, naphthazarins, fusaric acid, moniliformin, and sambutoxin. At present, to reduce the impact of Fusarium wilt, cultural practices, use of resistant cultivars, chemical fungicides, and biological control agents are used, among which wilt-resistant varieties provide the best prevention. However, pathogen races are likely to develop to overcome cultivar resistance (Khan et al. 2017). There are presently two registered biological control products for the management of Fusarium wilt in Canada: *Streptomyces griseoviridis* strain K61 (Mycostop®) and *Trichoderma harzianum* strain KRL-AG2 (Rootshield®). Studies of Fusarium wilt on tomato demonstrate that strains of *B. subtilis* have antagonistic effects against the pathogen. *B. subtilis* reduces the impact of *F. oxysporum* through production of fungitoxic compounds and induction of systemic resistance, particularly via lipid metabolism of the plant (Abd-Allah et al. 2007). Furthermore, *B. subtilis* can produce siderophores which take iron away from fungal pathogens; consequently, plant growth is promoted by the bacteria (Khan et al. 2017). *Bacillus* species colonize root tissues of tomato plants rapidly, resulting in reduced infection by *F. oxysporum* (Shafi et al. 2017). Colonization of plant roots directly contributes to successful biological control of soilborne diseases (Cavaglieri et al. 2005). Abd-Allah et al. (2007) found that a strain of *B. subtilis* isolated from the rhizosphere of tomato plants inhibited the development of *F. oxysporum*. They observed that there was a positive correlation between the amount of fusaric acid in tomato tissues and Fusarium wilt incidence (Abd-Allah et al. 2007). It appears that *B. subtilis* controlled the Fusarium wilt disease through the bioremediation of fusaric acid (Abd-Allah et al. 2007). Infection by *F. oxysporum* in tomato tissues results in rupture of the epidermal tissues (Abd-Allah et al. 2007), and the wilt pathogen also damages and colonizes the cortex tissues. In tomato plants pretreated with the formulated *B. subtilis*, it was observed that the wilt pathogen only colonized the epidermal tissues and did not destroy the cortex tissues (Abd-Allah et al. 2007). As described earlier, *B. subtilis* produces antibiotics and chitinase enzymes and induces the production of proteinase inhibitors, neutral lipids, fatty acids, and phytoalexins by the plant to control the wilt pathogen (Abd-Allah et al. 2007). A previous study reported that *B. subtilis* strain 174 elicited systemic resistance against Fusarium wilt in tomato plants and the levels of phenylalanine ammonia lyase (PAL), phenoloxidase (PO), and polyphenol oxidase (PPO) were enhanced in bacteria-treated tomato plants and Fusarium wilt was significantly inhibited by this strain (Akram and Anjum 2011).

1.4.4 Postharvest Fruit Rot Control on Tomato

Blue mold rot caused by *Penicillium* spp. is a common postharvest fungal disease of tomato fruits in greenhouses. Blue mold rot caused by *P. oxalicum* appeared only on mature tomato fruits and not on immature or young tomato fruits (Kwon et al. 2008). Typically, the pathogen causes infection through wound sites caused by damage during harvest. The first symptom of *Penicillium* is the waterlogged lesions, followed by softened and watery fruits (Kwon et al. 2008). Early in the growth stage, fungal growth is white, but it eventually turns to green (Fig. 1.9a). Mycelial growth of *Penicillium* spp. was suppressed by approximately 82% by *B. subtilis* UTB96 (Soleyman et al. 2014). In our research, disease incidence and severity of *Penicillium* sp. were inhibited by *B. subtilis* strain QST 713 following weekly applications to tomato plants (Punja et al. 2016). Production of antibiotic compounds in culture medium also inhibited colony growth of *Penicillium* and *Botrytis* species (Fig. 1.9b, c). It has been shown that iturin A produced by *B. subtilis* has antifungal properties, but some fungal species are not vulnerable to iturin (Klich et al. 1991). As for *P. italicum*, in vitro results showed that fungal development was greatly reduced by the presence of iturin A at a very low concentration of 4 μg/plate (Klich et al. 1991). Chitarra et al. (2002) isolated an iturin-like compound from strain *B. subtilis* YM 10-20 and found this heat-stable compound was able to permeabilize fungal spores and inhibit spore germination of *P. roqueforti*.

In our research, populations levels of *B. subtilis* strain QST713 on tomato fruits were increased significantly following application of Rhapsody to plants grown under commercial conditions at monthly intervals, but the populations gradually declined in the ensuing 4 weeks (Fig. 1.10a). Repeated applications increased the

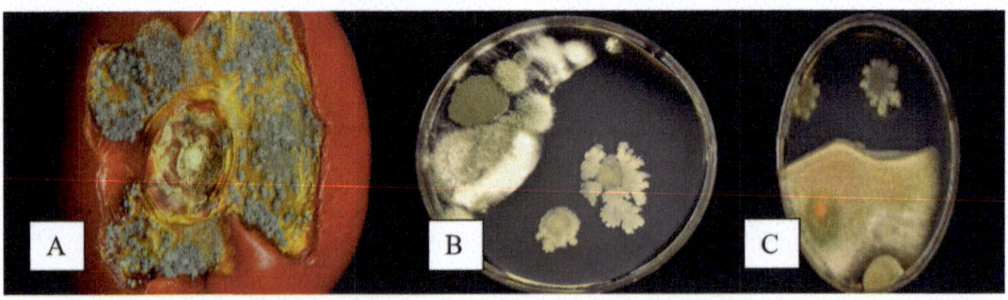

Fig. 1.9 (**a**) Penicillium fruit rot on tomato caused by *P. olsonii*. (**b**) Antagonistic activity of *B. subtilis* colony against *P. olsonii*. (**c**) Antagonistic activity of *B. subtilis* colony against *B. cinerea*

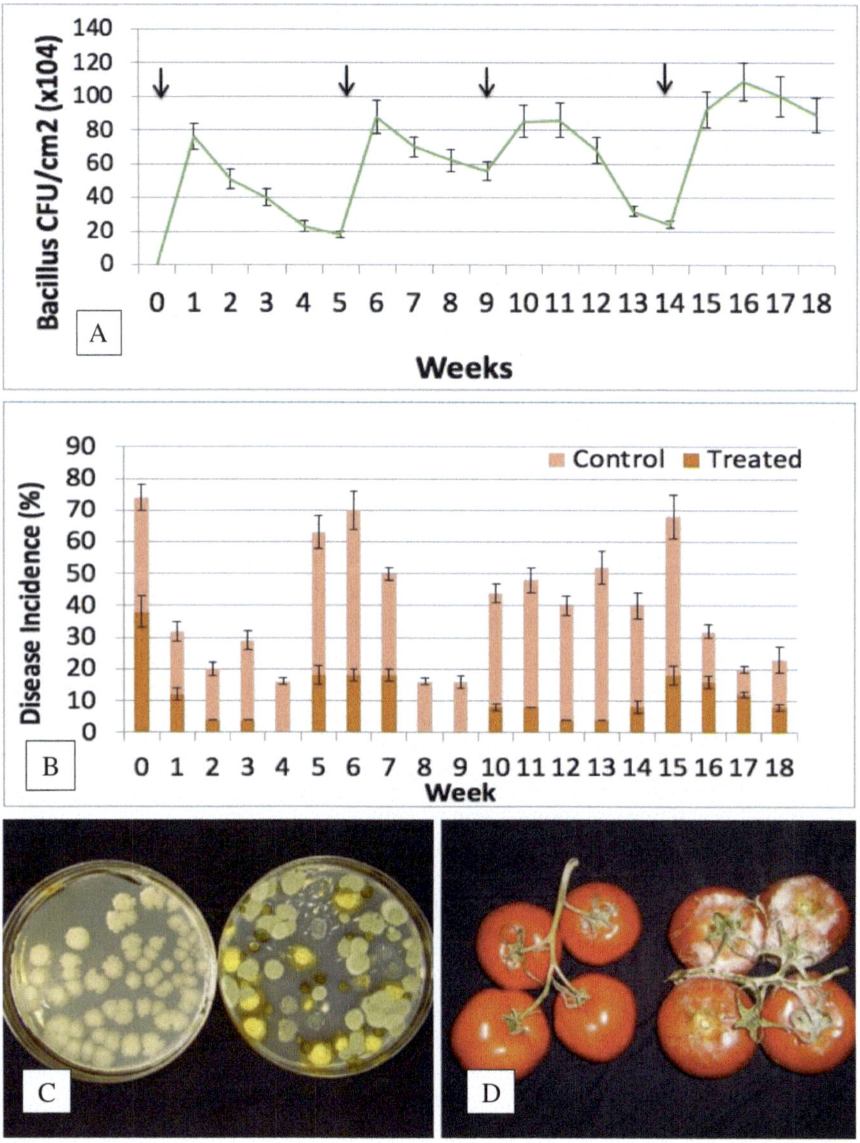

Fig. 1.10 (**a**) Population densities of *B. subtilis* strain QST713 on tomato fruit surfaces over an 18-week period following four applications of Rhapsody (arrows) at monthly intervals. (**b**) Disease incidence on tomato fruits after four applications of *B. subtilis*. Fruits were harvested and incubated at 25 °C for 7 days and disease incidence on control and treated fruits rated. (**c**) Population levels of *B. subtilis* (left plate) on tomato fruit surfaces compared to control fruit (right plate) which have colonies of *Penicillium* and *Aspergillus*. (**d**) Disease incidence on treated tomato fruits (left) compared to control fruits (right) after harvest and incubation for 7 days at 25 °C. Reproduced by permission from Punja et al. (2016)

population levels on the fruit. The effects of these applications on postharvest disease incidence, caused mainly by *Penicillium* species, are shown in Fig. 1.10b. Fruits from control and treated plants were incubated at 25 °C and high humidity for 7 days to promote disease. Disease incidence was significantly reduced by the *B. subtilis* treatments. The high populations of the bacteria were visualized by plating fruit surface washings onto agar medium, and they inhibited growth of *Penicillium* species on the fruit surface compared to control fruits (Fig. 1.10c) and reduced disease severity on the fruit (Fig. 1.10d).

1.5 Conclusions

The use of *B. subtilis* is a promising strategy to reduce a range of fungal diseases on cucumber and tomato plants grown under greenhouse conditions. *Bacillus* species colonize root tissues of plants rapidly and produce large numbers of endospores. They have been successfully formulated, and many stable and effective formulations have been developed for application in biocontrol of plant diseases. Many studies have shown that *B. subtilis* is a successful biological control agent against foliar, soilborne, or postharvest fungal diseases. In addition, disease severity can be effectively reduced by specific strains of *B. subtilis*, but the efficacy of strains may vary for control of different plant diseases. More than one mechanism may be used by *B. subtilis* to suppress the development of the pathogen. The three main mechanisms are competition with other microbes that could cause adverse effects on the plant; production of antibiotics, such as iturins, surfactins, and fengycins; and induction of host defense responses against pathogens. In some instances, the antagonistic effect of *B. subtilis* in vitro is not correlated with the ability of the bacterium to suppress disease development on the plant. *B. subtilis* can reduce infection by pathogens by limiting the germination of spores when seedlings, detached leaves, and plants are treated with the bacteria ahead of the time of inoculation of pathogens. *B. subtilis* seems to be a plant growth-promoting rhizobacteria based on reports of improved growth and yield of plants without the presence of pathogens. The colonization of *B. subtilis* on plant roots not only can promote plant growth but also can significantly reduce yield losses caused by the pathogens. Previous studies demonstrated that *B. subtilis* was more effective when alternated with certain fungicides than when applied alone. The application of *B. subtilis* offers an exciting opportunity to reduce the use of fungicides during greenhouse vegetable production.

References

Abd-Allah EF, Ezzat SM, Tohamy MR (2007) *Bacillus subtilis* as an alternative biologically based strategy for controlling Fusarium wilt disease in tomato: a histological study. Phytoparasitica 35:474–478

Agriculture and Agri-Food Canada (2006) Crop greenhouse tomato in Canada. Available from: http://publications.gc.ca/collections/collection_2009/agr/A118-10-24-2006E.pdf

Agriculture and Agri-Food Canada (2016) Crop profile for greenhouse tomato in Canada, 2014. Available from: http://publications.gc.ca/collections/collection_2016/aac-aafc/A118-10-24-2014-eng.pdf

Agriculture and Agri-Food Canada (2017) Statistical overview of the Canadian greenhouse vegetable industry 2016. Available from: http://www.agr.gc.ca/resources/prod/doc/pdf/st_ovrv_greenhouse_legumes_de_serre_2016-rev-eng.pdf

Akram W, Anjum T (2011) Quantitative changes in defense system of tomato induced by two strains of *Bacillus* against Fusarium wilt. Indian J Fund Appl Life Sci 1:7–13

Alabouvette C, Olivain C, Steinberg C (2006) Biological control of plant diseases: the European situation. Eur J Plant Pathol 114:329–341

Allard-Massicotte R, Tessier L, Lecuyer F, Lakshmanan V, Lucier J, Garneau D, Caudwell L, Vlamakis H, Bais HP, Beauregard PB (2016) *Bacillus subtilis* early colonization of *Arabidopsis thaliana* roots involve multiple chemotaxis receptors. MBio 7:e01664–e01616

Allen DJ, Ampofo JKO, Wortmann CS (1996) Pests, diseases, and nutritional disorders of the common bean in Africa: A field guide. Centro Internacional de Agricultura Tropical (CIAT). CIAT Publication no. 260. Technical Centre for Agricultural and Rural Cooperation (TCA), Wageningen, Cali, CO. 132 p.

Al-Tuwaigri MMY (2008) Biological control of Fusarium root rot of cucumber (*Cucumis sativus* L.) by rhizospheric isolates of *Bacillus subtilis* and *Trichoderma viride*. Egypt J Exp Biol 4:79–86

Asaka O, Shoda M (1996) Biocontrol of *Rhizoctonia solani* damping-off of tomato with *Bacillus subtilis* RB14. Appl Environ Microbiol 62:4081–4085

Baruzzi F, Quintien L, Morea M, Caputo L (2011) Antimicrobial compounds produced by *Bacillus* spp. and applications to food. In: Mendez-Vilas A (ed) Science against microbial pathogens: communicating current research and technical advances. Fomatex Microbiology Series 3, Formatex Research Center, Badajoz, pp 1102–111

Beauregard PB, Chai Y, Vlamakis H, Losick R, Kolter R (2013) *Bacillus subtilis* biofilm induction by plant polysaccharides. Natl Acad Sci 110:E1621–E1630

Becker OJ, Schwinn FJ (1993) Control of soil borne pathogens with living bacteria and fungi: status and outlook. Pestic Sci 37:355–363

Brown J (2002) Comparative genetics of avirulence and fungicide resistance in the powdery mildew fungi. In: Bélanger R, Bushnell WR, Dik AJ, Carver TLW (eds) The powdery mildews. A comprehensive treatise. APS Press, St. Paul, pp 56–65

Burr TJ, Caesar AJ (1984) Beneficial plant bacteria. CRC Crit Rev Plant Sci 2:1–20

Cavaglieri L, Orlando J, Rodriguez M, Chulze S, Etcheverry M (2005) Biocontrol of *Bacillus subtilis* against *Fusarium verticillioides* in vitro and at the maize root level. Res Microbiol 156:748–754

Cawoy H, Bettiol W, Fickers P, Ongena M (2011) *Bacillus*-based biological control of plant diseases. In pesticides in the modern world – pesticides use and management. Available from: https://www.intechopen.com/books/pesticides-in-the-modern-world-pesticides-use-and-management/bacillus-based-biological-control-of-plant-diseases

Cawoy H, Debois D, Franzil L, Pauw ED, Thonart P, Ongena M (2014) Lipopeptides as main ingredients for inhibition of fungal phytopathogens by *Bacillus subtilis/amyloliquefaciens*. Microb Biotechnol 8:281–295

Cerkauskas R (2012) Fusarium stem and root rot of greenhouse cucumber. Available from http://www.omafra.gov.on.ca/english/crops/facts/01-081.htm

Chitarra GS, Breeuwer P, Nout MJR, Aelst AC, Rombouts FM, Abee T (2002) An antifungal compound produced by *Bacillus subtilis* YM 10-20 inhibits germination of *Penicillium roqueforti* conidiospores. J Appl Microbiol 94:159–166
Conrath U, Beckers GJM, Flors V, Garcia-Agustin P, Jakab G, Mauch F, Newman MA, Pieterse CMJ, Poinssot B, Pozo MJ, Pugin A, Schaffrath U, Ton J, Wendehenne D, Zimmerli L, Mauch-Mani B, Prime A (2006) Priming: getting ready for battle. Mol Plant-Microbe Interact 19:1062–1071
Corrêa OS, Soria MA (2010) Potential of *Bacilli* for biocontrol and its exploitation in sustainable agriculture. Microbiol Monogr 18:197–209
Corrêa EB, Bettiol W, Sutton JC (2010) Biocontrol of root rot (*Pythium aphanidermatum*) and growth promotion with *Pseudomonas chlororaphis* 63-28 and *Bacillus subtilis* GB03 in hydroponic lettuce. Summa Phytopathol 36:275–281
Danchin A (2001) *Bacillus subtilis*. In: Encyclopedia of genetics. Academic Press, Cambridge, MA, pp 135–144
Davey ME, O'Toole GA (2000) Microbial biofilms: from ecology to molecular genetics. Microbiol Mol Biol Rev 64:847–867
Dey D, Calpas J, Morton D, Dyck D (2017) Commercial greenhouse vegetable production. Government of Alberta agriculture and forestry. Available from: http://www1.agric.gov.ab.ca/$department/deptdocs.nsf/all/agdex1443
Dickerson, GW (1996) Greenhouse vegetable production. Cooperative extension service. College of Agriculture and home economics. New Mexico State University. Available from: https://aces.nmsu.edu/pubs/_circulars/CR556.pdf
Earl AM, Losick R, Kolter R (2008) Ecology and genomics of *Bacillus subtilis*. Trends Microbiol 16:6
Fan HY, Ru JJ, Zhang YY, Wang Q, Li Y (2017) Fengycin produced by *Bacillus subtilis* 9407 plays a major role in the biocontrol of apple ring rot disease. Microbiol Res 199:89–97
Favrin RJ, Rahe JE, Mauza B (1988) *Pythium* spp. associated with crown rot of cucumbers in British Columbia greenhouses. Plant Dis 72:683–687
Ferguson G (2009) Gummy stem blight of greenhouse cucumber. Factsheet. Ontario ministry of agriculture, food and rural affairs. Available from: http://www.omafra.gov.on.ca/english/crops/facts/09-051w.htm
Foolad MR, Lin GY (2000) Genetic analysis of cold tolerance during vegetative growth in tomato, *Lycopersicon esculentum* Mill. Euphytica 122:105–111
Garcia-Gutierrez L, Zeriouh H, Romero D, Cubero J, de Vicente A, Perez-Garcia A (2013) The antagonistic strain *Bacillus subtilis* UMAF6639 also confers protection to melon plants against cucurbit powdery mildew by activation of jasmonate and salicylic acid-dependent defence responses. Microb Biotechnol 6:264–274
Gilardi G, Manker DC, Garibaldi A, Gullino ML (2008) Efficacy of the biocontrol agents *Bacillus subtilis* and *Ampelomyces quisqualis* applied in combination with fungicides against powdery mildew of zucchini. J Plant Dis Prot 115:208–213
Glazebrook J (2005) Contrasting mechanisms of defence against biotrophic and necrotrophic pathogens. Annu Rev Phytopathol 43:205–227
Grosch R, Junge H, Krebs B, Bochow H (1999) Use of *Bacillus subtilis* as a biocontrol agent. III Influence of *Bacillus subtilis* on fungal root diseases and on yield in soilless culture. J Plant Dis Prot 106:568–580
Haas D, Defago G (2005) Biological control of soil-borne pathogens by fluorescent pseudomonads. Nat Rev Microbiol 3:307–319
Harrison G (1996) Grey mould (*Botrytis*) in greenhouse tomato crops. Department of environment and primary industries. Available from: http://agriculture.vic.gov.au/agriculture/pests-diseases-and-weeds/plant-diseases/vegetable/tomato-diseases/grey-mould-botrytis-in-greenhouse-tomato-crops
Higgins D, Dworkin J (2011) Recent progress in *Bacillus subtilis* sporulation. FEMS Microbiol Rev 36:131–148

Hofte M, Bakker PAHM (2007) Competition for iron and induced systemic resistance by siderophores of plant growth promoting rhizobacteria. In: Microbi sideroph. Springer, Berlin/Heidelberg, pp 121–133
Hong SJ (2012) Study on the control of leaf mold, powdery mildew and gray mold for organic tomato cultivation. Korean J Organ Agric 20:655–668
Jahn M, Munger MH, McCreight DJ (2002) Breeding cucurbit crops for powdery mildew resistance. In: Belanger RR, Bushnell RW, Dik JA, Carver WLT (Eds) The powdery mildew. A comprehensive treatise, pp 239–248
Jarvis WR, Shoemaker RA (1978) Taxonomic status of *Fusarium oxysporum* causing foot and root tor of tomato. Phytopathology 68:1679–1680
Kaewkham T, Hynes RK, Siri B (2016) The effect of accelerated seed ageing on cucumber germination following seed treatment with fungicides and microbial biocontrol agents for managing gummy stem blight by *Didymella bryoniae*. Biocontrol Sci Technol 26:1048–1061
Keinath AP, Dubose VB (2004) Evaluation of fungicides for prevention and management of powdery mildew on watermelon. Crop Prot 23:35–42
Khabbaz SE, Abbasi PA (2013) Isolation, characterization, and formulation of antagonistic bacteria for the management of seedlings damping-off and root rot disease of cucumber. Can J Microbiol 60:25–33
Khan N, Maymon M, Hirsch AM (2017) Combating *Fusarium* infection using *Bacillus*-based antimicrobials. Microorganisms 5:75
Kim Y, Song J, Lee I, Yeo W, Yun B (2013) *Bacillus* sp. BS061 suppresses powdery mildew and gray mold. Mycobiology 41:108–111
Kipngeno P, Losenge T, Maina N, Kahangi E, Juma P (2015) Efficacy of *Bacillus subtilis* and *Trichoderma asperellum* against *Pythium aphanidermatum* in tomatoes. Biol Control 90:92–95
Klich MA, Lax AR, Bland JM (1991) Inhibition of some mycotoxigenic fungi by iturin A, a peptidolipid produced by *Bacillus subtilis*. Mycopathologia 116:77–80
Knapp S, Peralta E (2016) The Tomato (*Solanum lycopersicum* L., Solanaceae) and its botanical relatives. In: Causse M et al (eds) The tomato genome, compendium of plant genomes. Springer, Berlin/Heidelberg, pp 7–21
Krebs B, Hoding B, Kubart S, Workie MA, Junge H, Schmiedeknecht G, Grosch R, Bochow H, Hevesi M (1998) Use of *Bacillus subtilis* as biological agent. I Activities and characterization of *Bacillus subtilis* strains. J Plant Dis Prot 105:181–197
Křistková E, Lebeda A, Sedláková B, Duchoslav M, Dančák M (2003) Distribution of powdery mildew species on Cucurbitaceaous vegetables in the Czech Republic. Scientific works of the Lithuanian institute of horticulture and Lithuanian university of Agriculture. Hortic Veget Growing 22:31–41
Kwon JH, Shen SS, Jee HJ (2008) Occurrence of blue mold on tomato caused by *Penicillium oxalicum* in Korea. Plant Pathol J 24:87–89
Lebeda A (1983) The genera and species spectrum of cucumber powdery mildew in Czechoslovakia. Phytopathol Zeitschrift 108:71–79
Leclère V, Béchet M, Adam A, Guez JS, Wathelet B, Ongena M, Thonart P, Gancel F, Chollet-Imbert M, Jacques P (2005) Mycosubtilin overproduction by *Bacillus subtilis* BBG100 enhances the organism's antagonistic and biocontrol activities. Appl Environ Microbiol 71:4577–4584
Lin TC, Ishizaka M, Ishii H (2009) Acibenzolar-S-methyl induced systemic resistance against anthracnose and powdery mildew diseases on cucumber plants without accumulation of phytoalexins. J Phytopathol 157:40–50
Loper JE, Henkels MD (1997) Availability of iron to *Pseudomonas fluorescens* in rhizosphere and bulk soil evaluated with an ice nucleation reporter gene. Appl Environ Microbiol 63:99–105
Lugtenberg BJJ, Chin-A-Woeng TFC, Bloemberg GV (2002) Microbe-plant interactions: principles and mechanisms. Anton Leeuw Int J Gen Mol Microbiol 81:373–383
Manjula K, Podile AR (2005) Production of fungal cell wall degrading enzymes by a biocontrol strain of *Bacillus subtilis* AF 1. Indian J Exp Biol 43:892–896

Manker DC (2005) Natural products as green pesticides. In: Clark JM, Ohkawa H (eds) New discoveries in agrochemicals, ACS symposium series, vol 892, pp 283–294
Matheron ME, Porchas M (2000) Evaluation of fungicide performance for control of powdery mildew on lettuce in 2000. Online publication no. AZ1177 in: vegetable: College of agriculture report 2000, College of Agriculture, University of Arizona, Tucson, AZ, USA. Available from: https://www.researchgate.net/publication/260386900_Evaluation_of_Fungicide_Performance_for_Control_of_Powdery_Mildew_on_Lettuce_in_2000
McGrath MT (2001) Fungicide resistance in cucurbit powdery mildew: experiences and challenges. Plant Dis 85:236–245
McGrath MT (2017) Powdery mildew of cucurbits. Cooperative extension. Available from: http://vegetablemdonline.ppath.cornell.edu/factsheets/Cucurbits_PM.htm
Menzies J, Stan DLE (1996) Effect of inoculum density of *Pythium aphanidermatum* on the growth and yield of cucumber plants grown in recirculating nutrient film culture. Can J Plant Pathol 18:50–54
Merrill L, Dunbar J, Richardson J, Kuske CR (2006) Composition of *Bacillus* species in aerosols from 11 U.S. cities. J Forensic Sci 51:559–565
Mihalache G, Balaes T, Gostin I, Stefan M (2017) Lipopeptides produced by *Bacillus subtilis* as new biocontrol products against *fusariosis* in ornamental plants. Environ Sci Pollut Res 25:29784–29793
Morikawa M (2006) Beneficial biofilm formation by industrial bacteria *Bacillus subtilis* and related species. J Biosci Bioeng 101:1–8
Nagorska K, Bikowski M, Obuchowski M (2007) Multicellular behavior and production of a wide variety of toxic substances support usage of *Bacillus subtilis* as a powerful biocontrol agent. Acta Biochim Pol 54:495–508
Nakano MM, Zuber P (1998) Anaerobic growth of a "Strict Aerobe" (*Bacillus subtilis*). Annu Rev Microbiol 52:65–90
Nunez-Palenius HG, Hopkins D, Cantliffe DJ (2006) Powdery mildew of cucurbits in Florida. University of Florida. IFAS extension. Available from: http://edis.ifas.ufl.edu
Ongena M, Jacques P (2008) *Bacillus* lipopeptides: versatile weapons for plant disease biocontrol. Trends Microbiol 16:115–125
Ongena M, Jourdan E, Adam A, Paquot M, Brans A, Joris B, Arpigny JL, Thonart P (2007) Surfactin and fengycin lipopeptides of *Bacillus subtilis* as elicitors of induced systemic resistance in plants. Environ Microbiol 9:1084–1090
Paulitz TC, Belanger RR (2001) Biological control in greenhouse systems. Annu Rev Phytopathol 39:103–133
Pavlou GC, Vakalounakis DJ, Ligoxigakis EK (2002) Control of root and stem rot of cucumber, caused by *Fusarium oxysporum* f. sp. *radicis-cucumerinum*, by grafting onto resistant rootstocks. Plant Dis 86:379–382
Peet MM, Welles G (2005) Greenhouse tomato production. CAB international 2005. Tomatoes (ed. E. Heuvelink). 9, pp 257–304
Perez-Garcia A, Romero D, Fernandez-Ortuno D, Lopez-Ruiz F, Vicente A, Tores JA (2009) The powdery mildew fungus *Podosphaera fusca* (synonym *Podosphaera xanthii*), a constant threat to cucurbits. Mol Plant Pathol 10:153–160
Pieterse CMJ, León-Reyes A, van der Ent S, Wees SCM (2009) Networking by small-molecule hormones in plant immunity. Nat Chem Biol 5:308–316
Punja ZK, Parker M (2000) Development of Fusarium root and stem rot, a new disease on greenhouse cucumber in British Columbia, caused by *Fusarium oxysporum* f. sp. *radicis-cucumerinum*. Can J Plant Pathol 22:349–363
Punja ZK, Yip R (2003) Biological control of damping-off and root rot caused by *Pythium aphanidermatum* on greenhouse cucumbers. Can J Plant Pathol 25:411–417
Punja ZK, Rodriguez G, Tirajoh A (2016) Effects of *Bacillus subtilis* strain QST 713 and storage temperatures on post-harvest disease development on greenhouse tomatoes. Crop Prot 84:98–104

Roberts P, Kucharek T (2005) Florida plant disease management guide: watermelon. Electronic data information source of UF/IFAS extension, PDMG-V3-55. Available from: http://edis.ifas.ufl.edu/

Romero D, Pérez-García A, Rivera ME, Cazorla FM, de Vicente A (2004) Isolation and evaluation of antagonistic bacteria towards the cucurbit powdery mildew fungus *Podosphaera fusca*. Appl Microbiol Biotechnol 64:263–269

Romero D, de Vicente A, Zeriouh H, Cazorla FM, Fernández-Ortuño D, Torés JA, Pérez-García A (2007a) Evaluation of biological control agents for managing cucurbit powdery mildew on greenhouse-grown melon. Plant Pathol 56:976–986

Romero D, de Vicente A, Olmos JL, Dávila JC, Pérez-García A (2007b) Effect of lipopeptides of antagonistic strains of *Bacillus subtilis* on the morphology and ultrastructure of the cucurbit fungal pathogen *Podosphaera fusca*. J Appl Microbiol 103:969–976

Romero D, de Vicente A, Rakotoaly RH, Dufour SE, Veening JW, Arrebola E, Cazorla FM, Kuipers OP, Paquot M, Pérez-García A (2007c) The iturin and fengycin families of lipopeptides are key factors in antagonism of *Bacillus subtilis* toward *Podosphaera fusca*. Mol Plant-Microbe Interact 20:430–440

Sabaratnam S (2018) Gummy stem blight of greenhouse cucumber. British Columbia ministry of agriculture. Available from: https://www2.gov.bc.ca/assets/gov/farming-natural-resources-and-industry/agriculture-and-seafood/animal-and-crops/plant-health/phu-gummystemblight-greenhousecucumberss.pdf

Saber WI, Ghoneem KM, Al-Askar AA, Rashad YM, Ali AA, Rashad EM (2015) Chitinase production by *Bacillus subtilis* ATCC 11774 and its effect on biocontrol of *Rhizoctonia* diseases of potato. Acta Biol Hung 66:436–448

Salmeron J, Vernooij B, Lawton K, Kramer C, Frye C, Osterndorp M (2002) Powdery mildew control through transgenic expression of antifungal proteins, resistance genes and systemic acquired resistance. In: Bélanger RR, Bushnell WR, Dik AJ, Carver TLW (eds) The Powdery Mildews. A comprehensive treatise. APS Press, Minnesota, pp 169–199

Schisler DA, Slininger PJ, Behle RW, Jackson MA (2004) Formulation of *Bacillus* spp. for biological control of plant diseases. Phytopathology 94:1267–1271

Shafi J, Tian H, Ji M (2017) *Bacillus* species as versatile weapons for plant pathogens: a review. Biotechnol Biotechnol Equip 31:446–459

Shankar SM (2016) Epidemiology and management of damping-off of tomato (*Solanum esculentum* L.). Plant pathology department of N.M. college of Agriculture, Navsari Agricultural University. Available from: http://krishikosh.egranth.ac.in/handle/1/5810036276

Siripornvisal S (2010) Biocontrol efficacy of *Bacillus subtilis* BCB3-19 against tomato gray mold. Sci Tech J 10:37–44

Soleyman G, Masoud A, Siavash T (2014) Biological control of Alternaria rot of tomato by two bacterial strains, *Pseudomonas fluorescens* UTPF68, and *Bacillus subtilis* UTB96. Iranian J Plant Prot Sci 44:299–305

Stein T (2005) *Bacillus subtilis* antibiotics: structures, syntheses and specific functions. Mol Microbiol 56:845–857

Touré Y, Ongena M, Jacques P, Guiro A, Thonart P (2004) Role of lipopeptides produced by *Bacillus subtilis* GA1 in the reduction of grey mould disease caused by *Botrytis cinerea* on apple. J Appl Microbiol 96:1151–1160

Utkhede RS, Koch CA (2002) Chemical and biological treatments for control of gummy stem blight of greenhouse cucumbers. Eur J Plant Pathol 108:443–448

Utkhede RS, Koch CA, Menzies JG (1999) Rhizobacterial growth and yield promotion of cucumber plants inoculated with *Pythium aphanidermatum*. Can J Plant Pathol 21:265–271

Utkhede RS, Levesque CA, Dinh D (2000) *Pythium aphanidermatum* root rot in hydroponically grown lettuce and the effect of chemical and biological agents on its control. Can J Plant Pathol 22:138–144

Van Loon LC (2007) Plant responses to plant growth-promoting rhizobacteria. Eur J Plant Pathol 119:243–254

Wahab S (2009) Biotechnological approaches in the management of plant pests, diseases and weeds for sustainable agriculture. J Biopest 2:115–134

Wang Z (2016) Statistical overview of the Canadian greenhouse vegetable industry, 2015. Minister of Agriculture and Agri-Food Canada. Available from: http://www.agr.gc.ca/resources/prod/doc/pdf/1468861362193x-eng.pdf

Wang H, Shi Y, Wang D, Yao Z, Wang Y, Liu J, Zhang S, Wang A (2018) A biocontrol strain of *Bacillus subtilis* WXCDD105 used to control tomato *Botrytis cinerea* and *Cladosporium fulvum* and promote the growth of seedlings. Int J Mol Sci 19:1371

Weller DM (1988) Biocontrol of soilborne plant pathogens in the rhizosphere with bacteria. Annu Rev Phytopathol 26:379–407

Whipps JM (2001) Microbial interactions and biocontrol in the rhizosphere. J Exp Bot 52:487–511

Zalte A, Gade RM, Shitole AV, Belkar Y (2013) Management of tomato damping off by using plant growth promoting microorganisms. J Plant Dis Sci 8:200–203

Zhang J, Zhang J, Howell CR, Starr JL, Zhang JX (1996) Suppression of *Fusarium* colonization of cotton roots and Fusarium wilt by seed treatments with *Gliocladium virens* and *Bacillus subtilis*. Biocontrol Sci Technol 6:175–187

Zitter TA (1992) Gummy Stem Blight. Cooperative extension. New York State. Cornell University. Fact Sheet Page: 730.70. Available from: http://vegetablemdonline.ppath.cornell.edu/factsheets/Cucurbit_GSBlight.htm

Chapter 2
Bacillus Species: A Potential Plant Growth Regulator

Usha Rani, Shivesh Sharma, and Vivek Kumar

2.1 Introduction

Plant growth- and development-stimulating rhizobacteria are defined as root-settling or root-inhabiting bacteria, i.e., *Bacillus* that lead to either plant growth encouragement, regulation, or biocontrol of potential plant diseases. Presently, the world population is 7 billion, and we expect that this figure is going to reach 8 billion by 2022. With the increase in world population, there is going to be an increased industrialization and a growing demand for food. Increases in the use of synthetic chemical inputs in agriculture are expected due to this demand, and this will cause adverse impacts on the environment. To solve this problem, we must increase food production while simultaneously maintaining the sustainability of our soil. Therefore, we need to explore sustainable methods which would lessen the chemical inputs in the forms of pesticides and inorganic chemicals.

Agricultural chemical inputs can be managed by using soil microbes such as *Bacillus* spp. These bacterial strains are able to solubilize the insoluble P in soil, which is then uptaken by the plant roots (Ramani and Patel 2011; Tallapragada and Seshachala 2012). The biological health of soil is poor if it has small amounts or nearly no microorganisms. It is not considered fertile and does not support healthy plant growth. Therefore, poor soil should be replete with minute living organisms such as bacterial, fungal species, actinomycetes, protozoans, and algal species. Out of these microbes, bacterial species are the most dominant and common. The soil contains about 95% of bacterial biomass. Some healthy soils contain a huge number of bacterial cells (often around 10^8 to 10^9 cells/g soil). Interestingly, if we try to

U. Rani · V. Kumar (✉)
Himalayan School of Biosciences, Swami Rama Himalayan University, Dehradun, India

S. Sharma
Department of Biotechnology, MLN National Institute of Technology, Allahabad, India

M. T. Islam et al. (eds.), *Bacilli and Agrobiotechnology: Phytostimulation and Biocontrol*, Bacilli in Climate Resilient Agriculture and Bioprospecting,
https://doi.org/10.1007/978-3-030-15175-1_2

isolate bacterial cells from such rich soil, hardly will we be able to get 1% of the culturable bacteria (Su et al. 2004). On the other hand, in any ecologically stressed (biotic or abiotic) soils, the culturable bacteria number may be up to 10^4 cells/g of soil (Jacoby et al. 2017). Generally, the type and number of bacterial cells that are present in dissimilar or diverse soils are affected by soil environmental conditions such as pH, temperature, presence of various salts, heavy metals, moisture, and other inorganic and organic chemicals as well as by the types and number of flora and fauna found in that soils (Garbeva et al. 2004).

The varied communities of aerobic endospore-forming bacteria (AEFB), i.e., *Bacillus* spp., commonly occur in all types of farming fields, with different types of plants, and play a significant role in enhancing crop productivity by its direct or indirect functions (Grayston et al. 1998). Other physiological properties, viz., multilayered cell walls, endospore formation which are stress resistant, and excretion of varied peptide antibiotics, signal peptide molecules, and extra cellular enzymes, are omnipresent in case of these various *Bacillus* spp., and these traits help in proliferation and survival of the bacterial cells under various adverse climatic conditions for very long duration (Pirttijärvi et al. 2000). Many species of *Bacillus* and *Paenibacillus* are very well known to enhance plant development and growth. The chief means and approach for growth encouragement includes the manufacture of growth energizing phytohormones, solubilization and mobilization of insoluble phosphate present in soil, production of proteinaceous components such as siderophore, and demonstration of the phenomenon of antibiosis, i.e., antibiotics production. Besides these beneficial properties, several *Bacilli* are also involved in inhibition of plant ethylene production and stimulation of plant systemic resistance against several plant pathogens (Gutiérrez-Mañero et al. 2008; Idris et al. 2004; 2007a, b; Richardson et al. 2009). The disease-causing microbes negatively affect the plant growth and health and, therefore, are a major challenge to the production of food. Conventional approaches such as rotation of crops, breeding resistant plant varieties, and applying chemical pesticides seem to be inadequate for controlling plant root diseases of significant crop plants (Johri et al. 2003). Additionally, it seems unavoidable that lesser amounts of chemical pesticides will be employed and that more and more dependence will be rested on novel micro-biotechnological or genetic engineering usages, which chiefly includes the application of microbes as potential biocontrol of antagonistic agents. Biocontrol usage and research has increased recently, partly due to the change in public thought (Bale et al. 2008). There is also an urgent need to find suitable substitutes for harmful chemicals employed in plant disease control. There are several reports about the *Bacillus* and *Paenibacillus* species expressing antagonistic behaviors by the process of suppressing pathogens under in vivo and in vitro conditions (Govindasamy et al. 2010).

2.2 Phosphate Solubilization

Phosphate is the second most important nutrient for plants, and its uptake by plant roots is vital for optimal plant growth. There are several reports of phosphate solubilization by *Bacillus* spp. Kang et al. (2014) studied the beneficial aspects of *B. megaterium* strain mj1212 inoculated with mustard plants. It was observed that application of *Bacillus* sp. enhanced shoot and root length as well as plant fresh weight. The biochemical investigation showed that chlorophyll, fructose, glucose, sucrose, and many amino acids' content were higher in *B. megaterium* strain mj1212 inoculated plants, as compared to the uninoculated plants. Phosphate content was also higher in inoculated plants rather than control plants. In another study conducted by Swain et al. (2012), *Bacillus subtilis* thermotolerant strains (<50 °C) were isolated from cow dung. This strain also solubilized tricalcium phosphate, and this solubilization was associated with enzyme phosphatase production, especially the acid phosphatase (AcP). The inoculation of this bacterium with cowpea (*Vigna unguiculata* L.) resulted in increased root length, shoot height, and plant biomass as compared to the control plants. Wang et al. (2014) isolated phosphate-solubilizing bacteria *Bacillus thuringiensis* strain B1 from an acidic soil in China. Inoculation of B1 increased available P and peanut growth under acidic soil condition. Inoculation by strain B1 also considerably enhanced shoot length, branch number, hundred seed weight, and crude protein contents. Similar results on groundnut have also been reported by Maheswar and Sathiyavani (2012). As an endophytic bacterium, *Bacillus* sp. has been reported to solubilize P and enhance growth of banana plant (Matos et al. 2017). Kibrom et al. (2017) reported on the isolation and characterization of phosphate-solubilizing *Bacillus* spp. from various agroclimatic regions of Tigray, Ethiopia. The isolated bacteria enhanced P uptake as well as root and shoot length. Mohamed et al. (2018) isolated phosphate-solubilizing *Bacillus subtilis* and *Serratia marcescens* from the tomato plant rhizosphere. Inoculation of both bacteria in tomato plants enhanced phosphate uptake. Similarly, Turan et al. (2007) reported about the influence of *Bacillus* strain FS 3 on growth and development of tomato (*Lycopersicon esculentum* L.) plants and enhanced phosphate content. The inoculation of FS 3 increased plant height and root length. In another study conducted by Tahir et al. (2013), three phosphate-solubilizing bacterial strains, viz., *Azospirillum*, *Bacillus*, and *Enterobacter*, were isolated and characterized. Later these strains were identified based on 16SrRNA sequence analysis. Inoculation of these three strains improved wheat (*Triticum aestivum* L.) growth and phosphate uptake in grains. Jadhav (2016) studied the P solubilization and biocontrol aspects of *Bacillus licheniformis* isolated from the *Cajanus cajan* rhizosphere. It was found that *Bacillus licheniformis* solubilized a good amount of phosphate under laboratory conditions. Owing to its P-solubilizing attribute, this isolate may be utilized as a potential P-mobilizing biofertilizer. Ahmad et al. (2018) studied the effect of phosphate-solubilizing *Bacillus subtilis* Q3 and *Paenibacillus* sp. Q6 for enhancing cotton plant growth under alkaline soil conditions. The strains Q3 and Q6 enhanced plant growth and P uptake. From a unique rhizosphere of an aromatic plant,

Tallapragada and Seshachala (2012) isolated and characterized phosphate-solubilizing microbes from the rhizospheres of Piper betel. The isolated *Bacillus* sp. exhibited good amount of phosphate-solubilizing potential. Under semiarid conditions, the effect of salt-tolerant and phosphate solubilizers *Bacillus sphaericus* and *Burkholderia cepacia* were evaluated by Ramani and Patel (2011) on food and fodder crops. Both the bacterial isolates exhibited noteworthy effect under pot culture and field conditions.

Bahadir et al. (2018) discovered potential bioinoculants 440 *Bacillus* isolates from various sources. These were investigated qualitatively for P solubilization, and affirmative isolates were further tested for quantitative determination of P solubilization and production of organic acid. The six best phosphate solubilizers were again tested for production of phytohormone (IAA), seed germination under in vitro conditions, and pot experiments. All six best *Bacillus* strains produced a good amount of IAA, meaningfully improved root and shoot length, and noticeably enhanced plant growth and development. In another study, the effect of P-solubilizing *B. pumilus* was studied on cauliflower. *Bacillus* sp. not only improved P uptake but also enhanced cauliflower size and weight as compared to the control (Dipta et al. 2017). The phosphate-solubilizing *Bacillus* spp. employed several mechanisms for P solubilization. Figure 2.1 shows few important mechanisms for phosphate solubilization by *Bacillus* spp.

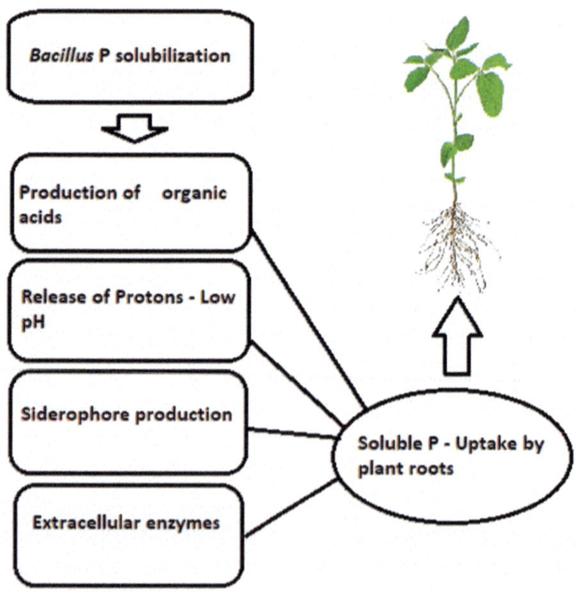

Fig. 2.1 Some of the P-solubilizing features of *Bacillus* spp.

2.3 Role of *Bacillus* in Stress Regulation

Tiwari et al. (2017) reported the role of *Bacillus amyloliquefaciens* (strain SN13) in remodeling the numerous abiotic stresses, viz., desiccation, heat, salt, drought, freezing, and cold, on a prevalent rice cv. Saryu 52 under hydroponic conditions. In addition to this, the rice seedlings were also supplied exogenously salicylic acid, abscisic acid, ethephon, and jasmonic acid to find out the role of *Bacillus* strain SN13 in phytohormone persuaded abiotic stresses. Phytohormone treatment and the abiotic stresses considerably influenced many biochemical and physiological plant parameters such as integrity of cell membrane and accumulation of osmolyte. Bacterial strain SN13 also modulated positively the stress responsible expressions of genes under numerous abiotic stress conditions and phytohormone treatments, which suggests its versatile role.

Zhou et al. (2016) examined the influence of *Bacillus megaterium* strain BOFC15 on *Arabidopsis* plants. Strain BOFC15 manufactured and excreted spermidine (Spd), which a type of polyamine (PA) that plays a crucial part in plant growth and development. The Spd making BOFC15 strain enhanced the plant tolerance against drought, which was related to changed cell ABA levels and stimulated an adaptive response. Several species of *Bacillus*, such as *B. subtilis*, *B. amyloliquefaciens*, *B. thuringiensis*, *B. licheniformis*, and *P. favisporus*, were applied to maize to study growth, antioxidant, and osmolyte status. Inoculation with *Bacillus* spp. augmented biomass of plant, relative water content (RWC), and lessened loss of leaf water. Inoculation with *Bacillus* spp. also resulted in increased sugar, proline, free amino acids, diminished leakage of electrolytes, and reduction of the antioxidant enzyme activities. Because of inoculation with the various spp. of *Bacillus*, there was an alleviation in drought stress in maize seedlings; therefore, bacterial inoculation resulted in physiological changes in plant (Vardharajula et al. 2011).

In Telangana State of India, Chari et al. (2018) isolated 44 *Bacillus* rhizospheric isolates, and of those isolates, 28 showed plant growth-promoting attributes. These 28 responsive rhizospheric isolates were tested for several abiotic stresses such as pH tolerance, salt, temperature, heavy metals (arsenic and cadmium), and drought tolerance. Out of 28, 4 *Bacillus* isolates demonstrated good growth at pH 4.0 to 12.0, 5 isolates exhibited tolerance to NaCl concentration from 1.5 to 20%, 6 isolates showed temperature tolerance from 20 °C to 50 °C, and, finally, 4 isolates exhibited water potential tolerance ranging from −0.05 Mpa to −0.73 Mpa. *Bacillus cereus* strain AR 156 triggered induced systemic resistance in the *Arabidopsis thaliana* plant by instantaneously triggering the salicylate- and jasmonate-/ethylene-dependent signaling pathways under abiotic stress. This plant growth-promoting bacterium also exhibited biocontrol effects against plant pathogen *Pseudomonas syringae* pv. tomato DC3000 (Niu et al. 2011).

In another interesting study conducted in Egypt (Sinai region) by Bochow et al. (2001), one potential rhizobacterium *Bacillus subtilis* strain FZR24, which was already registered as biological control agent, was tested in field as a salt-tolerant strain on two pepper and eggplant cultivars in saline soil conditions. Irrigation to both the cultivars was done by saline groundwater. It was observed that use of *Bacillus* strain FZB24 resulted in increased yield up to 430% in the pepper and 550% in eggplant cultivars, as compared to the uninoculated cultivars. In saline water-irrigated plots, other significant plant growth parameters were found to be increased by inoculation of *Bacillus* strain. Therefore, the bacterial inoculation resulted in 50 and 25% diminution in water salinity effect on pepper and eggplant yield, respectively.

Edkie et al. (2014) isolated *Bacillus subtilis* isolates from the rhizospheric zone of sugarcane variety, CoM 0265, which was saline tolerant. This strain was tested against commercially grown sugarcane cultivar, Co 86032, which was subjected to saline stress conditions under greenhouse studies. Plant parameters like root and shoot length, number of leaves, chlorophyll content, and cations such as Cu, Mn, Fe, and K showed noteworthy increases with *Bacillus* application. Interestingly, biochemical parameters such as malondialdehyde, reducing sugar, proline, proteins, catalase and ascorbate peroxidase contents, and minerals like Na and Zn increased under saline stress conditions, but inoculation with the *Bacillus* isolate resulted in noteworthy lessening in abovementioned components and, therefore, helped in protecting plants from saline stress. Figure 2.2 shows some of the stress regulating features of *Bacillus* spp.

Kumar et al. (2014) reported the abiotic as well as biotic stress regulating potential of *Bacillus* sp. isolated from different rhizospheric zones of India. It was found

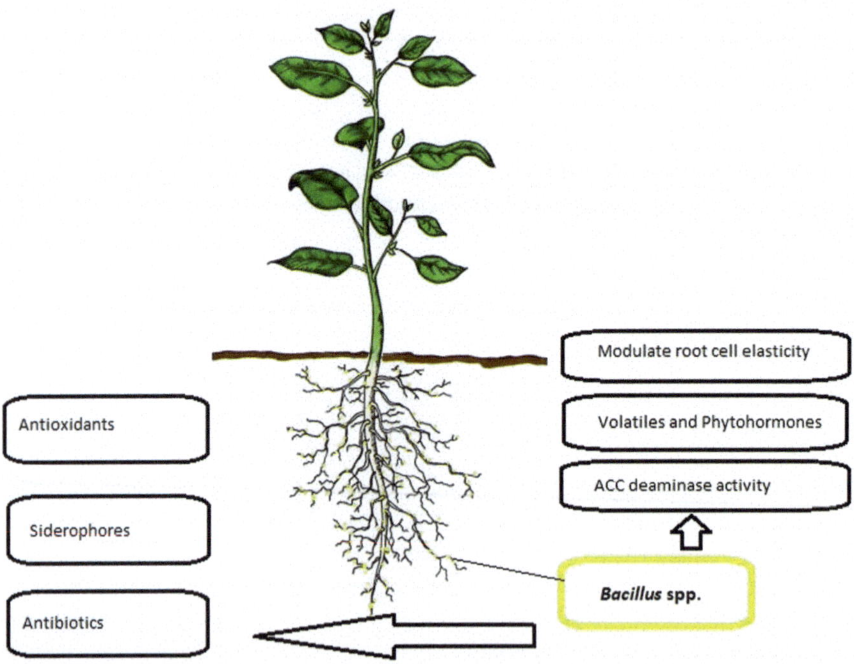

Fig. 2.2 Some of the stress regulating features of *Bacillus* spp.

that *Bacillus* sp. could control the growth of pathogenic fungal species such as *Sclerotium rolfsii*, *Botrytis ricini*, *Macrophomina phaseolina*, *Fusarium oxysporum*, and *Rhizoctonia solani*. On the other hand, this bacterium also demonstrated abiotic stress tolerance against salinity (NaCl 7%), higher temperature (50 °C), and drought (−1.2 MPa).

Gagné-Bourque et al. (2016) reported on the interaction of endophytic *Bacillus subtilis* strain B26 with timothy (*Phleum pratense* L.) on growth parameters, photosynthetic activity, water status, and overall metabolism under drought stress conditions. Under drought-induced and non-drought-induced conditions, *Bacillus* B26 effectively occupied the internal tissues of timothy plant and resulted in a positive influence on overall plant growth and development. After exposing the inoculated timothy plant to 8-week drought stress conditions, root and shoot biomass increased by 63.8 and 26.6%, also increment of 214.9 and 55.2% photosynthetic and stomatal conductance activity, respectively, as compared to the uninoculated timothy plants. Liu et al. (2013) studied the effect of *B. subtilis* inoculated with *Platycladus orientalis*. It was observed that the cytokinin hormone production was elevated which resulted in enhancement of ABA levels in shoots and increased the stomatal conductance.

Saikia et al. (2018) evaluated the effect of a consortium having ACC deaminase-producing activity *Bacillus subtilis* RJ46, *Ochrobactrum pseudogrignonense* RJ12, and *Pseudomonas* sp. RJ15 on alleviation of drought stress in *Pisum sativum* L. and *Vigna mungo* L. Inoculation with the consortium of three bacteria meaningfully increased germination of seed percentage, shoot and root length, and dry biomass. A raised level of reactive oxygen species (ROS) rummaging enzymes and osmolytes of cell, upsurge in relative water contents and leaf chlorophyll contents were detected after treatment with the consortium as compared to uninoculated plants under drought condition.

In addition, Armada et al. (2014) studied the influence of *Bacillus thuringiensis* on *Lavandula dentata* plant, and it was observed that production of IAA by bacterium resulted in higher proline and potassium contents and reduction in ascorbate peroxidase (APX) and glutathione reductase (GR) activities.

2.4 *Bacillus* as Biocontrol Agent

Utilization of microorganisms as a biological control agent to prevent the numerous diseases of plants is an attractive approach as compared to synthetic chemicals. The prosperous heterogeneity of microorganisms offers infinite potential for this alternative approach. Varieties of microbes are known for the biological removal of disease-causing agents in plants, but the majority of the research has employed strains of *Bacillus*, *Pseudomonas*, and *Trichoderma* (McSpadden and Driks 2004). It is a global challenge to generate and supply quality food to consumers which is devoid of undesirable levels of toxic chemicals. Recent strategies to prevent the loss due to plant pathogens depend upon the resistant cultivars or chemical pesticides, but some drawbacks are also associated with these approaches. The usage of chemicals is regulated because of health and environmental concerns. Secondly, after some time, pathogens develop resistance. Further, GMOs are highly controversial with the public (Fry 2008). Thus, eco-friendly strategies like the use of microbial strains as biopesticides provides an interesting alternative for the prevention of pests and plant diseases.

Traditionally plant growth-promoting rhizobacteria (PGPRs) are utilized as biocontrol agents, and their demand is increasing steadily (Lugtenberg and Kamilova 2009). Comprehensive studies have made a way for the commercialization and screening of various *Bacillus* strains that can be employed as biocontrol agents (McSpadden and Fravel 2002). The utilization and number of significant *Bacillus* strains are growing extensively. *Bacillus* strains have unique properties. They can replicate rapidly; most of them are resistant to different environmental conditions and possess a wide spectrum of biocontrol ability (Compant et al. 2005). Some strains of *Bacillus* species have the ability to intrude into the innermost plant tissues and play an essential role in plant growth promotion and plant protection. These *Bacillus* species mostly belong to free-living soil creatures, and most of the species include *B. subtilis* (Mishagi and Donndelinger 1990), *B. insolitus*, *B. pumilus* (McInroy and Kloepper 1995), *Paenibacillus polymyxa* (Shishido et al. 1999), *B. amyloliquefaciens* (Reva et al. 2002), *B. cereus* (Pleban et al. 1997), *B. megaterium* (McInroy and Kloepper 1995), and *B. licheniformis*. *B. endophyticus* a novel species has been also isolated from the cotton plant's inner tissues. US Environmental Protection Agency has registered eight species of microbes that can be utilized commercially as biocontrol agents (Cook et al. 1996). It includes three gram-positive bacteria (two *B. subtilis* and one *Streptomyces griseoviridis*). Numerous evidences are reported in literature that shows that *Bacillus* serves as an excellent biocontrol agent by utilizing various mechanisms like antibiosis, siderophore production, secretion of hydrolytic enzymes, etc. Different strains of *Bacillus* have been field-tested, and this has been directed toward the generation of products that can be utilized commercially as biocontrol agents. US EPA has provided a catalogue of biopesticides in which commercial formulations of distinct strains of *Bacillus* that can be employed as biocontrol agents are prescribed (McSpadden Gardener 2004). These products are accessible in distinct types of preparations such as a wettable powder, dry cakes, or liquid or suspension in a liquid, as it depends on the nature of relationship between the biocontrol strain and carrier molecule. Products such as Taegro, Sublilex, Companion, Serenade, and Kodiak are all dependent on the utilization of various strains of *B. subtilis* as a biocontrol agent. Kodiak is known for the eradication of root-borne disease-causing agents of soybean and cotton like *Aspergillus* spp., *Alternaria* spp., *Fusarium* spp., and *R. solani*. Moreover, Serenade (Agra Quest, Daius CA, USA) constituting *B. subtilis* strain QST713 is known to remediate the *Cercospora* leaf spot, early and late blight diseases related with different crop plants. Until recently, there was no data available regarding the genetic mechanisms behind the biocontrol potential of *B. subtilis* strain (Joshi and McSpadden Gardener 2006). *B. subtilis* produces volatile compounds that have an essential part in plant growth promotion and stimulation of defense mechanisms in plants. The *B. subtilis*, enzymatic, and endosporic products are extremely active against different fungal pathogens. Other than biocontrol ability of *Bacillus* species, they have essential role in plant growth promotion by stimulating the biosynthesis of numerous plant hormones such as indole-3-acetic acid (IAA) and gibberellic acid (GA3) (Chen et al. 2007; Harman 2011). Plant growth promotion by *Bacillus* species is observed when cell suspensions or spores of *B. subtilis* OTPBI were administered on tomato seed; it vigorously promoted the root and shoot growth, leaf area, and seedling vigor of tomato plant. High levels of plant defense-related enzymes and growth-promoting hormones (GA3 and IAA) were observed in treated plants as compared to nontreated plants (Chowdappa et al. 2013). Many researchers have analyzed that *B. subtilis* has the potential of plant growth promotion and improving yield by enhancing the nutrient uptake and stimulation of plant hormones and reducing ethylene production (Idris et al. 2007a, b). IAA plays a significant role in development of shoots and adventitious roots (Gardner 2009). Researchers analyzed hundreds of bacterial strains for their growth promotion and biocontrol ability and observed that *B. amyloliquefaciens* strain 54 has considerably enhanced the plant growth and increased resistance against the bacterial fruit blotch of cucurbitaceous crops (Jiang et al. 2015). *B. subtilis* strains NDSR7, MDSR14, and MDSR11 were evaluated for solubilization, and mobilization of Zn, soil biology, and crop growth ability, under laboratory conditions. The majority of the experiments conducted have proven that *Bacillus* strains have potential and they can induce resistance against a wide range of fungal and plant pathogens. *B. vallismortis* has the ability to control various bacterial and fungal infections (Park et al. 2001). *B. subtilis* strain FZB-G is able to stimulate phytohormone precursors which play significant roles in activation of defense mechanisms which leads to the generation of defense-related compounds (Gupta et al. 2000). Some enzymes with high lytic activity such as b-1,3-glucanase and chitinase are also produced by *Bacillus* strains. These enzymes can degrade the fungal cell wall very well. Several *Bacillus* species produce hydrolyzing enzymes like chitinase and glucanases that can effectively degenerate the pathogenic fungal cell wall (Leelasuphakul et al. 2006). Two types of chitinase enzymes are synthesized by *B. amyloliquefaciens* and can inhibit *F. oxysporum* growth (Wang et al. 2004). Chitinase enzymes with high chitinitic character are also produced *by B. subtilis* against various fungal pathogens (Chang et al. 2009). *B. thuringiensis* also shows the inhibition of *S. rolfsii* and several other fungal related diseases of soybean. Subspecies of *B. thuringiensis*, called as *colmeri*, produces chitinase which can block the maturation of fungal spores (Liu et al. 2010). Biocontrol agents based on *Bacillus* offer an effective, efficient, attractive, and eco-friendly approach toward the management of plant pests. The potential of *Bacillus* species to produce antibiotics that belong to multiple classes can be exploited to prevent wide range of plant diseases. But for successful implementation of biological agents, it is necessary to study their ecological behavior. Practical understanding is required regarding the diversity, ecological distribution, mode of action, and interacting environment in which biocontrol agent must perform their work, for the efficient acclimatization of these biological agents for sustainable agricultural approach.

2.5 *Bacillus* as a General Plant Growth Promoter

Seed germination, plant development, and plant growth are considerably manipulated by the available nutrients in the soil ecosystem. The plant roots absorb nitrogen (N) and phosphorus (P) from the soil because of the root transporter system, but the biologically available N and P forms are in limited amounts in the rhizospheric zone (De-Willigen 1986). The valuable and advantageous effect of *Bacillus* spp. in the crop improvement is very well known. *Bacillus* spp. uses various means (direct as well as indirect) to promote the plant growth. *Bacillus* spp. transform the intricate or difficult form of indispensable nutrients, such as N and P, into very simple and bioavailable forms that can be uptaken by the plant root system (Kuan et al. 2016; Shafi et al. 2017). Furthermore, nitrogen is a vital and significant constituent of nucleic acids, proteins, and other organic components in plants. The biological available form of nitrogen in soil is partial or inadequate, which slows down plant growth in its natural habitat system (Barker et al. 1974; De-Willigen 1986). There are some *Bacillus* spp. which release NH_3 from nitrogenous organic matter (Hayat et al. 2010). Oliveira et al. (1993) reported that some of the *Bacillus* spp. possess the nif gene and demonstrate nitrogenase enzyme activity. This enzyme helps in fixation of atmospheric nitrogen and delivers it to the plants to augment plant development, growth, and yield by fulfilling N requirements and delaying the process of senescence (Seldin et al. 1984; Ding et al. 2005). The *Bacillus* spp. also excretes the proteinaceous compound known as a siderophore, which helps in iron chelation from the rhizospheric zone of plant (Wilson et al. 2006). The iron-chelating compound siderophores bind Fe^{+++} in complex substances and reduce the ferric form into ferrous form, which can easily enter plant root system (Dertz et al. 2006).

Kumar et al. (2012) isolated seven soil bacterial isolates from bean rhizosphere in the Uttarakhand Himalayan region. These demonstrated excellent plant growth-promoting and biocontrol activities. On the bases of 16S rRNA gene sequence, the soil isolate was identified as *Bacillus* sp. and named BPR7. The rhizospheric strain BPR7 made phytohormone IAA, siderophores, enzyme phytase, organic acids, cyanogens, and ACC deaminase and was also able to solubilize numerous types of inorganic and organic phosphatic material as well as Zn and K. Interestingly, the *Bacillus* strain BPR7 demonstrated biocontrol activities and hindered the growth of many phytopathogens such as *Fusarium oxysporum*, *Fusarium solani*, *Macrophomina phaseolina*, *Sclerotinia sclerotiorum*, *Colletotrichum* sp., and *R. solani* under laboratory conditions. Gutiérrez-Mañero et al. (2008), reported that, *B. pumilus* and *B. licheniformis* isolated from alder (*Alnus glutinosa* [L.] Gaertn.) rhizosphere exhibited strong plant growth-promoting potential. Therefore, both the *Bacillus* bacteria demonstrated the property of potential PGPR. The bioassay facts displayed that the dwarf phenotype produced in alder seedlings by a chemical paclobutrazol (which inhibits biosynthesis of gibberellin or GA) was reversed effectively by applying bacterial extracts from media and by exogenous treatment of GA3.

Calvo et al. (2010) isolated 63 *Bacillus* strains from the native potato rhizospheric regions growing in the Andean highlands of Peru. The strains demonstrated strong biocontrol activity against *Rhizoctonia solani* and *Fusarium solani*. The antagonistic *Bacillus* strains were further verified for additional plant growth promotional aspects. It was observed that 81% produced a certain amount of phytohormone indole-3-acetic acid (IAA) and 58% could solubilize tricalcium phosphate (TCP). The phylogenetic examination divulged that the majority of soil isolates belonged to *B. amyloliquefaciens* sp., while other three potential strains may be novel putative *Bacillus* sp. Therefore, these *Bacillus* spp. may be uses as potential potato growth promoter.

From various locations in Pakistan (Punjab province), Shakeel et al. (2015) isolated 234 soil isolates from the rhizosphere of basmati super rice and basmati-385 varieties cultivated in clay loam and saline soils. Out of the total 234 rice rhizospheric isolates, 27 isolates solubilized Zn from zinc oxide, zinc carbonate, and zinc phosphate. The soil isolate SH10 exhibited maximum zone of Zn solubilization (24 mm) using zinc phosphate, and soil isolate SH17 showed maximum zone of Zn solubilization (14–15 mm) using zinc oxide and zinc carbonate. Strains SH10 and SH17 also solubilized P (38–46 mm) and K (47–55 mm) under laboratory conditions. Both the strains also exhibited biocontrol activities against potential plant pathogens *Fusarium moniliforme* and *Pyricularia oryzae* by 22–29% and manufactured many biological control factors in vitro conditions. Bacterial strains SH10 and SH17 also enhanced zinc uptake and translocation into the grains and augmented plant yield of super basmati rice and basmati-385 varieties by 18–47% and 22–49%, correspondingly. Using 16S rRNA gene analysis, both the SH10 and SH17 bacterial strains were identified as *Bacillus* sp. and *Bacillus cereus*.

Idris et al. (2002) studied many *Bacillus* strains belonging to *B. subtilis* or *B. amyloliquefaciens*. They were isolated from plant pathogen-contaminated soils and demonstrated PGPR properties. Soil isolate *B. amyloliquefaciens* could biodegrade extracellular compound phytate (myoinositol hexakisphosphate). The maximum phytase activity was found in isolate FZB45, and the diluted bacterial cultural filtrate of this isolate encouraged maize seedling growth in phytate presence and P limitation.

Lucas Garcia et al. (2004) studied the effect of inoculation of *B. licheniformis* on development and growth of tomato and pepper in three experiments. In the first experimental condition, the bacterium meaningfully enhanced plant height and the leaf surface area in both cultivars, and the effect was more on pepper as compared to the tomato. In a second experiment, the plant seedlings were grown in sand, and hydroponic systems were tested. Interestingly, the size and number of tomato fruit increased after application of bacterial inoculation in sand and in the hydroponic system. Moreover, the inoculated cultivars exhibited less disease infestation as compared to untreated plants. In the third experiment, the pepper yield was greater in inoculated plants as compared to the uninoculated plants. This *Bacillus* strain had appreciable colonization and survival potential, and it could be employed as a biofertilizer or as biological control means.

Bacillus strain ML3 was isolated from a soil sample, and it produced the phytohormone and IAA and NH_3 in the range of 174.72 and 0.66 μg ml^{-1}, correspondingly. This strain also produced siderophores and effectively solubilized tricalcium phosphate (TCP). The main functional nitrogenase gene nif H was also found in this strain. Based on morphological, biochemical, and partial 16S rRNA gene sequencing, this strain ML3 was identified as *B. licheniformis*. This strain has all properties to be utilized as an efficient plant growth promoter (Kayasth et al. 2013).

Agarwal and Agarwal (2013) isolated 28 bacterial strains from dissimilar tomato rhizospheric soils from Dehradun area of Uttarakhand, India. All of the soil isolates were characterized biochemically and tested for plant growth-promoting abilities such as indole acetic acid production, P solubilization, HCN production, siderophore excretion, and catalase activity. Out of 28 soil isolates, only 5 *Bacillus* isolates exhibited potential plant growth-promoting abilities. Inoculation of tomato plants with selected *Bacillus* sp. resulted in increment in shoot and root length of tomato seedlings as compared to the uninoculated plants. *Bacillus* sp. also enhanced the seed germination percentage as well as seed vigor index.

Ramírez and Kloepper (2010) have studied the consequence of inoculum concentration and soil phosphate-associated properties on plant growth encouragement by *B. amyloliquefaciens* strain FZB45, which produced phytase. A noteworthy synergism between soil phosphate and bacterial application was observed. *B. amyloliquefaciens* strain FZB45 encouraged plant growth and phosphate uptake, which confirms the role of enzyme phytase and less P uptake in uninoculated plants. This strain also produced IAA under laboratory conditions, but its role was not determined.

Sharma et al. (2013) isolated a bacterium from the rhizosphere of soybean plants grown at Directorate of Soybean Research, Indore, Madhya Pradesh, India, and this bacterium was identified as *Bacillus* sp. on the basis of morphological and biochemical tests as well as FAME profile. Studies on 16S rRNA gene revealed 98.7% similarity to *B. amyloliquefaciens*, and thereafter, it was labeled as strain sks_bnj_1 (AY 932823). This *Bacillus* strain owned manifold plant growth-promoting attributes such as IAA production; siderophore production; ACC deaminase activity; enzymes like phosphatases, phytases, and cellulases; Zn solubilization; and HCN production. This bacterium also exhibited biocontrol properties. Interaction of this strain with soybean increased the shoot root biomass as well as nutrient uptake as compared to the uninoculated control plants. In another study on *Bacillus* sp. with banana plant conducted by Cruz-Martín et al. (2015) in Cuba, it was observed that strain *B. pumilus* could fix atmospheric nitrogen and was able to grow in nitrogen-free culture media and produced IAA (28.9 μg ml^{-1}). Moreover, strain *B. pumilus* significantly enhanced the plant height and thickness of the stem, altered root architecture, and improved fresh and dry plant weights. Table 2.1 shows the dissimilar mechanism involved in PGP attributes by *Bacillus* spp.

Table 2.1 Various means and mechanisms of *Bacillus* spp. shows plant growth-promoting attributes

S. No	Functions	References
1	Phytohormone production, abiotic stress management, drought resistance	Gutiérrez-Mañero et al. (2008), Vardharajula et al. (2011), Armada et al. (2014), Kumar et al. (2014), Kang et al. (2015), Tiwari et al. (2017), Zhou et al. (2016), Chari et al. (2018)
2	Activation of salicylate- and jasmonate-/ethylene-dependent signaling pathways	Niu et al. (2011)
3	Salt tolerance, act as biocontrol agent, excretion of antibiotic compounds	Sailaja et al. (1997), Bochow et al. (2001), Joshi and McSpadden Gardener (2006), Chen et al. (2009), Shafi et al. (2017), Edkie et al. (2014), Kumar et al. (2014), Shafi et al. (2017)
4	Drought stress management and metabolic changes	Gagné-Bourque et al. (2016)
5	Production of cytokinin, PGPR properties, drought stress resistance, ACC deaminase activity	Liu et al. (2013), Armada et al. (2014), Saikia et al. (2018)
6	Identification of nif gene, nitrogen fixation	Oliveira et al. (1993), Ding et al. (2005), Seldin et al. (1984)
7	Siderophore production, induced systemic resistance	Bargabus et al. (2002), Kloepper et al. (2004), Dertz et al. (2006), Wilson et al. (2006), Choudhary and Johri (2008)
8	IAA production, enzyme production	Lucas Garcia et al. (2004), Calvo et al. (2010)
9	Zinc uptake and mobilization to grains	Shakeel et al. (2015)
10	Extracellular phytase production, P uptake	Idris et al. (2002), Ramírez and Kloepper (2010)
11	IAA, ammonia production, tricalcium phosphate solubilization	Kayasth et al. (2013), Cruz-Martín et al. (2015)
12	Indole acetic acid production, P solubilization, HCN production, siderophore excretion, and catalase activity	Kilian et al. (2000), Wilson et al. (2006), Yu et al. (2011), Agarwal and Agarwal (2013), Panhwar et al. (2013)
13	IAA production; siderophore production; ACC deaminase activity; enzymes like phosphatases, phytases, and cellulases; Zn solubilization; and HCN production	Idris et al. (2007a, b), Sharma et al. (2013), Ramesh et al. (2014), Ma et al. (2015), Kamran et al. (2017)

2.6 Conclusions

The application of *Bacillus* spp. as a potential plant growth promoter as well as a biocontrol agent has been reported by researchers all over the globe. People have tested this bacterium with several crops under different climatic conditions. Still, this bacterium is utilized as potential P solubilizer rather than as a comprehensive biopriming agent to be utilized as a complete plant growth promoter. Many researchers have isolated *Bacillus* sp. form soil and rhizospheric zone having beneficial plant growth attributes such as phytohormone (IAA) production, phosphate, potassium and Zn solubilization, enzymes like phytase, phosphatase, cellulose manufacturer, HCN and siderophores producer as well as biological control properties. In the various corners of the developed world, where agrochemicals are comparatively low-priced and reasonable, here the application of plant growth-promoting (PGP) *Bacillus* spp. plays a significant role toward the growth and progress of organic agriculture. Additionally, it is also rational and judicious to assume the enhanced employment of PGP *Bacillus* spp. in numerous phyto-remediation approaches. Numerous soil isolates of *Bacillus* spp. have been fostered as potential biocontrol agents to control the multiplication of plant pathogens. On the other hand, to be applied as effective and fruitful biocontrol agent, we have understood the ecological perspective of a *Bacillus* sp. In this context, better understanding of the assortment, dissemination, and physiological aspects of gram-positive *Bacillus* species will be very useful in identification and documentation of novel strains matching with the existing cropping systems. Before introducing a bacterium into the rhizosphere, we have to understand its ecological perspectives for its successful colonization. The genomic level mechanisms implicated in colonization of roots by *Bacillus* spp. are being studied by several workers. The development and refinement of technology and progression in metabolic and genomic techniques will offer novel possibilities for refining the process of biopriming agents' selection, their characterization, and regulation in association with plants. Improvement in techniques of functional genomics and proteomics will help us in controlling the gene functions for mass production, their consortium formulation (if required), and their application under field conditions. Transfer of genes responsible for drought tolerance, or metal tolerance in a *Bacillus* sp. having effective plant growth-promoting attributes, will allow it to be utilized under any required conditions. There would be a huge demand in the market for improved or engineered bacterium, and this will be a big advancement for organic farming as well as sustainable agriculture.

References

Agrawal DPK, Agrawal S (2013) Characterization of *Bacillus* sp. strains isolated from rhizosphere of tomato plants (*Lycopersicon esculentum*) for their use as potential plant growth promoting rhizobacteria. Int J Curr Microbiol Appl Sci 2(10):406–417

Ahmad M, Ahmad I, Hilger TH, Nadeem SM, Akhtar MF, Jamil M, Hussain A, Zahi ZA (2018) Preliminary study on phosphate solubilizing *Bacillus subtilis* strain Q3 and *Paenibacillus* sp. strain Q6 for improving cotton growth under alkaline conditions. PeerJ 6:e5122. https://doi.org/10.7717/peerj.5122

Armada E, Roldan A, Azcon R (2014) Differential activity of autochthonous bacteria in controlling drought stress in native Lavandula and Salvia plants species under drought conditions in natural arid soil. Microb Ecol 67:410–420

Bahadir PS, Liaqat F, Eltem R (2018) Plant growth promoting properties of phosphate solubilizing *Bacillus* species isolated from the Aegean Region of Turkey. Turk J Bot 42:183–196

Bale JS, van Lenteren JC, Bigler F (2008) Biological control and sustainable food production. Philos Trans R Soc Lond B Biol Sci 363(1492):761–776

Bargabus RL, Zidack NK, Sherwood JW, Jacobsen BJ (2002) Characterization of systemic resistance in sugar beet elicited by a non-pathogenic, phyllosphere colonizing *Bacillus mycoides* biological control agent. Physiol Mol Plant Pathol 61:289–298

Barker AV, Maynard DN, Mills HA (1974) Variations in nitrate accumulation among spinach cultivars. J Am Soc Horticult Sci 99:32–134

Bochow H, El-Sayed SF, Junge H, Stavropoulou A, Schmiedeknecht G (2001) Use of *Bacillus subtilis* as biocontrol agent. IV. Salt-stress tolerance induction by *Bacillus subtilis* FZB24 seed treatment in tropical vegetable field crops, and its mode of action. J Plant Dis Prot 108(1):21–30

Calvo P, Ormeño-Orrillo E, Martínez-Romero E, Zúñiga D (2010) Characterization of *Bacillus* isolates of potato rhizosphere from Andean soils of Peru and their potential PGPR characteristics. Braz J Microbiol 41(4):899–906

Chari KD, Reddy RS, Triveni S, Trimurtulu N, Rani CVD, Sreedhar M (2018) Isolation and characterization of abiotic stress tolerant plant growth promoting *Bacillus* spp. from different rhizospheric soils of Telangana. Biosci Biotechnol Res Asia 15(2):47–53

Chang WT, Chen ML, Wang SL (2009) An antifungal chitinase produced by Bacillus subtilis using chitin waste as a carbon source. World J Microbiol Biotechnol 26(5):945–950

Chen XH, Koumoutsi A, Scholz R (2007) Comparative analysis of the complete genome sequence of the plant growth-promoting bacterium *Bacillus amyloliquefaciens* FZB42. Nat Biotechnol 25(9):1007–1014

Chen XH, Koumaoutsi A, Scholz R, Borriss R (2009) More than anticipated-production of antibiotics and other secondary metabolites by *Bacillus amyloliquefaciens* FZB42. J Mol Microbiol Biotechnol 16:14–24

Choudhary DK, Johri BN (2008) Interactions of *Bacillus* spp. and plants-with special reference to induced systemic resistance (ISR). Microbiol Res 164:493–513

Chowdappa P, Mohan Kumar SP, Jyothi Lakshmi M, Upreti KK (2013) Growth stimulation and induction of systemic resistance in tomato against early and late blight by Bacillus subtilis OTPB1 or Trichoderma harzianum OTPB3. Biol Control 65(1):109–117

Compant S, Duffy B, Nowak J (2005) Use of plant growth promoting bacteria for biocontrol of plant diseases: principles, mechanisms of action, and future prospects. Appl Environ Microbiol 71(9):4951–4959

Cook RJ, Bruckart WL, Coulson JR, Goettel MS, Humber RA, Lumsden RD, Maddox JV, McManus ML, Moore L, Meyer SF, Quimby PC Jr, Stack JP, Vaughn JL (1996) Safety of microorganisms intended for pest and plant disease control: a framework for scientific evaluation. Biol Control 7:333–351

Cruz-Martín M, Mena E, Sánchez-García C, Roque B, Acosta-Suárez M, Pichardo T, Leiva-Mora M, Alvarado-Capó Y (2015) The effects of plant growth promoting *Bacillus pumilus* CCIBPC5 on 'Grande naine' (Musa AAA) plants in acclimatization stage. Biotecnología Vegetal 15(3):151–156

Dertz EA, Stintzi A, Raymond KN (2006) Siderophore-mediated iron transport in *Bacillus subtilis* and Corynebacterium glutamicum. J Biol Inorg Chem 11(8):1087–1097

De-Willigen P (1986) Supply of soil nitrogen to the plant during the growing season. In: Lambers H, Neeteson JJ, Stulen I (eds) Fundamental, ecological and agricultural aspects of nitrogen metabolism in higher plants. Martinus Nijhoff, Dordrecht, pp 417–432

Ding Y, Wang J, Liu Y, Chen S (2005) Isolation and identification of nitrogen-fixing bacilli from plant rhizospheres in Beijing region. J Appl Microbiol 99(5):1271–1281

Dipta B, Kirti S, Bhardwaj S, Gupta S, Kaushal R (2017) Phosphate solubilizing potential of *Bacillus pumilus* for the enhancement of cauliflower (*Brassica oleracea* var. botrytis l.). EM Int 23(3):1541–1548

Edkie GN, Jadhav S, Prasad DT (2014) *Bacillus subtilis* isolated from the sugarcane root rhizosphere: a potential bioinoculum to alleviate salinity stress in sugarcane cultivation. Asian J Agric Res 8:84–95

Fry W (2008) *Phytophthora infestans*: the plant (and R gene) destroyer. Mol Plant Pathol 9:385–4002

Gagné-Bourque F, Bertrand A, Claessens A, Aliferis KA, Jabaji S (2016) Alleviation of drought stress and metabolic changes in Timothy (*Phleum pratense* L.) colonized with *Bacillus subtilis* B26. Front Plant Sci. https://doi.org/10.3389/fpls.2016.00584

Garbeva P, van Veen JA, van Elsas JD (2004) Microbial diversity in soil: selection microbial populations by plant and soil type and implications for disease suppressiveness. Ann Rev Phytopathol 42:243–270

Gardner G (2009) Plant hormone signaling. Hortic Sci 44(1):222–223

Govindasamy V, Senthilkumar M, Magheshwaran V, Kumar U, Bose P, Sharma V, Annapurna K (2010) *Bacillus* and *Paenibacillus* spp.: potential PGPR for sustainable agriculture. In: Maheshwari DK (ed) Plant growth and health promoting bacteria. Springer, Berlin/Heidelberg, pp 333–356

Grayston SJ, Wang S, Campbell CD, Edwards AC (1998) Selective influence of plant species on microbial diversity in the rhizosphere. Soil Biol Biochem 30:369–378

Gupta V, Bochow H, Dolej S (2000) Ein das Pflanzenwachstum f€ordernder *Bacillus subtilis*-Stamm als potentieller Resistenzinduktor gegen die Fusarium-Welke an Tomaten [Plant growth-promoting *Bacillus subtilis* strain as potential inducer of systemic resistance in tomato against Fusarium wilt]. J Plant Dis Prot 107:145–154

Gutiérrez-Mañero FJ, Ramos B, Probanza A, Mehouachi J, Talon M (2008) The plant growth promoting rhizobacteria *Bacillus pumilus* and *Bacillus licheniformis* produce high amounts of physiologically active gibberelins. Physiol Plant 111:206–211

Harman GE (2011) Multifunctional fungal plant symbionts: new tools to enhance plant growth and productivity. New Phytol 189(3):647–649

Hayat R, Ali S, Amara U, Khalid R, Ahmed I (2010) Soil beneficial bacteria and their role in plant growth promotion: a review. Ann Microbiol 60:579–598

Idris EE, Makarewicz O, Farouk A, Rosner K, Greiner R, Bochow H, Richter T, Borriss R (2002) Extracellular phytase activity of *Bacillus amyloliquefaciens* FZB45 contributes to its plant-growth-promoting effect. Microbiology 148:2097–2109

Idris EE, Bochow H, Ross H, Borriss R (2004) Use of *Bacillus subtilis* as biocontrol agent. VI. Phytohormone action of culture filtrate prepared from plant growth promoting Bacillus amyloliquefaciens FZB24, FZB42, FZB45 and Bacillus subtilis FZB37. J Plant Dis Prot 111:583–597

Idris EE, Iglesias DJ, Talon M (2007a) Tryptophan dependent production of indole-3-acetic acid (IAA) affects level of plant growth promotion by *Bacillus amyloliquefaciens* FZB42. Mol Plant-Microbe Interact 20(6):619–626

Idris EES, Iglesias DJ, Talon M, Borriss R (2007b) Tryptophan-dependent production of Indole-3-Acetic Acid (IAA) affects level of plant growth promotion by *Bacillus amyloliquefaciens* FZB42. Mol Plant Microbe Interact 20:619–626

Jacoby R, Peukert M, Succurro A, Koprivova A, Kopriva S (2017) The role of soil microorganisms in plant mineral nutrition-current knowledge and future directions. Front Plant Sci 8:1617–1628

Jadhav RN (2016) Study of phosphate solubilization and antimicrobial activity of *Bacillus licheniformis* isolated from rhizosphere of *Cajanus cajan* cultivated in Marathwada region. Life Sci Int Res J 3(2):129–132

Jiang CH, Wu F, Yu ZY (2015) Study on screening and antagonistic mechanisms of *Bacillus amyloliquefaciens* 54 against bacterial fruit blotch (BFB) caused by *Acidovorax avenae* subsp. *citrulli*. Microbiol Res 170:95–104

Johri BN, Sharma A, Virdi JS (2003) Rhizobacterial diversity in India and its influence on soil and plant health. Adv Biochem Eng Biotechnol 84:49–89

Joshi R, McSpadden Gardener BB (2006) Identification and characterization of novel genetic markers associated with biological control activities in *Bacillus subtilis*. Biol Control 96:145–154

Kamran S, Shahid I, Baig DN, Rizwan M, Malik KA, Mehnaz S (2017) Contribution of zinc solubilizing bacteria in growth promotion and zinc content of wheat. Front Microbiol 8:2593–2601

Kang SM, Radhakrishnan R, You YH, Joo GJ, Lee IJ, Lee KE, Kim JH (2014) Phosphate solubilizing *Bacillus megaterium* mj1212 regulates endogenous plant carbohydrates and amino acids contents to promote mustard plant growth. Indian J Microbiol 54(4):427–433

Kang SM, Radhakrishnan R, Lee KE, You YH, Ko JH, Kim JH, Lee IJ (2015) Mechanism of plant growth promotion elicited by *Bacillus* sp. LKE15 in oriental melon. J Acta Agric Scand B Soil Plant Sci 65(7):637–647

Kayasth M, Kumar V, Gera R (2013) Exploring the potential of PGPR strain *Bacillus licheniformis* to be developed as multifunctional biofertilizer. Cent Eur J Exp Biol 2(1):12–17

Kibrom FG, Alemayehu W, Prakasam VR, Kiros W (2017) Isolation and characterization of efficient phosphate solubilizing *Bacillus* (PSB) from different agro-ecological zones of Tigray Soil, Ethiopia. Momona Ethiop J Sci 9(2):262–273

Kilian M, Steiner U, Krebs B, Junge H, Schmiedeknecht G, Hain R (2000) FZB24® *Bacillus subtilis* – mode of action of a microbial agent enhancing plant vitality. Pflanzenschutz Nachr Bayer 1:72–93

Kloepper JW, Ryu C-M, Zhang S (2004) Induced systemic resistance and promotion of plant growth by Bacillus spp. Phytopathology 94:1259–1266

Kuan KB, Othman R, Rahim KA, Shamsuddin ZH (2016) Plant growth-promoting rhizobacteria inoculation to enhance vegetative growth, nitrogen fixation and nitrogen remobilisation of maize under greenhouse conditions. PLoS ONE 11:e0152478. https://doi.org/10.1371/journal.pone.0152478

Kumar P, Dubey RC, Maheshwari DK (2012) *Bacillus* strains isolated from rhizosphere showed plant growth promoting and antagonistic activity against phytopathogens. Microbiol Res 167(8):493–499

Kumar GP, Ahmed SKMH, Desai S, Amalraj ELD, Rasul A (2014) In vitro screening for abiotic stress tolerance in potent biocontrol and plant growth promoting strains of *Pseudomonas* and *Bacillus* spp.. Int J Bacteriol 195946: 6p. https://doi.org/10.1155/2014/195946

Leelasuphakul W, Sivanunsakul P, Phongpaichit S (2006) Purification, characterization and synergistic activity of b1,3-glucanase and antibiotic extract from an antagonistic Bacillus subtilis NSRS 89-24 against rice blast and sheath blight pathogens. Enzym Microb Technol 38:990–997

Liu XY, Ruan LF, Hu ZF (2010) Genome-wide screening reveals the genetic determinants of an antibiotic insecticide in *Bacillus thuringiensis*. J Biol Chem 285(50):39191–39200

Liu F, Xing S, Ma H, Du Z, Ma B (2013) Cytokinin producing, plant growth promoting rhizobacteria that confer resistance to drought stress in *Platycladus orientalis* container seedlings Appl. Microbiol Biotechnol 97:9155–9164

Lucas Garcia JA, Probanza A, Ramos B, Ruiz Palomino M, Gutiérrez Mañero FJ (2004) Effect of inoculation of *Bacillus licheniformis* on tomato and pepper. Agronomie 24:169–176

Lugtenberg B, Kamilova F (2009) Plant-growth promoting rhizobacteria. Annu Rev Microbiol 63:541–556

Ma Y, Oliveira RS, Wu L, Luo Y, Rajkumar M, Rocha I, Freitas H (2015) Inoculation with metal mobilizing plant growth promoting rhizobacterium *Bacillus* sp. SC2b and Its role in rhizoremediation. J Toxicol Environmen Health A 78(13–14):931–944

Maheswar NU, Sathiyavani G (2012) Solubilization of phosphate by *Bacillus* Sps, from groundnut rhizosphere (*Arachis hypogaea* L). J Chem Pharm Res 4(8):4007–4011

Matos ADM, Gomes ICP, Nietsche S, Xavier AA, Gomes WS, Neto JADS, Pereira MCT (2017) Phosphate solubilization by endophytic bacteria isolated from banana trees. Anais da Academia Brasileira de Ciências 89(4):2945–2954

McInroy JA, Kloepper JW (1995) Survey of indigenous bacterial endophytes from cotton and sweet corn. Plant Soil 173:337–342

Mcspadden Gardener BB (2004) Ecology of Bacillus and Paenibacillus species in agricultural systems. Phytopathology 94:1252–1258

McSpadden Gardener BB, Driks A (2004) Overview of the nature and applications of biocontrol microbes: Bacillus spp. Phytopathology 94:1244

McSpadden Gardener BB, Fravel DR (2002) Biological control of plant pathogens: Research, commercialization and application in the USA. Plant Health Progress. https://doi.org/10.1094/PHP-2002-0510-01-RV

Mishagi IJ, Donndelinger CR (1990) Endophytic bacteria in symptom free cotton plants. Phytopathology 9:808–811

Mohamed EAH, Farag AG, Youssef SA (2018) Phosphate solubilization by *Bacillus subtilis* and *Serratia marcescens* isolated from tomato plant rhizosphere. J Environ Prot 9(3):266–277

Niu DD, Liu HX, Jiang CH, Wang YP, Wang QY, Jin HL, Guo JH (2011) The plant growth promoting rhizobacterium *Bacillus cereus* ar156 induces systemic resistance in Arabidopsis thaliana by simultaneously activating salicylate- and jasmonate/ethylene-dependent signaling pathways. Mol Plant Microbe Interact 24(5):533–542

Oliveira SS, Seldin L, Bastos MC (1993) Identification of structural nitrogen-fixation (nif) genes in *Bacillus polymyxa* and *Bacillus macerans*. World J Microbiol Biotechnol 9(3):387–389

Panhwar QA, Jusop S, Naher UA, Othman R, Razi MI (2013) Application of potential phosphate-solubilizing bacteria and organic acids on phosphate solubilization from phosphate rock in aerobic rice. Sci World J 2013: 10 p. 272409. https://doi.org/10.1155/2013/272409.

Park K, Ahn IP, Kim CH (2001) Systemic resistance and expression of the pathogenesis-related genes mediated by the plant growth-promoting rhizobacterium *Bacillus amyloliquefaciens* EXTN-1 against anthracnose disease in cucumber. Mycobiology 29(1):48–53

Pirttijärvi TS, Andersson MA, Salkinoja-Salonen MS (2000) Properties of *Bacillus cereus* and other bacilli contaminating biomaterial-based industrial processes. Int J Food Microbiol 60(2–3):231–239

Pleban S, Chernin L, Chet I (1997) Chitinolytic activity of an endophytic strain of *Bacillus cereus*. Lett Appl Microbiol 25:284–288

Ramani V, Patel HH (2011) Phosphate solubilization by *Bacillus sphaericus* and *Burkholderia cepacia* in presence of pesticides. J Agric Technol 7(5):1331–1337

Ramesh A, Sharma SK, Sharma MP, Yadav N, Joshi OP (2014) Inoculation of zinc solubilizing *Bacillus aryabhattai* strains for improved growth, mobilization and biofortification of zinc in soybean and wheat cultivated in Vertisols of central India. Appl Soil Ecol 73:87–96

Ramírez CA, Kloepper JW (2010) Plant growth promotion by *Bacillus amyloliquefaciens* FZB45 depends on inoculum rate and P-related soil properties. Biol Fertil Soil 46(8):835–844

Reva O, Smirnov VV, Petterson B, Priest FG (2002) *Bacillus endophyticus* sp. nov., isolated from the inner tissues of cotton plants (*Gossypium* sp.). Int J Syst Evol Microbiol 52:101–107

Richardson AE, Barea JM, McNeill AM, Prigent-Combaret C (2009) Acquisition of phosphorus and nitrogen in the rhizosphere and plant growth promotion by microorganisms. Plant Soil 321:305–339

Saikia J, Sarma RK, Dhandia R, Yadav A, Bharali R, Gupta VK, Saikia R (2018) Alleviation of drought stress in pulse crops with ACC deaminase producing rhizobacteria isolated from acidic soil of Northeast India. Sci Rep 8:3560

Sailaja PR, Podile AR, Reddanna P (1997) Biocontrol strain *Bacillus subtilis* AF1 rapidly induces lipoxygenase in groundnut (*Arachis hypogaea* L.) compared to crown rot pathogen *Aspergillus niger*. Eur J Plant Pathol 104:125–132

Seldin L, Van Elsas JD, Penido EGC (1984) *Bacillus azotofixans* sp. nov. a nitrogen-fixing species from Brazilian soils and grass roots. Int J Syst Bacteriol 34(4):451–456

Shafi J, Tian H, Ji M (2017) *Bacillus* species as versatile weapons for plantpathogens: a review. Biotechnol Biotechnological Equip. https://doi.org/10.1080/13102818.2017.1286950

Shakeel M, Rais A, Hassan MN, Hafeez FY (2015) Root associated *Bacillus* sp. improves growth, yield and zinc translocation for basmati Rice (*Oryza sativa*) varieties. Front Microbiol. https://doi.org/10.3389/fmicb.2015.01286

Sharma SK, Ramesh A, Johri BN (2013) Isolation and characterization of plant growth-promoting *Bacillus amyloliquefaciens* strain sks_bnj_1 and its influence on rhizosphere soil properties and nutrition of Soybean (*Glycine max* L. Merrill). J Virol Microbiol 2013:446006. https://doi.org/10.5171/2013.446006

Shishido M, Breuil C, Chanway CP (1999) Endophytic colonization of spruce by plant growth promoting rhizobacteria. FEMS Microbiol Ecol 29:191–196

Su J, Wu Y, Ma X, Zhang G, Feng H, Zhang Y (2004) Soil microbial counts and identification of culturable bacteria in an extreme by arid zone. Folia Microbiol 49(4):423–439

Swain MR, Laxminarayana K, Ray RC (2012) phosphorus solubilization by thermotolerant *Bacillus subtilis* isolated from cow dung microflora. Agric Res 1(3):273–279

Tahir M, Mirza MS, Zaheer A, Dimitrov MR, Smidt H, Hameed S (2013) Isolation and identification of phosphate solubilizer *Azospirillum*, *Bacillus* and *Enterobacter* strains by 16SrRNA sequence analysis and their effect on growth of wheat (*Triticum aestivum* L.). Aust J Crop Sci 7(9):1284–1292

Tallapragada P, Seshachala U (2012) Phosphate-solubilizing microbes and their occurrence in the rhizospheres of Piper betel in Karnataka. India Turk J Biol 36:25–35

Tiwari S, Prasad V, Chauhan PS, Lata C (2017) *Bacillus amyloliquefaciens* confers tolerance to various abiotic stresses and modulates plant response to phytohormones through osmoprotection and gene expression regulation in rice. Frontiers Plant Sci 8. https://doi.org/10.3389/fpls.2017.01510

Turan M, Ataoğlu N, Şahin F (2007) Effects of *Bacillus* FS-3 on growth of tomato (*Lycopersicon esculentum* L.) plants and availability of phosphorus in soil. Plant Soil Environ 53(2):58–64

Vardharajula S, Ali SJ, Grover M, Reddy G, Bandi V (2011) Drought tolerant plant growth promoting *Bacillus* spp.: effect on growth, osmolytes, and antioxidant status of maize under drought stress. J Plant Interact 6(1):1–14

Wang J, Liu J, Wang X (2004) Application of electrospray ionization mass spectrometry in rapid typing of fengycin homologues produced by Bacillus subtilis. Lett Appl Microbiol 39(1):98–102

Wang T, Liu MQ, Li HX (2014) Inoculation of phosphate solubilizing bacteria *Bacillus thuringiensis* B1 increases available phosphorus and growth of peanut in acidic soil. Acta Agric Scand B Soil Plant Sci 64(3):252–259

Wilson MK, Abergel RJ, Raymond KN, Arceneaux JE, Byers BR (2006) Siderophores of *Bacillus anthracis*, *Bacillus cereus*, and *Bacillus thuringiensis*. Biochem Biophys Res Commun 348(1):320–325

Yu X, Ai C, Xin L, Zhou G (2011) The siderophore-producing bacterium, *Bacillus subtilis* CAS15, has a biocontrol effect on *Fusarium* wilt and promotes the growth of pepper. Eur J Soil Biol 47(2):138–145

Zhou C, Ma Z, Zhu L, Xiao X, Xie Y, Zhu J, Wang J (2016) Rhizobacterial strain *Bacillus megaterium* BOFC15 induces cellular polyamine changes that improve plant growth and drought resistance. Int J Mol Sci 17(6):976–986

Chapter 3
Bacilli in the Biocontrol of Mycotoxins

Subbaiah Chalivendra and Jong Hyun Ham

3.1 Introduction

Mycotoxins are low molecular weight (<900 daltons), secondary metabolites produced by fungi belonging mainly to *Aspergillus*, *Fusarium*, and *Penicillium* spp. Mycotoxicosis is a term applied to the noninfectious and noncontagious toxic effects of mycotoxins on humans or animals. Many of the mycotoxigenic fungi are necrotrophic or hemibiotrophic plant pathogens with a broad host range. Thus, contaminated food and feed is the main source of mycotoxicosis, although contamination can also occur through skin contact and inhalation. Mycotoxicosis has been a threat for millennia, particularly far more serious in the past when our understanding of fungal secondary metabolism was only rudimentary (Peraica et al. 1999). For example, contamination of wheat by toxins from the ergot fungus, *Claviceps purpurea*, was the cause of the notorious "St. Anthony's fire" of the Middle Ages in Europe. It was characterized by an intense burning sensation progressing into mutilation and massive deaths (ergotism). In 1891, cardiac beriberi in animals fed with imported rice reached epidemic proportions in Japan. This "yellow rice syndrome" was due to contamination with the mycotoxin citreoviridin, an inhibitor of mitochondrial ATP synthetase, secreted by *Penicillium citreonigrum*. Alimentary toxic aleukia (characterized by nausea, vomiting, diarrhea, leukopenia, hemorrhaging, skin inflammation, and sometimes death) was another example of mycotoxicosis outbreak that occurred during the World War II in Russia and associated with the *Fusarium* toxin T-2. Many mycotoxicoses, including those mentioned above, have been contained or completely eliminated with an increased knowledge of their

S. Chalivendra (✉) · J. H. Ham
Department of Plant Pathology and Crop Physiology, Louisiana State University AgCenter, Baton Rouge, LA, USA
e-mail: schalivendra@agcenter.lsu.edu

M. T. Islam et al. (eds.), *Bacilli and Agrobiotechnology: Phytostimulation and Biocontrol*, Bacilli in Climate Resilient Agriculture and Bioprospecting,
https://doi.org/10.1007/978-3-030-15175-1_3

etiology and stringent regulation of food and feed safety (Peraica et al. 1999). Yet, potential outbreak of other mycotoxic diseases is a continuing threat (Table 3.1). For example, a rapid spread of hepatitis in 2004 that led to many deaths in Kenya was found to be a case of aflatoxicosis, caused by aflatoxin-contaminated maize (Probst et al. 2007). In 2005, 19 different pet foods contaminated with aflatoxin were recalled in the USA, following the death of at least 100 dogs (Leung et al. 2006). Dogs are among the most sensitive to aflatoxins. Exposure to even as low as 0.5–1 mg of aflatoxin/kg of body weight kills them within days (Böhm and Razzai-Fazeli 2005). The preferential colonization of food and feed crops (e.g., cereals and oilseed crops) by mycotoxigenic fungi and the potential exacerbation of the problem by ongoing global climate change (Medina et al. 2015; de Sousa and Warren 2018) demand robust solutions in order to minimize public health crises and migration of key crops from their traditional growing areas. Nearly 25% of global crop production is affected by mycotoxin contamination (Marin et al. 2013), leading to an annual loss of 5 billion dollars in the USA and Canada alone according to the Food and Agriculture Organization.

Of all mycotoxins of serious concern, aflatoxin B_1 (AF), made by *Aspergillus flavus* and *A. parasiticus*, is the most dangerous crop contaminant due to its very potent carcinogenicity. It was the cause of the notorious epidemic of turkey X disease in the 1960s. AF is strictly regulated in the USA (20 ppb in food and 0.5 ppb in milk), and so, growers spend >500 million dollars/year in testing and disposal of contaminated grain. It poses a more serious threat to food safety and security in developing countries due to poor or lack of regulatory systems. Although not as dangerous as AF, fumonisins (FUMs) are among the most prevalent food and feed contaminants in many countries. FUMs are produced by *F. verticillioides* and classified as probable carcinogens due to their suspected involvement in esophageal cancer. Unlike *A. flavus*, *F. verticillioides* can also grow as an endophyte and cause stalk and ear rot diseases in maize (Blacutt et al. 2018, for a review). Genetic variation for resistance to aspergillus and fusarium ear rots exists in maize, the globally most cultivated crop with a remarkable genetic diversity. However, incorporation of genetic resistance to major mycotoxigenic fungi into crops still remains an unrealized goal. The complex genetic basis of resistance and the strong influence of environmental factors have been barriers for accurate QTL localization for resistance to either of the pathogens. For example, AF or FUM accumulation is sporadic and exacerbated by drought and insect damage at the preharvest stage or by high humidity during postharvest storage (Bennett and Klich 2003). Germplasm screens targeted to clearly defined phenotypes and not vulnerable to environmental factors are important to exploit the natural variation (Chalivendra et al. 2017). Chemical fungicides, even if effective, are not a sustainable solution due to concerns of environmental toxicity and resistance development when heavily used. Converse to their effect on fungal growth, certain classes of fungicides (e.g., strobilurins) have often been found to promote mycotoxin synthesis (Ellner 2005; Malandrakis et al. 2013; Kim et al. 2015). Biocontrol by atoxigenic *A. flavus* strains (e.g., Afla-Guard) or nonpathogenic *F. oxysporum* strains (e.g., FUSAclean or Biofox C) is the only AF or FUM control measure currently available to growers (Abbas et al. 2017; Fravel et al. 1998). However, the use of

Table 3.1 Major mycotoxigenic fungi and mycotoxins

Mycotoxigenic species	Mycotoxins made	Commodities contaminated	Species affected	Clinical effects
Aspergillus flavus and *A. parasiticus*	Aflatoxins	Maize, peanuts, cottonseed, coconut, tree nuts, figs and dairy, and poultry products (pre- and postharvest)	Humans, cattle, sheep, swine, dogs, cats, and young birds	Intestinal bleeding, hepatocarcinoma (liver cancer), Aspergillosis (lung disease)
Claviceps purpurea, C. africana, C. fusiformis and *C. sorghi*	Ergot alkaloids	Rye, sorghum, pearl millet, and pasture grasses	Humans, cattle, and sheep	Hallucinations, feeling of itching and burning, gangrene, loss of limbs, abortion
Fusarium verticillioides, F. proliferatum, A. niger, A. welwitschiae	Fumonisins	Maize, coffee, and grapes	Humans, swine, and horses	Pulmonary edema, leukoencephalomalacia, esophageal cancer, neural tube defects, liver damage, reduced growth
Aspergillus ochraceus, A. carbonarius, A. niger, Penicillium verrucosum, P. carbonarius	Ochratoxins	Cereal grains, dry beans, meat products, coffee, grapes, and dried fruit	Humans, swine, pets, and poultry	Kidney and liver damage, cancer, and abortion
Fusarium, Myrothecium, Trichoderma, Trichothecium, Cephalosporium, Verticimonosporium, and *Stachybotrys* spp.	Trichothecenes	Wheat, barley, oats, and maize	Humans, swine, dairy cattle, poultry, and horses	Feed refusal, diarrhea, vomiting, skin disorders, reduced growth
F. graminearum, F. culmorum, F. cerealis, F. equiseti, F. verticillioides, and *F. incarnatum*	Deoxynivalenol, Zearalenone	Maize, barley, oats, wheat, rice, and sorghum	Humans, swine, dairy, and cattle	Enlargement of uterus, abortion, early puberty in girls, malformation of testicles and ovaries
Stenocarpella maydis, S. macrospora	Diplodiol, Diploidiatoxin	Maize	Poultry, farm animals	Diplodiasis (neuromycotoxicosis): ataxy, paralysis, and hepatic damage and death
Aspergillus, Penicillium and *Byssochlamys spp.*	Patulin	Apples, cherries, blueberries, plums, bananas, strawberries, grapes, fruit products, and sea food	Humans	Group 3 carcinogen; weight loss, gastric, intestinal, and renal dysfunction, sperm malformation, and abortion

intraspecific biocontrol strains has raised two key concerns at least in the case of *A. flavus*: (1) atoxigenic strains are effective against only a limited range of toxigenic strains whose populations keep changing periodically in the soil (Huang et al. 2011; Damann 2015), and (2) the requirement of physical contact for the biocontrol effect (touch inhibition) (Huang et al. 2011) may lead to the evolution of super toxigenic strains via genetic recombination. In contrast, interspecific biocontrol agents (*BCA*) preclude recombination and generally are active against a broad range of fungal strains, including mycotoxigenic fungi. Many of these organisms are isolated from plants growing in disease suppressive soils or native colonizers of plant hosts (Gond et al. 2015; Shrestha et al. 2016). Here, we review the development, mode of action, and prospects of biocontrol *Bacilli* (*BCB*), particularly relevant to mycotoxigenic fungi that colonize maize and other important crops.

3.2 *Bacillus*-Based Biocontrol Systems

Although fungi (e.g., *Trichoderma*, *Metarhizium*, and *Beauveria*) constitute popular BCAs, bacterial species comprise a much larger range of microbial biocontrol products. *Bacilli* are the most extensively studied as well as commercially formulated bacterial BCAs (Cawoy et al. 2011). The genetic diversity, ability to sporulate and survive in harsh environments, rapid colonization facilitated by flagellar and non-flagellar motility, nonpathogenic lifestyle (other than *B. cereus*, which causes food poisoning in humans; Bottone 2010), and a broad range of pathogenic fungal targets aided by diverse modes of action are some of the reasons why the genus constitutes a majority of commercially viable biocontrol formulations. The most popular BCB, *B. thuringiensis*, is mostly known for its insecticidal activity of toxic crystal proteins. Transgenic crops expressing Bt genes for the insecticidal crystal proteins have played an important role in enhancing food security as well as safety (more in Sect. 3.6). In addition to those listed in Table 3.2, many closely related genera also display antagonism against fungal pathogens including mycotoxigenic species. For example, *Paenibacillus polymyxa* isolates are potent antagonists of *Fusarium* spp. (Abd El Daim et al. 2015; Cawoy et al. 2015; Shi et al. 2017). Similarly, *Lysinibacillus* spp. showed significant anti-aflatoxigenic activity in vitro and in maize and peanut seeds (Chalivendra et al. 2018; Wang et al. 2013) and an ability to biodegrade AF (Adebo et al. 2016) as well as zearalenone (Wang et al. 2018).

3.3 Modes of Biocontrol by *Bacilli*

Although fungal growth inhibition is the predominant criterion used in laboratory screens to select biocontrol species, *Bacillus* BCAs act via diverse effects on the pathogen as well as the host in the suppression of pathogenicity (reviewed in Cawoy et al. 2011; Shafi et al. 2017). These include inhibition of hyphal growth, induction

Table 3.2 *Bacilli* with biocontrol activity on mycotoxigenic fungi

Bacillus species	Targeted fungus	Mode of action	References
B. amyloliquefaciens *B. licheniformis* *B. megaterium* *B. methylotrophicus* *B. subtilis*	*A. flavus*	Loss of membrane and cell wall integrity, germination inhibition, transcriptional inhibition of AF biosynthesis, AF degradation	Afsharmanesh et al. (2018), Al-Saad et al. (2016), Chalivendra et al. (2018), Farzaneh et al. (2016), Gong et al. (2014), Kong et al. (2014), Raksha Rao et al. 2017
B. amyloliquefaciens *B. subtilis*	*A. parasiticus*	Chitinolytic activity	Wang et al. (2013)
B. safensis *B. amyloliquefaciens* *B. subtilis*	*A. westerdijkiae*	Not investigated	Einloft et al. (2017)
B. subtilis	*A. ochraceus* *A. carbonarius*	Not investigated	Shi et al. (2014a)
B. amyloliquefaciens	*A. niger*	Not investigated	Raut et al. (2014)
B. subtilis	*F. oxysporum*	Cell lysis	Swain et al. (2008)
B. amyloliquefaciens *B. mojavensis* *B. subtilis*	*F. verticillioides (also known as F. moniliforme)*	Cell vacuolization, lysis and loss of cell wall integrity	Blacutt et al. (2016), Gond et al. (2015), Pereira et al. (2007)
B. velezensis *B. amyloliquefaciens*	*F. graminearum*	Production of antifungal volatiles and LPs	Shi et al. (2014b), Chen et al. (2018), Lim et al. (2017), Palazzini et al. (2016)
B. amyloliquefaciens	*F. solani*	Production of antifungal LPs	Li et al. (2014)

of cell death, competition for space and nutrients, direct effects on mycotoxin biosynthesis and stability, and/or promotion of host immunity and fitness. Considerable progress has been made in unraveling the molecular mechanisms underlying these effects during the past decades.

3.3.1 Antifungal Activity

The most common mode of action of BCAs is antibiosis. The extent of pathogen growth inhibition varies with the *Bacillus* strain and the pathogen. In many BCA-fungal interactions, the pathogen recovers once the bacterium is removed (i.e., fungistasis), while in others, the pathogen is killed after its exposure to the BCA (fungicidal effect). Understanding the molecular basis of these BCA-pathogen interactions would facilitate the discovery of new biocides that are robust and appropriate to a specific pathogen (DeFilippi et al. 2018).

Co-cultivation is used as the primary screen to detect antifungal activities. Therefore, it is important that the assays are done under a broad range of conditions that are particularly relevant to the field environment in order to maximize the discovery of robust BCA candidates. For example, 20 out of 59 *Bacillus* spp. isolated from maize fields inhibited *A. flavus* and *A. parasiticus* at 0.98 water activity. However, when water activity of the medium was lowered to 0.955, only three isolates were effective (Bluma and Etcheverry 2006). *Bacillus* strains from maize fields showed much lower inhibition of *A. flavus* and *F. verticillioides* growth in potato dextrose broth than in maize kernel extract medium, suggesting nutritional regulation of antifungal activity (Palumbo et al. 2007). An apparent lack of fungistasis when co-cultured in popular synthetic media may occasionally be due to unmet nutritional needs resulting in the poor growth of a test *Bacillus* isolate (Chalivendra et al. 2018). Conversely, some *Bacilli* strains were shown to stimulate the proliferation of *Fusarium* spp. and promote disease under certain abiotic stress conditions, suggesting that stress-adapted bacteria may supply growth-stabilizing compounds to the fungus (Cray et al. 2016). In addition, strains that require nutrient-rich media to show any significant antifungal activity may raise concerns because of their poor adaptability to the adverse phyllosphere environment. In such cases, cell-free extracts (crude or enriched) prepared under conducive growth conditions may be a suitable alternative to the field application of live cultures.

3.3.2 *Inhibition of Mycotoxin Biosynthesis*

Mycotoxigenic fungal pathogens pose a unique problem in that factors governing virulence may be different from those that control toxin biosynthesis. For example, the amount of FUMs or AF in the grain often does not always correspond to the severity of ear rot or fungal biomass in infected seeds (e.g., Cotty et al. 1994; Santiago et al. 2015). It is not surprising that BCA also may differ in their mode of biocontrol activity. However, the possibility of distinctive BCA effects on mycotoxigenic fungi has not been examined until recently. In an assay designed to measure both antibiosis and antimycotoxigenic activities, 29 *Bacilli* isolates from rice phyllosphere (RABs) were tested for their effects in *A. flavus* (Chalivendra et al. 2018; Fig. 3.1). Only 15 RABs showed significant growth inhibition, whereas 26 of them (~90%) showed significant anti-aflatoxigenic activity, ten of which had negligible or no effect on colony growth (Chalivendra et al. 2018). It follows that the primary BCA screens for (myco)toxigenic pathogens need to assay for both growth and toxin inhibition to capture all potentially useful strains. Identification of isolates with different activities may allow the formulation of a cocktail with complementary strains for robust mycotoxin control. For example, BCB that show significant antimycotoxigenic activity and negligible antibiosis are safe to be combined with atoxigenic intraspecific fungal strains or chemical fungicides in mitigating mycotoxin production (e.g., Hervás et al. 1997).

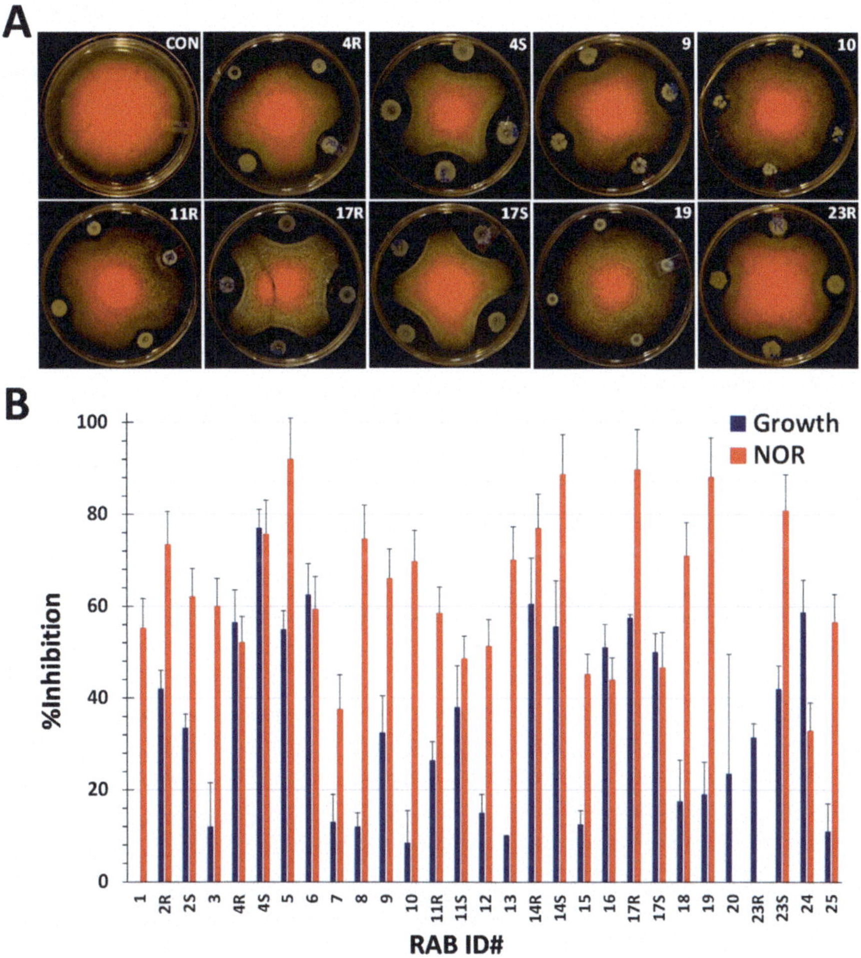

Fig. 3.1 Assay to screen putative BCAs for both fungistatic and antimycotoxigenic activities in the same plate. (Reproduced from Chalivendra et al. 2018). Assay details are presented in the original publication

3.4 Mechanisms of Antibiosis and Antimycotoxigenicity

In co-cultivation assays, the size of the growth inhibition zone is used to score the efficacy of BCA and to discard the negatives. A study by Bluma and Etcheverry (2006) is an exception, where the authors found that 56 of 59 *Bacilli* isolates from maize fields needed contact for antagonism in *A. flavus*, while only ~5% of isolates formed inhibition zones. A growth inhibitory zone in co-cultivation assays is a strong indication of diffusible antifungal activities as the basis of biocontrol. *Bacilli* are known to synthesize and secrete a variety of antifungal metabolites, such as lipopeptides, polyketides, hydrolases, and siderophores, thus offering an ex vivo method of pathogen control, in addition to biological formulations. Purified compounds or crude culture supernatants of these BCB show antifungal/toxigenic activities that are equivalent or even superior to live bacteria (e.g., Blacutt et al. 2016; Chalivendra et al. 2018). The molecular basis of antibiosis induced by many of these secreted compounds is fairly understood (Ongena and Jacques 2008). Recent studies show that antifungal compounds from BCB strains (which also inhibit fungal growth) repress transcript levels of AF biosynthesis genes (Al-Saad et al. 2016; Kong et al. 2014). On the other hand, the mechanism(s) of antimycotoxigenic activity independent of antibiosis or mycotoxin degradation is yet to be explored.

3.4.1 Cyclic Lipopeptides (CLPs)

CLPs are the predominant class of *Bacillus* antifungals. As suggested by their name, they are rings of amino acids with fatty acid side chains. Remarkable heterogeneity has been observed among the CLPs, although they are as short as 7–10 amino acids. In addition to amino acid substitutions, the variability is contributed by the nature, length, and configuration of the fatty acid chain. CLPs provide a broad spectrum and environmentally stable antimicrobial activity due to their structural diversity and cyclic configuration. Their production is often modulated by the presence and the type of pathogens (DeFilippi et al. 2018). CLPs belong to three families, surfactin, iturin, and fengycin, and a single *Bacillus* strain can make one or more of them (Ongena and Jacques 2008). Each of these family members has both common and specific biological activities. Surfactins are poor antagonists of some pathogenic fungi, but they facilitate swarming movement and rapid colonization of plant surfaces as well as the rhizosphere, in addition to inducing host systemic resistance (Leclère et al. 2006; Cawoy et al. 2015). Fengicyns inhibit most filamentous fungi, although were shown to be more effective on *F. verticillioides* than on *A. flavus* (Hu et al. 2007; Farzaneh et al. 2016). Fengicyns disrupt membrane integrity leading to organelle dysfunction, particularly a loss of mitochondrial membrane potential, resulting in oxidative stress, chromatin condensation, and hyphal cell death (Zhang and Sun 2018). These subcellular effects are manifested as hyphal bulging, curling, or emptying. Iturins are strong antifungal CLPs. Bacillomycin D, an iturin, is strongly antagonistic to *A. flavus* (Moyne et al. 2001; Gong et al. 2014) as well as *F. graminearum* (Sun et al. 2018). Similar to surfactins, iturins may help BCB to colonize plant surfaces by promoting biofilm formation (Xu et al. 2013).

3.4.2 Non-peptide Antifungals

BCBs also produce non-peptide antibiotics, but their role in the biocontrol of mycotoxigenic fungi has not been sufficiently investigated. Chitinolytic activity in some *Bacillus* spp. has been found to be a factor in the biocontrol of *A. parasiticus* and AF mitigation (Wang et al. 2013). Volatile compounds made by some strains of *B.*

amyloliquefaciens are reported to restrict the growth and spore germination in *F. oxysporum* (Yuan et al. 2012). However, the chemical nature or the mode of action of these volatile compounds has not been investigated.

3.5 Mycotoxin Degradation

Some BCB eliminate mycotoxins by metabolizing them into less toxic or nontoxic by-products (e.g., Ciegler et al. 1966; Adebo et al. 2016; Verheecke et al. 2016; Vanhoutte et al. 2016, and references therein), in addition to repressing fungal growth or toxin biosynthesis. Many species of *Bacillus*, including *B. subtilis* and *B. licheniformis* or their culture supernatants, have been shown to degrade AF to nontoxic products (reviewed in Raksha Rao et al. 2017; Verheecke et al. 2016). However, neither the enzymes nor the degradation products have been investigated in these studies. Recently, Afsharmanesh et al. (2018) presented genetic evidence for the involvement of an oxidoreductase in AF degradation by *B. subtilis*. The enzyme, encoded by the *bacC* gene, participates in the biosynthesis of the antifungal peptide bacilysin, suggesting that the pathway provides a robust biocontrol of AF. Only a few microorganisms are known to degrade and detoxify FUMs (Vanhoutte et al. 2016), and *Bacilli* are among them (Camilo et al. 2000). Based on an increase in the medium pH, it was hypothesized that the degradation was by the removal of the free amino group.

3.6 Induced Systemic Resistance (ISR)

A fifth mechanism by which *Bacilli* may inhibit mycotoxigenic fungi is by enhancing the host plant defense through the stimulation of its immune machinery. Surfactins are shown to induce significant protective effect against diseases caused by fungal pathogens, while fengycins are poor inducers of ISR (Cawoy et al. 2014). Treatment of maize roots with a suspension of *B. subtilis* strain endophytic to maize seeds (but not with lipopeptide extract) induced defense genes (Gond et al. 2015). However, there has been no report on the induction of ISR by *Bacilli* against a specific mycotoxigenic fungus.

3.7 Plant Growth Promotion

In addition to ISR, many BCBs also promote plant growth (e.g., Shrestha et al. 2016), by producing phytohormones that are appropriate to growing conditions (Park et al. 2017). For example, plants treated with *B. aryabhattai* showed greater tolerance to heat stress and produced more abscisic acid, the stress hormone that controls stomatal closure (Park et al. 2017). Heat and drought stress are known to exacerbate mycotoxin synthesis and also insect damage (Bennett and Klich 2003).

Species such as *B. megaterium* and *Paenibacillus polymyxa* can be used as biofertilizers, since they can fix nitrogen and solubilize phosphate (Anand and Chanway 2013; Ding et al. 2005).

3.8 Insect Vector Control

Transgenic Bt crops expressing *cry* genes from *B. thuringiensis* are protected against not only insect pests but pathogenic microbes vectored by those insects. Bt corn, although not very effective in aflatoxin mitigation, has been shown to reduce FUMs and deoxynivalenol contamination (e.g., Bowers 2013). *A. flavus* and *A. parasiticus* invade the seed through silk and are vectored mainly by insects that are not controlled by Bt corn (Abbas et al. 2013). Lower mycotoxin levels in Bt corn have been shown from the feeding damage caused by the European corn borer but not the corn earworm. *Bacillus thuringiensis* may have additional biocontrol effects. For example, seed treatment with a chitinase from *B. thuringiensis* controlled *Fusarium* infection in soybean (Gomaa 2012).

3.9 Summary

Being sporulating Gram-positive bacteria, *Bacilli* are resistant to temperature, dryness, UV radiation, and organic solvents. This stability is extended to their culture supernatants or many antifungal compounds secreted by *Bacilli*. For example, culture filtrates from other microbial BCAs such as *Pseudomonas* spp. were unstable at temperatures and pH conditions tested, in contrast with the filtrates from *B. subtilis*. The cell-free extract from the BCB maintained antifungal activity even after 30 days of exposure to extreme temperatures and pH values or sterilization by autoclaving (Petatán-Sagahón et al. 2011). Although commercial formulations specifically targeted to mycotoxigenic fungi are not currently available, there has been an increased interest in developing such products to promote food safety currently under threat from global warming.

References

Abbas HK, Zablotowicz RM, Weaver MA, Shier WT, Bruns HA, Bellaloui N, Accinelli C, Abel CA (2013) Implications of Bt traits on mycotoxin contamination in maize: overview and recent experimental results in Southern United States. J Agric Food Chem 61(48):11759–11770

Abbas HK, Accinelli C, Shier WT (2017) Biological control of aflatoxin contamination in U.S. Crops and the use of bioplastic formulations of aspergillus flavus biocontrol strains to optimize application strategies. J Agric Food Chem 65(33):7081–7087

Abd El Daim IA, Häggblom P, Karlsson M, Stenström E, Timmusk S (2015) Paenibacillus polymyxa A26 Sfp-type PPTase inactivation limits bacterial antagonism against Fusarium graminearum but not of F. culmorum in kernel assay. Front Plant Sci 6(368)

Adebo OA, Njobeh PB, Mavumengwana V (2016) Degradation and detoxification of AFB1 by *Staphylococcus warneri*, *Sporosarcina* sp. and *Lysinibacillus fusiformis*. Food Control 68:92–96

Afsharmanesh H, Perez-Garcia A, Zeriouh H, Ahmadzadeh M, Romero D (2018) Aflatoxin degradation by *Bacillus subtilis* UTB1 is based on production of an oxidoreductase involved in bacilysin biosynthesis. Food Control 94:48–55

Al-Saad LA, Al-Badran AI, Al-Jumayli SA, Magan N, Rodríguez A (2016) Impact of bacterial biocontrol agents on aflatoxin biosynthetic genes, aflD and aflR expression, and phenotypic aflatoxin B_1 production by *Aspergillus flavus* under different environmental and nutritional regimes. Int J Food Microbiol 217:123–129

Anand R, Chanway C (2013) N_2-fixation and growth promotion in cedar colonized by an endophytic strain of *Paenibacillus polymyxa*. Biol Fertil Soils 49:235–239

Bennett JW, Klich M (2003) Mycotoxins. Clin Microbiol Rev 16(3):497–516

Blacutt AA, Mitchell TR, Bacon CW, Gold SE (2016) *Bacillus mojavensis* RRC101 lipopeptides provoke physiological and metabolic changes during antagonism against *Fusarium verticillioides*. Mol Plant-Microbe Interact 29:713–723

Blacutt AA, Gold SE, Voss KA, Gao M, Glenn AE (2018) *Fusarium verticillioides*, Advancements in understanding the toxicity, virulence, and niche adaptations of a model mycotoxigenic pathogen of maize. Phytopathology 108:312–326

Bluma RV, Etcheverry MG (2006) Influence of *Bacillus* spp. isolated from maize agroecosystem on growth and aflatoxin B1 production by *Aspergillus* section *Flavi*. Pest Manag Sci 62:242–251

Böhm J, Razzai-Fazeli E (2005) Effects of mycotoxins on domestic pet species. In: Diaz D (ed) The mycotoxin blue book. Nottingham University Press, Nottingham, pp 77–91

Bottone EJ (2010) *Bacillus cereus*, a volatile human pathogen. Clin Microbiol Rev *23*:382–398

Bowers EL (2013) Fumonisin content in conventional versus Bt maize and implications of fumonisin contamination on fuel ethanol yield and DDGS quality. Graduate theses and dissertations. https://lib.dr.iastate.edu/etd/13596.

Camilo SB, Ono CJ, Ueno Y, Hirooka EY (2000) Anti-*Fusarium moniliforme* activity and fumonisin biodegradation by corn and silage microflora. Braz Arch Biol Technol 43:159–164

Cawoy H, Bettiol W, Fickers P, Ongena M (2011) *Bacillus*-based biological control of plant diseases. In: Stoytcheva M (ed) Pesticides in the modern world – pesticides use and management. InTech, Rijeka, pp 274–302

Cawoy H, Mariutto M, Henry G, Fisher C, Vasilyeva N, Thonart P et al (2014) Plant defense stimulation by natural isolates of *Bacillus* depends on efficient surfactin production. Mol Plant Microbe Interact 27:87–100

Cawoy H, Debois D, Franzil L, De Pauw E, Thonart P, Ongena M (2015) Lipopeptides as main ingredients for inhibition of fungal phytopathogens by *Bacillus subtilis/amyloliquefaciens*. Microb Biotechnol 8:281–295

Chalivendra SC, DeRobertis C, Chang PK, Damann KE (2017) Cyclopiazonic acid is a pathogenicity factor for *Aspergillus flavus* and a promising target for screening germplasm for ear rot resistance. Mol Plant Microbe Interact 30:361–373

Chalivendra SC, DeRobertis C, Reyes Pineda J, Ham H-J, Damann KE (2018) Rice phyllosphere *Bacillus* species and their secreted metabolites suppress *Aspergillus flavus* growth and aflatoxin production in vitro and in maize seeds. Toxins 10:159

Chen L, Heng J, Qin S, Bian K (2018) A comprehensive understanding of the biocontrol potential of *Bacillus velezensis* LM2303 against *Fusarium* head blight. PLoS ONE *13*(*6*):e0198560

Ciegler A, Lillehoj EB, Peterson RE, Hall HH (1966) Microbial detoxification of aflatoxin. Appl Microbiol 14:934–938

Cotty PJ, Bayman DS, Egel DS, Elias KS (1994) Agriculture, aflatoxins and *Aspergillus*. In: Powell K (ed) The genus *Aspergillus*. Plenum Press, New York, pp 1–27

Cray JA, Connor MC, Stevenson A, Houghton JDR, Rangel DEN, Cooke LR et al (2016) Biocontrol agents promote growth of potato pathogens, depending on environmental conditions. Microb Biotechnol 9:330–354

Damann KE (2015) Atoxigenic *Aspergillus flavus* biological control of aflatoxin contamination, what is the mechanism? World Mycotoxin J 8:235–244

de Sousa A, Warren H (2018) Climate change is messing with your dinner. Bloomberg

DeFilippi S, Groulx E, Megalla M, Mohamed R, Avis TJ (2018) Fungal competitors affect production of antimicrobial Lipopeptides in *Bacillus subtilis* strain B9–5. J Chem Ecol 44(4):374–383

Ding Y, Wang J, Liu Y, Chen S (2005) Isolation and identification of nitrogen-fixing *Bacilli* from plant rhizospheres in Beijing region. J Appl Microbiol 99(5):1271–1281

Einloft TC, de Oliveira PB, Veras FF, Welke JE, Mallmann CA, Dilkin P, Dionello RG (2017) Effect of *Bacillus* spp. on *Aspergillus westerdijkiae* growth, sporulation and ochratoxin A production in green-coffee medium. Food Sci Technol 37:24–27

Ellner FM (2005) Results of long-term field studies into the effect of strobilurin containing fungicides on the production of mycotoxins in several winter wheat varieties. Mycotoxin Res 21:112–115

Farzaneh M, Shi Z-Q, Ahmadzadeh M, Hu L-B, Ghassempour A (2016) Inhibition of the *Aspergillus flavus* growth and Aflatoxin B1 contamination on pistachio nut by Fengycin and Surfactin-producing *Bacillus subtilis* UTBSP1. Plant Pathol J 32:209–215

Fravel DR, Connick WJ, Lewis JA (1998) Formulation of microorganisms to control plant diseases. In: Burges HD (ed) Formulation of microbial biopesticides. beneficial microorganisms, nematodes and seed treatments. Springer, Dordrecht, pp 188–191

Gomaa EZ (2012) Chitinase production by *Bacillus thuringiensis* and *Bacillus licheniformis*: Their potential in antifungal biocontrol. J Microbiol 50(1):103–111

Gond SK, Bergen MS, Torres MS, White JF Jr (2015) Endophytic *Bacillus* spp. produce antifungal lipopeptides and induce host defence gene expression in maize. Microbiol Res 172:79–87

Gong Q, Zhang C, Lu F, Zhao H, Bie X, Lu Z (2014) Identification of bacillomycin D from *Bacillus subtilis* fmbJ and its inhibition effects against *Aspergillus flavus*. Food Control 36:8–14

Hervás A, Landa B, Jiménez-Díaz RM (1997) Influence of chickpea genotype and *Bacillus* sp. on protection from Fusarium wilt by seed treatment with nonpathogenic *Fusarium oxysporum*. Eur J Plant Pathol 103(7):631–642

Hu LB, Shi ZQ, Zhang T, Yang ZM (2007) Fengycin antibiotics isolated from B-FS01 culture inhibit the growth of *Fusarium moniliforme* Sheldon ATCC 38932. FEMS Microbiol Lett 272:91–98

Huang C, Jha A, Sweany R, DeRobertis C, Damann KE Jr (2011) Intraspecific aflatoxin inhibition in Aspergillus flavus is Thigmoregulated, independent of vegetative compatibility group and is strain dependent. PLoS ONE 6:e23470

Kim JH, Chan KL, Mcconnell LL (2015) Augmenting the efficacy of fungal and mycotoxin control via chemosensitization. Outlooks Pest Manag 26:172–176

Kong Q, Chi C, Yu J, Shan S, Li Q, Guan B, Nierman WC, Bennett JW (2014) The inhibitory effect of Bacillus megaterium on aflatoxin and cyclopiazonic acid biosynthetic pathway gene expression in Aspergillus flavus. Appl Microbiol Biotechnol 98:5161–5172

Leclère V, Marti R, Bechet M, Fickers P, Jacques P (2006) The lipopeptides mycosubtilin and surfactin enhance spreading of *Bacillus subtilis* strains by their surface-active properties. Arch Microbiol 186:475–483

Leung MCK, Diaz-Llano G, Smith TK (2006) Mycotoxins in pet food, a review on worldwide prevalence and preventative strategies. J Agric Food Chem 54(26):9623–9635

Li B, Li Q, Xu Z, Zhang N, Shen Q, Zhang R (2014) Responses of beneficial *Bacillus amyloliquefaciens* SQR9 to different soilborne fungal pathogens through the alteration of antifungal compounds production. Front Microbiol 5(636)

Lim SM, Yoon M-Y, Choi GJ, Choi YH, Jang KS, Shin TS et al (2017) Diffusible and Volatile Antifungal Compounds Produced by an Antagonistic *Bacillus velezensis G341* against Various Phytopathogenic Fungi. Plant Pathol J 33(5):488–498

Malandrakis AA, Vattis KN, Doukas EG, Markoglou AN (2013) Effect of phenylpyrrole-resistance on fitness parameters and ochratoxin production in *Aspergillus carbonarius*. Int J Food Microbiol 165:287–294
Marin S, Ramos AJ, Cano-Sancho G, Sanchis V (2013) Mycotoxins: occurrence, toxicology, and exposure assessment. Food Chem Toxicol 60:218–237
Medina Á, Rodríguez A, Magan N (2015) Climate change and mycotoxigenic fungi, Impacts on mycotoxin production. Curr Opin Food Sci 5:99–104
Moyne AL, Shelby R, Cleveland T, Tuzun S (2001) Bacillomycin D, an iturin with antifungal activity against *Aspergillus flavus*. J Appl Microbiol 90:622–629
Ongena M, Jacques P (2008) *Bacillus* lipopeptides, versatile weapons for plant disease biocontrol. Trends Microbiol 16:115–125
Palazzini JM, Dunlap CA, Bowman MJ, Chulze SN (2016) *Bacillus velezensis* RC 218 as a biocontrol agent to reduce Fusarium head blight and deoxynivalenol accumulation: genome sequencing and secondary metabolite cluster profiles. Microbiol Res 192:30–36
Palumbo JD, O'Keeffe TL, Abbas HK (2007) Isolation of maize soil and rhizosphere bacteria with antagonistic activity against *Aspergillus flavus* and *Fusarium verticillioides*. J Food Prot 70:1615–1621
Park Y-G, Mun B-G, Kang S-M, Hussain A, Shahzad R, Seo C-W et al (2017) *Bacillus aryabhattai* SRB02 tolerates oxidative and nitrosative stress and promotes the growth of soybean by modulating the production of phytohormones. PLoS ONE 12(3):e0173203
Peraica M, Radić B, Lucić A, Pavlović M (1999) Toxic effects of mycotoxins in humans. Bull World Health Organ 77:754–766
Pereira P, Nesci A, Etcheverry M (2007) Effects of biocontrol agents on *Fusarium verticillioides* count and fumonisin content in the maize agroecosystem: Impact on rhizospheric bacterial and fungal groups. Biol Control 42(3):281–287
Petatán-Sagahón I, Anducho-Reyes MA, Silva-Rojas HV, Arana-Cuenca A, Tellez-Jurado A, Cárdenas-Álvarez IO, Mercado-Flores Y (2011) Isolation of bacteria with antifungal activity against the phytopathogenic fungi *Stenocarpella maydis* and *Stenocarpella macrospora*. Int J Mol Sci 12:5522–5537
Probst C, Njapau H, Cotty PJ (2007) Outbreak of an acute aflatoxicosis in Kenya in 2004, identification of the causal agent. Appl Environ Microbiol 73:2762–2764
Raksha Rao K, Vipin AV, Hariprasad P, Anu Appaiah KA, Venkateswaran G (2017) Biological detoxification of Aflatoxin B1 by *Bacillus licheniformis* CFR1. Food Control 71:234–241
Raut I, Calin M, Stancu M, Roseanu A, Badea Doni M, Oancea F, Jecu L (2014) Inhibition of toxigenic *Aspergillus niger* by microbial metabolites. Rev Chim 65:779–783
Santiago R, Cao A, Butrón A (2015) Genetic factors involved in Fumonisin accumulation in maize kernels and their implications in maize agronomic management and breeding. Toxins 7(8):3267–3296
Shafi J, Tian H, Ji M (2017) Bacillus species as versatile weapons for plant pathogens: a review. Biotechnol Biotechnol Equip 31(3):446–459
Shi L, Liang Z, Li J, Hao J, Xu Y, Huang K et al (2014a) Ochratoxin A biocontrol and biodegradation by *Bacillus subtilis* CW 14. J Sci Food Agric 94(9):1879–1885
Shi C, Yan P, Li J, Wu H, Li Q, Guan S (2014b) Biocontrol of *Fusarium graminearum* growth and deoxynivalenol production in wheat kernels with bacterial antagonists. Int J Environ Res Public Health 11(1):1094–1105
Shi L, Du N, Shu S, Sun J, Li S, Guo S (2017) Paenibacillus polymyxa NSY50 suppresses Fusarium wilt in cucumbers by regulating the rhizospheric microbial community. Sci Rep 7:41234
Shrestha BK, Karki HS, Groth DE, Jungkhun N, Ham JH (2016) Biological control activities of rice-associated Bacillus sp. Strains against sheath blight and bacterial panicle blight of rice. PLoS ONE 11(1):e0146764
Sun J, Li W, Liu Y, Lin F, Huang Z, Lu F, Bie X, Lu Z (2018) Growth inhibition of *Fusarium graminearum* and reduction of deoxynivalenol production in wheat grain by bacillomycin D. J Stored Prod Res 75:21–28

Swain RC, Ray RC, Nautiyal CS (2008) Biocontrol efficacy of *Bacillus subtilis* strains isolated from cow dung against postharvest yam (*Dioscorea rotundata* L.) pathogens. Curr Microbiol 57:407–411

Vanhoutte I, Audenaert K, De Gelder L (2016) Biodegradation of Mycotoxins: Tales from Known and Unexplored Worlds. Front Microbiol 7:561

Verheecke C, Liboz T, Mathieu F (2016) Microbial degradation of aflatoxin B1: Current status and future advances. Int J Food Microbiol 237:1–9

Wang K, Yan P-s, Ding Q-l, Wu Q-x, Wang Z-b, Peng J (2013) Diversity of culturable root-associated/endophytic bacteria and their chitinolytic and aflatoxin inhibition activity of peanut plant in China. World J Microbiol Biotechnol 29(1):1–10

Wang JQ, Yang F, Yang PL, Liu J, Lv ZH (2018) Microbial reduction of Zearalenone by a new isolated *Lysinibacillus* sp. ZJ-2016-1. World Mycotoxin J. In Press

Xu Z, Shao J, Li B, Yan X, Shen Q, Zhang R (2013) Contribution of Bacillomycin D in *Bacillus amyloliquefaciens* SQR9 to antifungal activity and biofilm formation. Appl Environ Microbiol 79(3):808–815

Yuan J, Raza W, Shen Q, Huang Q (2012) Antifungal activity of *Bacillus amyloliquefaciens* NJN-6 volatile compounds against *Fusarium oxysporum* f. sp. *cubense*. Appl Environ Microbiol 78(16):5942–5944

Zhang L, Sun C (2018) Fengycins, cyclic lipopeptides from marine *Bacillus subtilis* strains, kill the plant-pathogenic fungus *Magnaporthe grisea* by inducing reactive oxygen species production and chromatin condensation. Appl Environ Microbiol 84(18):e00445–e00418

Chapter 4
Bacillus subtilis and Its Effect on the Postharvest of Fruit and Flowers

Fabián Heriberto Rivera Chávez, Hortencia Gabriela Mena Violante, Martin Arriaga Rubi, Jaime Bravo, María del Rosario Abraham Juárez, Silvia Edith Valdés Rodríguez, and Victor Olalde Portugal

4.1 Introduction

We live in an era of science and technological advances in all the fields of knowledge, including those associated with agriculture. These advances in the past 60 years have created disease-resistant, high-yielding, and profitable crops and have developed more effective agricultural inputs. However, famines still arise, and the indiscriminate use of synthetic inputs has had a negative impact on the environment (Unfao 2010; Bhardwaj et al. 2014).

Famine has fostered the development of technologies that take a holistic advantage of nature. Studying the ecological side of agriculture has allowed us to use its resources more effectively, such as rhizospheric microorganisms, including plant growth-promoting rhizobacteria (PGPRs), like *Bacillus subtilis* that has the capacity to form thermoresistant spores and biofilms (Rivera Pérez 2009). Originally, these bacteria were only associated with the production of root growth, but there is

F. H. Rivera Chávez · S. E. Valdés Rodríguez · V. Olalde Portugal (✉)
Departamento de Biotecnología y Bioquímica, Centro de Investigación y Estudios Avanzados Campus Guanajuato, Irapuato, Guanajuato, México

H. G. Mena Violante
Departamento de Biotecnología, Centro de Interdisciplinario de Investigación para el Desarrollo Integral Regional IPN Unidad Michoacán, Jiquilpan, Michoacán, México

M. Arriaga Rubi
Facultad de Ciencias Agrícolas, Universidad Autónoma del Estado de México, Toluca, Estado de México, México

J. Bravo
Instituto Tecnológico de Tepic, Tepic, Nayarit, México

M. d. R. Abraham Juárez
Departamento de alimentos, División Ciencias de la Vida, Universidad de Guanajuato, Irapuato, Guanajuato, México

M. T. Islam et al. (eds.), *Bacilli and Agrobiotechnology: Phytostimulation and Biocontrol*, Bacilli in Climate Resilient Agriculture and Bioprospecting,
https://doi.org/10.1007/978-3-030-15175-1_4

new evidence that these rhizospheric bacteria can function as biofertilizers or biocontrol agents that promote the growth of more vigorous plants, increase their yield, and improve their response to biotic stress (Tejera Hernández et al. 2013). Recent findings have revealed that the inoculation of agricultural crops with *B. subtilis* can improve the final quality of their products, having a positive impact on attributes, like size and firmness (Olalde-Portugal and Mena-Violante 2008). This impact may involve the activity of hormone-like substances that affect plant growth, flowering, and fruit ripening as well as involve nutritional mechanisms and protection against pathogens. This paper addresses the potential of applying *B. subtilis* as a biofertilizer, a biocontrol agent, and a biostimulant to improve the quality and shelf life of agricultural products in the context of sustainable agricultural production.

4.1.1 The Challenge of Sustainable Agricultural Production

The world is currently experiencing a food crisis in which approximately 30% of the population suffers some form of malnourishment. More than half of the diseases in the world are associated with nutrient deficiency (Unfao 2010). The nutritional requirements of humanity initiated the development of a massive farm production model based mainly on the use of great expanses of land, the intensive application of fertilizers, the improvement of crops, and the application of pest-control substances with the objective of assuring a bountiful harvest which eventually gave way to the well-known green revolution (Bhardwaj et al. 2014; Glick 2012).

Data from the FAO (2000) suggest that by 2030, the agricultural production in developing countries will be 70% higher than it was in 1995–1997. Around 80% of this production will come from intensive farming systems that will require an increase in the amount of arable land. Around 120 million of hectares will be turned into agricultural soil (Unfao 2010). Innovations in fertilization, plant nutrition, and irrigating systems, as well as the control of diseases and pathogens, have allowed high yields in crops over the last 50 years. However, these technologies have had a highly negative impact on the environment (Pingali 2012).

The ecological study of agriculture is crucial for discovering the best production methods with the least negative impact on the environment. A deeper understanding of the ecological nature of the agricultural systems will make it possible to improve their capacity to benefit mankind. One of the biotechnological alternatives for sustainable agricultural production is the use of beneficial soil microorganisms functioning as biofertilizers, particularly from the rhizosphere, because these microorganisms produce metabolites during their life cycle that produce positive effects on the crops and improve the soil quality (Bhardwaj et al. 2014). The rhizosphere is the region of the soil surrounding and affected by the root, wherein the root can give shelter to more than 1×10^{11} cells of microorganisms per gram of soil and can contain more than 30,000 prokaryotic species (Egamberdieva et al. 2008). The group of microbial communities that inhabit the rhizosphere is called the microbiome, and its interactions may determine the health and efficacy of a crop that grows in an agroecosystem (Mendes et al. 2013).

Hiltner (1904) suggested that the volume of the rhizosphere or soil adherent to the plant root is richer in free-living microorganisms than the rest of the surrounding soil where the root has less influence (reviewed by Lugtenberg and Kamilova 2009). This phenomenon could be due to the fact that an important percentage of carbon fixed by the plants (5–21% depending on the plant) is secreted into the environment through radicular exudates. The plant root secretes a complex mix of metabolites into the soil, including sugars, peptides, phenols, and hormones that are used by the different microorganisms in the rhizosphere. This concentration of metabolites may explain why the concentration of microorganisms in the rhizosphere is 10 to 100 times denser than in the rest of the soil (Lugtenberg and Kamilova 2009). These microbial associations result in a wide range of benefits involving biochemical and physiological changes that occur in the host plants (Gunes et al. 2015). The use of microorganisms in different crops promotes a significant increase in the production of fruits, vegetables, and grains (Glick 2012; Bergottini et al. 2015).

Field and greenhouse experiments have shown yield increase when the beneficial microorganisms are inoculated into the soil, such as plant growth-promoting rhizobacteria (PGPRs). However, improvements occur not only in biomass production but also in terms of crop protection and the quality of agricultural products (Olalde-Portugal and Mena-Violante 2008; Ordookhani et al. 2010).

4.1.2 Plant Growth-Promoting Rhizobacteria (PGPRs)

Plant growth-promoting rhizobacteria (PGPRs) are organisms associated with plant roots, and they stimulate plant growth, facilitate nutrient assimilation, and provide protection against infectious agents (Bais et al. 2004). These microorganisms live in a mutualistic relationship with their host. Initially it was observed that their presence promoted plant tissue growth, hence the name (Kloepper et al. 1989). Several studies have suggested that there are multiple mechanisms through which PGPRs can promote plant growth. Some of these microorganisms have the ability to produce molecules comparable to auxins (Ahmed and Hasnain 2014), cytokinins (Liu et al. 2013), gibberellins (Kang et al. 2012), and abscisic acid (Cohen et al. 2015) that act as plant growth hormones.

It has been established that the presence of PGPRs enhances the efficacy of nutrient absorption into plants. This can be done by modifying kinetic transport, modifying root structure, or promoting the bioavailability of nutrients (Bashan et al. 1989; Bais et al. 2004; Glick 2012). It has been shown that some PGPRs have the ability to process molecules containing nitrogen, phosphorus, and iron into forms that can be assimilated by plants (Antoun 2013).

Nitrogen fertilizers are among the most indispensable and costly of agricultural inputs. Only 45% or less of applied nitrogen is used by the plants, while the remainder is lost through gaseous emissions or through filtrations causing a serious contamination issue. An alternative would be to utilize a group of bacteria having the ability to fix nitrogen, capturing the atmospheric nitrogen in its nonreactive state N_2 and changing it into ammonia NH_3. This conversion is carried out by the enzyme nitrogenase and is known as biological nitrogen fixation (Bhattacharjee et al. 2008). Also, soluble phosphate availability can be a limiting factor on plant growth. Some PGPRs solubilize phosphorus from organic and inorganic sources to facilitate phosphorus absorption and bioavailability through the release of low-molecular weight organic acids functioning as chelating agents (Antoun 2013; Baldan et al. 2015).

Additionally, some PGPRs produce siderophores, low-molecular weight compounds with chelating potential containing a high affinity for metal ions, particularly Fe^+, which bind to form a Fe-siderophore that can be assimilated by the plant or bacteria. This assimilation leads to an increase in the amount of iron available in the internal tissues constituting a nutritional benefit and an indirect defense tool against pathogens (Aguado-Santacruz et al. 2012; Kloepper et al. 1989).

4.2 *Bacillus subtilis*

B. subtilis—one of the 65 species of the genus *Bacillus*—is a soil microorganism widely distributed in several habitats. The successful colonization of *B. subtilis* is due to its capacity to form thermoresistant endospores as a response to several environmental stress factors, such as nutritional deprivation and the lack of moisture (Rivera Pérez 2009), and it is dispersed by wind. *B. subtilis* has a high reproduction rate and produces extracellular hydrolytic enzymes and antibiotic substances (Earl et al. 2008). The use of fluorescent antibodies to distinguish the vegetative state of the spore in different soil samples revealed that *B. subtilis* is found predominantly in this state when it is associated with organic materials found in decomposition (Norris and Wolf 1961). Experiments proved *B. subtilis* to be a saprophytic organism. Soil samples saturated with twice as much volume of sterile organic matter were inoculated with *B. subtilis* spores, wherein spore germination and the proliferation of the vegetative form of the inoculum were observed for several days, until final sporulation occurred due to nutrient deficiency (Vilain et al. 2006).

For a long time, it was taken for granted that *B. subtilis* was an obligate aerobe. Sequencing of its genome in 1968 showed genes possibly associated with the synthesis of nitrate reductase enzymes (Kunst et al. 1997), suggesting that it can grow in anaerobic conditions using nitrate instead of oxygen as an electron acceptor (Folmsbee et al. 2004; Kunst et al. 1997). Additionally, *B. subtilis* has been found to complete its vegetative cycle under anaerobic conditions like those present in the gastrointestinal tract of animals (Hong et al. 2005; Tam et al. 2006).

4.2.1 Biofilms Formed by Bacillus subtilis

A biofilm is a bacterial population in which bacteria adhere to each other and to a surface through the excretion of various polymers, generating an array of several attached layers (Costerton and Lewandowski 1995). Initially during biofilm formation, bacteria change their phenotype according to their proximity to the surface. As the process continues, fastened cells interact as microcolonies with similar cells and those of other species (Fletcher 1991). Different bacterial biofilms respond to specific microenvironmental conditions with different growth patterns (Costerton et al. 1987) that may be pure or mixed cultures of cellular aggregates on different surfaces under controlled or natural conditions (Andrews and Harris 2000). Biofilms have aroused interest in the study of plant-microorganism interaction. It is well known that most of naturally free-living bacteria are associated with different surfaces in the form of multicellular clusters known as biofilms (Branda et al. 2005). Biofilms growth offers some benefits to its constituents, including improved resistance to severe weather, access to nutrients, and protection (Davey and O'Toole 2000).

B. subtilis also has the ability to form biofilms, since it can develop into a functional cooperative community with specialized cells differentiated from an isogenic progenitor population and produce the necessary molecules to form the biofilm matrix (Vlamakis et al. 2008). This biofilm matrix is generally made of exopolysaccharides and the TasA protein that polymerizes into amyloid fibrils-like structures (Beauregard et al. 2013).

One of *B. subtilis*' characteristics that holds great significance is the capacity to colonize on the roots of a large number of plants and to grow in a mutualistic relationship with them. There is evidence correlating the capacity of *B. subtilis* to form biofilms with root colonization, promotion of growth and protection against pathogens (Beauregard et al. 2013; Cairns et al. 2014).

4.2.2 Bacillus subtilis *as a Biofertilizer*

Crop yield and the quality of agricultural products depend directly on the quality of the soil that provides the plant with nutrients and support. In many cases, low crop productivity has been associated with a poor handling of the soil. Loss of arable land is mainly caused by excessive exploitation of arable lands without the addition of organic material and the increase in salinity due to the use of fertilizers and the occurrence of droughts (Yuan et al. 2007). For more than 60 years, the application of soil endemic microorganisms, particularly those associated with the rhizosphere, has been proposed as an alternative to chemical fertilizer use. The supporters of sustainable agriculture and biosecurity programs are becoming very interested in this method to resolve infertile soil limitations (Bhardwaj et al. 2014).

Biofertilizers contain live microorganisms which have the ability to enrich native soil flora. Moreover, several studies have shown that biofertilizers enhance the texture and other features of soil where the crops grow (Bhardwaj et al. 2014). Thus, biofertilizers are defined as substances that contain live microorganisms that colonize in the rhizosphere or internal part of the plant when they are applied to a seed, a root surface, or soil. Additionally, some microorganisms have the capacity to release chemical compounds that may regulate the plant as well as antibiotics (Vessey 2003).

Phosphorus is the most indispensable nutrient for the optimal growth of a plant after nitrogen (Rajankar et al. 2007; Tejera Hernández et al. 2013). Most of the phosphorus present in soil cannot be assimilated by plants with only 0.1–0.4% of phosphorus being present in organic form (Corrales Ramírez, et al. 2014). Phosphorus forms metal compounds with iron and aluminum in very acidic soils, and it forms compounds with calcium carbonate in very alkaline soils. Some microorganisms are known to have the capacity to solubilize phosphorus from various sources through several methods (Toro et al. 1997). These methods mainly include reducing the pH of the soil by means of organic acid excretion which dissolves phosphorus-rich minerals or chelate cations joined to phosphorus, thereby releasing it (He et al. 2007). The genus *Bacillus* is one of the most studied bacteria regarding its capacity to solubilize phosphorus (Rajankar et al. 2007). Several species that belong to this genus have the ability to solubilize phosphates that are associated with promoting growth of the bacteria in different economically important crops, such as corn and wheat (Egamberdiyeva et al. 2004). Various studies have proven that several *B. subtilis* strains have the capacity to solubilize phosphorus in vitro (Qureshi et al. 2012). Likewise, field studies suggest that *B. subtilis* can enhance phosphorus mobilization in the soil by functioning independently or along with other microorganisms like *Pseudomonas aeruginosa* or arbuscular mycorrhizal fungi. In these cases, excretion of organic acids is the main mechanism of solubilization that occurs with cation exchange allowing insoluble phosphate to become soluble and available to the plant. Citric acid, lactic acid, succinic acid, and propionic acid are among the most common excreted organic acids (Adesemoye et al. 2008; Tejera Hernández et al. 2013; Toro et al. 1997).

Most plants absorb iron, an essential nutrient, as a ferrous ion (Fe^{2+}), but the ferric ion (Fe^{3+}) is the most abundant in the soil that can be precipitated as ion-oxide forms easily. Most plants exude compounds that can interact with Fe^{3+} so they can assimilate it. The "chelating" compounds deposit the Fe^{3+} on a root surface where it is reduced to Fe^{2+} and then absorbed. The siderophores link to Fe^{3+} and allow it to be absorbed through the plasmalemma (Vessey 2003). The production of siderophores is considered a secondary mechanism through which PGPRs stimulate plant growth, since they facilitate iron ions bioavailability in the rhizosphere. Additionally, siderophores are believed to inhibit the growth of certain pathogens (Kloepper et al. 1980). In a study conducted by Díaz Peralta et al. (2012), *B. subtilis* strains were found to produce hydroxamate siderophores also identified in other bacterial species classified as promoting plant growth.

Consequently, *B. subtilis* strains can be supplementary components to the conventional handling of soil fertilization along with crop rotation, residue reincorporation, and pathogen biocontrol (Sahoo et al. 2013). Several studies documented the positive effects of PGPRs on crops caused by an increase in the content of certain plant nutrients. Karlidag et al. (2007) reported that the inoculation of apple trees (*Malus domestica* L. cv. Granny Smith) with *B. subtilis* had a significant impact on the nutrient content (i.e., P, K, Ca, Fe, Mn, Cu, and Zn) in addition to increasing growth and yield. Karakurt et al. (2011) tested strains of *Bacillus* OSU-142 (N_2-fixing) and M3 (N2-fixing and phosphate solubilizing) in isolation and in combination with organically grown primocane raspberry (cv. Heritage) and found an increase in the crop growth and yield. Additionally, the contents of nitrogen, phosphorus, and calcium in raspberry leaves inoculated with both strains, and the contents of iron and manganese on raspberry leaves treated with M3 increased.

Published information about nutrient content of agricultural products coming from plants inoculated with PGPR is scarce. In this regard, Dursun et al. (2010) sprayed PGPR on tomato (*Lycopersicon esculentum* L.) and cucumber (*Cucumis sativus* L.) and found that the application of *B. subtilis* BA-142 increased the mineral content (i.e., N, P, Mg, Ca, Na, K, Cu, Mn, Fe, and Zn) in fruits of both crops. The results showed that there are positive nutritional effects on plants fertilized with *B. subtilis* causing an increase of fruit nutritional quality.

4.2.3 **Bacillus subtilis** *as a Biostimulator*

Over the past few years, an important group of secondary metabolites synthesized by several microorganisms have aroused interest among biotechnologists. They are molecules classified as plant growth hormones due to their similarity to those compounds produced by plants during their growth. Even though the early reports about these metabolites date back to the 1960s, they have drawn attention recently because of their results when applied to sustainable agricultural models. Several species of microorganisms have been reported to synthesize indole-3-acetic acid, gibberellin, zeatin, and abscisic acid (Karadeniz et al. 2006).

About three decades ago, *B. subtilis* strains were reported to produce chemical compounds promoting plant growth with most of the studies documenting the presence of indoleacetic acid and abscisic acid (Araújo et al. 2005). Abscisic acid inhibits root elongation setting up a negative correlation between tissue growth and the endogenous content of abscisic acid. On the other hand, indoleacetic acid has been associated with secondary roots and root hair proliferation. Consequently, it has been hypothesized that the production of these substances from *B. subtilis* contributes to plant growth, especially when observing root length, increase of woody tissue density, greater vigor, and flower and fruit production (Díaz Peralta et al. 2012).

The angiosperms, also known as "flowering plants," are defined in botany as plants with seed(s) and flowers having whorls or arrangement of sepals, petals, stamens, and carpels. While the carpels contain the ovules, the pollen makes contact with the carpels stigmatic surface instead of directly with the ovule, as in gymnosperms. The main role of the floral organ is the production of seeds through sexual reproduction in order to perpetuate and propagate the species. This has led to the growth of attractive crowns and modified edible ovaries, known as fruits (Ferrara Sarmiento et al. 2013).

The use of flowers has gained great interest in the scientific and commercial fields to such an extent that strategies have been searched to control the development of these organs through techniques ranging from the application of chemical compounds similar to plant hormones to genetic engineering (Klee and Giovannoni 2011). The economic interest has led to the overexploitation and indiscriminate use of uncontrolled nutritional systems.

In recent works, the application of *B. subtilis* strains has altered flower growth as well as plant life span. These alterations have been associated with better nutrition and plant stimuli that are typical of the interaction with these microorganisms. Ornamental flowers of the genus *Lilium* are produced under intensive systems due to their high demand. It was observed that when these plants were inoculated with *B. subtilis* along with mycorrhizal fungi *Glomus fasciculatum* and a minimum dose of phosphorus, plants were more vigorous, taller, heavier with greater stem diameter, and had a more intense color. In addition, flowers from the treated plants bloomed faster and lived significantly longer (Rubí Arriaga et al. 2012).

B. subtilis application has also been reported to influence the quality of some agricultural products. Mena-Violante and Olalde-Portugal (2007) showed that fruit and pericarp firmness increased in the latest ripening stages when tomato plant roots were inoculated with *B. subtilis*. The changes in firmness involved alterations in the components of the primary cell wall, such as cellulose, pectin, and glucans, that normally occurred during ripening (Brownleader et al. 1999) and are associated with ethylene production. Ethylene is the hormone that controls the expression of ripening genes (Alexander and Grierson 2002), and it is engaged in the promotion of plant growth (Glick et al. 1998). Although these changes occur in the roots, they could be associated with signaling pathways that affect the fruit growth and ripening (Mena-Violante and Olalde-Portugal 2007). Changes associated with tomato texture (*Solanum lycopersicum* Mill.) in plants inoculated with *Bacillus subtilis* BEB-13bs were also studied. The results showed that the fruit pericarp firmness at the light red ripening stage was significantly higher in fruits from inoculated plants. In addition, the treatment with *B. subtilis* significantly reduced the activity of the polygalacturonase enzyme (PG) in the fruit at the light red ripening stage. Finally, it was reported that the expression pattern of Aco, the gene that encodes for the aminocyclopropane carboxylic acid oxidase—enzyme regulating ethylene synthesis during ripening—showed a significant decrease in the transcript accumulation in red fruit from inoculated plants. The expression pattern changes of this gene associated with ripening, along with the PG enzymatic activity, show the influence of *B. subtilis* on the ripening process. The fact that only the bacteria on the roots produced these effects on the fruit suggested that the participation of signals somehow influenced the ethylene biosynthetic pathway. However, further research is needed to study this possibility.

Another biostimulating effect associated with the fruit quality was documented by Karakurt et al. (2011), who reported that the foliar application of *B. subtilis* promoted the biomass growth in cherry trees and significant changes on the fruit chemical composition, suggesting that the fruit quality could be indirectly affected by bacteria through the production of substances comparable to phytohormones.

4.2.4 **Bacillus subtilis** *as a Postharvest Biocontrol Agent*

Fruits and vegetables are considered living beings even after harvest since they continue breathing and release energy in the form of heat and water. This makes them perishable products that are susceptible to the loss of desirable features that can lead to the loss of commercial value (Giovannoni 2004).

Various technologies have been developed to preserve the organoleptic features (i.e., color, flavor, texture) of fruits and vegetables, such as controlled atmosphere refrigeration systems and radiation at different wavelengths (Ayala Gil 2011). The use of synthetic chemicals to inhibit pathogen growth has contributed to a significant decrease in sprouting and reduced the loss of product (Kim et al. 2015). However, over the long term, it has been negatively observed that the excessive use of these technologies can raise product prices, eventually lose their efficacy, may be toxic for people, and may contribute to environmental deterioration. These drawbacks have promoted the search of less dangerous and more effective alternatives. In this regard, biocontrol of postharvest decay in agricultural products could provide an effective alternative technology to chemical control.

B. subtilis is one of the most commonly used bacteria for disease biocontrol (Ongena and Jacques 2008) and, along with other *Bacillus* species, represents half of the commercially available biopesticides for postharvest disease control worldwide (Fravel 2005). The capacity of *B. subtilis* to form spores makes it one of the best candidates for the development of effective biopesticides (Ongena et al. 2007) in addition to being able to produce metabolites with strong antifungal properties, high degradability, and being environmentally friendly (Chen et al. 2008).

B. subtilis sequencing revealed that a large portion of its genome (4%) produces secondary metabolites. These secondary metabolites include antibiotic peptides that inhibit fungi and bacteria growth (Emmert et al. 2004), volatile compounds (Yuan et al. 2012), and several types of lipopeptides (Bais et al. 2004; Hossain et al. 2015) associated with *B. subtilis* biocontrol activity. In addition, it has been documented that *B. subtilis* produces catabolic enzymes (proteases, kinases, and glucanases) and other components that are toxic to phytopathogens (Ashwini and Srividya 2013).

Since the 1980s, beneficial effects have been observed regarding the application of organisms to different fruits, such as peaches, nectarines, apricots, and plums. For example, spraying with *Pseudomonas* and *Bacillus* bacterial cultures demonstrated antagonist activity against the fungal pathogen *Monilinia fructicola* (Pusey and Wilson 1984). In this work, the treatment with the strain B-3 of *B. subtilis* affected the development of brown rot caused by fungi in all the treated cases. Jiang et al. (2001) reported that *B. subtilis* and its cell-free extracts inhibited the occurrence of *Peronophithora litchi*. The use of *B. subtilis* to control pathogens in several crops has been reported recently (Bais et al. 2004; Choudhary and Johri 2009), and some strains have been proposed as biocontrol agents for the development of bioproducts to control postharvest fruit decay caused by fungi (Arrebola et al. 2010; Casals et al. 2010; Obagwu and Korsten 2003). However, they are not currently commercially available. For instance, Obagwu and Korsten (2003) isolated *B. subtilis* strains from a citrus surface and assayed their capacity to inhibit the occurrence of *Penicillium digitatum* and *Penicillium atalicum.* They found that a treatment of *B. subtilis* combined with sodium bicarbonate was as effective as the mixture of commercial fungicides use to control these pathogens. In a study where *B. subtilis* was applied to avocado plantations to control fungus-causing stalk rot that infected plants from flowering, Demoz and Korsten (2006) demonstrated the capacity of *B. subtilis* to colonize different surfaces of the plant, such as the flower and stomata, and to reduce the incidence of infections associated with fruit rot.

There is a considerable interest in using *B. subtilis* strains that produce antibiotic lipopeptides (e.g., iturine and surfactin) (Bais et al. 2004). Meticulous studies have shown that *B. subtilis* CPA-8 in culture (i.e., cells, spores, and antifungal metabolites), cells, and cell-free supernatant are effective to control brown rot on stone fruit mainly through the production of fengycin lipopeptides (Yánez-Mendizábal et al. 2012).

An orange fruit study associated with *B. subtilis* showed that these bacteria were capable of reducing the postharvest incidence of *Penicillium crustosum* by 25% (Arrebola et al. 2010). Raw extracts of the same *B. subtilis* strain were used to reduce the incidence and decay of *Penicillium digitatum* in mandarin oranges (Leelasuphakul et al. 2008). The biocontrol observed in *B. subtilis* may be due to the production of volatile compounds that can affect the production of mycelium, such as ketones, organic acids, alcohols, sulfur-nitrogen compounds, and esters, including more than 21 different types of compounds (Arrebola et al. 2010).

A postharvest study on the *Malus domestica* "Golden Delicious" apple reported that strains from different *Bacillus* species, including *B. subtilis*, showed antagonist activity against gray mold caused by *Botrytis mali*, preventing its growth and reducing the lesion diameter (Jamalizadeh et al. 2009). Dimkić et al. (2013) showed that the ethyl acetate extracts from the cell-free supernatants of two *B. subtilis* strains were active against several apple fungal pathogens after the harvest, in vitro and in vivo. The mass spectrometry analysis of the extracts confirmed the presence of surfactin. Kim et al. (2015) demonstrated the capacity of the *B. subtilis* strain HM1 as a control agent for apple anthracnose caused by *Colletotrichum acutatum* fungi, one of the phytopathogens that most affects postharvest fruits. *B. subtilis* HM1 exhibited a wide spectrum of antagonistic properties to several phytopathogenic fungi, and the production of lipopeptides was identified to attribute to the inhibitory properties of this strain. The authors reported that the application of *B. subtilis* could prevent up to 80.7% of infection caused by *Colletotrichum acutatum*, while the cell-free supernatant only showed 69.4% efficacy on anthracnose control. Interestingly, three compounds associated with the phytopathogenic fungi inhibition were identified in the supernatant: iturine A, fengycin, and surfactin.

Despite much research, the mechanisms of *B. subtilis* to function as a biocontrol agent are still not completely clear, and it is suggested that this antimicrobial activity may be due to the production of mycolytic enzymes. Srivastava et al. (2012) reported that the *B. subtilis* JN032305 strain isolated from the chili rhizosphere produced three mycolytic enzymes: chitinase, glucanase, and cellulase, showing a wide spectrum antagonistic property against bacteria and phytopathogenic fungi.

Interestingly, the application of bacteria to influence the quality of agricultural products regarding pre- and postharvest disease is just beginning to be documented. Feliziani et al. (2015) carried out research in which several alternative treatments were applied and demonstrated that the routine pre- and postharvest application of the treatments reduced the loss of desirable features by 33%, such as color and turgor in strawberry (*Fragaria* × *ananassa*).

4.2.5 **Bacillus subtilis** *as an Agricultural Product Quality Promoter*

The quality of fruits and vegetables is defined as their degree of excellence (Abbott 1999) and centers around organoleptic properties (i.e., texture, color, flavor), bioactive substances content (i.e., carotenoids and dietary fiber), and essential nutritional compounds (i.e., proteins and vitamins), in addition to the absence of unwanted attributes (i.e., pesticides and heavy metals) (Schreiner 2007).

The quality of agricultural products is affected by several pre- and postharvest factors. The preharvest factors that affect the quality of products already harvested include biological factors (i.e., pathologic and entomological), physiological (i.e., nutritional imbalance and ripeness), and cultural (i.e., fertilization and growth regulators) (Mattheis and Fellman 1999). In this regard, the PGPR can be considered as preharvest agent that influences the yield and quality of fruits, vegetables, and other agricultural products (Mena-Violante and Olalde-Portugal 2007; Mena-Violante et al. 2009).

The quality of fruits and vegetables is highly dependent on their ripening stage. Fruit ripening consists of a number of biochemical and structural changes that make the fruit more attractive to seed dispersal vectors (Brummell 2006). It is a highly complex and coordinated process that involves different metabolic pathways, involving pigment biosynthesis (Andersen et al. 2004), sugars, acids, and volatile compounds that alter the flavor (Darbellay et al. 2004), degradation of the cell wall and the middle lamella (Osorio et al. 2013), as well as the synthesis and action of regulation hormones (Given et al. 1988). Color is the main indicator of fruit ripeness, and it depends on pigment accumulation, such as carotenoids and flavonoids, that have been attributed to nutraceutical properties (Andersen et al. 2004; Wang et al. 2014).

Reports on organoleptic fruit properties caused by the inoculation of crops with PGPR are scarce, but one of these studies was carried out by Mena-Violante and Olalde-Portugal (2007). They demonstrated that the *B. subtilis* BEB-13bs strain had positive effects on tomato quality (*Solanum lycopersicum*), particularly on the size and texture. Interestingly, firmer fruits from plants inoculated with *B. subtilis* were obtained. It was suggested that the effects on the texture (i.e., firmness) of fruits could be associated with the changes in ethylene production. In this regard, it is known that the plant hormone regulates ripeness (Alexander and Grierson 2002) and is also involved in the promotion of plant growth by the PGPR (Glick et al. 1998). Although these changes occur in the roots, they could be associated with signal pathways that affect the growth and ripeness of fruit.

Mena-Violante et al. (2009) reported that during the latest ripening stages, pericarp, and whole fruit firmness increased when the tomato roots were inoculated with *B. subtilis*. Texture changes include alterations in the primary wall components, such as cellulose, pectin, and glucan, that normally occur during ripening (Brownleader et al. 1999). Texture is a very important quality parameter, because it generally determines the shelf life of fruit (Brandy 1987; Manning 1996; Paniagua et al. 2014). Low quality in fruit is mainly caused by excessive softening (Giovannoni 2004). The firmest fruits are expected to be more resistant to decay caused by microorganisms and consequently have a longer shelf life. In this regard, the authors found that the decay percentage of fruit after 10 days in storage was significantly reduced in fruits from plants treated with *B. subtilis*.

Some research showed that applying *B. subtilis* as a biofertilizer or as a biocontrol agent had a positive impact on the shelf life of agricultural products. Pusey and Wilson (1984) reported that the treatment of coffee with the *B. subtilis* B-1849 strain reduced fruit decay. Jiang et al. (2001) reported that *B. subtilis* as well as its cell-free extracts inhibited the growth of *Peronophithora litchi*. In addition to controlling infection, fruits were stored up to 30 days at 5 °C without acquiring unwanted alterations when the microorganisms or its extracts were applied after harvest.

Similarly, in research carried out at the Ecological Biochemistry Laboratory of CINVESTAV-IPN Irapuato, we obtained results similar to those reported by Jiang et al. (2001). After spraying with *B. subtilis* strains and storing at commercial refrigeration temperatures, strawberries exhibited better preservation of physical features for a longer period of time compared to the controls.

It is important to note that other positive effects of plant roots inoculated with *B. subtilis* have been observed on fruit quality. Datta et al. (2011) found through a field study that C2 and C25 bacterial strains identified as *Bacillus* species isolated from the rhizosphere of the chili cultivar "Suryamukhi" promoted a significant increase in fruit weight, thus demonstrating that fruit quality benefited in terms of size. On the other hand, Erturk et al. (2012) showed that *Bacillus* species not only increased the yield and growth of the strawberry "Fern" but also positively influenced fruit quality features, such as average weight, diameter, soluble solids content, and vitamin C content. Pırlak and Köse (2009) studied the effects of several PGPR strains containing biofertilizing and biocontrolling features on the strawberry cultivar "Selva," and they reported that the inoculation of strawberry roots with the *Bacillus* OSU-142 strain resulted in increased yields and soluble solids content as well as affected sugar content.

4.3 Conclusions

Bacillus subtilis is a rhizospheric bacterium with potential for applications in sustainable agricultural production due to its capacity to form endospores and biofilms and to its properties as a biofertilizer, a biostimulant, and a biocontrol agent. This potential of *B. subtilis* lies within mechanisms involving plant growth regulators' synthesis, secondary metabolites, and mycolytic enzymes.

B. subtilis is known to promote plant growth and to increase yield as well as to prevent and control disease. In addition, these bacteria can improve the quality of agricultural products, like flowers and fruits. Several quality features in these products, such as size, color, firmness, and shelf life, can be positively affected by the application of *B. subtilis*.

References

Abbott JA (1999) Quality measurement of fruits and vegetables. Postharvest Biol Technol 15(3):207–225

Adesemoye AO, Obini M, Ugoji EO (2008) Comparison of plant growth-promotion with pseudomonas aeruginosa and bacillus subtilis in three vegetables. Braz J Microbiol 39:423–426

Aguado-Santacruz GA, Moreno-Gómez B, Jiménez-Francisco B, García-Moya E, Preciado-Ortiz RE (2012) Impacto de los sideróforos microbianos y fitosideróforos en la asimilación de hierro por las plantas: Una síntesis. Rev Fitotec Mex 35(1):9–21

Ahmed A, Hasnain S (2014) Auxins as one of the factors of plant growth improvement by plant growth promoting rhizobacteria. Pol J Microbiol 63(3):261–266

Alexander L, Grierson D (2002) Ethylene biosynthesis and action in tomato: a model for climacteric fruit ripening. J Exp Bot 53(377):2039–2055. https://doi.org/10.1093/jxb/erf072

Andersen ØM, Fossen T, Torskangerpoll K, Fossen A, Hauge U (2004) Anthocyanin from strawberry (Fragaria ananassa) with the novel aglycone, 5-carboxypyranopelargonidin. Phytochemistry 65(4):405–410. https://doi.org/10.1016/j.phytochem.2003.10.014

Andrews JH, Harris RF (2000) The ecology and biogeography of microorganisms on plant surfaces. Annu Rev Phytopathol 38:145–180

Antoun H (2013) Plant-growth-promoting rhizobacteria. In: Brenner's encyclopedia of genetics, vol. 5. Elsevier. https://doi.org/10.1016/B978-0-12-374984-0.01169-4

Araújo FF, Henning AA, Hungria M (2005) Phytohormones and antibiotics produced by Bacillus subtilis and their effects on seed pathogenic fungi and on soybean root development. World J Microbiol Biotechnol 21(8–9):1639–1645. https://doi.org/10.1007/s11274-005-3621-x

Arrebola E, Sivakumar D, Korsten L (2010) Effect of volatile compounds produced by Bacillus strains on postharvest decay in citrus. Biol Control 53(1):122–128. https://doi.org/10.1016/j.biocontrol.2009.11.010

Ashwini N, Srividya S (2013) Potentiality of Bacillus subtilis as biocontrol agent for management of anthracnose disease of chilli caused by Colletotrichum gloeosporioides OGC1. 3 Biotech 4(2):127–136. https://doi.org/10.1007/s13205-013-0134-4

Ayala Gil ME (2011) Efecto de la luz ultravioleta y la temperatura en la concentración de flavonoides del fruto de fresa (Fragaria x ananassa)

Bais HP, Fall R, Vivanco JM, Bais Pal H, Fall R, Vivanco JM et al (2004) Biocontrol of Bacillus subtilis against infection of Arabidopsis roots by Pseudomonas syringae is facilitated by biofilm formation and surfactin production. Plant Physiol 134(January):307–319. https://doi.org/10.1104/pp.103.028712.zobacterium

Baldan E, Nigris S, Romualdi C, D'Alessandro S, Clocchiatti A, Zottini M et al (2015) Beneficial bacteria isolated from grapevine inner tissues shape arabidopsis thaliana roots. Plos One 10(10):e0140252. https://doi.org/10.1371/journal.pone.0140252
Bashan Y, Levanony H, Mitiku G (1989) Changes in proton efflux intact wheat root induced by Azospirillum brasilense Cd. Can J Microbiol 35:691–697
Beauregard PB, Chai Y, Vlamakis H, Losick R, Kolter R (2013) Bacillus subtilis biofilm induction by plant polysaccharides. Proc Natl Acad Sci USA 110(17):E1621–E1630. https://doi.org/10.1073/pnas.1218984110
Bergottini VM, Otegui MB, Sosa DA, Zapata PD, Mulot M, Rebord M et al (2015) Bio-inoculation of yerba mate seedlings (Ilex paraguariensis St. Hill.) with native plant growth-promoting rhizobacteria: a sustainable alternative to improve crop yield. Biol Fertil Soils 51(6):749–755. https://doi.org/10.1007/s00374-015-1012-5
Bhardwaj D, Ansari MW, Sahoo RK, Tuteja N (2014) Biofertilizers function as key player in sustainable agriculture by improving soil fertility, plant tolerance and crop productivity. Microbial Cell Factories 13(1):66. Retrieved from http://www.pubmedcentral.nih.gov/articlerender.fcgi?artid=4022417&tool=pmcentrez&rendertype=abstract
Bhattacharjee RB, Singh A, Mukhopadhyay SN (2008) Use of nitrogen-fixing bacteria as biofertiliser for non-legumes: Prospects and challenges. Appl Microbiol Biotechnol 80(2):199–209. https://doi.org/10.1007/s00253-008-1567-2
Branda SS, Vik S, Friedman L, Kolter R (2005) Biofilms: the matrix revisited. Trends Microbiol 13(1):20–26. https://doi.org/10.1016/j.tim.2004.11.006
Brandy CJ (1987) Fruit ripening. Annu Rev *38*:155–178
Brownleader MD, Jackson P, Mobasheri A, Pantelides AT, Sumar S, Trevan M, Dey PM (1999) Molecular aspects of cell wall modifications during fruit ripening. Crit Rev Food Sci Nutr 39(2):149–164. https://doi.org/10.1080/10408399908500494
Brummell DA (2006) Primary cell wall metabolism during fruit ripening. N Z J For Sci 36(June 2005):99–111
Cairns LS, Hobley L, Stanley-Wall NR (2014) Biofilm formation by Bacillus subtilis: new insights into regulatory strategies and assembly mechanisms. Mol Microbiol 93(4):587–598. https://doi.org/10.1111/mmi.12697
Casals C, Teixidó N, Viñas I, Cambray J, Usall J (2010) Control of Monilinia spp. on stone fruit by curing treatments. Part II: the effect of host and Monilinia spp. variables on curing efficacy. Postharvest Biol Technol 56(1):26–30. https://doi.org/10.1016/j.postharvbio.2009.11.009
Chen H, Xiao X, Wang J, Wu L, Zheng Z, Yu Z (2008) Antagonistic effects of volatiles generated by Bacillus subtilis on spore germination and hyphal growth of the plant pathogen, Botrytis cinerea. Biotechnol Lett 30(5):919–923. https://doi.org/10.1007/s10529-007-9626-9
Choudhary DK, Johri BN (2009) Interactions of Bacillus spp. and plants – with special reference to induced systemic resistance (ISR). Microbiol Res 164(5):493–513
Cohen AC, Bottini R, Piccoli P (2015) Role of abscisic acid producing PGPR in sustainable agriculture. In: Bacterial metabolites in sustainable agroecosystem, pp 258–282
Corrales Ramírez LC, Sánchez Leal LC, Arévao Galvez Z, Moreno Burbano VE (2014) Bacillus: género bacteriano que demuestra ser un importante solubilizador de fosfato Bacillus: a genus of bacteria that exhibits important phosphate. NOVA 12(21):165–178
Costerton JW, Lewandowski Z (1995) Microbial biofilms. Annu Rev Microbiol 49:711–745
Costerton JW, Cheng KJ, Geesey GG, Lass TI, Nickel JC, Dasgupta M, Merrie TJ (1987) Bacterial biofilms in nature and disease. Annu Rev Microbiol 41:435–464
Darbellay C, Luisier J-L, Villettaz J-C, Amadó R, Azodanlou R (2004) Changes in flavour and texture during the ripening of strawberries. Eur Food Res Technol 218(2):167–172. https://doi.org/10.1007/s00217-003-0822-0
Datta M, Palit R, Sengupta C, Pandit MK, Banerjee S (2011) Plant growth promoting rhizobacteria enhance growth and yield of chilli (Capsicum annuum L.) under field conditions. Aust J Crop Sci 5(3):531–536

Davey ME, O'toole GA (2000) Microbial biofilms: from ecology to molecular genetics. Microbiol Mol Biol Rev 64(4):847–867. https://doi.org/10.1128/MMBR.64.4.847-867.2000

Demoz BT, Korsten L (2006) Bacillus subtilis attachment, colonization, and survival on avocado flowers and its mode of action on stem-end rot pathogens. Biol Control 37(1):68–74. https://doi.org/10.1016/j.biocontrol.2005.11.010

Díaz Peralta K, Araya T, Valenzuela S, Sossa K, Martínez M, Peña-Cortés H, Sanfuentes E (2012) Production of phytohormones, siderophores and population fluctuation of two root-promoting rhizobacteria in Eucalyptus globulus cuttings. World J Microbiol Biotechnol 28(5):2003–2014. https://doi.org/10.1007/s11274-012-1003-8

Dimkić I, Živković S, Berić T, Ivanović Ž, Gavrilović V, Stanković S, Fira D (2013) Characterization and evaluation of two Bacillus strains, SS-12.6 and SS-13.1, as potential agents for the control of phytopathogenic bacteria and fungi. Biol Control 65(3):312–321. https://doi.org/10.1016/j.biocontrol.2013.03.012

Dursun A, EkIncI M, Dönmez MF (2010) Effects of foliar application of plant growth promoting bacterium on chemical contents, yield and growth of tomato (Lycopersicon esculentum L.) and cucumber (Cucumis sativus L.). Pak J Bot 42(5):3349–3356

Earl AM, Losick R, Kolter R (2008) Ecology and genomics of Bacillus subtilis. Trends Microbiol 16(6):1–11

Egamberdieva D, Kamilova F, Validov S, Gafurova L, Kucharova Z, Lugtenberg B (2008) High incidence of plant growth-stimulating bacteria associated with the rhizosphere of wheat grown on salinated soil in Uzbekistan. Environ Microbiol 10(1):1–9. https://doi.org/10.1111/j.1462-2920.2007.01424.x

Egamberdiyeva D, Juraeva D, Poberejskaya S, Myachina O, Teryuhova P, Seydalieva L, Aliev A (2004) Improvement of wheat and cotton growth and nutrient uptake by phosphate solubilizing bacteria. In: 26th Southern conservation tillage conference, pp 58–66

Emmert EAB, Klimowicz AK, Michael G, Handelsman J, Thomas MG (2004) Genetics of Zwittermicin A production by bacillus cereus. Appl Environ Microbiol 70(1):104–113. https://doi.org/10.1128/AEM.70.1.104

Erturk Y, Ercisli S, Cakmakci R (2012) Yield and growth response of strawberry to plant growth-promoting Rhizobacteria inoculation. J Plant Nutr 35(6):817–826. https://doi.org/10.1080/01904167.2012.663437

FAO (2000) FAO statement on biotechnology. http://www.fao.org/biotech/fao-statement-on-biotechnology/es/, reviewed May 2018

Feliziani E, Landi L, Romanazzi G (2015) Preharvest treatments with chitosan and other alternatives to conventional fungicides to control postharvest decay of strawberry. Carbohydr Polym 132:111–117. https://doi.org/10.1016/j.carbpol.2015.05.078

Ferrara Sarmiento O, Hernández Najarro F, Díaz Montesinos MG (2013) Las plantas con flores. In: La biodiversidad en Chiapas, pp 121–125. https://doi.org/10.1002/9780470561119.socpsy001017

Fletcher M (1991) The physiological activity of bacteria attached to solid surfaces. Adv Microb Physiol 32:53–85

Folmsbee MJ, Mcinerney MJ, Nagle DP (2004) Anaerobic growth of Bacillus mojavensis and Bacillus subtilis requires Deoxyribonucleosides or DNA. Appl Environ Microbiol 70(9):5252–5257

Fravel DR (2005) Commercialization and implementation of biocontrol. Annu Rev Phytopathol 43:337–359. https://doi.org/10.1146/annurev.phyto.43.032904.092924

Giovannoni JJ (2004) Genetic regulation of fruit development and ripening. Plant Cell 16:170–180. https://doi.org/10.1105/tpc.019158.Fruit

Given NK, Venis MA, Gierson D (1988) Hormonal regulation of ripening in the strawberry, a non-climacteric fruit. Planta 174(3):402–406

Glick BR (2012) Plant growth-promoting bacteria: mechanisms and applications. Scientifica 2012:963401. https://doi.org/10.6064/2012/963401

Glick B, Penrose D, Li J (1998) A model for the lowering of plant ethylene concentrations by plant growth-promoting bacteria. J Theor Biol 190(1):63–68. https://doi.org/10.1006/jtbi.1997.0532
Gunes A, Karagoz K, Turan M, Kotan R, Yildirim E, Cakmakci R, Sahin F (2015) Fertilizer efficiency of some plant growth promoting rhizobacteria for plant growth. Res J Soil Biol 7(2):28–45. https://doi.org/10.3923/rjsb.2015.28.45
He ZL, Bian W, Zhu J (2007) Screening and identification of microorganisms capable of utilizing phosphate adsorbed by goethite. Commun Soil Sci Plant Anal 33(5–6):647–663. https://doi.org/10.1081/CSS-120003057
Hiltner L (1904) Uber neuere Erfahrungen und Probleme auf dem Gebiete der Bodenbakteriologie unter besonderden berucksichtigung und Brache. Arb Dtsch Landwirtsch Gesellschaft 98:59–78
Hong HA, Le HD, Cutting SM (2005) The use of bacterial spore formers as probiotics. FEMS Microbiol Rev 29(4):813–835. https://doi.org/10.1016/j.femsre.2004.12.001
Hossain MJ, Ran C, Liu K, Ryu C-M, Rasmussen-Ivey CR, Williams MA et al (2015) Deciphering the conserved genetic loci implicated in plant disease control through comparative genomics of Bacillus amyloliquefaciens subsp. plantarum. Front Plant Sci 6(August):1–14. https://doi.org/10.3389/fpls.2015.00631
Jamalizadeh M, Etebarian HR, Aminian H, Alizadeh A (2009) Evaluation of Bacillus spp. as potential biocontrol agent for postharvest gray mold control on golden delicious apple in Iran. J Plant Protect Res 49(4). https://doi.org/10.2478/v10045-009-0064-7
Jiang YM, Zhu XR, Li YB (2001) Postharvest control of litchi fruit rot by Bacillus subtilis. LWT Food Sci Technol 34(7):430–436. https://doi.org/10.1006/fstl.2001.0758
Kang SM, Khan AL, Hamayun M, Hussain J, Joo GJ, You YH et al (2012) Gibberellin-producing Promicromonospora sp. SE188 improves Solanum lycopersicum plant growth and influences endogenous plant hormones. J Microbiol *50*(6):902–909. https://doi.org/10.1007/s12275-012-2273-4
Karadeniz A, Topcuoğlu SF, Inan S (2006) Auxin, gibberellin, cytokinin and abscisic acid production in some bacteria. World J Microbiol Biotechnol 22(10):1061–1064. https://doi.org/10.1007/s11274-005-4561-1
Karakurt H, Kotan R, Dadaşoğlu F, Aslantaş R, Şahİn F (2011) Effects of plant growth promoting rhizobacteria on fruit set , pomological and chemical characteristics, color values, and vegetative growth of sour cherry (Prunus cerasus cv. Kütahya). Turk J Biol 35:283–291. https://doi.org/10.3906/biy-0908-35
Karlidag H, Esitken A, Turan M, Sahin F (2007) Effects of root inoculation of plant growth promoting rhizobacteria (PGPR) on yield, growth and nutrient element contents of leaves of apple. Sci Hortic 114(1):16–20. https://doi.org/10.1016/j.scienta.2007.04.013
Kim HM, Lee K-J, Chae J-C (2015) Postharvest biological control of Colletotrichum acutatum on apple by Bacillus subtilis HM1 and the structural identification of antagonists. J Microbiol Biotechnol 25(11):1954–1959
Klee HJ, Giovannoni JJ (2011) Genetics and control of tomato fruit ripening and quality attributes. Annu Rev Genet 45:41–59. https://doi.org/10.1146/annurev-genet-110410-132507
Kloepper JW, Leong J, Teintze M, Schroth M (1980) Enhanced plant growth by siderophores produced by plant growth-promoting rhizobacteria
Kloepper JW, Lifshitz R, Zablotowicz RM (1989) Free-living bacterial inocula for enhancing crop productivity. Trends Biotechnol 7(2):39–44. https://doi.org/10.1016/0167-7799(89)90057-7
Kunst F, Ogasawara N, Moszer I, Albertini AM, Alloni G, Azevedo V et al (1997) The complete genome sequence of the gram-positive bacterium Bacillus subtilis. Nature 390(6657):249–256. https://doi.org/10.1038/36786
Leelasuphakul W, Hemmanee P, Chuenchitt S (2008) Growth inhibitory properties of Bacillus subtilis strains and their metabolites against the green mold pathogen (Penicillium digitatum Sacc.) of citrus fruit. Postharvest Biol Technol 48(1):113–121. https://doi.org/10.1016/j.postharvbio.2007.09.024

Liu F, Xing S, Ma H, Du Z, Ma B (2013) Cytokinin-producing, plant growth-promoting rhizobacteria that confer resistance to drought stress in Platycladus orientalis container seedlings. Appl Microbiol Biotechnol 97(20):9155–9164. https://doi.org/10.1007/s00253-013-5193-2

Lugtenberg B, Kamilova F (2009) Plant-growth-promoting rhizobacteria. Annu Rev Microbiol 63:541–556. https://doi.org/10.1146/annurev.micro.62.081307.162918

Manning K (1996) Soft fruit. In: Biochemistry of fruit ripening, pp 347–377

Mattheis JP, Fellman JK (1999) Preharvest factors influencing flavor of fresh fruit and vegetables. Postharvest Biol Technol 3(15):227–232

Mena-Violante HG, Olalde-Portugal V (2007) Alteration of tomato fruit quality by root inoculation with plant growth-promoting rhizobacteria (PGPR): Bacillus subtilis BEB-13bs. Sci Hortic 113(1):103–106

Mena-Violante H, Cruz-Hernández A, Paredes-López O, Gómez-Lim M, Olalde-Portugal V (2009) Fruit texture related changes and enhanced shelf-life through tomato root inoculation with Bacillus subtilis BEB-13BS. Agrociencias 43:559–567

Mendes R, Garbeva P, Raaijmakers JM (2013) The rhizosphere microbiome: significance of plant beneficial, plant pathogenic, and human pathogenic microorganisms. FEMS Microbiol Rev 37(5):634–663. https://doi.org/10.1111/1574-6976.12028

Norris J, Wolf J (1961) A study of antigens of the aerobic spore-forming bacteria. J Appl Bacteriol 24:42–56

Obagwu J, Korsten L (2003) Integrated control of citrus green and blue molds using Bacillus subtilis in combination with sodium bicarbonate or hot water. Postharvest Biol Technol 28(1):187–194. https://doi.org/10.1016/S0925-5214(02)00145-X

Olalde-Portugal V, Mena-Violante HG (2008) Rhizosphere microorganisms and their effect on fruit quality. In: Postharvest biology and technology of fruits, vegetables, and flowers, pp 360–373

Ongena M, Jacques P (2008) Bacillus lipopeptides: versatile weapons for plant disease biocontrol. Trends Microbiol 16(3):115–125. https://doi.org/10.1016/j.tim.2007.12.009

Ongena M, Jourdan E, Adam A, Paquot M, Brans A, Joris B et al (2007) Surfactin and fengycin lipopeptides of Bacillus subtilis as elicitors of induced systemic resistance in plants. Environ Microbiol 9(4):1084–1090. https://doi.org/10.1111/j.1462-2920.2006.01202.x

Ordookhani K, Khavazi K, Moezzi A, Rejali F (2010) Influence of PGPR and AMF on antioxidant activity, lycopene and potassium contents in tomato. Afr J Agric Res 5(10):1108–1116. https://doi.org/10.5897/AJAR09.183

Osorio S, Scossa F, Fernie AR (2013) Molecular regulation of fruit ripening. Front Plant Sci 4(June):198. https://doi.org/10.3389/fpls.2013.00198

Paniagua C, Posé S, Morris VJ, Kirby AR, Quesada MA, Mercado JA (2014) Fruit softening and pectin disassembly: an overview of nanostructural pectin modifications assessed by atomic force microscopy. Ann Bot 114(6):1375–1383. https://doi.org/10.1093/aob/mcu149

Paul Schreiner R (2007) Effects of native and nonnative arbuscular mycorrhizal fungi on growth and nutrient uptake of 'pinot noir' (*Vitis vinifera* L.) in two soils with contrasting levels of phosphorus. Appl Soil Ecol 36(2–3):205–215

Pingali PL (2012) Green revolution: impacts, limits, and the path ahead. Proc Natl Acad Sci 109(31):12302–12308. https://doi.org/10.1073/pnas.0912953109

Pırlak L, Köse M (2009) Effects of plant growth promoting rhizobacteria on yield and some fruit properties of strawberry. J Plant Nutr 32(7):1173–1184. https://doi.org/10.1080/01904160902943197

Pusey PL, Wilson CL (1984) Postharvest biological control of stone fruit brown rot by bacillus subtilis. Plant Dis 68(9):753–756

Qureshi MA, Ahmad ZA, Akhtar N, Iqbal A, Mujeeb F, Shakir MA (2012) Role of phosphate solubilizing bacteria (PSB) in enhancing P availability and promoting cotton growth. J Anim Plant Sci 22(1):204–210

Rajankar PN, Tambekar DH, Wate SR (2007) Study of phosphate solubilization efficiencies of fungi and bacteria isolated from saline belt of Purna river basin. Res J Agric Biol Sci 3(6):701–703

Rivera Pérez JA (2009) Análisis del proteoma extracelular de cepas de Bacillus subtilis de interés agronómico

Rubí Arriaga M, Gonzalez Huerta A, Olalde Portugal V, Reyes Reyes BG, Castillo González AM, Pérez López D, de J, Aguilera Gómez LI (2012) Contribución de fósfor al mejoramiento de la calidad en Lilium y la relación con Glomus fasciculatum y Bacillus subtilis. Revsita Mecivana de CIencias Agrícolas 3(1):125–139

Sahoo RK, Bhardwaj D, Tuteja N (2013) Biofertilizers: a sustainable eco-friendly agricultural approach to crop improvement. In: Plant acclimation to environmental stress, pp 403–432

Srivastava N, Chauhan AS, Sharma B (2012) Isolation and characterization of some phytochemicals from Indian traditional plants. Biotechnol Res Int 2012:549850. https://doi.org/10.1155/2012/549850

Tam NKM, Uyen NQ, Hong HA, Le H, Hoa TT, Serra CR et al (2006) The intestinal life cycle of Bacillus subtilis and close relatives the intestinal life cycle of Bacillus subtilis and close relatives. J Bacteriol 188(7):2692–2700. https://doi.org/10.1128/JB.188.7.2692

Tejera Hernández B, Heydrich Pérez M, Rojas Bodía M (2013) Aislamiento de bacillus solubilizadores de fosfatos asociados al cultivo del arroz. Agronomía Mesoamaricana 24(2):357–364

Toro M, Azcón R, Barea JM (1997) Improvement of arbuscular mycorrhiza development by inoculation of soil with phosphate-solubilizing rhizobacteria to improve rock phosphate bioavailability (32P) and nutrient cycling. Appl Environ Microbiol 63(11):4408–4412

Unfao (2010) The state of food insecurity in the World

Vessey J (2003) Plant growth promoting rhizobacteria as biofertilizers. Plant Soil 255(2):571–586. https://doi.org/10.1023/A:1026037216893

Vilain S, Luo Y, Hildreth MB, Brözel VS (2006) Analysis of the life cycle of the soil Saprophyte Bacillus cereus in liquid soil extract and in soil. Appl Environl Microbiol 72(7):4970–4977. https://doi.org/10.1128/AEM.03076-05

Vlamakis H, Aguilar C, Losick R, Kolter R (2008) Control of cell fate by the formation of an architecturally complex bacterial community. Genes Dev 22(7):945–953. https://doi.org/10.1101/gad.1645008

Wang Y, Zhang D, Liu Y, Wang D, Liu J, Ji B (2014) The protective effects of berry-derived anthocyanins against visible light-induced damage in human retinal pigment epithelial cells. J Sci Food Agric. https://doi.org/10.1002/jsfa.6765

Yánez-Mendizábal V, Zeriouh H, Viñas I, Torres R, Usall J, de Vicente A et al (2012) Biological control of peach brown rot (Monilinia spp.) by Bacillus subtilis CPA-8 is based on production of fengycin-like lipopeptides. Eur J Plant Pathol 132(4):609–619. https://doi.org/10.1007/s10658-011-9905-0

Yuan BC, Li ZZ, Liu H, Gao M, Zhang YY (2007) Microbial biomass and activity in salt affected soils under arid conditions. Appl Soil Ecol 35(2):319–328. https://doi.org/10.1016/j.apsoil.2006.07.004

Yuan J, Raza W, Shen Q, Huang Q (2012) Antifungal activity of bacillus amyloliquefaciens NJN-6 volatile compounds against Fusarium oxysporum f. sp. cubense. Appl Environ Microbiol 78(16):5942–5944. https://doi.org/10.1128/AEM.01357-12

Chapter 5
Plant Growth Promotion by ACC Deaminase-Producing Bacilli Under Salt Stress Conditions

Gustavo Santoyo, Adrian Equihua, Aurora Flores, Edgardo Sepulveda, Eduardo Valencia-Cantero, Juan M. Sanchez-Yañez, Luzmaria R. Morales, M. Govindappa, and Sergio de los Santos-Villalobos

5.1 Introduction

Agriculture is a dominant activity in worldwide economic, social, and environmental development, contributing to 80% of the food consumed globally (FAO 2018). However, this activity around the world faces several soil adversities, which can negatively impact food production, causing economic losses to producers. Soil degradation in agroecosystems is a serious problem that tends to be ignored, even when it has accelerated rates around the world (Borrelli et al., 2015; Dotterweich, 2013; Ligonja and Shrestha 2015). This event has negative impacts on the future of food safety by reducing soil fertility, which causes biotic and abiotic stress for crops (Dercon et al. 2012). Some examples of abiotic stress include extreme environmental conditions, such as high or low temperatures or pH, low nutrient availability in soils, flooding, prolonged periods of drought, and high concentrations of metals and salts (Glick 2014; Santoyo et al. 2016). Among them, salinity in soils is recognized as one of the factors that most affect agricultural production, and it is estimated that 20% of cultivated areas are affected by this stress condition (Flowers 2004). For

G. Santoyo (✉) · A. Equihua · A. Flores · E. Valencia-Cantero · J. M. Sanchez-Yañez · L. R. Morales
Instituto de Investigaciones Químico Biológicas, Universidad Michoacana de San Nicolás de Hidalgo, Morelia, Michoacán, Mexico
e-mail: gsantoyo@umich.mx

E. Sepulveda
CONACYT, Department of Microbiology, Centro de Investigación Científica y de Educación Superior de Ensenada, CICESE, Ensenada, Mexico

M. Govindappa
Dayananda Sagar College of Engineering, Bengaluru, Karnataka, India

S. de los Santos-Villalobos
CONACYT-Instituto Tecnológico de Sonora, Ciudad Obregón, Sonora, Mexico

M. T. Islam et al. (eds.), *Bacilli and Agrobiotechnology: Phytostimulation and Biocontrol*, Bacilli in Climate Resilient Agriculture and Bioprospecting,
https://doi.org/10.1007/978-3-030-15175-1_5

example, salinity has caused a loss of about 65% of wheat yield in moderately saline soils (Shafi et al. 2010); due to that abiotic stress affects almost all aspects of plant development including germination, vegetative growth, and reproductive development (Foolad 2004). Thus, the global food production is negatively affected by high concentrations of salt in agroecosystems, due to (i) natural biogenesis of agricultural soil located in arid and semiarid regions and/or (ii) application of high rate of synthetic fertilizers, irrigation, and backflow of seawater (Siddikee et al. 2010).

In different regions of the world, saline soils represent most of the arable and cultivable areas, so several strategies have been implemented to mitigate its effects and generate good yields. For example, tolerant plants to higher salinity concentrations by using genetic modification have been widely reported (Roy et al. 2014). As well as other techniques such as the selection of genotypes resistant to salt stress (first strategy) has shown that it may be another viable alternative and that it could be more accepted in certain legislations of countries where the cultivation of GMOs (second strategy) may be prohibited or highly restricted for cultivation (James 2015). A third strategy is the use of microorganisms that form beneficial interactions with plants, in particular, the plant growth-promoting bacteria (PGPB) (Lugtenberg and Kamilova 2009, Rojas-Solis et al. 2018). Thus, various bacterial genera from the rhizosphere, phyllosphere, or plant endosphere have been isolated and characterized by their ability to promote the growth of plants under salt stress conditions (Ghosh et al. 2003; Shrivastava and Kumar 2015). In this chapter we will focus on bacteria of the Bacilli group, mainly the *Bacillus* genus, within which there are several species that stand out for their multiple direct and indirect mechanisms of plant growth promotion, in addition to resistance to several types of environmental stress, fast-growing or duplication rate, and competent colonization, among others (Santoyo et al. 2012).

5.2 Salt Stress in Plants and Ethylene Biosynthesis

Ethylene is a hormone that is produced by the vast majority of plants and that occurs in various concentrations depending on the environmental conditions where the plants develop and grow (Glick 2014). Ethylene at optimal concentration (10 g L^{-1}) can induce seed germination and elongation of the roots and the formation of primordia in stems and roots and initiate the stages such as flowering. In fruits, it can induce ripening and degradation. Also, it may be part of the produced volatiles that are part of the compounds important in fruit aroma (Lynch and Brown 1997; Choudhary 2017). However, at a higher concentration (25 g L^{-1}), this hormone induces defoliation, inhibition of root elongation, leaf senescence and abscission, and chlorophyll destruction (Singh et al. 2015). Thus, it is determinant to control or regulate the ethylene production by roots for normal growth and development of the plants.

In plants, ethylene is synthesized in three steps: methionine is converted to S-adenosyl-methionine by S-AdoMet synthetase; then 1-aminocyclopropane-1-carboxylic acid (ACC) is synthesized from S-AdoMet by ACS (ACC synthase); and finally, ethylene is produced through the oxidation of ACC by ACO (ACC oxidase) (Lin et al. 2009).

Ethylene can be found in low concentrations in various plant tissues under stress-free conditions. The regulation of ethylene synthesis occurs at different steps of the biosynthetic pathway. In Arabidopsis, tobacco and cotton, expression of genes coding for ACSs was found to be increased under salt stress (Achard et al. 2006, Cao et al. 2006, and Peng et al. 2014a, b). Interestingly, in Arabidopsis it was also found that a moderate low salinity pretreatment alleviated salt stress induction of four ACSs (ACS2, ACS6, ACS7, and ACS8) (Shen et al. 2014). ACO is also regulated by salinity. In cotton, several ACOs were found to be upregulated after salt treatment (Peng et al. 2014b).

Ethylene synthesis is affected by several factors including temperature, light, nutrition, gravity, and the presence of various types of biological stresses, i.e., the plant growth under salt stress, which improves plant tolerance to high salinity (Peng et al. 2014a, b). Thus, ethylene and its precursor (ACC) are induced by salinity in plant species; in fact, ethylene is known as the "hormone of stress" (Arshad et al. 2008). Ethylene is not only produced in response to salt stress but as a generalized response caused by multiple types of stress. Besides, saline stress in plants causes a series of physiological responses, i.e., salinity in plants induces generation of reactive oxygen species (ROS), including superoxide anion (O_2^-), singlet oxygen ($^1O_2$), and hydrogen peroxide (H_2O_2), and causes cellular damage in the plant system (Arshad et al. 2008; Long et al. 2015; Peng et al. 2014a, b). In a recent work, it is proposed that salt stress can also block water absorbing by an osmotic stressful effect and a direct cell wall synthesis inhibition (Fig. 5.1). The previous stresses caused by the salt ends with a slow cell development and a shortage in root length (Long et al. 2015).

Thus, salinity causes a stress on the plant, which leads to an increase in the production of ethylene, causing the abscission of leaves, petals, and flowers. It can also cause yellowing of leaves, senescence of various organs, and premature death of the plant (Zahir et al. 2009). Ethylene synthesis pathways in plants have been reviewed in quite a lot of detail in various works and have been known in detail for years (Yang and Hoffman 1984, Gamalero and Glick 2012). Briefly, the enzyme ACC synthase converts the S-adenosylmethionine (SAM) to 1-aminocyclopropane-1-carboxylic acid (ACC) and 5′-methylthioadenosine (MTA). ACC is then converted to ethylene by the enzyme ACC oxidase. Indeed, it has been proposed that, while ethylene plays a positive role in the early stage of self-adjustment for survival under high-salinity stress, after self-adjustment has been achieved, excessive ethylene in plants will inhibit plant growth and development, which is disadvantageous for plants to survive under high-salinity stress (Tao et al. 2015).

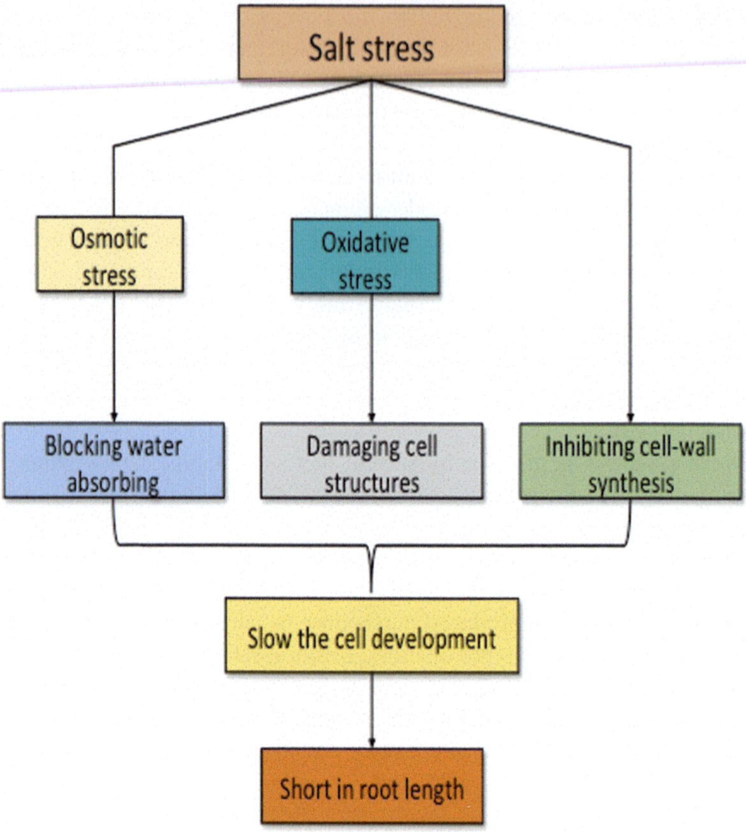

Fig. 5.1 Proposed model for root development inhibition under salt stress. See text for details. (Modified from Long et al. 2015)

5.3 Bacterial ACC Deaminase

The enzyme 1-aminocyclopropane-1-carboxylate deaminase (ACC deaminase) was first discovered and purified from an edaphic microorganism (*Pseudomonas* sp. ACP). This strain showed the ability to convert ACC to ammonia and α-ketobutyrate (Honma and Shimomura 1978). Thus, plants – under stress conditions, i.e., saline – respond by increasing the production of ethylene, causing various physiological changes that allow it to adapt and survive, such as tissue abscission and senescence. It is here where the bacterial enzyme ACC deaminase acts by degrading the plant ACC, the direct precursor of ethylene, generating α-ketobutyrate and ammonia, so the ethylene accumulation under stress conditions is avoided. Therefore, the bacterial enzyme ACC deaminase helps the plant to reduce the abiotic stress, promoting its growth and survival (Glick 2014). Glick and colleagues (1998) proposed the pioneering model on the action of the enzyme ACC deaminase as a relevant factor for growth promotion in plants. In general, the ACC deaminase-containing PGPB associated with plants act as a sink for ACC, generally causing an increase in the length of the roots and shoots, as well as a better resistance to the growth inhibition by the ethylene-inducing stresses.

In a more recent model (Fig. 5.2), Glick (2014) proposes that phytohormone indole-3-acetic acid (IAA), produced by the plant and the associated PGPB, plays an essential role during the promotion of plant growth. The roots of the plant exude various compounds to the rhizosphere, including sugars, organic acids, and amino acids, such as tryptophan. PGPB can assimilate tryptophan, which is an essential precursor of the IAA synthesis. Then, the PGPB that produce IAA (in addition to ACC deaminase) can induce the transcription of auxin response factors, promoting plant growth and transcription of the ACC synthase as well. In conclusion, PGPB that contain ACC deaminase and produce IAA can generate a cross talk in the plant between IAA and ACC deaminase. On the other hand, ACC deaminase lowers ethylene levels, while IAA stimulates plant growth (Duca et al. 2014; Nascimento et al. 2018).

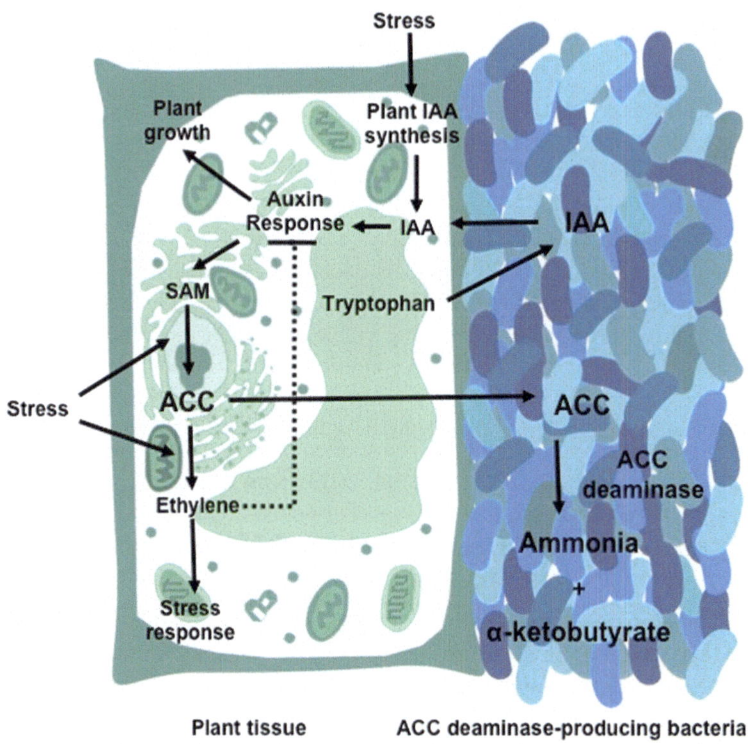

Fig. 5.2 Bacilli that both contains the enzyme 1-aminocyclopropane-1-carboxylate deaminase (ACC deaminase) and synthesize the phytohormone indole-3-acetic acid (IAA) may induce plant growth. The scheme is only showing the ACC deaminase enzyme and was modified from Glick (2014)

5.4 *Bacillus* Genus as Plant Growth-Promoting Bacteria

The *Bacillus* genus was first reported by Cohn in 1872 (Kokcha et al. 2012), who described it as heat-resistant, endospore-producing bacteria; at present, this genus includes over 336 species (Alcaraz et al. 2010). *Bacillus* is widely distributed worldwide (cultivable population from log 3 to log 6 per gram fresh weight of soil) (Vargas-Ayala et al. 2000) due to their ability to form endospores, a structure that provides them the ability to live in several habitats, both water and terrestrial ecosystems, and even in environments under extreme conditions (Tejera-Hernández et al. 2011). Regarding the agricultural sustainability, few researches have to be carried out to understand the diversity and dynamics of *Bacillus* in agroecosystems under stress conditions and how the crop with *Bacillus* interaction is modulated by extreme soil conditions.

Bacillus, among other PGPB, offers vital ecosystemic services, such as (i) social and ecological sustainability, (ii) adaptation and mitigation to climate change, (iii) biotechnological resource for humanity, (iv) cycling of water and nutrients, and (v) food security, mainly by nutrient cycling (van der Heijden et al. 2008), and improving the plant growth by avoiding the establishment of phytopathogenic agents (Compant et al. 2005), and the production of phytohormones, solubilization of nutrients, and activity of enzyme such as ACC deaminase (Hayat et al. 2010)

In the last decade, *Bacillus* strains have been reported to influence crop growth and yield under abiotic stress conditions, i.e., in a field experiment (Electric conductivity = 5.2 dS m^{-1}), Upadhyay and Singh (2015) reported a maximum root dry weight and shoot biomass after inoculation of wheat with *Bacillus aquimaris* SU44 and *B. aquimaris* SU8, after 60 and 90 days, respectively.

These traits make the Bacilli members excellent candidates for generating bio-inoculants, since, in addition to the aforementioned previous advantages, the spores can be stored for a long time, remaining viable until their inoculation in the field, which can survive even under adverse conditions such as saline stress (Villarreal-Delgado et al. 2018). Therefore, for several years Bacilli have been highlighted as effective bio-inoculants due to their consistent field results (Glick, and Skof 1986).

5.5 Presence of ACC Deaminase in Soil Microorganisms

The ACC deaminase activity has been reported in all three domains, i.e., Eukarya, Bacteria, and Archaea. This finding is corroborated by identifying the *acdS* gene in the genomes of soil microorganisms and endophytes, as well as the activity of the ACC deaminase enzyme, and is relatively frequent (Blaha et al. 2006; Nascimento et al. 2014). Plant fungi such as *Trichoderma asperellum* contain the ACC deaminase enzyme. *T. asperellum* has been reported as beneficial to plants, since it has phytopathogen biocontrol activity and plant growth-promoting traits (Viterbo et al. 2010). Bacterial groups such as *Rhizobiaceae* (*Rhizobium*, *Sinorhizobium*, and

Agrobacterium), *Phyllobacteriaceae* (*Phyllobacterium* and *Mesorhizobium*), and *Azospirillum* also contain the ACC deaminase enzyme (Nascimento et al. 2014). Other genera of PGPB bacteria that have been studied also exhibit deaminase ACC activities, including *Achromobacter*, *Burkholderia*, *Ralstonia*, *Pseudomonas*, and *Enterobacter* (Blaha et al. 2006; Duan et al. 2013; Wang et al. 2000; de los Santos-Villalobos et al. 2013).

In a recent work on the evolution and phylogeny of the *acdS* gene (ACC deaminase enzyme gene), the authors detect its presence in bacterial groups as diverse as *Actinobacteria*, *Deinococcus/Thermus*, *Alphaproteobacteria*, *Betaproteobacteria*, *Gammaproteobacteria*, and *Firmicutes*. Surprisingly, *acdS* genes were found in a wide range of plant and human pathogenic microorganisms, suggesting a different role in these organisms. The authors also reported the presence of Lrp-like regulatory proteins, such as AcdR, which is a common regulatory mechanism ACC deaminase expression in Proteobacteria (Nascimento et al. 2014).

5.6 Plant Growth Promotion by Bacilli Expressing ACC Deaminase

Bacilli are one of the most abundant bacterial groups in agricultural ecosystems; it is somehow expected to contain the *acdS* gene and, therefore, the ACC deaminase enzyme (Santoyo et al. 2012; Nascimento et al. 2014). The first report of the isolation and characterization of plant growth-promoting Bacilli with the ability to catabolize the ACC was carried out by Ghosh et al. (2003). In that work, the authors isolated plant growth-promoting bacteria from southeastern Wisconsin soils based on the unique ability of the isolates to use the ACC as the sole source of nitrogen. Thus, they isolated three Bacilli, including the species *Bacillus circulans* DUC1, *B. firmus* DUC2, and *B. globisporus* DUC3, which exhibited beneficial abilities by stimulating the root elongation in *Brassica campestris* (canola) seedlings under gnotobiotic conditions.

Other recent work has shown that several species of Bacilli show activities that promote plant growth associated with the ability of strains to use ACC. For example, Xu et al. (2014) carried out a screening in bacterial communities of *Bacillus* within the seeds of four commercial tomato varieties (*Lycopersicum esculentum* Mill.), by 16S rRNA gene PCR-RFLP (restriction fragment length polymorphism), in order to identify PGP traits under gnotobiotic experiments and greenhouse conditions. Thus, authors identified the strain *B. subtilis* HYT-12-1, which showed ACC deaminase activity, among other PGP mechanisms.

In another work, searching for plant growth-promoting endophytes associated with the medicinal plant *Lonicera japonica*, which grows in eastern China, several bacterial strains were identified, *Bacillus* and *Paenibacillus*. Such strains showed a promising ACC deaminase activity, as well as induction of root and shoots length and increment of chlorophyll of wheat (*Triticum aestivum* cv. "Zhoumai 18"). Also, some strains showed antagonism against the phytopathogens *Magnaporthe grisea*, *Fusarium oxysporum*, and *Alternaria alternata* (Zhao et al. 2015).

Two other endophytic strains of Bacilli, *B. subtilis* LK14 and LK15, were isolated from the medicinal plant *Moringa peregrina*, which grows in the arid regions of Arabia. The inoculation of one of the strains, LK14, significantly increased the shoot and root biomass and chlorophyll contents of tomato (*Solanum lycopersicum*) plants (compared to uninoculated plants). Interestingly, such strain exhibited a significant production of IAA, as well as ACC deaminase activity. Although the authors did not perform inoculation experiments under saline stress conditions, it can be concluded that using endophytic, Bacilli strains can be bio-prospective for plant growth promotion of crops in marginal lands (Khan et al. 2016).

5.7 Plant Growth Promotion by Bacilli Under Saline Stress

The previous works show a high potential to use plant growth-promoting Bacilli with ACC deaminase activity; however, there is no clear relationship between the ability to use ACC and the induction of growth in plants, in addition to the fact that the experiments were not carried out under stressful conditions, either salinity or other environmental stress, a situation where the ACC deaminase enzyme can have a potential to reduce ethylene levels in plants, facilitating survival and increasing plant growth. Therefore, the use of mutants in the *acdS* gene was necessary to confirm a relationship between ACC deaminase activity and the promotion of plant growth. Thus, Dr. Glick's group generated an *acdS* mutant in the bacterium *Pseudomonas* sp. UW4 (previously known as *Enterobacter cloacae* UW4), which showed a significant decrease in ACC deaminase activity and root elongation of canola plants. Thus, this work confirmed the importance of the ACC deaminase and validated the model where it is proposed that PGPB induce plant growth by lowering ethylene levels in plants, including ethylene inhibition of root elongation (Li et al. 2000).

In a recent work, Yaish et al. (2015) isolated and characterized several endophytic Bacilli (*Paenibacillus xylanexedens* PD-R6) of date palm (*Phoenix dactylifera* L.) with ACC deaminase activity, among other mechanisms such as the production of indole-3-acetic acid (IAA). Some strains were also able to chelate ferric iron (Fe^{3+}); solubilize phosphorus (PO43+), zinc (Zn2+), and potassium (K^{+}); and produce ammonia. The PD-R6 strain increased the root length of canola plants, either under normal growth conditions or salinity, but interestingly, it was possible to observe an increase in ACC activity and production of IAA in response to the increase in salt (NaCl) in the growth medium. The authors conclude that the isolated endophytic bacterium *Paenibacillus xylanexedens* PD-R6 can alter ethylene and IAA levels and also facilitate nutrient uptake in roots and therefore have the potential role to promote the growth of date palm trees growing under salinity stress.

The ability to produce IAA and the ACC deaminase activity displayed by Bacilli is a desirable feature in PGPB. Thus, Chinnaswamy et al. (2018) isolated to fast-growing, endophytic strain of *Bacillus megaterium* (NMp082) from root nodules of *Medicago polymorpha*. This species, apart from exhibiting to produce IAA and ACC deaminase activity, contained *nifH* and *nodD* genes with a 100% identity to those of *Ensifer meliloti*, which suggest an unusual event of lateral gene transfer. The authors also reported that *B. megaterium* NMp082 was not able to form effective nodules, but it induced nodule-like unorganized structures in alfalfa roots. Interestingly, *B. megaterium* NMp082 induced tolerance to salt stress in alfalfa and *Arabidopsis* plants and showed good traits to tolerate salt stress, water deficiency, and the presence of different heavy metals.

5.8 Plant Growth Promotion by Naturally Halotolerant Bacilli

Isolating strains that are naturally tolerant to high salt concentrations (halotolerant) is an excellent strategy to identify plant growth-promoting Bacilli strains, since usually these strains contain deaminase ACC activity. In countries such as Iran where 25% of arable land have high concentrations of salt, it is desirable to identify halotolerant strains as bioinoculating potentials that allow their survival in such stressful conditions and carry out effective plant growth-promoting actions. Recently, a halotolerant strain of *Bacillus mojavensis* K78 was identified in Iranian rhizospheric soils, which contains ACC deaminase activity and was able to increase dry root weight and shoots, chlorophyll content, and nutrient intake in low wheat plants conditions of salt stress. Additionally, strain K78 improved the water content of wheat grown under stress, improving the osmotic balance of plant cells (Pourbabaee et al. 2016).

Other halotolerant strains of *Bacillus* with ACC deaminase activity have been reported. For example, the species *B. aryabhattai* strain RS341 showed more than 40% increase in root elongation and dry weight in canola seedlings, when compared with uninoculated salt-stressed plants (Siddikee et al. 2010). More recently the same group reported that the inoculation of ACC deaminase-producing halotolerant *B. aryabhattai* RS341 at 120 mM of NaCl significantly increased the seed germination and decreased seed ACC content. Importantly, the ethylene emission of salt stress exposed canola seedlings was reduced with the inoculation of strain *B. aryabhattai* RS34, compared to uninoculated salt stress control. The authors concluded that amelioration of salt stress inhibitory effect on the canola seed was attributed to the modulation in ethylene emission (and induction of hydrolytic enzymes) by bioinoculation of ACC deaminase-producing halotolerant strain RS341 (Siddikee et al. 2015). Table 5.1 shows relevant works of ACC deaminase-producing Bacilli species that promote plant growth under salt stress conditions.

Table 5.1 Bacilli species producing ACC deaminase and containing other direct and indirect mechanisms to induce plant growth

Bacilli species and strain	Plant host/ endophyte or rhizospheric origin	Plant growth-promoting effect	References
Bacillus megaterium NMp082	*Medicago polymorpha*/ Endophyte	Promote plant growth in *Medicago* spp.	Chinnaswamy et al. (2018)
		Induce tolerance to salt stress in alfalfa and Arabidopsis plants	
B. circulans DUC1	Alfalfa, soybean, tomato, corn/	Promote plant growth in canola seedlings	Ghosh et al. (2003)
Bacillus firmus DUC2	Rhizospheric		
B. globisporus DUC3		Low ethylene levels	
		Root elongation is enhanced	
B. aryabhattai RS341	Halophytic plants/ Rhizospheric	Increase root elongation and dry weight in salt-stressed canola seedlings	Siddikee et al. (2010)
B. aryabhattai	Halophytic plants/ Rhizospheric	Alleviate salt stress effect in canola seed germination	Siddikee et al. (2015)
RS341			
B. subtilis LK14	*Moringa peregrina*/ Endophyte	Improve the growth of tomato plants	Khan et al. (2016)
B. subtilis LK15			
		Produce IAA	
		Phosphate solubilization	
		ACC deaminase activity	
B. mojavensis K78	Wheat plants/ Rhizospheric	Facilitate plant growth promotion in the presence of inhibitory levels of salt in wheat plants	Pourbabaee et al. (2016)
		ACC deaminase activity	
B. subtilis HYT-12-1	Tomato seeds/ Endophyte	Promote plant growth in tomato plants	Xu et al. (2014)
		ACC deaminase activity	
		Fix Nitrogen	
B. megaterium QM B1551	Date palm (*Phoenix dactylifera* L.)/ Endophyte	Some strains have ACC deaminase activity and produce IAA and ammonia (except *Bacillus megaterium* QM B1551)	Yaish et al. (2015)
B. endophyticus 2DT			
B. oleronius ATCC			
B. thuringiensis Bt407			
B. anthracis Ames			
B. atrophaeus 170	*Lonicera japonica*/ Endophyte	Promote plant growth in wheat plants. Strain 124 shows capacity for siderophore production and phosphate solubilization	Zhao et al. (2015)
B. megaterium 124			
B. subtilis 132			
B. subtilis 135			

5.9 Conclusions and Perspectives

The Bacilli have great potential versus other groups of PGPB, for example, the capacity to sporulate, the high tolerance to saline stress, fast-growing rate and elevated competence to colonize niches in the rhizosphere are common traits in newly isolates around the world, as well as its wide distribution in different latitudes, noticing the good capacities to promote plant growth of plants through ACC deaminase activity, without ruling out other mechanisms that allow an additional benefit. For example, Bacilli can enhance the efficiency of water use and nutrient uptake, as well as maintaining K^+/Na^+ ratio in plant cells (Nadeem et al. 2007).

It is noteworthy that there is a lack of knowledge regarding the level of participation of the ACC deaminase with respect to other mechanisms of stimulation of growth and development in plants; this is, in part, due to the lack of mutant strains and double mutants (i.e., ACC deaminase gene plus other genes that encode mechanisms such as production of siderophores, solubilization of nutrients, and production of osmoprotective compounds) that have delayed their analysis in the plant-bacteria interaction. Therefore, it is essential that this area of genetic analysis be developed further to allow the evaluation of the cross talk between functions and bacterial mechanisms of PGP. Finally, we propose that more work needs to be done on the interaction between various species of Bacilli and other PGPB, since it has been observed, in a few recent works, the additive and/or synergistic activity of Bacillus species with other beneficial bacterial species or fungi (Armada et al. 2016; Kumar et al. 2016). Finally, whether a bioinoculant containing different PGPB is developed, the presence of one or more species of the Bacilli group is essential, since such species have enormous potential to benefit plant health, particularly, those agroecosystems with problems of salinity.

Acknowledgments G.S. thanks to the Coordinación de la Investigación Científica (Project: 2018–2019) of the Universidad Michoacana de San Nicolás de Hidalgo for the financial support to research projects.

References

Achard P, Cheng H, De Grauwe L, Decat J, Schoutteten H, Moritz T et al (2006) Integration of plant responses to environmentally activated phytohormonal signals. Science 311:91–94. https://doi.org/10.1126/science.1118642

Alcaraz LD, Moreno-Hagelsieb G, Eguiarte LE, Souza V, Herrera-Estrella L, Olmedo G (2010) Understanding the evolutionary relationships and major traits of *Bacillus* through comparative genomics. BMC Genomics 11:332. https://doi.org/10.1186/1471-2164-11-332

Armada E, Probanza A, Roldán A, Azcón R (2016) Native plant growth promoting bacteria *Bacillus thuringiensis* and mixed or individual mycorrhizal species improved drought tolerance and oxidative metabolism in *Lavandula dentata* plants. J Plant Physiol 192:1–12

Arshad M, Shaharoona B, Mahmood T (2008) Inoculation with *Pseudomonas* spp. containing ACC-deaminase partially eliminates the effects of drought stress on growth, yield, and ripening of pea (*Pisum sativum* L.). Pedosphere 18(5):611–620

Blaha D, Prigent-Combaret C, Mirza MS, Moënne-Loccoz Y (2006) Phylogeny of the 1-aminocyclopropane-1-carboxylic acid deaminase-encoding gene *acdS* in phytobeneficial and pathogenic Proteobacteria and relation with strain biogeography. FEMS Microbiol Ecol 56:455–470

Borrelli P, Märker M, Schütt B (2015) Modelling post-tree-harvesting soil erosion and sediment deposition potential in the turano river basin (Italian central apennine). Land Degrad Dev 26(4):356–366

Cao WH, Liu J, Zhou QY, Cao YR, Zheng SF, Du BX et al (2006) Expression of tobacco ethylene receptor NTHK1 alters plant responses to salt stress. Plant Cell Environ 29:1210–1219. https://doi.org/10.1111/j.1365-3040.2006.01501.x

Chinnaswamy A, Coba de la Peña T, Stoll A, de la Peña Rojo D, Bravo J, Rincón A, Pueyo JJ (2018) A nodule endophytic *Bacillus megaterium* strain isolated from *Medicago polymorpha* enhances growth, promotes nodulation by *Ensifer medicae* and alleviates salt stress in alfalfa plants. Ann Appl Biol 172:295–308. https://doi.org/10.1111/aab.12420

Choudhary DK (2017) In: Varma A, Tuteja N (eds) Plant-microbe interaction: an approach to sustainable agriculture. Springer, New Delhi

Compant S, Duffy B, Nowak J, Clément C, Barka EA (2005) Use of plant growth-promoting bacteria for biocontrol of plant diseases: principles, mechanisms of action, and future prospects. Appl Environ Microbiol 71:4951–4959. https://doi.org/10.1128/AEM.71.9.4951-4959.2005

de los Santos-Villalobos S, de Folter S, Délano-Frier JP, Gómez-Lim MA, Guzmán-Ortiz DA, Peña-Cabriales JJ (2013) Growth promotion and flowering induction in mango by *Burkholderia* and *Rhizobium* inoculation: morphometric, biochemical and molecular events. J Plant Growth Regul 32(3):615–627

Dercon G, Mabit L, Hancock G, Nguyen ML, Dornhofer R, Bacchi OOS, Benmansour M, Bernard C, Froehlich W, Golosov VN, Haciyakupoglu S, Hai PS, Klik A, Li Y, Lobb DA, Onda Y, Popa N, Rafiq M, Ritchie JC, Schuller P, Shakhashiro A, Wallbrink P, Walling DE, Zapata F, Zhang X (2012) Fallout radionuclide-based techniques for assessing the impact of soil conservation measures on erosion control and soil quality: an overview of the main lessons learnt under an FAO/IAEA Coordinated Research Project. J Environ Radioact 107:78–85

Dotterweich M (2013) The history of human-induced soil erosion: geomorphic legacies, early descriptions and research, and the development of soil conservation-A global synopsis. Geomorphology 201:1–34

Duan J, Jiang W, Cheng Z, Heikkila JJ, Glick BR (2013) The complete genome sequence of the plant growth-promoting bacterium *Pseudomonas* sp. UW4. PLoS One 8(3):e58640

Duca D, Lorv J, Patten CL, Rose D, Glick BR (2014) Indole-3-acetic acid in plant–microbe interactions. Antonie Van Leeuwenhoek 106(1):85–125

FAO (2018) World food situation. Retrieved from http://www.fao.org/worldfoodsituation/csdb/en/

Flowers T (2004) Improving crop salt tolerance. J Exp Bot 55:307–319

Foolad MR (2004) Recent advances in genetics of salt tolerance in tomato. Plant Cell Tissue Organ Cult 76:101–111

Gamalero E, Glick BR (2012) Ethylene and abiotic stress tolerance in plants. In: Environmental adaptations and stress tolerance of plants in the era of climate change. Springer, New York, pp 395–412

Ghosh S, Penterman JN, Little RD, Chavez R, Glick BR (2003) Three newly isolated plant growth-promoting bacilli facilitate the seedling growth of canola, *Brassica campestris*. Plant Physiol Biochem 41(3):277–281

Glick BR (2014) Bacteria with ACC deaminase can promote plant growth and help to feed the world. Microbiol Res 169(1):30–39

Glick BR, Skof YC (1986) Environmental implications of recombinant DNA technology. Biotechnol Adv 4:261–277

Glick BR, Penrose DM, Li J (1998) A model for the lowering of plant ethylene concentrations by plant growth promoting bacteria. J Theor Biol 190:63–68

Hayat R, Ali S, Amara U (2010) Soil beneficial bacteria and their role in plant growth promotion: a review. Ann Microbiol 60:579–598. https://doi.org/10.1007/s13213-010-0117-1

Honma M, Shimomura T (1978) Metabolism of 1-aminocyclopropane-1-carboxylic acid. Agric Biol Chem 42:1825–1831

James C (2015) Global status of commercialized biotech/GM crops: 2014. ISAAA brief, 49

Khan AL, Halo BA, Elyassi A, Ali S, Al-Hosni K, Hussain J et al (2016) Indole acetic acid and ACC deaminase from endophytic bacteria improves the growth of *Solanum lycopersicum*. Electron J Biotechnol 21:58–64

Kokcha S, Mishra AK, Lagier JC, Million M, Leroy Q, Raoult D, Fournier PE (2012) Non contiguous-finished genome sequence and description of Bacillus timonensis sp. nov. Stand Genomic Sci 6(3):346

Kumar M, Mishra S, Dixit V, Kumar M, Agarwal L, Chauhan PS, Nautiyal CS (2016) Synergistic effect of *Pseudomonas putida* and *Bacillus amyloliquefaciens* ameliorates drought stress in chickpea (*Cicer arietinum* L.). Plant Signal Behav 11(1):e1071004

Li J, Ovakim DH, Charles TC, Glick BR (2000) An ACC deaminase minus mutant of *Enterobacter cloacae* UW4 No longer promotes root elongation. Curr Microbiol 41(2):101–105

Ligonja PJ, Shrestha RP (2015) Soil erosion assessment in kondoa eroded area in Tanzania using universal soil loss equation, geographic information systems and socioeconomic approach. Land Degrad Dev 26(4):367–379

Lin Z, Zhong S, Grierson D (2009) Recent advances in ethylene research. J Exp Bot 60:3311–3336. https://doi.org/10.1093/jxb/erp204

Long W, Zou X, Zhang X (2015) Transcriptome Analysis of canola (*Brassica napus*) under salt stress at the germination stage. PLoS One 10(2):e0116217

Lugtenberg B, Kamilova F (2009) Plant-growth-promoting rhizobacteria. Annu Rev Microbiol 63:541–556

Lynch J, Brown KM (1997) Ethylene and plant responses to nutritional stress. Physiol Plant 100(3):613–619

Nadeem SM, Zahir ZA, Naveed M, Arshad M (2007) Preliminary investigations on inducing salt tolerance in maize through inoculation with rhizobacteria containing ACC deaminase activity. Can J Microbiol 53(10):1141–1149

Nascimento FX, Rossi MJ, Soares CR, McConkey BJ, Glick BR (2014) New insights into 1-ami nocyclopropane-1-carboxylate (ACC) deaminase phylogeny, evolution and ecological significance. PLoS One 9(6):e99168

Nascimento FX, Rossi MJ, Glick BR (2018) Ethylene and 1-Aminocyclopropane-1-carboxylate (ACC) in plant–bacterial interactions. Front Plant Sci 9:114

Peng J, Li Z, Wen X, Li W, Shi H, Yang L et al (2014a) Salt-induced stabilization of EIN3/EIL1 confers salinity tolerance by deterring ROS accumulation in *Arabidopsis*. PLoS Genet 10:e1004664. https://doi.org/10.1371/journal.pgen.1004664

Peng Z, He S, Gong W, Sun J, Pan Z, Xu F et al (2014b) Comprehensive analysis of differentially expressed genes and transcriptional regulation induced by salt stress in two contrasting cotton genotypes. BMC Genomics 15:760. https://doi.org/10.1186/1471-2164-15-760

Pourbabaee AA, Bahmani E, Alikhani HA, Emami S (2016) Promotion of wheat growth under salt stress by halotolerant bacteria containing ACC deaminase. J Agri Sci Technol (JAST) 18(3):855–864

Rojas-Solís D, Zetter-Salmón E, Contreras-Pérez M, del Carmen Rocha-Granados M, Macías-Rodríguez L, Santoyo G (2018) *Pseudomonas stutzeri* E25 and *Stenotrophomonas maltophilia* CR71 endophytes produce antifungal volatile organic compounds and exhibit additive plant growth-promoting effects. Biocatal Agric Biotechnol 13:46–52

Roy SJ, Negrão S, Tester M (2014) Salt resistant crop plants. Curr Opin Biotechnol 26:115–124

Santoyo G, Orozco-Mosqueda MDC, Govindappa M (2012) Mechanisms of biocontrol and plant growth-promoting activity in soil bacterial species of *Bacillus* and *Pseudomonas*: a review. Biocontrol Sci Tech 22(8):855–872

Santoyo G, Moreno-Hagelsieb G, del Carmen Orozco-Mosqueda M, Glick BR (2016) Plant growth-promoting bacterial endophytes. Microbiol Res 183:92–99

Shafi M, Bakht J, Khan MJ, Khan MA, Anwar S (2010) Effect of salinity on yield and ion accumulation of wheat genotypes. Pak J Bot 42(6):4113–4121

Shen X, Wang Z, Song X, Xu J, Jiang C, Zhao Y, Ma C, Zhang H (2014) Transcriptomic profiling revealed an important role of cell wall remodeling and ethylene signaling pathway during salt acclimation in Arabidopsis. Plant Mol Biol 86(3):303–317

Shrivastava P, Kumar R (2015) Soil salinity: a serious environmental issue and plant growth promoting bacteria as one of the tools for its alleviation. Saudi J Biol Sci 22(2):123–131

Siddikee MA, Chauhan PS, Anandham R, Han GH, Sa T (2010) Isolation, characterization, and use for plant growth promotion under salt stress, of ACC deaminase-producing halotolerant bacteria derived from coastal soil. J Microbiol Biotechnol 20(11):1577–1584

Siddikee MA, Sundaram S, Chandrasekaran M, Kim K, Selvakumar G, Sa T (2015) Halotolerant bacteria with ACC deaminase activity alleviate salt stress effect in canola seed germination. J Korean Soc Appl Biol Chem 58(2):237–241

Singh RP, Shelke GM, Kumar A, Jha PN (2015) Biochemistry and genetics of ACC deaminase: a weapon to "stress ethylene" produced in plants. Front Microbiol 6(937):2015. https://doi.org/10.3389/fmicb.2015.00937

Tao J-J, Chen H-W, Ma B, Zhang W-K, Chen S-Y, Zhang J-S (2015) The role of ethylene in plants under salinity stress. Front Plant Sci 6:1059. https://doi.org/10.3389/fpls.2015.01059

Tejera-Hernández B, Rojas-Badía MM, Heydrich-Pérez M (2011) Potencialidades del género *Bacillus* en la promoción del crecimiento vegetal y el control de hongos fitopatógenos. Revista CENIC Ciencias Biológicas 42:131–138. Disponible en línea: http://www.redalyc.org/pdf/1812/181222321004.pdf

Upadhyay SK, Singh DP (2015) Effect of salt-tolerant plant growth-promoting rhizobacteria on wheat plants and soil health in a saline environment. Plant Biol (Stuttg) 17(1):288–293. https://doi.org/10.1111/plb.12173

Van der Heijden M, Bardgett R, van Straalen N (2008) The unseen majority: soil microbes as drivers of plant diversity and productivity in terrestrial ecosystems. Ecol Lett 11:296–310. https://doi.org/10.1111/j.1461-0248.2007.01139.x

Vargas-Ayala R, Rodríguez-Kábana R, Morgan-Jones G, Mclnroy JA, Kloepper JW (2000) Shifts in soil microflora induced by velvet bean (*Mucuna deeringiana*) in cropping systems to control root-knot nematodes. Biol Control 17:11–22. https://doi.org/10.1006/bcon.1999.0769

Villarreal-Delgado MF, Villa-Rodríguez ED, Cira-Chávez LA, Estrada-Alvarado MI, Parra-Cota FI, de los Santos-Villalobos S (2018) The genus *Bacillus* as a biological control agent and its implications in the agricultural biosecurity. Mex J Phytopathol 36:95–130. https://doi.org/10.18781/R.MEX.FIT.1706-5

Viterbo A, Landau U, Kim S, Chernin L, Chet I (2010) Characterization of ACC deaminase from the biocontrol and plant growth-promoting agent *Trichoderma asperellum* T203. FEMS Microbiol Lett 305:42–48

Wang C, Knill E, Glick BR, Défago G (2000) Effect of transferring 1-aminocyclopropane-1-carboxylic acid (ACC) deaminase genes into *Pseudomonas fluorescens* strain CHA0 and its *gacA* derivative CHA96 on their growth-promoting and disease-suppressive capacities. Can J Microbiol 46(10):898–907

Xu M, Sheng J, Chen L, Men Y, Gan L, Guo S, Shen L (2014) Bacterial community compositions of tomato (*Lycopersicum esculentum* Mill.) seeds and plant growth promoting activity of ACC deaminase producing *Bacillus subtilis* (HYT-12-1) on tomato seedlings. World J Microbiol Biotechnol 30(3):835–845

Yaish MW, Antony I, Glick BR (2015) Isolation and characterization of endophytic plant growth-promoting bacteria from date palm tree (*Phoenix dactylifera* L.) and their potential role in salinity tolerance. Antonie Van Leeuwenhoek 107(6):1519–1532

Yang SF, Hoffman NE (1984) Ethylene biosynthesis and its regulation in higher plants. Annu Rev Plant Physiol Plant Mol Biol 35:155–189

Zahir ZA, Ghani U, Naveed M, Nadeem SM, Asghar HN (2009) Comparative effectiveness of *Pseudomonas* and *Serratia* sp. containing ACC-deaminase for improving growth and yield of wheat (*Triticum aestivum* L.) under salt-stressed conditions. Arch Microbiol 191(5):415–424

Zhao L, Xu Y, Lai XH, Shan C, Deng Z, Ji Y (2015) Screening and characterization of endophytic *Bacillus* and *Paenibacillus* strains from medicinal plant *Lonicera japonica* for use as potential plant growth promoters. Braz J Microbiol 46(4):977–989

Chapter 6
Bacillus subtilis-Mediated Abiotic Stress Tolerance in Plants

Oksana Lastochkina

6.1 Introduction

Abiotic stresses have a significant negative impact on agricultural industry by inhibiting plant growth and decreasing crop productivity worldwide (Pereira 2016). Losses of major crops from adverse environmental stress factors like drought, salinity, extreme temperature, heavy metals, UV radiation, various oxidative stresses, etc. have reached 50–82% and pose serious threats to agriculture and food security (FAO 2016; Benedetto et al. 2017). According to the Food and Agriculture Organization of the United Nations (FAO), the world's population is rapidly increasing. To ensure food security by the year 2050, the world food production of major agricultural crops should be increased by 70% and in developing countries by 100% (FAO 2014, 2016). The chemicals traditionally applied for plant protection can cause significant damage to the environment and can contaminate food products with toxic compounds. The emerging acuteness of food and environmental problems as well as the search for energy-saving natural "green" technologies and the production of organic products raise the urgency of searching for novel, efficient, and environment - friendly ways to cope with the threat of global abiotic stresses to agriculture and to increase plant productivity. Several strategies have been developed in order to solve these issues including plant genetic engineering (Dandekar 2000; Key et al. 2008) and plant growth-promoting bacteria (PGPB) that activate various physiological features of host plant metabolism without causing adverse effects on plants, humans, and the environment (Dimkpa et al. 2009; Baez-Rogelio et al. 2016; Numan et al. 2018). Since PGPB inoculation is more easily manipulated

O. Lastochkina (✉)
Institute of Biochemistry and Genetics, Ufa Federal Research Centre of the Russian Academy of Sciences, Ufa, Russia

Bashkir Research Institute of Agriculture, Ufa Federal Research Centre of the Russian Academy of Sciences, Ufa, Russia

M. T. Islam et al. (eds.), *Bacilli and Agrobiotechnology: Phytostimulation and Biocontrol*, Bacilli in Climate Resilient Agriculture and Bioprospecting,
https://doi.org/10.1007/978-3-030-15175-1_6

compared to transgenic plants, the application of PGPB in agriculture has gained recent attention (Ma 2017).

The PGPB are group of beneficial bacteria either free-living in the soil or colonizing the rhizosphere, phyllosphere, or plant tissue interior (endophytes) that can enhance plant growth and induce systemic resistance in host plants to a broad spectrum of pathogens (Beneduzi et al. 2012) and abiotic stresses (Dimkpa et al. 2009; Sayed et al. 2011; Cherif et al. 2015; Van Loon 2007; Baez-Rogelio et al. 2016). PGPB include the strains in the genera *Acinetobacter*, *Alcaligenes*, *Arthrobacter*, *Azospirillum*, *Azotobacter*, *Bacillus*, *Beijerinckia*, *Burkholderia*, *Erwinia*, *Enterobacter*, *Rhizobium*, *Flavobacterium*, and *Serratia* (Bashan and de-Bashan 2005; Maksimov et al. 2015). However, there is particular interest in the genus *Bacillus* spp., viz., *B. subtilis* for development of natural fertilizers and plant protection products for application in agriculture, horticulture, and forestry (Sarma et al. 2018; Maksimov et al. 2015). A large number of findings in the literature indicate that *B. subtilis* is involved in the regulation of plant growth, development, and response to both biotic (pathogens, pests) (Van Loon 2007; Akram et al. 2013) and abiotic stress factors such as drought, salinity, high and low temperatures, toxic metals, UF radiation, etc. (Lastochkina et al. 2017a; Moon et al. 2017; Kuramshina et al. 2018; Martins et al. 2018; Numan et al. 2018), therefore increasing plant productivity. So, the protective effect of *B. subtilis* under different abiotic stresses was revealed for various plant species including wheat (Turan et al. 2012; Çakmakçı et al. 2017; Lastochkina et al. 2017a), maize (Rojas-Tapias et al. 2012; Moon et al. 2017), strawberry (Karlidag et al. 2010, 2013), soybean (Martins et al. 2018), lettuce (Yildirim et al. 2011), cabbage (Turan et al. 2014), brahmi (Bharti et al. 2013), sugar beet (Pusenkova et al. 2015, 2016), sweet basil (Sayed et al. 2011), and many others. Of special interest are endophytic *B. subtilis*, living inside plant tissues, which allow them to be less dependent on external environmental factors (compared to rhizosphere and phyllosphere strains) while exhibiting economically "useful" features (Sessitsch et al. 2012; Maksimov and Khairullin 2016; Pandey et al. 2017). Wherein, endophytes once implanted in plant tissues can contribute to the formation of their long-term protection against adverse environmental factors throughout the entire growing season (Sessitsch et al. 2012; Cherif et al. 2015; Pusenkova et al. 2016). A number of reports indicate the positive effect of *B. subtilis* on the postharvest physiology of various fruits and vegetables, enhancing their resistance against postharvest diseases and unfavorable storage conditions which is manifested in extending their marketing life with maintaining nutritional qualities (Jiang et al. 2001).

The mechanism of *B. subtilis*-induced physiological action on the host plants are numerous, diverse, and often specific (Van Loon 2007; Berg 2009; Pérez-García et al. 2011; Turan et al. 2012; Numan et al. 2018). Conceptually, *B. subtilis* may use direct and indirect mechanisms to promote plant growth and increase plant stress tolerance. This is associated with the synthesis of wide range of biologically active substances such as antibacterial and insecticidal components, siderophores and chelators, hormones and enzymes, production of 1-aminocyclopropane-1-carboxylate (ACC) deaminase (Saikia et al. 2018), exopolysaccharides (EPS) (Upadhyay et al.

2011), ethylene level reduction in plants, nitrogen fixation, increased availability of macro-/microelements (Pandey et al. 2017), and the launch of host plant's systemic resistance and tolerance (Van Loon 2007; Akram et al. 2013; Lastochkina et al. 2017a; Moon et al. 2017; Kuramshina et al. 2018; Martins et al. 2018; Numan et al. 2018). The activation of the defense mechanisms throughout the plant can be carried out in different ways. Two common pathways are induced systemic resistance (ISR) and systemic acquired resistance (SAR) (Ryals et al. 1996; Van Loon 2007; Pieterse et al. 2012). It has been discovered that *B. subtilis* is triggered by both SAR and ISR in plants which indicates the prospects of work on the development of biological products on the basis of *B. subtilis*, especially endophytes, for the protection of agricultural plants against stresses both biogenic and abiogenic nature (Maksimov and Khairullin 2016; Pandey et al. 2017). Furthermore, PGPB-induced physical and chemical changes in plants result in increased tolerance against abiotic stresses which have been described as induced systemic tolerance (IST) (Yang et al. 2009; Arya et al. 2018). The role of PGPB in IST has been revealed and studied very recently (Sarma et al. 2018). However, it is not yet clear how *B. subtilis* specifically regulates the protective system of host plants under abiotic stresses and how *B. subtilis*-induced plant defense signal system interacts with the classical signaling pathways specifically induced by such signaling molecules as salicylic acid (SA), jasmonates (JAs), and ethylene. The effectiveness of the same strain of *B. subtilis* can vary depending not only from environmental conditions but also on many factors such as plant species, its ecological and geographic origin, and varietal features (Maksimov and Khairullin 2016; Lastochkina et al. 2017b). Therefore, one of the main reasons development of plant protection products including *B. subtilis* has been restrained is due to lack of systemic studies of the mechanisms of relationships in the systems "*B. subtilis* – host plant – abiotic stress." Thus, for fully using the potential of *B. subtilis* as inoculants in sustainable agriculture, it's extremely important to understand their underlying mechanisms in physiological action on host plants under different adverse environmental factors.

In this chapter, the published findings have been summarized to show the influence of *B. subtilis* strains on plant growth, development, and productivity. The current state of knowledge of the physiological and biochemical mechanisms of *B. subtilis*-induced plant tolerance under different abiotic stress conditions is also discussed.

6.2 Role of *B. subtilis* in Plant Growth and Development

B. subtilis plays an important role in enhancing plant growth and development through a wide variety of interrelated direct and indirect mechanisms (Bloemberg and Lugtenberg 2001). Conceptually, direct ways of employing those bacterial traits that result in the direct promotion of plant growth include improving the bioavailability of macro-/micronutrients (nitrogen fixation, phosphate solubilization, sequestering iron), producing or changing the concentrations of phytohormones (auxins, cytokinins (CKs), gibberellins, ABA, JAs, SA), and regulating ethylene levels in plant (Oliveira et al. 2003; Jha et al. 2012; Ishak et al. 2016). Transcriptome analysis of the bacteria *B. subtilis* OKB105 in response to *Oryza sativa* seedlings showed that many of the 176 differentially expressed genes are involved in the interaction between bacteria and plants. A total of 176 genes (3.8%) of the *B. subtilis* OKB105 transcriptome showed significantly altered expression levels in response to rice seedlings. Among these, the 52 genes were upregulated, the majority of which (including *araA*, *ywkA*, *yfls*, *mtlA*, *ydgG*) are involved in metabolism, nutrients transport, and stress responses; the 124 genes were downregulated (including *cheV*, *fliL*, *spmA*, *tua*) and involved in chemotaxis, motility, sporulation, and teichuronic acid biosynthesis (Xie et al. 2015). Many studies demonstrated the ability of *B. subtilis* to facilitate nutrient uptake or increase nutrient availability acting as a biofertilizer (Ramesh et al. 2014). Inoculation of wheat and soybean with *B. subtilis* strains MDSR7, MDSR11, and MDSR14 significantly increased the availability of zinc which enhances the plant growth and assimilation of zinc in seeds (Ramesh et al. 2014). Ahmad et al. (2017) discovered that endophytic *B. subtilis* 330-2 significantly increased rice and maize growth (ranging 14–37%) and triggered their tolerance against biotic/abiotic stresses differentially expressed in 114 genes, among which 32% and 10% were involved in metabolism (e.g., *gltA*, *pabA*, and *ggt*) and transportation of nutrients (e.g., *fhu*, *glpT*, and *gltT*), respectively.

6.2.1 Production of Plant Growth Regulators

Production of phytohormones is one of the most important mechanisms of *B. subtilis*-induced plant growth promotion (PGP). The liquid culture filtrates of *B. subtilis* were found to contain the natural plant growth regulators such as auxins, CKs, ABA, and gibberellins (Sivasakthi et al. 2013; Ishak et al. 2016) which were beneficial for effective photosynthesis, growth, and plasma membrane integrity. Gupta et al. (2000) found a culture filtrate of *B. subtilis* FZB-G capable of triggering phytohormones' precursor, which plays an important role in signal transduction and activation of defense genes resulting in the production of defense-related compounds. Many studies have demonstrated that *B. subtilis* promotes plant growth through the production of auxins (Moon et al. 2017; Ahmad et al. 2017), CKs (Arkhipova et al. 2005), and gibberellins (Bottini et al. 2004) or by regulating the high levels of endogenous ethylene in the host plants (Glick 2014). Among the PGP traits, indolic compounds such as indole-3-acetic acid (IAA) play a vital role in plant growth and development being the main plant auxin which is responsible for the division, expansion, and differentiation of plant cells and tissues (Pandey et al. 2017). It has been shown that *B. subtilis* FZB24 producing IAA promotes root system development which improves plant ability to absorb more water and nutrients (Sorty et al. 2016). The rhizobacteria *B. subtilis* LDR2 increased IAA content in wheat under salt stress and enhanced plant tolerance (Barnawal et al. 2017). Most studies have shown that IAA biosynthesis is greatly influenced by L-TRP precursor that is believed to be the primary precursor for formation of IAA in microorganisms (Sessitsch et al. 2012). Several different IAA biosynthesis pathways may also be used and a single bacterial strains sometime containing more than one pathway (Pandey et al. 2017). For example, in the endophyte metagenome, all corresponding genes for IAA production have been detected for three pathways: the indole-3-acetamide, indole-3-pyruvate, and tryptamine (Sessitsch et al. 2012). Interestingly, IAA production is reported to be more common among the endophytic than among epiphytic bacteria. Thus, from the tested 363 epiphytic and 373 endophytic soybean bacteria, the proportion of IAA-producing endophytic bacteria was 34% and for epiphytes 21% (Kuklinski-Sorbal et al. 2004). Of the 25 carrot endophyte isolates, 9 (36%) had IAA activity from 2 to 15 $\mu g/mL^{-1}$ (Blagova et al. 2014). Endophytic strain *B. subtilis* 330-2 was revealed to produce a considerable amount of IAA (24.13 $\mu g\ mL^{-1}$) and significantly enhance the growth of plants in comparison to the non-inoculated control. So, compared with the control, the *B. subtilis* 330-2-inoculated rice/maize seedlings showed 12–34% enhancement in seedling elongation and biomass accumulation (Ahmad et al. 2017). The endophytic strain *B. subtilis* 10-4 that increases wheat plant growth (seed germination, length of roots and shoots, plant biomass) produced 5.8 $\mu g/mL^{-1}$ of IAA in the presence of precursor L-TRP (Lastochkina et al. 2017a). *B. subtilis* OTPB1 considerably enhanced growth (shoots, roots, seedling vigor, leaf area) and induced systemic resistance in tomato plants against early and late blight. Higher levels of IAA and GA3 and defense-related enzymes (peroxidase (PO), polyphenol oxidase (PPO), and superoxide dismutase (SOD)) were detected in inoculated plants (Chowdappa et al. 2013). The other important plant hormone, which is produced by *B. subtilis* is CKs increasing plant cellular activities including cell division and leaf growth (Arkhipova et al. 2005, 2007; Liu et al. 2013). For example, Arkhipova et al. (2005) demonstrated that the production of CK by *B. subtilis* enhanced the rate of CK production in shoots and roots of lettuce and stimulated host plant growth. Accumulation of CKs in inoculated plants was associated with an increasing of the plant shoots and roots weight of approximately 30% (Arkhipova et al. 2005). The CK-producing *B. subtilis* can increase host plant growth under abiotic stress conditions as well (Arkhipova et al. 2005, 2007; Liu et al. 2013). Liu et al. (2013) reported that the oriental *Thuja* seedlings inoculated with CK-producing *B. subtilis* strains were more resistant to stress due to drought. The production of GA has also been indicated for different species of PGPB including *Bacillus* ssp. The exact mechanism which promotes plant growth by gibberellins has not been indicated. It is assumed that increasing the root growth, specifically the density of root hairs by gibberellins, enhances the nutrient and water uptake by host plants (Mohamed and Gomaa 2012). Ethylene is another essential hormone for plant growth and development regulating cell growth, seed germination, leaf senescence, flower, and fruit ripening, but it has different effects on growth depending on its concentration in root tissues (Reid 1981). High concentrations of ethylene can be harmful since it causes defoliation and cellular processes leading to inhibition of stem and root growth as well as premature senescence, all of which resulted in yield reduction. The production of ethylene increases under stress conditions adversely effecting plant growth and is regulated by ACC synthase (Glick 2014). Under different types of stresses, such as drought, cold, flooding, infections with pathogens, and the presence of toxic metals, plants respond by synthesizing ACC synthase, which is the precursor for ethylene (Chen et al. 2002; Chen and Gallie 2010; Glick et al. 2007; Glick 2014). Research findings indicated that the primary mechanism used by PGPB that decrease ethylene is the production of ACC deaminase (EC 4.1.99.4) enzyme, which is able to degrade ACC synthase into α-ketobutyrate and ammonia (Grichko and Glick 2001). Thus, ACC deaminase production may reduce abiotic stress by balancing production of plant ethylene level since elevated ethylene levels inhibit cell division, DNA synthesis, and growth of roots and shoots (Burg 1973). For many species of *Bacillus* sp., including *B. subtilis*, ACC deaminase activity was found (Saikia et al. 2018). Various studies have demonstrated that inoculation of plants with ACC deaminase producing *B. subtilis* increased their growth and resistance to environmental stresses (Barnawal et al. 2013; Treesubsuntorn et al. 2017; Saikia et al. 2018). There are many reports on plant health improvement through inoculating ACC deaminase-containing bacterial strains during drought, flooding, salinity, heavy metal stress, etc. But some of the investigations did not directly describe the causal relationship of the bacterial ACC deaminase enzyme in stress resistance. There might exist some other bacterial determinants for stress alleviation in affected plants (Saikia et al. 2018). In other studies, the endophytic PGPB from sunflower plants were shown to produce such hormones as JAs, ABA (Forchetti et al. 2007), and SA (García-Gutiérrez et al. 2013). Since the influence of JAs and SA on plant growth and resistance is well known (Shakirova et al. 2012), the growth-promoting effect of bacteria under normal and stress conditions is usually ascribed to the ability of bacteria to synthesize these compounds (Sayed et al. 2011; Fravel 2005; Chourdhary and Johri 2009; Wenhao et al. 2012).

Indirect mechanisms refer to bacterial traits that inhibit the functioning of one or more plant pathogenic organisms in both fungi and bacteria, including the synthesis of a range of different antibiotics, cell wall-degrading enzymes (chitinases, cellulases, glucanases, proteases, lipases), lipopeptides (surfactin, iturin, fengycin, etc.), siderophores, VOCs, lowering plant ethylene levels, ACC deaminase, quorum quenching, activation of SAR, ISR responses, and IST in plants against abiotic stresses (Bowen and Rovira 1999; Van Loon 2007; Glick 2014). To date, numerous examples of *B. subtilis* strains with indirect PGP mechanisms have been reported. Jensen et al. (2000) demonstrated that inoculation of bean with *B. subtilis* strain GB03 suppressed root rot disease of plants and significantly increased their DW and yields in comparison to non-inoculated control both in greenhouse and field experiments. In another study, inoculation with *B. subtilis* ME488 suppressed several plant pathogens growth in vitro reducing the cucumber and pepper diseases caused by *Fusarium oxysporum* and *Phytophthora capsici*, respectively, as well as increased germination in seeds and plant growth in comparison to controls (Chung et al. 2008).

6.2.2 Production of Lytic and Oxidative Enzymes

Growth enhancement through enzymatic activity is another mechanism used by *B. subtilis*. The lytic enzymes synthesized by *Bacillus* spp. have proved to be very active in degrading fungal cell walls (Leelasuphakul et al. 2006). Inoculation of rice plants with *B. subtilis* AUBS1 induced systemic resistance of host plant against sheath blight through increasing the levels of β-1-3-glucanases and thaumatin (Jayaraj et al. 2004a, b). The defense-related activities of the enzymes have been proven in various plant species and against various plant pathogens (Thilagavathi et al. 2007). Endophytic *B. subtilis* 330-2 that enhanced plant growth under abiotic/biotic stresses secreted a complex of hydrolytic enzymes including β-1,3-glucanase, β-1,4-glucanase, and proteases, which possibly degrade the contents of the fungal cell wall, such as β-1,3-glucan and glucosidic bonds (Ahmad et al. 2017). *B. subtilis* also produced many defense-related oxidative enzymes such as PO, PPO, and phenylalanine ammonia-lyase (PAL) (Jayaraj et al. 2004a, b) which induced lignin and oxidative phenolic compounds playing an important role in contraction defense-related obstacle by producing structural changes in the cellular defense system against plant pathogens (Thilagavathi et al. 2007). Thus, PAL has an important role in the lignin formation and production of flavonoids, which play a key role in the phenylpropanoid biosynthetic pathway (Sailaja et al. 1998). Ramyabharathi et al. (2012) showed that *B. subtilis* EPCO16 induced defense-related proteins and increased accumulation of phenolics as well as activities of PAL and CAT in tomato plants infected by *F. oxysporum f.* sp. *lycopersici*. *B. subtilis* strain AUBS1 induced systemic resistance in rice plants against sheath blight significantly increasing the level of PO, PAL, and PR proteins (Jayaraj et al. 2004a, b).

6.2.3 Production of Siderophores

The ability of *B. subtilis* to produce siderophores is another important PGP trait of bacteria with significant influence on plant growth (Pieterse et al. 2014; Pandey et al. 2017). Siderophore-secreting PGP bacteria inside the plant tissues help in the transport of Fe^{3+} inside the plant cell and contribute to the plant growth and productivity through synthesis of ATP, DNA precursor, and the heme. The ability to produce siderophores confers competitive advantages to endophytic bacteria for the colonization of plant tissues and the exclusion of other microorganisms from the same ecological niche (Pandey et al. 2017). *B. subtilis* strain B28 producing siderophore, protease, and hydrogen cyanide (HCN) reduced Fusarium wilt of chickpea in vitro and significantly increased plant growth parameters such as length, FW, and DW in comparison to controls in greenhouse experiments (Karimi et al. 2012). In another study, an endophytic *B. subtilis* 330-2 producing siderophore demonstrated its antagonistic activity against various pathogens along with regulation of plant growth (Ahmad et al. 2017).

6.2.4 Production of Malic Acid (MA)

The interaction between plants and PGPB, including *B. subtilis*, is a complex and reciprocal process. This relationship can be initiated through the action of such compounds as MA, which the plants exudate to attract rhizobacteria and results in subsequent migration of microorganisms toward the roots, movement of bacteria along the roots, and adherence (Rudrappa et al. 2008). L-MA secreted from *Arabidopsis* roots selectively signals and recruits the beneficial rhizobacteria *B. subtilis* FB17 (Rudrappa et al. 2008). Rekha et al. (2018) demonstrated that *B. subtilis* strain RR4 is chemotactic to MA and can induce MA biosynthesis in roots of rice plants. Significant differential expressions of the genes of MA biosynthesis – malate synthase (*OsMS*) – and the transporter gene, aluminum-activated malate transporter (*OsALMT*), caused 1.8-fold and 0.58-fold changes, respectively, in *B. subtilis* RR4 inoculated roots. The gene expression pattern of *OsMS* corroborated the data obtained by high-performance liquid chromatography, which showed elevated MA levels in roots (1.52-fold), whereas the levels of MA in root exudates were not altered significantly although expression of *OsALMT* was reduced (Rekha et al. 2018).

6.2.5 Production of Lipopeptides

The results of recent year's research indicate that *Bacillus* spp. produce a variety of bioactive antibiotic metabolites, including lipopeptides, which play an important role in antifungal activity and PGP (Raaijmakers et al. 2010). It has been shown that these antimicrobial compounds are active against a wide range of microorganisms, including bacteria, fungi, oomycetes, and viruses (Raaijmakers et al. 2010). Romero et al. (2007) reported that iturin and fengycin families of lipopeptides play a major role in the antagonism of *B. subtilis* against *Podosphaera fusca*. The surfactin- and fengycin- type lipopeptides can act as bacterial determinants in plant cells by initiating the immune response through the stimulation of ISR (Ongena et al. 2007). Lipopeptides of *B. subtilis* S499 proved to be effective triggers of SAR in tobacco and bean plants (Ongena et al. 2007). Comparison of the level of production of antibiotic substances with different strains showed the advantages of *B. subtilis* species. For example, concentrations of lipopeptides in the culture medium of the most active strain of *B. subtilis* IB-54 exceeded those of active *P. ehimensis* and *P. polymyxa* by 10–100 times. Differences between the *B. subtilis* strains were also observed in both individual activity level and in the spectrum of the lipopeptides formed. It has been established that the main components of the *B. subtilis* lipopeptide complex are iturin, surfactin, and related compounds (Romero et al. 2007; Ahmad et al. 2017). It was identified that antifungal peptides produced by *B. subtilis* AU195 show similarity with bacillomycin (group iturin A) and exhibit strong antagonistic activity against *Aspergillus flavus* and other plant pathogenic fungi. The iturin and fengycin from *B. subtilis* strains UMAF6614, UMAF6616, UMAF6639, and UMAF8561 were involved in the suppression of cucurbit disease caused by *Podosphaera fusca* (Romero et al. 2007). The analysis of culture filtrate, along with the recovery of inhibitory components (fengycin, surfactin, and iturin A or bacillomycin) from the melon leaves inoculated with *B. subtilis* UMAF6614 and *B. subtilis* UMAF663, strongly supported the evidence of in situ production of these antimicrobials. Bais et al. (2004) reported that protective action of surfactin produced by *B. subtilis* against infection caused by *Pseudomonas syringae* in *Arabidopsis thaliana* and suggested that surfactin was necessary both for root colonization and protecting plants against the pathogen. Ahmad et al. (2017) observed that endophytic *B. subtilis* 330-2 enhanced rice and maize plant growth (ranging 14–37%) and tolerance to biotic/abiotic stresses regulates the expression of different NRPSs, including surfactin (*srfAA*), fengycin (*fen*), and bacillaene (*bae*). In addition to NRPS, *B. subtilis* 330-2 triggered the production of secondary metabolites, including antibiotics such as macrolactin (*mln*), difficidin (*dfn*), iturin A (*ituA*), bacillibactin (*dhbF*), penicillin-binding protein 2B (*pbpB*), and beta-lactamase (*penP*), which are important manifestations of a plant's defense mechanism and can cope with competing microorganisms and inhibit the growth of phytopathogenic fungi or bacteria. It should be noted that in antibiotic production, 10% of all 114 genes differentially expressed by *B. subtilis* 330-2 were involved (Ahmad et al. 2017).

6.2.6 Emission of Volatile Organic Compounds (VOCs)

Many types of bacteria, including *B. subtilis* can regulate plant growth from a distance without any contact, suggesting the possibility that they emit invisible VOCs which promote or inhibit growth of plants. It is reported that about 350 bacterial species produces around 846 different potential VOCs with 5431 synonyms (Lemfack et al. 2014). The bacterial VOCs can interact with phytohormones by taking part in morphogenetic processes, resulting in plant growth stimulation (Ryu et al. 2003; Zhang et al. 2007; Ortíz-Castro et al. 2014). Recently, VOCs emitted by *Bacillus* spp. were suggested to help increase plant growth through regulating the synthesis of phytohormones and metabolism (Zhang et al. 2007; Tahir et al. 2017). Auxin regulation by VOCs from *B. subtilis* GBO3 resulted in the initiation of *Arabidopsis* growth promotion. The accumulation of auxin in the leaves decreased while it increased in the roots after exposure to VOCs, perhaps due to the initiation of basipetal auxin transport (Zhang et al. 2007). Tahir et al. (2017) demonstrated that *B. subtilis* SYST2 produced VOCs along with regulating the endogenous phytohormone levels and altering the expression of genes involved in auxin, CKs, gibberellin, expansin, and ethylene biosynthesis and also increased the photosynthesis rate resulting in enhanced tomato growth both in vitro and pot experiments. It should be noted that 2 of the 11 individual VOCs emitted by *B. subtilis* SYST2 (albuterol and 1,3-propanediol) were found to be the key growth-promoting factors (Tahir et al. 2017). *B. subtilis* GBO3 emits VOCs and was shown to increase *Arabidopsis* photosynthetic activity and chlorophyll content in leaves (Xie et al. 2009). In another study, VOCs emitted by *B. subtilis* GBO3 enhanced photosynthesis by increasing chlorophyll content (Xie et al. 2014) and by altering the expression of genes relating to chloroplasts, suggesting a significant relationship between the increase in photosynthetic rates and growth promotion (Zhang et al. 2008). It is interesting that some of the VOCs emitted from *B. subtilis* GB03 are bacterial determinants involved in IST (Ryu et al. 2004). VOCs emitted by *B. subtilis* SYST2 increased plant biomass and decreased the ethylene content in tomato seedlings in comparison to control (water) and *E. coli* DH5α (a non-VOC producer) both in vitro and pots, which could due to the fact that these VOCs repressed the *ACO1* transcriptional gene (Tahir et al. 2017). These results support findings that bacterial VOCs altered the transcription of genes related to ethylene biosynthesis (*ACO2*, *ACS4*, *ACS12*, *SAM-2*), and genes involved in the ethylene response (*CHIB*, *ERF1*, *GST*) (Kwon et al. 2010). It was revealed that *B. subtilis* GB03-emitted VOCs regulate ethylene biosynthesis enzymes as well as ethylene biosynthesis-related genes (*ERF1*, *GST2*, and *CHIB*). It also regulates the jasmonic acid (JA) and SA-mediated defense mechanisms (Shafi et al. 2017).

6.2.7 Production of Polyamines

Polyamines produced by *B. subtilis* can also play an important physiological role in plant growth and development as they are involved in various processes such as cell division and differentiation, nucleic acid replication, translation, transcription, protein synthesis, and membrane stability (Kusano et al. 2008). A common polyamine spermidine has attracted widespread interest as it is essential for eukaryotic cell viability and is correlated with lateral root development. Spermidine also plays a protective role against various environmental factors leading to oxidative, osmotic, and acidic stresses as well as pathogen attacks. Imai et al. (2004) revealed that spermidine synthase genes are essential for survival of *Arabidopsis*. A double mutant of two spermidine synthase encoded genes *SPDS1* and *SPDS2* exhibit a lethal defect in embryo development (Imai et al. 2004). Hummel et al. (2002) reported involvement of the polyamines spermidine and spermine in root development at low temperature in the subantarctic cruciferous species *Pringlea antiscorbutica*. It was revealed that a decrease in spermidine and spermine levels results in reduced lateral root development (Hummel et al. 2002). Xie et al. (2014) demonstrated that *B. subtilis* strain *OKB105* producing spermidine significantly enhanced root growth in tobacco. The analysis of *B. subtilis* OKB105 culture filtrates demonstrated that a common polyamine spermidine is the pivotal PGP compound. It was revealed the genes *yecA* (encoding a putative amino acid/polyamine permease) and *speB* (encoding agmatinase) are involved in the secretion or biosynthesis of polyamine in *B. subtilis* OKB105. In addition, OKB105-induced expression of the expansin genes (*Nt-EXPA1*, *Nt-EXPA2*) inhibited the expression of the ethylene biosynthesis gene *ACO1* (Xie et al. 2014). In another study, the transcription profiles of *B. subtilis* showed that genes associated with acetylation, transportation, and biosynthesis of polyamine were differentially expressed when growing in alkaline or acidic conditions (Wilks et al. 2009). Moreover, some evidence on the relationships between polyamines and biofilm formation has been proposed. Polyamine depletion in *B. subtilis* resulted in deficiency of biofilm formation, and the defect was restored by adding agmatine or spermidine into the medium (Burrell et al. 2010). It is known that biofilm formation by *B. subtilis* is a complex process and plays an important role in host plant protection. Bais et al. (2004) reported that upon root colonization, *B. subtilis* (strain 6051) forms a stable, extensive biofilm and secretes surfactin, which act together to protect plants against attack by pathogenic bacteria. It was shown that a wild-type *B. subtilis* 6051 can form biofilm-like structures on the *Arabidopsis* roots and protect plants from *Pseudomonas syringae* pathogen attacks (Bais et al. 2004). It is reported that plant protection by *B. subtilis* depends on widely conserved genes, including genes for regulation and matrix production which are required for biofilm formation (Chen et al. 2012).

6.2.8 Induced Systemic Resistance and Tolerance

Triggering the mechanisms of systemic resistance and tolerance in plants by bacterial microbes including *B. subtilis* is the main cause of disease suppression by crop plants (Ahn et al. 2002) leading to increase in yield. The activation of *B. subtilis* induced defense mechanisms throughout the plant can be carried out in different ways. Two common pathways are ISR and SAR, which have been significantly appreciated over the past decade (Ryals et al. 1996; Van Loon 2007; Pieterse et al. 2012). It has been discovered that *B. subtilis* triggered both SAR and ISR in plants. Because of this, development of biological control products based on *B. subtilis* for the protection of agricultural plants against environmental stresses shows great promise (Maksimov and Khairullin 2016; Pandey et al. 2017). The role of *B. subtilis* in increasing host plant tolerance to abiotic stresses has only been revealed and studied very recently and described as IST (Yang et al. 2009; Arya et al. 2018; Sarma et al. 2018). The whole influence of *B. subtilis*-mediated conditioned responses in plants, whether on a physical, biochemical, or molecular level, can lead to protection against abiotic stresses. Only a few reports on PGPB as elicitors of tolerance to abiotic stresses such as salinity, drought, nutrient deficiency or excess have been published (Lastochkina et al. 2017a; Sarma et al. 2018; Arya et al. 2018). The interaction between *B. subtilis* and plants leads to the activation of local and systemic defenses controlled by plant signaling hormones such as SA, JA, and ethylene (Koornneef and Pieterse 2008). Some researchers reported the key target affected by *B. subtilis* is the plant signaling pathways, regulated by SA and JA, leading to development of a protective responses to stresses (Van Loon 2007; Chourdhary and Johri 2009; Niu et al. 2011; García-Gutiérrez et al. 2013; Veselova et al. 2015). However, it is not yet clear how *B. subtilis* regulates the protective system of host plants under abiotic stresses and how the *B. subtilis*-induced plant defense signal system interacts with classical signaling pathways specifically induced by such signaling molecules such as SA and JA. On the one hand, it is considered that the influence of *B. subtilis* is similar to the action of "weak" pathogens on plants, and on the other hand, they themselves produce metabolites with hormonal and signaling functions (auxins, CKs, ethylene, gibberellins, ABA, SA, JAs) (Dodd et al. 2010; Veselova et al. 2015).

Various species of *Bacillus* spp. as well as strains of *B. subtilis* often possess one or more of the above mentioned traits, although the bacterial activities may be different and may depend on many factors. It should be noted that *Bacillus* species are considered to be the most effective because they have the ability to produce spores that allow them to survive even in adverse environmental conditions (Francis et al. 2010). In fact, no single organism including *B. subtilis* has the ability to make use of all the available mechanisms that could be used to promote plant growth both under normal and stress conditions.

6.3 *B. subtilis*-Mediated Plant Stress Tolerance

In many regions, agricultural production is adversely affected by various abiotic stresses such as drought, salinity, extreme temperatures, and toxic metals resulting in significant reduction of crop productivity (FAO 2014, 2016; Pereira 2016). Abiotic stresses are often interrelated, either individually or in combination. They induce general signaling pathways for the regulation of cellular responses aimed at adaptation, and therefore plants can have similar morphological, physiological-biochemical, molecular-genetic changes (Potters et al. 2007; Huang et al. 2008; Munns and Tester 2008). Many studies have shown that various strains of *B. subtilis* contribute to protection of different host plants against diverse abiotic stresses leading to the increase of plant growth and productivity/yield (Turan et al. 2012; Ahmad et al. 2017; Egamberdieva et al. 2017; Martins et al. 2018; Wani et al. 2018). Investigations on the mechanisms mediated by *B. subtilis* on plant stress tolerance are of the most urgent problems in modern plant biology due to *B. subtilis*-inducing responses against stresses that could be diverse and often specific (Niu et al. 2011; Ahmad et al. 2017).

6.3.1 Drought

Drought is one of the major abiotic stresses worldwide leading to inhibition of plant growth and reduction the crop yield (Araus et al. 2008; FAO 2014, 2016). In response to water deficiency, plants rapidly modulate a series of physiological reactions in order to economize the water use. This includes stomata closure, ABA accumulation, photosynthetic rate reduction, expression of aquaporins and vacuolar H+-pyrophosphatases for maintaining cell turgor through osmotic adjustments, the accumulation of compatible osmolytes, and reactive oxygen species (ROS) enzymes which resulted in cell integrity, functionality, and survival of plants (Araus et al. 2002; Ruan et al. 2010; Cramer et al. 2011; Krasensky and Jonak 2012). Numerous studies demonstrated that *B. subtilis* contributes to alleviate drought stress in different plants by several mechanisms (Zhang et al. 2010; Wang et al. 2012; Martins et al. 2018; Saikia et al. 2018) (Table 6.1). For example, Gagné-Bourque et al. (2016) demonstrated that under drought stress endophytic *B. subtilis* B26 successfully colonized the internal tissues of timothy (*Phleum pratense* L.) positively impacting plant growth and metabolism. Exposure of inoculated plants to drought stress (8 week) led to significant increases in biomass of shoots (by 26.6%) and roots (63.8%), photosynthetic activity (by 55.2%), and stomatal conductance (214.9%) compared to non-inoculated plants grown under similar stress conditions (Gagné-Bourque et al. 2016). It has been proposed that *B. subtilis*-induced drought stress tolerance in host plants is mediated by several mechanisms such as phytohormonal and ACC deaminase activity, production of ROS-scavenging antioxidant enzymes, accumulation of osmolytes and soluble sugars, regulation of the expression of drought-responsive genes, etc. (Gagné-Bourque et al. 2015; Barnawal et al. 2017; Saikia et al. 2018). Phytohormones produced by plants such as IAA, ABA, CKs, gibberellins, and ethylene are important for their growth and development under normal conditions and play a major role under abiotic stresses including drought (Shakirova et al. 2012). The ability of *B. subtilis* to synthesize plant hormones such as IAA, ABA, CKs, gibberellins, ethylene, SA, and JAs was shown in many studies (Barnawal et al. 2017; Ahmad et al. 2017; Lastochkina et al. 2017a). Various plant species inoculated with IAA-producing PGPB represent an increase in root initiation and growth which result in elevated water and nutrient uptake and plant tolerance to water deficit (Dimkpa et al. 2009; Egamberdieva and Kucharova 2009). Moon et al. (2017) showed that auxin production by different *Bacillus* strains, including *B. subtilis*, is associated to enhance the vegetative growth parameters of *Zea mays* L. under water stress conditions. For drought-tolerant bacterial strains *B. subtilis* AB-21, *B. subtilis* AB-61, *B. aryabhattai* AB-51, *B. toyonensis* AM-25, *B. licheniformis* AM-21, and *B. pumilus* AB-33 isolated from a semiarid region located in Northeastern Pakistan at highest water stress (14% field capacity (FC)), significant response for FW (25%) and DW (45%) was recorded with *B. subtilis* AB-21 in comparison to the control. For shoot length, mixed culture combination C-3 (*B. licheniformis* AM-21, *B. pumilus* AB-33, and *B. subtilis* AB-61) recorded a 46% increase over the control at 14% FC. On the other hand, *B. subtilis* AB-21 and *B. aryabhattai* AB-51 recorded 15% and 50% increases for root length and number of roots at 14% and 21% FC, respectively (Moon et al. 2017). The bacterial strain *B. subtilis* UFGS1 can mitigate drought stress in soybean and may improve water efficiency under certain conditions (Martins et al. 2018). It is known that drought causes oxidative stress in plants associated with increased production of ROS (superoxide anion radicals (O^{2-}), hydrogen peroxide (H_2O_2), hydroxyl radicals (OH), singlet oxygen (O^{12}), and alkoxy radicals), which potentially force negative impact on the integrity of the plant cells membrane structures. Antioxidant defense system involves both enzymatic (SOD, PO, CAT, ascorbate peroxidase, APX, glutathione reductase, GR) and non-enzymatic (ascorbic acid, cysteine, glutathione) components serving to prevent accumulation of ROS and alleviate the oxidative damage occurring during drought stress (Kaushal and Wani 2015). Current

Table 6.1 Examples of *B. subtilis*-mediated abiotic stress tolerance in plants

Stress	Microbe inoculation	Plant	Effect on plants and tolerance strategy	Reference
Drought	*B. subtilis* GB03	Arabidopsis	Elevated accumulation of osmoprotectants Cho, GlyBet	Zhang et al. (2010)
Drought	*B. subtilis* RMPB44	Maize	Increasing Pro, sugars, and free amino acids, decreasing electrolyte leakage, and reducing the activity of antioxidant enzymes (APX, CAT, and glutathione peroxidase)	Vardharajula et al. (2011)
Drought	*B. subtilis* SM21	*Cucumis sativus*	Production of monodehydroascorbate, Pro, and antioxidant enzymes	Wang et al. (2012)
Drought	*B. subtilis LDR2* (rhizobacteria)	Trigonella	Enhancing nodulation and arbuscular mycorrhizal fungi colonization in the plants resulting in improved nutrient uptake and plant growth Reducing levels of ACC	Barnawal et al. (2013)
Drought	*B. subtilis* B26 (endophyte)	*Brachypodium distachyon*	Upregulation of expression of drought-responsive genes (*DREB2B-like*, *DHN3-like*, *LEA-14-A-like*), modulation of the DNA methylation process (*MET1B-like*, *CMT3-like*, *DRM2-like* genes), and an increase in the soluble sugars and starch content of leaves	Gagné-Bourque et al. (2015)
Drought	*B. subtilis* UFGS1 (rhizobacteria)	Soybean	Increase fresh and dried biomass Maintaining the photosynthetic rates and higher stomatal conductance and transpiration	Martins et al. (2018)

(continued)

Table 6.1 (continued)

Stress	Microbe inoculation	Plant	Effect on plants and tolerance strategy	Reference
Drought	*B. subtilis* B26 (endophyte)	Timothy (*Phleum pratense* L.)	Increasing biomass of shoots (by 26.6%) and roots (63.8%) Increasing photosynthesis (by 55.2%), stomatal conductance (214.9%), and the levels of several sugars (sucrose, fructans) and amino acids (asparagine, glutamic acid, glutamine, GABA)	Gagné-Bourque et al. (2016)
Drought	*B. subtilis* RJ46 (rhizobacteria) *B. subtilis* RJ46 in consortium with *Ochrobactrum pseudogrignonense* RJ12 and *Pseudomonas* sp. RJ15	*Vigna mungo* L., *Pisum sativum* L.	Increasing seed germination, lengths of root and shoot, DW, RWC, intension of root recovery Production of ROS-scavenging enzymes, cellular osmolytes, chlorophyll, producing ACC deaminase, decreasing the ACC accumulation and downregulated ACC oxidase gene expression	Saikia et al. (2018)
Water stress	*B. subtilis* AB21, *B. subtilis* AB61 (rhizobacterium)	*Zea mays*	Auxin synthesis, phosphate solubilization, ammonia production	Moon et al. (2017)
Drought Salinity	*B. subtilis* LDR2 (rhizobacteria)	Wheat	Enhancing photosynthetic efficiency and IAA content, reducing ABA and ACC content, enhancing *TaDREB2* gene expression, and modulating expression of a regulatory component (CTR1) of ethylene signaling pathway and DREB2 transcription factor	Barnawal et al. (2017)

(continued)

Table 6.1 (continued)

Stress	Microbe inoculation	Plant	Effect on plants and tolerance strategy	Reference
Salinity	*B. subtilis* FZB24 (rhizobacteria)	Eggplant, pepper	Increasing yield up to 550% (eggplants) and 430% (pepper), reduction of salinity effect on plant yield on 25%–50% Regulation of auxin synthesis	Bochow et al. (2001)
Salinity	*B. subtilis*	Wheat	Restricting sodium uptake and stimulates plant growth	Ashraf et al. (2004)
Salinity	*B. subtilis* (rhizobacterium)	*Bradyrhizobium japonicum*	Producing various plant hormones for growth improvement	Han and Lee (2005)
Salinity	*B. subtilis FZB24*	Artichoke (*Cynara scolymus* L.)	Stimulating the vegetative growth, improving gas exchange leading to the increase photosynthesis rate	Saleh et al. (2005)
Salinity	*B. subtilis*	*Lactuca sativa*	Stimulation of shoot biomass CK-signaling	Arkhipova et al. (2007)
Salinity	*B. subtilis* GB03	*Arabidopsis thaliana*	Tissue-specific regulation of sodium transporter HKT1 Decreasing root transcriptional expression of a high-affinity K^+ transporter (AtHKT1) decreasing root Na^+ import	Zhang et al. (2008)
Salinity	*B. subtilis* EY2 (rhizobacteria)	Strawberry	Increasing RWC and N content of leaves	Karlidag et al. (2010)
Salinity	*B. subtilis* FZB 24	Lettuce	Raised the root vigor, yield Significantly increasing SOD and CAT activities, decreasing MDA, enhancing the stability of the cell membrane, improving photosynthesis by increasing the net photosynthetic rate and the stomatal conductance	Liu et al. (2010)

(continued)

Table 6.1 (continued)

Stress	Microbe inoculation	Plant	Effect on plants and tolerance strategy	Reference
Salinity	*B. subtilis* EY2 (rhizobacteria)	Lettuce	Increasing growth, head weight and height, root diameter, plant nutrient element contents Increasing chlorophyll content	Yildirim et al. (2011)
Salinity	*B. subtilis* (rhizobacterium)	Sweet basil	Positive effects on growth, oil %, oil yield and nutrient uptake	Sayed et al. (2011)
Salinity	*B. subtilis* SU47 (rhizobacteria)	Wheat (*Triticum aestivum* L.)	Increase in dry biomass and total soluble sugars and Pro content and reduced activity of antioxidant enzymes	Upadhyay et al. (2012)
Salinity	*B. subtilis*	Radish (*Raphanus sativus*)	Increasing in fresh and dry masses of roots and leaves, photosynthetic pigments, Pro, total free amino acids, crude protein contents compared to non-inoculated ones under salinity Increasing phytohormones contents (IAA, GA3) and the contents of N, P, K^+, Ca^{2+}, and Mg^{2+} but decreased ABA and Na^+ and Cl^- content	Mohamed and Gomaa (2012)
Salinity	*B. subtilis EY2* (rhizobacteria)	Strawberry	Significantly increased the growth, chlorophyll content, RWC, nutrient element content, and yield lowered electrolyte leakage	Karlidag et al. (2013)
Salinity	*B. subtilis* BERA71	*Acacia gerrardii*	Stimulating growth of roots and shoots, nodulation, nutrient uptake	Hashem et al. (2016)
Salinity	*B. subtilis* NUU4 (endophyte)	Chickpea (*Cicer arietinum* L.)	Improving symbiotic performance of host plant with rhizobia, decreasing H_2O_2 concentration, and increasing Pro content	Egamberdieva et al. (2017)

(continued)

Table 6.1 (continued)

Stress	Microbe inoculation	Plant	Effect on plants and tolerance strategy	Reference
Salinity	*B. subtilis*10-4 (endophyte)	Wheat (*Triticum aestivum* L.)	Increasing seed germination and biomass and length of shoots and roots Increasing RWC, SA content, and Pro accumulation and decreasing MDA and hydrolysis of statolite starch SA-signaling pathway	Lastochkina et al. (2017a)
Salinity	*B. subtilis* BERA 71	Chickpea	Enhancing plant biomass and synthesis of photosynthetic pigments; reducing the levels of ROS and lipid peroxidation; increasing the activities of ROS-scavenging antioxidant enzymes (SOD, PO, GR), the levels of non-enzymatic antioxidants (ascorbic acid, glutathione), and the total phenol content; and enhancing the membrane stability	Abd-Allah et al. (2018)
Salinity Heat Drought Cold Oxidative	*B. subtilis* 330-2	Rice, maize	Production IAA, siderophores, lytic enzymes, and solubilized different sources of organic/inorganic phosphates and zinc, expression 114 genes among which 32% and 10% were involved in metabolism (e.g., gltA, pabA, and ggt) and transportation of nutrients (e.g., fhu, glpT, and gltT), respectively	Ahmad et al. (2017)

(continued)

Table 6.1 (continued)

Stress	Microbe inoculation	Plant	Effect on plants and tolerance strategy	Reference
Arid climate stresses (water shortage, soil salinity)	*B. subtilis*	Wheat	Increasing the concentration of Crude protein, N, P, and total soluble carbohydrates in grains Enhancing Pro, soluble carbohydrates, Chl a, Chl b, Chl a + Chl b; Chl a/Chl b, carotenoids	Haggag et al. (2017)
Variable environmental factors		Primocane-fruiting raspberries	Increasing the number of new raspberry canes, plant generative organs, crop productivity	Belyaev et al. (2017)
Cold	*B. subtilis* OSU142 (rhizobacteria)	Wheat, barley	Enhancing root and shoot DW, leaf total chlorophyll content, stomatal conductance, leaf RWC	Turan et al. (2012)
Fungicide stress	*B. subtilis* BC8		Secreting active biomolecules	Shahid and Khan (2018)
Alkaline stress (high pH)	*B. subtilis* OSU142	Grapevine rootstocks (*Vitis* spp.)	Improving vegetative growth, leaf physiology and nutrient acquisition	Karaca and Sabir (2017)
Toxic metal (Pb)	*B. subtilis* QM3	Wheat	Growth promotion Increasing the activity of antioxidant enzymes CAT, PO, APX, SOD and decreasing the concentration of metal in the roots	Hao et al. (2015)
Chromium salts ($CrCl_3$, K_2CrO_4 and $K_2Cr_2O_7$)	*Bacillus sp.* AMP2	*Triticum aestivum*	Reduction in the Cr uptake of seedlings Increased the levels of acid phosphatase and PA	Naseem et al. (2016)
Cr toxicity	*B. subtilis* MAI3	Soybean	Increasing photosynthetic pigments, antioxidants, reduction MDA production	Wani et al. (2018)
Cd, Ni toxicity	*B. subtilis* strains (endophytes)	*Sinapis alba* L.	Reducing oxidative stress manifestation	Kuramshina et al. (2018)

(continued)

Table 6.1 (continued)

Stress	Microbe inoculation	Plant	Effect on plants and tolerance strategy	Reference
Cd toxicity	*B. subtilis*	Rice (*Oryza sativa* L.)	Reducing Cd accumulation in roots, shoots, and grains, higher Ca and Mg accumulation Increasing DW, phosphate solubilization, IAA production, control ethylene level by ACC deaminase	Treesubsuntorn et al. (2017)

evidences indicate the effect of *B. subtilis* on the activity of PA, CAT, SOD, and APX involved in the utilization of ROS. The decrease in the adverse effect of oxidative stress by *B. subtilis* is shown in many crops (Vardharajula et al. 2011; Wang et al. 2012; Saikia et al. 2018).

Plants' adaptation to drought stress is associated with the accumulation of compatible solute/osmolytes like proline (Pro), sugars, betaines, polyamines, quaternary ammonium compounds, amino acids, and water stress-related proteins like dehydrins (Chen and Jiang 2010; Krasensky and Jonak 2012). Increased levels of osmolytes maintain cell water status and protect membrane and protein damage against stresses (Sandhya et al. 2011; Krasensky and Jonak 2012). Various studies revealed that *B. subtilis* produces osmolytes in response to drought and stimulates plant growth by increasing the root cells' osmotic potential (Paul et al. 2008; Dimkpa et al. 2009). For instance, the beneficial soil bacterium *B. subtilis* GB03 increases *Arabidopsis* tolerance to osmotic stress by elevating accumulation of such osmoprotectants as choline (Cho) and glycine betaine (GlyBet). It was reported that *B. subtilis* GB03 enhances *Arabidopsis* Cho and GlyBet synthesis associated with enhanced plant tolerance to osmotic stress (Zhang et al. 2010). Osmoregulation in plants through accumulation of soluble sugars (sucrose, glucose, fructose) is also the mechanism for maintaining homeostasis in plants under drought stress (Valliyodan and Nguyen 2006), and their metabolism plays a significant role in drought and cold stress tolerance (Valliyodan and Nguyen 2006; Livingston et al. 2009). Inoculation of *Brachypodium distachyon* plants with endophytic *B. subtilis* B26 increased root and shoot weights and accelerated growth rate and seed yield as compared to control plants under drought stress (Gagné-Bourque et al. 2015). Inoculated plants exposed to stress for 8 days had 2.9-fold (glucose) and 1.4-fold (fructose) more total soluble and starch contents in stressed inoculated plants (Gagné-Bourque et al. 2015). In another study, Gagné-Bourque et al. (2016) revealed that *B. subtilis* B26 improves timothy growth under drought stress through accumulating more osmolytes (total carbohydrates and soluble sugars) in roots and shoots. Inoculation of timothy with *B. subtilis* B26 most notably increased sucrose and fructan concentrations in shoots under non-stressed and drought-stressed conditions over a period of 8 weeks of drought, while glucose increased in plants after 4 weeks of drought. The results show that *B. subtilis* contributed to increase the sugar biosynthesis to allow for better osmotic adjustment and thus alleviates the adverse effect of drought on the host plant (Gagné-Bourque et al. 2016). The bacterium *B. subtilis* B26 had a significant impact on plant metabolism and resulted in higher levels of several sugars (sucrose, fructans) and an increase of key amino acids such as asparagine, glutamic acid, and glutamine in shoots and roots of inoculated plants compared to non-inoculated ones. Also, in the presence of *B. subtilis* B26, the accumulation of nonprotein gamma-aminobutyric acid (GABA) in shoots of stressed plants and in roots of stressed and unstressed plants was increased (Gagné-Bourque et al. 2016).

It was discovered that the protective effects of the bacterium on plants were linked to the upregulation of the drought-response genes (*DREB2B*-like, *DHN3*-like, and *LEA-14-A*-like) and modulation of the DNA methylation genes (*MET1B*-like, *DRM2*-like, and *CMT3*-like) regulating the process (Gagné-Bourque et al. 2015). It should be noted that a single inoculation of *Brachypodium* with *B. subtilis* B26 affected the whole growth cycle of the plant, accelerating its growth rates, shortening its vegetative period, and alleviating drought stress effects. It was revealed that endophytic *B. subtilis* B26 efficiently colonized the plant and was recovered from roots, stems, and blades as well as seeds of *Brachypodium*, indicating that the bacterium is able to migrate, spread systemically inside the plant, and establish itself in the aerial plant tissues and organs and is vertically transmitted to seeds. Interestingly, the presence of *B. subtilis* B26 in the seed led to systemic colonization of the next generation of *Brachypodium distachyon* (Brachypodium) plants (Gagné-Bourque et al. 2015). *B. subtilis* B26 isolated from switchgrass confers drought resistance in Brachypodium through the upregulation of the drought-responsive gene expression, modulation of the process of DNA methylation, an increase in the content of sugars and starch in leaves (Gagné-Bourque et al. 2015), as well as the production of phytohormones and lipopeptide toxins (Gagne-Bourgue et al. 2013).

Some works demonstrated that bacterial consortium comprising *B. subtilis* could be effective bio-formulators for crop health improvement in drought-affected agricultural fields. For example, Saikia et al. (2018) evaluated the effect consortium of ACC deaminase producing rhizobacteria (*B. subtilis* RJ46, *Ochrobactrum pseudogrignonense* RJ12, *Pseudomonas* sp. RJ15) on drought stress alleviation in *Vigna mungo* L. and *Pisum sativum* L. plants. It was revealed that inoculation of plants with the consortium significantly increased seed germination, lengths of root and shoot, and their DW and offered drought stress tolerance by regulating plant ethylene levels. Particularly, the consortium treatment decreased the ACC accumulation and downregulated ACC oxidase gene expression. Besides, increased production of ROS-scavenging enzymes and cellular osmolytes, leaf chlorophyll content, relative water content (RWC), and root recovery intension was observed after consortium treatment in comparison to non-inoculated plants under drought stress (Saikia et al. 2018). ACC deaminase-containing *B. subtilis* LDR2 protected *Trigonella* plants under severe drought stress conditions through the reduction of the levels of ACC (responsible for stress ethylene generation) alleviating ethylene-induced damage (Barnawal et al. 2013). The findings about drought-induced changes in biochemical markers such as reduced chlorophyll concentration, increased Pro content, and higher lipid peroxidation were monitored and clearly indicated the protective effects of *B. subtilis* LDR2 under drought. Furthermore, LDR2 enhanced nodulation and arbuscular mycorrhizal fungi colonization in the *Trigonella* resulting in improved nutrient uptake and plant growth under drought (Barnawal et al. 2013). Vardharajula et al. (2011) reported *Bacillus* spp. improved maize plant growth and tolerance to drought stress via enhanced production of Pro, amino acids, and soluble sugars. Ahmad et al. (2017) demonstrated that genes cystathionine beta-lyase and cysteine desulfhydrase associated with plants tolerance to drought were differentially expressed in endophytic *B. subtilis* 330-2.

Interestingly, the effectiveness of the same strain of *B. subtilis* can vary depending from both environmental conditions and many other factors such as plant species, its ecological and geographic origin, and varietal features (Maksimov and Khairullin 2016; Lastochkina et al. 2017b). It was previously reported that endophytic strain *B. subtilis* 26D differently manifested drought tolerance on wheat plants belonging to different ecological groups (ecotypes) in the same growth conditions (Lastochkina et al. 2017b). It was revealed that under the drought stress, the influence of *B. subtilis* on the drought tolerance of wheat varieties from different ecotypes was observed at the initial stages of ontogenesis, which, apparently, was associated with different adaptation strategies used against drought stress by plants. Thus, inoculation with *B. subtilis* promoted germination of the seeds of the variety Saratovskaya-55 (the steppe Volga ecotype) under the conditions of the *simulated* drought and, in opposite scenario, had an inhibitory effect on the germination of seeds of the variety Omskaya-35 (the forest-steppe West Siberian ecotype) (Lastochkina et al. 2017b). Different strategies used by *B. subtilis* 26D as protective mechanisms in different ecotypes will allow discovering similar and different drought tolerance mechanisms and lead to a significant progress in identifying the effective protective components improvement pathway and management of this fundamental process.

6.3.2 Salinity

Salinity can severely limit crop growth and productivity (Munns and Tester 2008). In plants, salinity promotes oxidative stress and water deficit (that also accompanied by oxidative effect) (Hasegawa et al. 2000) and causes a deficiency of essential nutrients, such as K^+, and toxicity of Na^+ inside the plants, which consequently inhibits plant photosynthesis, lipid metabolism, protein synthesis, and biomass accumulation (Munns and Tester 2008; Asgari et al. 2012). To prevent the damaging effect of salt stress and maintain optimal water status, plants actively express genes coding for proteins with functions of molecular chaperones, proteases, and protease inhibitors. Under stress conditions, plants begin to accumulate previously absent low molecular weight organic compounds with osmoprotective properties such as amino acids, sugars, and betaines (Szabados and Savoure 2009; Chen and Jiang 2010). A general metabolic adaptation which enables plants to cope with water or osmotic stress involves an increased synthesis of osmoprotectant Pro. It should be noted that changes in Pro content along with the role in osmoregulation may also protect the structure of different biomolecules and membranes (Hare et al. 1998) or act as free radical scavengers that protect DNA from damaging effects of ROS (Ashraf and Foolad 2007). The protective effect of *B. subtilis* under salinity has been shown for various plant species, including lettuce (Liu et al. 2010), wheat (Ashraf et al. 2004; Lastochkina et al. 2017a), grasses (Chen and Jiang 2010), sweet basil (Sayed et al. 2011), radish (Mohamed and Gomaa 2012), *Acacia gerrardii* (Hashem et al. 2016), and many others (Table 6.1). The findings indicated that *B. subtilis* is able to induce plant tolerance to salinity through production of various plant hormones (Han and Lee 2005). Inoculation of plants with *B. subtilis* increased plant growth, yield, and nutrient uptake, especially under salt stress conditions, by influencing production of phytohormones (auxin, CKs, giberellines) and by enzymatic lowering of plant ethylene levels (Bochow et al. 2001; Ashraf et al. 2004; Saleh et al. 2005). It has been shown that PGPR strain *B. subtilis* LDR2 increase IAA content in wheat under salinity and enhance plant tolerance (Barnawal et al. 2017). Inoculation of eggplant and pepper with rhizobacterium *B. subtilis* FZB24 (registered as biocontrol agent) in the plots irrigated with saline groundwater, the yield increased up to 550% in eggplants, and up to 430% in pepper in comparison to non-inoculated ones (Bochow et al. 2001). Also, *B. subtilis* FZB24 caused 25% and 50% reduction in salinity effect on the yield of pepper and eggplants, respectively, and consequently resulted in a remarkable salt stress tolerance induction, which varied its degree according to the used plant species. The authors hypothesized for the mode of action of *B. subtilis* FZB24, which acts as plant growth and health promoter, and stress tolerance inducer, that the given bacterial production of auxin and auxin precursors during root colonization induces a push in the plant auxin synthesis with changing regulation of the appropriate mechanisms (Bochow et al. 2001). In the model experiments, the pretreatment of seedlings with millimolar amounts of auxin precursors (tryptophan, indole-3-pyruvic acid or indole-3-acetic aldehyde), 75% growth reduction in non-inoculated seedlings under salt stress could be compensated completely after 1 week. That was not observed to the same degree after pre-application of IAA. Thus, the presented model experiment could support the hypothesis of salt stress tolerance induction in *B. subtilis* FZB24-inoculated plants (Bochow et al. 2001). In another study, *B. subtilis* 10-4 producing IAA and siderophores significantly increased wheat plant growth and biomass under both nonsaline and saline conditions (Lastochkina et al. 2017a). Endophytic strain *B. subtilis* NUU4 (producing IAA, HCN, siderophores, cell wall-degrading enzymes and demonstrating antagonistic activity against *F. oxysporum*, *F. solani*, *F. culmorum*, *B. cinerea*, and *A. alternata*) improved symbiotic performance of host plant (*Cicer arietinum* L.) with rhizobia and nutrient uptake which resulted in significant increase in growth parameters (roots and shoots) and yield under salinity compared to non-inoculated control plants (Egamberdieva et al. 2017). Similar findings were reported by Hashem et al. (2016), stating that *B. subtilis* stimulated growth of roots and shoots and nodulation and nutrient uptake of *Acacia gerrardii* under salinity. The bacteria *B. subtilis* also increased contents of such phytohormones as IAA, gibberellins, and the contents of N, P, K^+, Mg^{2+}, and Ca^{2+} but decreased ABA contents and Na^+ and Cl^- content which may contribute to the activation of processes associated with a decrease of the salt effect (Mohamed and Gomaa 2012). An improved N, P, K, and Mg uptake in chickpea plants inoculated with *B. subtilis* in the combination with *M. ciceri* under saline soil conditions was observed (Egamberdieva et al. 2017). Along with the best plant growth-promoting treats, *B. subtilis* NUU4 showed good biocontrol capacity against chickpea root rot caused by *F. solani* under saline soil conditions (Egamberdieva et al. 2017). The results indicate that a combined inoculation of *B. subtilis* NUU4 and *M. ciceri* IC53 was also effective in terms of chickpea growth promotion, stress tolerance, nodulation, pod formation, and yield compared to the un-inoculated control and the treatment of *M. ciceri* IC53 alone (Egamberdieva et al. 2017). A combined inoculation of chickpea with *B. subtilis* NUU4 and *M. ciceri* IC53 decreased H_2O_2 concentrations and increased Pro contents compared to the non-inoculated plants indicating an alleviation of adverse effects of salt stress. The content of H_2O_2 in the chickpea leaves was 7.87 μM/g FW, while the single inoculation of plants with *M. ciceri* IC53 decreased H_2O_2 by 18% and inoculation with composition of *M. ciceri* IC53 and *B. subtilis* NUU4 by 29% (Egamberdieva et al. 2017). *B. subtilis* EY2 significantly increased the growth, chlorophyll content, nutrient element content, and yield of strawberry plants under natural field salinity conditions (Karlidag et al. 2013). *Raphanus sativus* seeds inoculated with *B. subtilis* caused a significant increase in fresh and dry masses of roots and leaves, photosynthetic pigments, Pro, total free amino acids, and crude protein contents in plants compared to non-inoculated ones under salinity (Mohamed and Gomaa 2012). In another study, the inoculated chickpea with *B. subtilis* NUU4 and in combination with *M. ciceri* IC53 showed 51% and 26% higher photosynthetic pigments compared to the non-inoculated plants and the single inoculation only with *M. ciceri* IC53, respectively. Increased contents of chlorophyll pigments in chickpea leaves were observed in plants co-inoculated with *B. subtilis* and *M. ciceri* (Egamberdieva et al. 2017). Liu et al. (2010) demonstrated that *B. subtilis* raised the root vigor of lettuce under salt stress and improved photosynthesis by increasing the net photosynthetic rate and stomatal conductance.

Salinity, like other stresses, significantly increases the level of ROS in plants and causes an increase of MDA (the final product of membrane lipid peroxidation) (Yazici et al. 2007; Koca et al. 2006). Therefore, leaf MDA content, representing the degree of cell membrane damage is usually used to evaluate plant tolerance to salinity and drought (Soleimanzadeh and Soleimanzadeh 2010). Exposure to salt (2% NaCl) stress reduced the level of stress-induced lipid peroxidation (MDA) in wheat plants inoculated by *B. subtilis* 10-4 (Lastochkina et al. 2017a). These results indicate weakening of oxidative stress in *B. subtilis*-treated plants and assume the protective effect of these bacteria is related to their ability to modulate the activity of oxidative enzymes and control hydrogen peroxide level (Lastochkina et al. 2017a). Hashem et al. (2016) found that endophytic *B. subtilis* isolated from *Acacia ger-*

rardii plant tissue reduced production of H_2O_2 and enhanced plant growth of *Acacia gerrardii* under salinity. In another study, *B. subtilis* significantly increased the activity of SOD and CAT of lettuce while decreasing the MDA content under salt stress. This indicated that *B. subtilis* enhanced the stability of the cell membrane (Liu et al. 2010). Endophytic *B. subtilis* 10-4 increases the water storage capacity of wheat leaves and decreases the level of salinity-induced hydrolysis of statolitic starch in the root cells of seedlings. This indicates an increase of the barrier properties in wheat plants that interfere with the penetration of toxic sodium ions under the influence of *B. subtilis* 10-4 (Lastochkina et al. 2017a). Also, it was revealed that *B. subtilis* 10-4 also increased SA content in seedlings (before stress) and contributed to decreasing the level of stress-induced SA accumulation. Activation of defense reactions in wheat plants under salinity by *B. subtilis* might be connected with the ability of bacteria to increase the endogenous SA level that plays an important role in the induction of systemic tolerance in plants.

6.3.3 Toxic Metals

Pollution of soils with heavy metals (HM) has become a common phenomenon throughout the world due to the increase in the scale of anthropogenic activities. The plants growing on HM-contaminated soils are characterized by a decreased growth rates and yields (Ma et al. 2016). Some of the direct toxic effects caused by high concentrations of HM include inhibition of cytoplasmic enzymes and damage to cellular structures due to the development of oxidative stress (Jadia and Fulekar 2009). An example of an indirect toxic effect is the replacement by HM of cations in the functional groups of various bioorganic compounds (Sytar et al. 2016). Moreover, the negative influence of HM on the activity of soil microorganisms can indirectly affect plant growth as well. The decrease in the number of beneficial soil microorganisms due to the high concentration of metal can lead to a decrease in the decomposition of organic matter, which in turn can cause a decrease in the content of certain nutrients in the soil. PGPB, including *B. subtilis*, are able to also grow in HM-contaminated environment and protect plants against toxicity of HM (Hao et al. 2015; Chaudhary and Khan 2018) (Table 6.1). Beneficial microorganisms have different mechanisms for the HM tolerance formation through mobilization, immobilization, and transformation to less toxic forms. Such mechanisms include exclusion, extrusion, active removal, biotransformation, biosorption, precipitation, or bioaccumulation of metals both in external and intracellular spaces of plants (Ma et al. 2016). These processes can influence the solubility and bioavailability of a metal for the plant, thereby altering its toxic effect (Rajkumar et al. 2008). Naseem et al. (2016) demonstrated that inoculation of wheat with *Bacillus* sp. AMP2 caused reduction in the Cr uptake of seedlings both at 10 and 20 μg mL^{-1} different chromium salts ($CrCl_3$, K_2CrO_4, and $K_2Cr_2O_7$) when compared with non-inoculated plants. Moreover, increased levels of acid phosphatase and PA were recorded in bacterial inoculated plants when compared to the control (Naseem et al. 2016). In another study, inoculation of wheat plants with *B. subtilis* QM3 promoted their growth in the presence of Pb (Hao et al. 2015). Herewith, in plant tissues increasing the activity of such antioxidant enzymes as CAT, PA, APX, and SOD decreased the concentration of metal in the roots in comparison with non-inoculated control plants (Hao et al. 2015). It was shown that endophytic strains of *B. subtilis* 26D and 11VM improve *S. alba* tolerance to the toxic effect of Cd and Ni and reduce manifestation of oxidative stress in the presence of higher levels of metal ions in the aboveground part of plants (Kuramshina et al. 2018). Treesubsuntorn et al. (2017) showed that *B. subtilis* inoculation of rice (*Oryza sativa* L.) can highly reduce Cd accumulation in every part of rice roots and shoots (45 days) and grains (120 days). *B. subtilis* can effectively absorb Cd compared to *B. cereus*, which might be the main mechanism to reduce Cd transportation in rice plants. Interestingly, plants that were inoculated with bacterial species individually harbored higher calcium (Ca) and magnesium (Mg) accumulation; *B. subtilis*-inoculated plants had the highest levels of Ca and Mg compared to plants inoculated with *B. cereus*. Moreover, *B. subtilis* could increase the DW of the rice plant and protect them from Cd stress due to the ability to produce IAA, solubilize phosphate, and control ethylene levels by ACC deaminase activity (Treesubsuntorn et al. 2017).

6.3.4 Variable Abiotic Stresses

During the growing season, plants can be subjected to one or various sequences of environmental stress factors such as fluctuating air temperature, relative humidity, and winter and spring frosts, which can all seriously influence the plant biological parameters and crop productivity. Belyaev et al. (2017) observed the effects of the preplanting treatment of primocane fruiting raspberry root system with *Bacillus* strains (*B. subtilis* RCAM B-10641, *B. amyloliquefaciens* RCAM B-10642, and *B. licheniformis* RCAM B-10562) on plant growth and crop productivity in 2012–2015 growing seasons which differed by environmental conditions (Western Siberia, Russia). It was revealed that preplanting treatment by bacterial strains increased the number of new raspberry canes, generative organs, and crop productivity under variable environmental factors. Moreover, the *Bacillus* strains acted as the standard humic fertilizer. Inoculation by *Bacillus* sp. including *B. subtilis* RCAM B-10641 decreased the negative effects of abiotic stresses on plants in all years of the research. It is important to note that from all investigated strains, *B. subtilis* RCAM B-10641 was shown to have the best results in adaptation of primocane fruiting raspberry plants to variable environmental factors in Western Siberia (Belyaev et al. 2017). Ahmad et al. (2017) demonstrated that different genes associated with plants tolerance to heat (elongation factor Tu; aspartokinase II; and dihydroorotase, pyrC), salinity (cardiolipin synthase, ywiE; glutaminase-1, ybgJ; phosphoglyceratemutase, gpmI; 4-hydroxy-3-methylbut-2-enyl diphosphate reductase, ispH and glutamate synthase (NADPH/NADH), gltA), drought (cystathionine beta-lyase and cysteine desulfhydrase), and cold stresses (sporulation cortex protein, coxA) were differentially expressed in *B. subtilis* 330-2 (Table 6.1). Interestingly, NADH-dependent butanol dehydrogenase A (yugJ) and formate dehydrogenase (yjgC) are commonly involved in tolerance to heat, salinity, and oxidative stresses. Likewise, the glutamate symporter (gltT) and S-adenosylmethionine synthase (metK) are commonly involved in salinity and drought stress. These results clearly suggest the beneficial role of endophytic *B. subtilis* 330-2 in enhancing the rice and maize plant's tolerance to different abiotic stresses.

6.4 The Potential of *B. subtilis* in Reducing Postharvest Losses

Abiotic stresses exert negative impacts on plants not only during the growing season but also during postharvest handling and storage. According to the FAO (2015) about 30% of cereals, 45% of fruits and vegetables, and 45% of roots and tubers are lost (FAO 2015). Postharvest food loss is substantial qualitative and quantitative loss of food through the supply chain, from harvesting to its consumption or other end uses (Hodges et al. 2010). Food loss is highest during storage due to pathogens (bacteria, molds, and insects), environmental conditions (rain, humidity, heat, and frost), sprouting, and quality loss (water loss, saccharification, etc.) (Buchholz et al. 2018). Recent studies suggest that *B. subtilis* has great potential to increase fruit (vegetables) set, quality, and tolerance against both postharvest biotic (diseases) and abiotic stresses such as temperature extremes, O_2 and CO_2 levels, mechanical injury associated with loading of product for transportation, transportation, unloading, throughout packaging, and processing lines (Miller 2003, Lastochkina et al. 2019). A number of reports indicate the positive effect of *B. subtilis* on postharvest physiology of various fruits and vegetables due to enhancing their resistance against postharvest diseases and unfavorable storage conditions which is manifested in extending their marketing life with maintaining nutritional qualities (Jiang et al. 2001, Lastochkina et al. 2019). The application of beneficial microbial inoculants that reduce adverse stress factors on the field could also have positive effects on crops during storage. For example, when peach fruits were treated with *B. subtilis* CF-3, the proportion of good quality fruits after storage for 36 days at 10 °C was 65%, which was more than 30% higher in comparison to the non-treated control group. The CF-3 strain was characterized by good UV, pH, and temperature stability (Gao et al. 2016). In another study, the application of *B. subtilis* prevented the fading of the *Litchi* fruits during storage (30 days at 5 °C) and maintained a high level of their consumer properties (in comparison to the control), specifically, by total dissolved solids content, titratable acidity, and ascorbic acid content. Herewith, the treatment with bacterial cells of *B. subtilis* did not alter the taste of fruit (Jiang et al. 2001). Despite the advantages of *B. subtilis* as eco-friendly strategy for preventing food losses during storage, the knowledge about the mechanisms of their action on postharvest physiology and preservation under abiotic stress are limited and require detailed future investigations.

6.5 Commercialization of *B. subtilis*

To date, several strains of *B. subtilis* are commercially available in the form of formulated products which are used as bio-fertilizers and plant protection biological products against biotic/abiotic stresses. For example, several *B. subtilis*-based products have recently been commercialized, including Quantum-400 (*B. subtilis* GB03), Serenade (*B. subtilis* QST713), Kodiak (*B. subtilis* improved GB03 strain), Subtilex (*B. subtilis* MBI600), Fitosporin-M (*B. subtilis* 26D), Bisolbisan (*B. subtilis* Ч-13), Baksis (*B. subtilis* 63-Z), Bactofit (*B. subtilis* ИПМ215), Alirin-B (*B. subtilis* B-10), Gamair (*B. subtilis* M-22), Vitaplan (*B. subtilis* BKM-B-2604D+ *B. subtilis* BKM-B-2605D), and others. Numerous studies have shown various stages in the process of commercialization: isolation of antagonist strains, screening, fermentation methods, mass production, formulation viability, toxicology, industrial relations, quality control, and field efficiency. The commercial success of beneficial strains requires economical and viable market demand, consistent and wide range of activities, safety and stability, a long shelf life, low costs, and easy availability of materials for a career (Nandakumar et al. 2001). Additionally, one of the main constraints to development and commercialization of plant protection products on the basis of *B. subtilis* is the lack of systemic research of the mechanisms of interactions in the systems "*B. subtilis* – host plant – environmental stress". Furthermore, the success in commercialization of perspective *B. subtilis* strains depends on the linkages between the scientific organizations and industries.

6.6 Conclusion and Future Perspectives

Summarizing the data available in the literature, we can conclude that *B. subtilis* plays an important role in regulation of plant growth, development, and tolerance to abiotic stresses which undoubtedly expands the significance of their practical application with the aim of increasing the yield and productivity of agricultural crops. For fully using the potential of *B. subtilis* as inoculants in sustainable agricultural productivity, it's extremely important to understand their underlying mechanisms in physiological action on host plants under different adverse environmental factors. Therefore, it is crucial to understand the mechanism of action of these bacteria in order to promote manufacturing of *B. subtilis*-based products which will support green technology in agriculture, increase crop productivity, and maintain long-term soil fertility and sustainability in a clean environment.

Acknowledgements This work was partially supported by RFBR (No. 19-016-00035) and RF Presidential Grants for Young Scientists (No. MK-643.2019.11).

References

Abd-Allah EF, Alqarawi AA, Hashem A, Radhakrishnan R, Al-Huqail AA, Al-Otibi FON, Malik JA, Alharbi RI, Egamberdieva D (2018) Endophytic bacterium *Bacillus subtilis* (BERA 71) improves salt tolerance in chickpea plants by regulating the plant defense mechanisms. J Plant Interact 13(1):37–44. https://doi.org/10.1080/17429145.2017.1414321

Ahmad Z, Wu J, Chen L, Dong W (2017) Isolated *Bacillus subtilis* strain 330-2 and its antagonistic genes identified by the removing PCR. Sci Rep 7(1):1777

Ahn I-P, Park K, Kim C-H (2002) Rhizobacteria-induced resistance perturbs viral disease progress and triggers defense-related gene expression. Mol Cell 13(2):302–308

Akram W, Anjum T, Ali B, Ahmad A (2013) Screening of native *Bacillus* strains to induce systemic resistance in tomato plants against *Fusarium* wilt in split root system and its field applications. Int J Agric Biol 15:1289–1294

Araus JL, Slafer GA, Reynolds MP, Royo C (2002) Plant breeding and drought in C3 cereals: what should we breed for? Ann Bot 89:925–940

Araus JL, Slafer GA, Royo C, Serret MD (2008) Breeding for yield potential and stress adaptation in cereals. Crit Rev Plant Sci 27(6):377–412

Arkhipova TN, Veselov SU, Melentiev AI, Martynenko EV, Kudoyarova GR (2005) Ability of bacterium *Bacillus subtilis* to produce cytokinins and to influence the growth and endogenous hormone content of lettuce plants. Plant Soil 272:201–209. https://doi.org/10.1007/s11104-004-5047-x

Arkhipova TN, Prinsen E, Veselov SU, Martinenko EV, Melentiev AI, Kudoyarova GR (2007) Cytokinin producing bacteria enhance plant growth in drying soil. Plant Soil 292:305–315. https://doi.org/10.1007/s11104-007-9233-5

Arya B, Komala BR, Sumalatha NT, Surendra GM, Gurumurthy PR (2018) PGPR induced systemic tolerance in plant. Int J Curr Microbiol Appl Sci 7:453–462

Asgari HR, Cornelis W, Van Damme P (2012) Salt stress effect on wheat (*Triticum aestivum* L.) growth and leaf ion concentrations. Int J Plant Prod 6(2):195–208

Ashraf M, Foolad MR (2007) Roles of glycine betaine and proline in improving plant abiotic stress resistance. Environ Exp Bot 59(2):206–216

Ashraf M, Hasnain S, Berge O, Mahmood T (2004) Inoculation wheat seedlings with exopolysaccharide-producing bacteria restricts sodium uptake and stimulates plant growth under salt stress. Biol Fertil Soils 40:157–162

Baez-Rogelio A, Morales-García YE, Quintero-Hernández V, Muñoz-Rojas J (2016) Next generation of microbial inoculants for agriculture and bioremediation. Microb Biotechnol 10(1). https://doi.org/10.1111/1751-7915.12448

Bais HP, Fall R, Vivanco JM (2004) Biocontrol of *Bacillus subtilis* against infection of *Arabidopsis* roots by *Pseudomonas* syringae is facilitated by biofilm formation and surfactin production. Plant Physiol 134:307–319

Barnawal D, Maji D, Bharti N, Chanotiya CS, Kalra A (2013) ACC deaminase-containing *Bacillus subtilis* reduces stress ethylene-induced damage and improves mycorrhizal colonization and rhizobial nodulation in *Trigonella foenum-graecum* under drought stress. J Plant Growth Regul 32(4):809–822

Barnawal D, Bharti N, Pandey SS, Pandey A, Chanotiya CS, Kalra A (2017) Plant growth-promoting rhizobacteria enhance wheat salt and drought stress tolerance by altering endogenous phytohormone levels and *TaCTR1/TaDREB2* expression. Physiol Plant 161(4):502–514. https://doi.org/10.1111/ppl.12614

Bashan Y, de-Bashan LE (2005) Bacteria/plant growth-promotion. In: Hillel D (ed) Encyclopedia of soils in the environment. Elsevier, Oxford

Belyaev AA, Shternshis MV, Chechenina NS, Shpatova TV, Lelyak AA (2017) Adaptation of primocane fruiting raspberry plants to environmental factors under the influence of *Bacillus strains* in Western Siberia. Environ Sci Pollut Res 24(8):7016–7022

Benedetto NA, Corbo MR, Campaniello D, Cataldi MP, Bevilacqua A, Sinigaglia M, Flagella Z (2017) The role of plant growth promoting bacteria in improving nitrogen use efficiency for sustainable crop production: a focus on wheat. AIMS Microbiol 3(3):413–434

Beneduzi A, Ambrosini A, Passaglia L (2012) Plant growth-promoting rhizobacteria (PGPR): their potential as antagonists and biocontrol agents. Genet Mol Biol 35(4):1044–1051

Berg G (2009) Plant-microbe interactions promoting plant growth and health: perspectives for controlled use of microorganisms in agriculture. Appl Microbiol Biotechnol 84(1):11–18

Bharti N, Yadav D, Barnawal D, Maji D, Kalra A (2013) Exiguo bacterium oxidotolerans, a halotolerant plant growth promoting rhizobacteria, improves yield and content of secondary metabolites in *Bacopa monnieri* (L.) Pennell under primary and secondary salt stress. World J Microbiol Biotechnol 29:379–387

Blagova DK, Sarvarova ER, Khairullin RM (2014) Isolation and characterization of bacterial carrot endophytes (*Daucus carota* L. var. Sativus). Bull Orenburg State Univ 174(13):10–12

Bloemberg GV, Lugtenberg BJ (2001) Molecular basis of plant growth promotion and biocontrol by rhizobacteria. Curr Opin Plant Biol 4(4):343–350

Bochow H, El-Sayed SF, Junge H, Stavropoulou A, Schmiedeknecht G (2001) Use of *Bacillus subtilis* as biocontrol agent. IV. Salt-stress tolerance induction by *Bacillus subtilis* FZB24 seed treatment in tropical vegetable field crops, and its mode of action. Z Pflanzenkrankh. Pflanzenschutz 108(1):21–30

Bottini R, Cassan F, Picolli P (2004) Gibberellin production by bacteria and its involvement in plant growth promotion. Appl Microbiol Biotechnol 65:497–503

Bowen GD, Rovira AD (1999) The rhizosphere and its management to improve plant growth. Adv Agron 66:1–102

Buchholz F, Kostic T, Sessitsch A, Mitter B (2018) The potential of plant microbiota in reducing post-harvest food loss. Microb Biotechnol 11(6):971–975. https://doi.org/10.1111/1751-7915.13252

Burg SP (1973) Ethylene in plant growth. Proc Natl Acad Sci U S A 70:591–597

Burrell M, Hanfrey CC, Murray EJ, Stanley-Wall NR, Michael AJ (2010) Evolution and multiplicity of arginine decarboxylases in polyamine biosynthesis and essential role in *Bacillus subtilis* biofilm formation. J Biol Chem 285:39224–39238

Çakmakçı R, Turan M, Kıtır N, Güneş A, Nikerel E, Özdemir BS, Yıldırım E, Olgun M, Topçuoğlu B, Tüfenkçi Ş, Karaman MR, Tarhan L, Mokhtari NEP (2017) The role of soil beneficial bacteria in wheat production: a review. Agricultural and biological sciences "wheat improvement, management and utilization". https://doi.org/10.5772/67274

Chaudhary K, Khan S (2018) Role of plant growth promoting bacteria (PGPB) for bioremediation of heavy metals: an overview. Biostimulation Remediation Technologies for Groundwater Contaminants 22. https://doi.org/10.4018/978-1-5225-4162-2.ch006

Chen J-F, Gallie DR (2010) Analysis of the functional conservation of ethylene receptors between maize and *Arabidopsis*. Plant Mol Biol 74(4-5):405–421. https://doi.org/10.1007/s11103-010-9686-4

Chen H, Jiang J-G (2010) Osmotic adjustment and plant adaptation to environmental changes related to drought and salinity. Environ Rev 18:309–319

Chen YF, Randlett MD, Findell JL, Schaller GE (2002) Localization of the ethylene receptor ETR1 to the endoplasmic reticulum of *Arabidopsis*. J Biol Chem 277:19861–19866. https://doi.org/10.1074/jbc.M201286200

Chen Y, Yan F, Chai Y, Liu H, Kolter R, Losick R, Guo JH (2012) Biocontrol of tomato wilt disease by *Bacillus subtilis* isolates from natural environments depends on conserved genes mediating biofilm formation. Environ Microbiol 15:848–864

Cherif H, Marasco R, Rolli E, Ferjani R, Fusi M, Soussi A et al (2015) Oasis desert farming selects environment-specific date palm root endophytic communities and cultivable bacteria that promote resistance to drought. Environ Microbiol Rep 7:668–678

Chourdhary D, Johri B (2009) Interactions of *Bacillus* spp. and plants – with special reference to induced systemic resistance (ISR). Microbiol Res 164:493–513

Chowdappa P, Kumar SM, Lakshmi MJ et al (2013) Growth stimulation and induction of systemic resistance in tomato against early and late blight by *Bacillus subtilis* OTPB1 or *Trichoderma harzianum* OTPB3. Biol Control 65(1):109–117

Chung S, Kong H, Buyer JS, Lakshman DK, Lydon J, Kim SD, Roberts DP (2008) Isolation and partial characterization of *Bacillus subtilis* ME488 for suppression of soilborne pathogens of cucumber and pepper. Appl Microbiol Biotechnol 80(1):115–123

Cramer GR, Urano K, Delrot S, Pezzotti M, Shinozaki K (2011) Effects of abiotic stress on plants: a systems biology perspective. BMC Plant Biol 11:163. https://doi.org/10.1186/1471-2229-11-163

Dandekar AM (2000) Genetic engineering to improve quality, productivity and value of crops. Calif Agric 54(4):49–56. https://doi.org/10.3733/ca.v054n04p49

De Jensen CE, Meronuck R, Percich JA (2000) Efficacy of *Bacillus subtilis* and two rhizobium strains for the management of bean root rot in Minnesota. Annu Rep Bean Improv Coop 43:33–34

Dimkpa C, Weinand T, Asch F (2009) Plant rhizobacteria interactions alleviate abiotic stress conditions. Plant Cell Environ 32(12):1682–1694

Dodd IC, Zinovkina NY, Safronova VI, Belimov AA (2010) Rhizobacterial mediation of plant hormone status. Ann Appl Biol 57:361–379

Egamberdieva D, Kucharova Z (2009) Selection for root colonising bacteria stimulating wheat growth in saline soils. Biol Fertil Soils 45(6):563–571

Egamberdieva D, Wirth SJ, Shurigin VV, Hashem A, Abd Allah EF (2017) Endophytic bacteria improve plant growth, symbiotic performance of chickpea (*Cicer arietinum* L.) and induce suppression of root rot caused by *Fusarium solani* under salt stress. Front Microbiol 8:1887. https://doi.org/10.3389/fmicb.2017.01887

FAO (2014). Available from: http://www.fao.org/news/story/en/item/273303/icode

FAO (2015) Food losses and waste. URL http://www.fao.org/food-loss-and-food-waste/en/

FAO (2016) Available online: http://www.fao.org/3/a-i6030e.pdf (2016)

Forchetti G, Masciarelli O, Alemano S, Alvarez D, Abdala G (2007) Endophytic bacteria in sunflower (*Helianthus annuus* L.) isolation, characterization, and production of jasmonates and abscisic acid in culture medium. Appl Microbiol Biotechnol 76:1145–1152

Francis I, Holsters M, Vereecke D (2010) The gram-positive side of plant-microbe interactions. Environ Microbiol 12:1–12. https://doi.org/10.1111/j.1462-2920.2009.01989.x

Fravel D (2005) Commercialization and implementation of biocontrol. Annu Rev Phytopathol 43:337–359

Gagne-Bourgue F, Aliferis KA, Seguin P, Rani M, Samson R, Jabaji S (2013) Isolation and characterization of indigenous endophytic bacteria associated with leaves of switchgrass (*Panicum virgatum* L.) cultivars. J Appl Microbiol 114(3):836–853

Gagné-Bourque F, Mayer BF, Charron JB, Vali H, Bertrand A, Jabaji S (2015) Accelerated growth rate and increased drought stress resilience of the model grass *Brachypodium distachyon* colonized by *Bacillus subtilis* B26. PLoS One 10(6):e0130456. https://doi.org/10.1371/journal.pone.0130456

Gagné-Bourque F, Bertrand A, Claessens A, Aliferis KA, Jabaji S (2016) Alleviation of drought stress and metabolic changes in timothy (*Phleum pratense* L.) colonized with *Bacillus subtilis* B26. Front Plant Sci 7:584. https://doi.org/10.3389/fpls.2016.00584

Gao H, Xu X, Dai Y, He H (2016) Isolation, identification and characterization of *Bacillus subtilis* CF-3, a bacterium from fermented bean curd for controlling postharvest diseases of peach fruit. Food Sci Technol Res 22(3):377–385. https://doi.org/10.3136/fstr.22.377

García-Gutiérrez L, Zeriouh H, Romero D, Cubero J, Vicente A, Pérez-García A (2013) The antagonistic strain *Bacillus subtilis* UMAF6639 also confers protection to melon plants against cucurbit powdery mildew by activation of jasmonate – and salicylic acid-dependent defense responses. Microb Biotechnol 6(3):264–274

Glick BR (2014) Bacteria with ACC deaminase can promote plant growth and help to feed the world. Microbiol Res 169(1):30–39

Glick BR, Todorovic B, Czarny J, Cheng Z, Duan J, McConkey B (2007) Promotion of plant growth by bacterial ACC deaminase. Crit Rev Plant Sci 26:227–242. https://doi.org/10.1080/07352680701572966

Grichko VP, Glick BR (2001) Amelioration of flooding stress by ACC deaminase containing plant growth-promoting bacteria. Plant Physiol Biochem 39(1):11–17

Gupta V, Bochow H, Dolej S et al (2000) Plant growth-promoting *Bacillus subtilis* strain as potential inducer of systemic resistance in tomato against *Fusarium* wilt. J Plant Dis Prot 107:145–154

Haggag WM, Tawfik MM, Abouziena HF, Abd El Wahed MSA, Ali RR (2017) Enhancing wheat production under arid climate stresses using bio-elicitors. Gesunde Pflanzen 69(3):149–158

Han HS, Lee KD (2005) Physiological responses of soybean-inoculation of Bradyrhizobium japonicum with PGPR in saline soil conditions. Res J Agric Biol Sci 1(3):216–221

Hao Y, Wu H, Liu Y, Hu Q (2015) Mitigative effect of *Bacillus subtilis* QM3 on root morphology and resistance enzyme activity of wheat root under lead stress. Adv Microbiol 5:469–478

Hare PD, Cress WA, Van Staden J (1998) Dissecting the roles of osmolyte accumulation during stress. Plant Cell Environ 21:535–553

Hasegawa PM, Bressan RA, Zhu JK, Bohnert HJ (2000) Plant cellular and molecular responses to high salinity. Annu Rev Plant Physiol Plant Mol Biol 51:463–499. https://doi.org/10.1146/annurev.arplant.51.1.463

Hashem A, Abd Allah EF, Alqarawi A, Al-Huqail AA, Wirth S, Egamberdieva D (2016) The interaction between arbuscular mycorrhizal fungi and endophytic bacteria enhances plant growth of *Acacia gerrardii* under salt stress. Front Plant Sci 7:1089. https://doi.org/10.3389/fmicb.2016.01089

Hodges RJ, Buzby JC, Bennett B (2010) Postharvest losses and waste in developed and less developed countries: opportunities to improve resource use. J Agric Sci 149:37–45

Huang D, Wu W, Abrams SR, Cutler AJ (2008) The relationship of drought-related gene expression in *Arabidopsis thaliana* to hormonal and environmental factors. J Exp Bot 59:2991–3007

Hummel I, Couée I, El Amrani A, Martin-Tanguy J, Hennion F (2002) Involvement of polyamines in root development at low temperature in the subantarctic cruciferous species *Pringlea antiscorbutica*. J Exp Bot 53:1463–1473

Imai A, Matsuyama T, Hanzawa Y, Akiyama T, Tamaoki M, Saji H, Shirano Y, Kato T, Hayashi H, Shibata D (2004) Spermidine synthase genes are essential for survival of *Arabidopsis*. Plant Physiol 135:1565–1573

Ishak Z, Mohd Iswadi MK, Russman Nizam AH, Ahmad Kamil MJ, Ernie Eileen RR, Wan Syaidatul A, Ainon H (2016) Plant growth hormones produced by endophytic *Bacillus subtilis* strain LKM-BK isolated from cocoa. Malays Cocoa J 9(1):127–133

Jadia CD, Fulekar MH (2009) Phytoremediation of heavy metals: recent techniques. Afr J Biotechnol 8(6):921–928

Jayaraj J, Anand A, Muthukrishnan S (2004a) Pathogenesis-related proteins and their roles in resistance to fungal pathogens. Fungal disease resistance in pants: biochemistry, molecular biology, and genetic engineering. Haworth Press, New York

Jayaraj J, Yi H, Liang G et al (2004b) Foliar application of *Bacillus subtilis* AUBS1 reduces sheath blight and triggers defense mechanisms in rice. J Plant Dis Prot 111(2):115–125

Jha B, Gontia I, Hartmann A (2012) The roots of the halophyte *Salicornia brachiata* are a source of new halotolerant diazotrophic bacteria with plant growth-promoting potential. Plant Soil 356:265–277

Jiang YM, Chen F, Li YB, Liu SX (2001) A preliminary study on the biological control of postharvest diseases of Litchi fruit. J Fruit Sci 14(3):185–186

Karaca U, Sabir A (2017) Sustainable mitigation of alkaline stress in grapevine rootstocks (vitis spp.) by plant growth-promoting rhizobacteria. Erwerbs-Obstbau:1–10. https://doi.org/10.1007/s10341-017-0361-7

Karimi K, Amini J, Harighi B, Bahramnejad B (2012) Evaluation of biocontrol potential of '*pseudomonas*' and '*bacillus*' spp. against *fusarium* wilt of chickpea. Aust J Crop Sci 6(4):695

Karlidag H, Esitken A, Yildirim E, Donmez MF, Turan M (2010) Effects of plant growth promoting bacteria (PGPB) on yield, growth, leaf water content, membrane permeability andionic composition of strawberry under saline conditions. J Plant Nutr 34:34–46

Karlidag H, Yildirim E, Turan M, Pehluvan M, Donmez F (2013) Plant growth-promoting rhizobacteria mitigate deleterious effects of salt stress on strawberry plants (*Fragaria*×*ananassa*). Hortscience 48(5):563–567

Kaushal M, Wani SP (2015) Plant-growth-promoting rhizobacteria: drought stress alleviators to ameliorate crop production in drylands. Ann Microbiol:1–8

Key S, Ma JK-C, Drake PMW (2008) Genetically modified plants and human health. J R Soc Med 101(6):290–298. https://doi.org/10.1258/jrsm.2008.070372

Koca H, Ozdemir F, Turkan I (2006) Effect of salt stress on lipid peroxidation and superoxide dismutase and peroxidase activities of *Lycopersicon esculentum* and *L. pennellii*. Biol Plant 50:745–748

Koornneef A, Pieterse CMJ (2008) Cross talk in defense signaling. Plant Physiol 146:839–844

Krasensky J, Jonak C (2012) Drought, salt, and temperature stress-induced metabolic rearrangements and regulatory networks. J Exp Bot 63(4):1593–1608

Kuklinski-Sorbal J, Araujo WL, Mendes R, Geraldi IO, Pizzirani-Kleiner AA, Azevedo JL (2004) Isolation and characterisation of soybean-associated bacteria and their potential for plant growth promotion. Environ Microbiol 6(12):1244–1251

Kuramshina ZM, Smirnova YV, Khairullin RM (2018) Cadmium and nickel toxicity for sinapis alba plants inoculated with endophytic strains of *Bacillus subtilis*. Russ J Plant Physiol 65(2):269–277

Kusano T, Berberich T, Tateda C, Takahashi Y (2008) Polyamines: essential factors for growth and survival. Plant 228:367–381

Kwon YS, Ryu C-M, Lee S, Park HB, Han KS, Lee JH et al (2010) Proteome analysis of *Arabidopsis* seedlings exposed to bacterial volatiles. Planta 232:1355–1370. https://doi.org/10.1007/s00425-010-1259-x

Lastochkina O, Pusenkova L, Yuldashev R, Babaev M, Garipova S, Blagova D, Khairullin R, Aliniaeifard S (2017a) Effects of *Bacillus subtilis* on some physiological and biochemical parameters of Triticum aestivum L. (wheat) under salinity. Plant Physiol Biochem 121:80–88

Lastochkina OV, Yuldashev RA, Pusenkova LI (2017b) The effect of bacteria *Bacillus subtilis* 26D on drought resistance of soft spring wheat cultivars of the West Siberian forest-steppe and Volga steppe ecotypes in the early stages of ontogenesis. Proc RAS Ufa Sci Cent 3(1):99–102

Lastochkina O, Seifikalhor M, Aliniaeifard S, Baymiev A, Pusenkova L, Garipova S, Kulabuhova D, Maksimov I (2019) Bacillus spp.: efficient biotic strategy to control postharvest diseases of fruits and vegetables. Plants 8(4):97

Leelasuphakul W, Sivanunsakul P, Phongpaichit S (2006) Purification, characterization and synergistic activity of β-1,3-glucanase and antibiotic extract from an antagonistic *Bacillus subtilis* NSRS 89-24 against rice blast and sheath blight. Enzym Microb Technol 38(7):990–997

Lemfack MC, Nickel J, Dunkel M, Preissner R, Piechulla B (2014) mVOC: a database of microbial volatiles. Nucleic Acids Res 42:D744–D748. https://doi.org/10.1093/nar/gkt1250

Liu W, Chen D, Yin H, Song M, Guo S (2010) *Bacillus subtilis* for salt stress relief in vegetable cultivation. Acta Hortic 856:237–242. https://doi.org/10.17660/ActaHortic.2010.856.33

Liu F, Xing S, Ma H, Du Z, Ma B (2013) Cytokinin-producing, plant growth-promoting rhizobacteria that confer resistance to drought stress in *Platycladus orientalis* container seedlings. Appl Microbiol Biotechnol 97(20):9155–9164. https://doi.org/10.1007/s00253-013-5193-2

Livingston DP, Hincha DK, Heyer AG (2009) Cell fructan and its relationship to abiotic stress tolerance in plants. Mol Life Sci 66(13):2007–2023. https://doi.org/10.1007/s00018-009-0002-x

Ma Y (2017) Beneficial bacteria for disease suppression and plant growth promotion. Plant-microbe interactions in agro-ecological perspectives. Springer, Singapore. https://doi.org/10.1007/978-981-10-5813-4_26

Ma Y, Rajkumar M, Zhang C, Freitas H (2016) Beneficial role of bacterial endophytes in heavy metal phytoremediation. J Environ Manag 174:14–25

Maksimov I, Khairullin R (2016) The role of Bacillus bacterium in formation of plant defence: mechanisms and reactions. The handbook of microbial bioresources.

Maksimov IV, Veselova SV, Nuzhnaya TV, Sarvarova ER, Khairullin RM (2015) Plant growth promoting bacteria in regulation of plant resistance to stress factors. Russ J Plant Physiol 62(6):715–726

Martins SJ, Rocha GA, de Melo HC, de Castro Georg R, Ulhôa CJ, de Campos Dianese É, Oshiquiri LH, da Cunha MG, da Rocha MR, de Araújo LG, Vaz KS, Dunlap CA (2018) Plant-associated bacteria mitigate drought stress in soybean. Environ Sci Pollut Res 25(14):13676–13686. https://doi.org/10.1007/s11356-018-1610-5

Miller AR (2003) Harvest and handling injury: physiology, biochemistry, and detection. In: Postharvest physiology and pathology of vegetables. Marcel Dekker Inc., New York

Mohamed HI, Gomaa EZ (2012) Effect of plant growth promoting *Bacillus subtilis* and Pseudomonas fluorescens on growth and pigment composition of radish plants (*Raphanus sativus*) under NaCl stress. Photosynthetica 50(2):263–272. https://doi.org/10.1007/s11099-012-0032-8

Moon S, Asif R, Basharat A (2017) Phylogenetic diversity of drought tolerant *Bacillus* spp. and their growth stimulation of *Zea mays* L. under different water regimes. Res J Biotechnol 12(10):38–46

Munns R, Tester M (2008) Mechanisms of salinity tolerance. Annu Rev Plant Biol 59:651–681

Nandakumar R, Babu S, Viswanathan R, Sheela J, Raguchander T (2001) A new bio-formulation containing plant growth promoting rhizobacterial mixture for the management of sheath blight and enhanced grain yield in rice. Biocontrol 46(4):493–510

Naseem S, Yasin M, Faisal M, Ahmed A (2016) Comparative study of plant growth promoting bacteria in minimizing toxic effects of chromium on growth and metabolic activities in wheat (*Triticum aestivum*). J Chem Soc Pak 38(3):509–516

Niu DD, Liu HX, Jiang CH, Wang YP, Wang QY, Jin HL, Guo JH (2011) The plant growth-promoting rhizobacterium *Bacillus cereus* AR156 induces systemic resistance in *Arabidopsis thaliana* by simultaneously activating salicylate- and jasmonate/ethylene-dependent signaling pathways. Mol Plant Microbe Interact 24(5):533–542

Numan M, Bashir S, Khan Y, Mumtaz R, Shinwari ZK, Khan AL, Khan A, AL-Harrasi A (2018) Plant growth promoting bacteria as an alternative strategy for salt tolerance in plants: a review. Microbiol Res 209:21–32. https://doi.org/10.1016/j.micres.2018.02.003

Oliveira ALM, Urquiaga S, Baldani JI (2003) Processos e mecanismos envolvidos na influência de microrganismos sobre o crescimento vegetal. Embrapa Agrobiologia Documentos 161:1–5

Ongena M, Jourdan E, Adam A, Paquot M, Brans A, Joris B, Arpigny JL, Thonart P (2007) Surfactin and fengycin lipopeptides of *Bacillus subtilis* as elicitors of induced systemic resistance in plants. Environ Microbiol 9:1084–1090

Ortíz-Castro R, Contreras-Cornejo HA, Macías-Rodríguez L, López-Bucio J (2014) The role of microbial signals in plant growth and development. Plant Signal Behav 4:701–712. https://doi.org/10.4161/psb.4.8.9047

Pandey PK, Singh MC, Singh SS, Kumar AK, Pathak MM, Shakywar RC, Pandey AK (2017) Inside the plants: endophytic bacteria and their functional attributes for plant growth promotion. Int J Curr Microbiol App Sci 6(2):11–21

Paul MJ, Primavesi LF, Jhurreea D, Zhang Y (2008) Trehalose metabolism and signaling. Annu Rev Plant Biol 59:417–441

Pereira A (2016) Plant abiotic stress challenges from the changing environment. Front Plant Sci 7:1123

Pérez-García A, Romero D, Vicente A (2011) Plant protection and growth stimulation by microorganisms: biotechnological applications of *Bacilli* in agriculture. Curr Opin Biotechnol 22:187–193

Potters G, Pasternak TP, Guisez Y, Palme KJ, Jansen MAK (2007) Stress-induced morphogenic responses: growing out of trouble. Trends Plant Sci 12:98–105

Pieterse CM, Van der Does D, Zamioudis C, Leon-Reyes A, Van Wees SC (2012) Hormonal modulation of plant immunity. Annu Rev Cell Dev Biol 28:489–521

Pieterse CMJ, Zamioudis C, Berendsen RL, Weller D, Van Wees SCM, Bakker PAHM (2014) Induced systemic resistance by beneficial microbes. Annu Rev Phytopathol 52:347–375

Pusenkova LI, Il'yasova EY, Maksimov IV, Lastochkina OV (2015) Enhancement of adaptive capacity of sugar beet crops by microbial biopreparations under biotic and abiotic stresses. Agric Biol 50(1):115–123

Pusenkova LI, Il'yasova EY, Lastochkina OV, Maksimov IV, Leonova SA (2016) Changes in the species composition of the rhizosphere and phyllosphere of sugar beet under the impact of biological preparations based on endophytic bacteria and their metabolites. Eurasian Soil Sci 49(10):1136–1144

Raaijmakers JM, De Bruijn I, Nybroe O, Ongena M (2010) Natural functions of lipopeptides from Bacillus and Pseudomonas: more than surfactants and antibiotics. FEMS Microbiol Rev 34(6):1037–1062

Rajkumar M, Ma Y, Freitas H (2008) Characterization of metal resistant plant-growth promoting *Bacillus weihenstephanensis* isolated from serpentine soil in Portugal. J Basic Microbiol 48:500–508

Ramesh A, Sharma SK, Sharma MP et al (2014) Inoculation of zinc solubilizing *Bacillus aryabhattai* strains for improved growth, mobilization and biofortification of zinc in soybean and wheat cultivated in vertisols of central India. Appl Soil Ecol 73:87–96

Ramyabharathi S, Meena B, Raguchander T (2012) Induction of chitinase and β-1,3-glucanase PR proteins in tomato through liquid formulated *Bacillus subtilis* EPCO 16 against *Fusarium* wilt. J Today Biol Sci Res Rev 1(1):50–60

Reid MS (1981) The role of ethylene in flower senescene. Acta Hortic 261:157–169

Rekha K, Baskar B, Srinath S, Usha B (2018) Plant-growth-promoting rhizobacteria *Bacillus subtilis* RR4 isolated from rice rhizosphere induces malic acid biosynthesis in rice roots. Can J Microbiol 64(1):20–27. https://doi.org/10.1139/cjm-2017-0409

Rojas-Tapias D, Moreno-Galván A, Pardo-Díaz S, Obando M, Rivera S, Bonilla R (2012) Effect of inoculation with plant growth promoting bacteria (PGPB) on amelioration of saline stress in maize (*Zea mays*). Appl Soil Ecol 61:264–272

Romero D, de Vicente A, Rakotoaly RH, Dufour SE, Veening J-W, Arrebola E, Cazorla FM, Kuipers OP, Paquot M, PérezGarcía A (2007) The iturin and fengycin families of lipopeptides are key factors in antagonism of *Bacillus subtilis* toward *Podosphaera fusca*. Mol Plant-Microbe Interact 20:430–440

Ruan YL, Jin Y, Yang YJ, Li GJ, Boyer JS (2010) Sugar input, metabolism, and signaling mediated by invertase: roles in development, yield potential, and response to drought and heat. Mol Plant 3(6):942–955

Rudrappa T, Czymmek KJ, Paré PW, Bais HP (2008) Root secreted malic acid recruits beneficial soil bacteria. Plant Physiol 148:1547–1556

Ryals JA, Neuenschwander UH, Willits MG, Molina A, Steiner HY, Hunt MD (1996) Systemic acquired resistance. Plant Cell 8(10):1809–1819

Ryu C-M, Farag MA, Hu C-H, Reddy MS, Wei H-X, Pare PW et al (2003) Bacterial volatiles promote growth in Arabidopsis. Proc Natl Acad Sci U S A 100:4927–4932. https://doi.org/10.1073/pnas.0730845100

Ryu CM, Mohamed AF, Hu CH, Munagala SR, Kloepper JW, Pare PW (2004) Bacterial volatiles induce systemic resistance in *Arabidopsis*. Plant Physiol 134:1017–1026

Saikia J, Sarma RK, Dhandia R, Yadav A, Bharali R, Gupta VK, Saikia R (2018) Alleviation of drought stress in pulse crops with ACC deaminase producing rhizobacteria isolated from acidic soil of Northeast India. Sci Rep 8:3560. https://doi.org/10.1038/s41598-018-21921-w

Sailaja P, Podile A, Reddanna P (1998) Biocontrol strain of *Bacillus subtilis* AF 1 rapidly induces lipoxygenase in groundnut (*Arachis hypogaea* L.) compared to crown rot pathogen *Aspergillus niger*. Eur J Plant Pathol 104(2):125–132

Saleh SA, Heuberger H, Schnitzler WH (2005) Alleviation of salinity effect on artichoke productivity by *Bacillus subtilis* FZB24, supplemental Ca and micronutrients. J Appl Bot Food Qual 79:24–32

Sandhya V, Shaik ZA, Minakshi G, Gopal R, Venkateswarlu B (2011) Drought-tolerant plant growth promoting *Bacillus* spp.: effect on growth, osmolytes, and antioxidant status of maize under drought stress. J Plant Interact 6(1):1–14

Sarma BK, Yadav KS, Singh DP, Singh HB (2018) Rhizobacteria mediated induced systemic tolerance in plants: prospects for abiotic stress management. Bact Agrobiol Stress Manag:225–238

Sayed SA, Atef AS, Soha E (2011) Response of three sweet basil cultivars to inoculation with *Bacillus subtilis* and arbuscular mycorrhizal fungi under salt stress conditions. Nat Sci 9(6):31–36

Sessitsch A, Hardoim P, Döring J, Weilharter A, Krause A, Woyke T, Mitter B, Hauberg-Lotte L, Friedrich F, Rahalkar M, Hurek T, Sarkar A, Bodrossy L, van Overbeek L, Brar D, van Elsas JD, Reinhold-Hurek B (2012) Functional characteristics of an endophyte community colonizing rice roots as revealed by metagenomic analysis. MPMI 25(1):28–36

Shafi O, Tian H, Ji M (2017) *Bacillus* species as versatile weapons for plant pathogens: a review. Biotechnol Biotechnol Equip 31(3):446–459. https://doi.org/10.1080/13102818.2017.1286950

Shahid M, Khan MS (2018) Cellular destruction, phytohormones and growth modulating enzymes production by *Bacillus subtilis* strain BC8 impacted by fungicides. Pestic Biochem Physiol 149:8–19

Shakirova FM, Avalbaev AM, Bezrukova MV, Fatkhutdinova RA, Maslennikova DR, Yuldashev RA, Allagulova CR, Lastochkina OV (2012) Phytohormones and abiotic stress tolerance in plants. Springer, Berlin/Heidelberg

Sivasakthi S, Kanchana D, Usharani G, Saranraj P (2013) Production of plant growth promoting substance by *Pseudomonas fluorescens* and *Bacillus subtilis* isolated from paddy rhizosphere soil of Cuddalore district, Tamil Nadu, India. Int J Microbiol Res 4(3):227–233

Soleimanzadeh N, Soleimanzadeh H (2010) Effect of VA-mycorrhiza on growth and yield of sunflower (*Helianthus annuus* L.) at different phosphorus level. World Acad Sci Eng Technol 71:414–417

Sorty AM, Meena KK, Choudhary K, Bitla UM, Minhas PS, Krishnani KK (2016) Effect of plant growth promoting bacteria associated with halophytic weed (*Psoralea corylifolia* L.) on germination and seedling growth of wheat under saline conditions. Appl Biochem Biotechnol 180:872–882. https://doi.org/10.1007/s12010-016-2139-z

Sytar O, Brestic M, Taran N, Zivcak M (2016) Plants used for biomonitoring and phytoremediation of trace elements in soil and water. Plant metal interaction emerging remediation techniques. Elsevier.

Szabados L, Savoure A (2009) Proline: a multifunctional amino acid. Trends Plant Sci 15:89–97

Tahir HAS, Gu Q, Wu H, Raza W, Hanif A, Wu L, Colman MV, Gao X (2017) Plant growth promotion by volatile organic compounds produced by *Bacillus subtilis* SYST2. Front Microbiol 8:171. https://doi.org/10.3389/fmicb.2017.00171

Thilagavathi R, Saravanakumar D, Ragupathi N et al (2007) A combination of biocontrol agents improves the management of dry root rot (*Macrophomina phaseolina*) in greengram. Phytopathol Mediterr 46(2):157–167

Treesubsuntorn C, Dhurakit P, Khaksar G, Thiravetyan P (2017) Effect of microorganisms on reducing cadmium uptake and toxicity in rice (*Oryza sativa* L.). Environ Sci Pollut Res:1–12

Turan M, Gulluce M, Şahin F (2012) Effects of plant-growth-promoting rhizobacteria on yield, growth, and some physiological characteristics of wheat and barley plants. Commun Soil Sci Plant Anal 43(12):1658–1673

Turan M, Ekinci M, Yıldırım E, Güneş K, Karagöz K, Kotan R, Dursun A (2014) Plant growth-promoting rhizobacteria improved growth, nutrient, and hormone content in cabbage (*Brassica oleracea*) seedlings. Turk J Agric For 38:327–333

Upadhyay SK, Singh JS, Singh DP (2011) Exopolysaccharide-producing plant growth promoting rhizobacteria under salinity condition. Pedosphere 2:214–222

Upadhyay SK, Singh JS, Saxena AK, Singh DP (2012) Impact of PGPR inoculation on growth and antioxidant status of wheat under saline conditions. Plant Biol 14(4):605–611

Valliyodan B, Nguyen HT (2006) Understanding regulatory networks and engineering for enhanced drought tolerance in plants. Curr Opin Plant Biol 9(2):189–195. https://doi.org/10.1016/j.pbi.2006.01.019

Van Loon LC (2007) Plant responses to plant growth-promoting rhizobacteria. Eur J Plant Pathol 119:243–254

Vardharajula S, Zulfikar AS, Grover M, Reddy G, Bandi V (2011) Drought-tolerant plant growth promoting *Bacillus* spp., effect on growth, osmolytes, and antioxidant status of maize under drought stress. J Plant Interact 6:1–14

Veselova SV, Nuzhnaya TV, Maksimov IV (2015) Jasmonic acid: biosynthesis, functions and role in plant development. Nova Science Publishers, New York

Wang C-J, Yang W, Wang C, Gu C, Niu D-D, Liu H-X, Wang Y-P, Guo J-H (2012) Induction of drought tolerance in cucumber plants by a consortium of three plant growth-promoting rhizobacterium strains. PLoS One 7(12):e52565

Wani PA, Wahid S, Singh R, Kehinde AM (2018) Antioxidant and chromium reductase assisted chromium (VI) reduction and Cr (III) immobilization by the rhizospheric *Bacillus* helps in the remediation of Cr (VI) and growth promotion of soybean crop. Rhizosphere 6:23–30

Wenhao X, Zhao L, Xu X, Yonghua Q, Guanghui Y (2012) Mutual information flow between beneficial microorganisms and the roots of host plants determined the biofunctions of biofertilisers. Am J Plant Sci 3:1115–1120

Wilks JC, Kitko RD, Cleeton SH, Lee GE, Ugwu CS, Jones BD, BonDurant SS, Slonczewski JL (2009) Acid and base stress and transcriptomic responses in *Bacillus subtilis*. Appl Environ Microbiol 75:981–990

Xie X, Zhang H, Pare P (2009) Sustained growth promotion in arabidopsis with long-term exposure to the beneficial soil bacterium *Bacillus subtilis* (GB03). Plant Signal Behav 4:948–953. https://doi.org/10.4161/psb.4.10.9709

Xie S, Wu H, Zang H, Wu L, Zhu Q, Gao X (2014) Plant growth promotion by spermidine-producing *Bacillus subtilis* OKB105. Mol Plant-Microbe Interact 27:655–663. https://doi.org/10.1094/MPMI-01-14-0010-R

Xie S, Wu H, Chen L, Zang H, Xie Y, Gao X (2015) Transcriptome profiling of *Bacillus subtilis* OKB105 in response to rice seedlings. BMC Microbiol 15:21. https://doi.org/10.1186/s12866-015-0353-4

Yang J, Kloepper JW, Ryu CM (2009) Rhizosphere bacteria help plants tolerate abiotic stress. Trends Plant Sci 14:1–4

Yazici I, Tuerkan I, Sekmen AH, Demiral T (2007) Salinity tolerance of purslane (*Portulaca oleracea* L.) is achieved by enhanced antioxidative system, lower level of lipid peroxidation and proline accumulation. Environ Exp Bot 61:49–57

Yildirim E, Turan M, Ekinci M, Dursun A, Cakmakci R (2011) Plant growth promoting rhizobacteria ameliorate deleterious effect of salt stress on lettuce. Sci Res Essays 6:4389–4396

Zhang H, Kim M, Krishnamachari V, Payton P, Sun Y, Grimson M et al (2007) Rhizobacterial volatile emissions regulate auxin homeostasis and cell expansion in Arabidopsis. Planta 226:839–851. https://doi.org/10.1007/s00425-007-0530-2

Zhang H, Xie X, Kim MS, Kornyeyev DA, Holaday S, Paré PW (2008) Soil bacteria augment *Arabidopsis* photosynthesis by decreasing glucose sensing and abscisic acid levels in planta. Plant J 56:264–273. https://doi.org/10.1111/j.1365-313X.2008.03593.x

Zhang H, Murzello SY, Kim M-S, Xie X, Jeter RM, Zak JC, Dowd SE, Paré PW (2010) Choline and osmotic-stress tolerance induced in *Arabidopsis* by the soil microbe *Bacillus subtilis* (GB03). Mol Plant-Microbe Interact 23(8):1097–1104. https://doi.org/10.1094/MPMI-23-8-1097

Chapter 7
Exploring the Utility of *Aneurinibacillus* as a Bioinoculant for Sustainable Crop Production and Environmental Applications

Chandandeep Kaur, G. Selvakumar, and A. N. Ganeshamurthy

7.1 Introduction

Aerobic endospore-forming bacteria (AEFB) were mostly accommodated in the genera *Bacillus* and *Paenibacillus* in the past. But with subsequent taxonomic rearrangements, a number of new *Bacillus*-derived genera were proposed and have gained acceptance (Heyndrickx et al. 1997). The bacterium *Bacillus aneurinilyticus* was originally discovered as a thiamine decomposer by (Aoyama 1952) and later re-characterized based on its distinct phenotypic characteristics and chemotaxonomic profiles by (Shida et al. 1994). Subsequent to the reclassification of the *B. aneurinilyticus* cluster by (Shida et al. 1996), a novel genus, namely, *Aneurinibacillus* came into existence. The genus *Aneurinibacillus* is a Gram-positive aerobic endospore-forming bacterium belonging to phylum *Firmicutes*, class *Bacilli*, order *Bacillales* and family *Paenibacillaceae*. It is an aerobic bacterium with ellipsoidal spores, and the cells are motile by peritrichous flagella. The major menaquinone is MK-7; and the major cellular fatty acids are $C_{15:0\ iso}$ (41.9–66.8), $C_{17:0\ iso}$ (1–23.8), $C_{16:0}$ (1.8–8.5) and $C_{16:0\ iso}$ (0.5–6.6). The G + C content ranges from 42.9 to 46.7 mol% (Shida et al. 1994; Goto et al. 2004). *Aneurinibacillus* is recognized by the presence of unique S-layer proteins. The species *A. thermoaerophilus* carries a square surface layer (S-layer) array composed of identical glycoprotein monomers at its outermost cell envelope component (Schaffer et al. 1999). At present the genus *Aneurinibacillus* comprises of the species *A. migulanus* (Takagi et al. 1993), *A. aneurinilyticus* (Shida et al. 1996; Heyndrickx et al. 1997), *A. thermoaerophilus* (Meier-Stauffer et al. 1996; Heyndrickx et al. 1997), *A. danicus* (Shida et al. 1996; Goto et al. 2004), *A. terranovensis* (Allan et al. 2005), *A. soli* (Lee et al. 2014), *A. tyrosinisolvens* (Tsubouchi et al. 2015) and *A. sediminis* (Subhash et al. 2017). Most described *Aneurinibacillus* species are of soil origin

C. Kaur · G. Selvakumar (✉) · A. N. Ganeshamurthy
ICAR- Indian Institute of Horticultural Research, Bengaluru, India

M. T. Islam et al. (eds.), *Bacilli and Agrobiotechnology: Phytostimulation and Biocontrol*, Bacilli in Climate Resilient Agriculture and Bioprospecting,
https://doi.org/10.1007/978-3-030-15175-1_7

and thereby tend to have an influence on plant growth. This chapter attempts to put together the available information on the genus *Aneurinibacillus* and its utility in the agricultural and related environments.

7.2 Disease Suppression and Plant Growth Promotion by *Aneurinibacillus* Species

The AEFB strains belonging to the genera *Bacillus*, *Brevibacillus* and *Paenibacillus* are highly diverse and widely prevalent in the agro-environment. They play an important role in disease management and prevention of pathogen resistance to fungicides (Seddon et al. 2000; Schmitt and Seddon 2004). *Aneurinibacillus migulanus* previously known as *Bacillus brevis* is well known for controlling plant disease development by the production of antimicrobial peptide gramicidin S (Edwards and Seddon 2001; Belbahri et al. 2015). It has membranolytic activity against pathogenic bacteria as well as fungi like *Fusarium oxysporum*, *Botrytis cinerea* and several phytopathogenic oomycete species (Maget-Dana and Ptak 1995; Epand and Vogel 1999; Tsukagoshi et al. 1970; Kondejewski et al. 1996; Alenezi et al. 2016). *A. migulanus* strain Nagano is effective in the biological control of *Dothistroma* needle blight on pine in forest nurseries (Alenezi et al. 2016). Apart from antibiosis, the Gramicidin S produced by *A. migulanus* has biosurfactant property, which helps in suppressing spore germination by increasing the rate of evaporation from plant surfaces and reducing the moisture content (Seddon et al. 1997, 2000).

Two *A. migulanus* strains, viz. DSM 5759 and DSM 5668, were studied in detail for the presence and synthesis of multiple antimicrobial peptides. Apart from membranolytic activity, other mechanisms have also been discovered in *A. migulanus* Nagano and NCTC 7096 strains. The genomes of Nagano and NCTC 7096 strains were found to have biosynthetic gene clusters, which leads to synthesis of two bacteriocins and a microcin cluster. Bacteriocins are known to supress the growth of competent bacterial strains because of their proteinaceous toxin nature. The whole genome mining of these strains also revealed the presence of the lasso and linaridin peptides which act as effective antimicrobial agents. The linaridin peptide's cluster leads to biosynthesis of phosphonate, which is more robust against oomycete pathogens such as *Pythium* and *Phytophthora* (Jackson et al. 2000). In the initial testing of anti-oomycete activity, the *A. migulanus* strain Nagano proved to have better bioactive potential against plant pathogens than *A. migulanus* NCTC 7096 though it possesses only 8 biosynthetic gene clusters compared to 11 clusters in *A. migulanus* NCTC 7096. The whole genome annotation of *A. migulanus* NCTC7096 and Nagano strains also showed the presence of single siderophore cluster and the *fur* gene, which encodes an iron regulator (Alenezi et al. 2015). Also, this strain has extensive antimicrobial properties against fungal pathogens. The antifungal activity has been attributed to a novel peptide, the elucidation of which will be advantageous from the biological control point of view (Alenezi et al. 2015). The draft genome sequence of *A. migulanus* displayed the presence of gramicidin S (GS) genes in both the strains, i.e. Nagano and NCTC 7096, with a difference of two amino acids that were designated as T3421 and P419S. Three-dimensional models suggested that these substitutions declined the stability of the *A. migulanus* NCTC 7096 GS synthase protein. Alternatively, it is proposed that the T3421 and P419S substitutions lead to synthesis of two diverse gramicidin homologs in the Nagano strain, but this fact requires further verification. The genomes of both *A. migulanus* NCTC 7096 and Nagano also revealed the presence of a single chitinase gene, but components of chitin regulatory system were not found (Alenezi et al. 2016).

Previously it was thought that the biocontrol potential of *A. migulanus* Nagano was due to the production of GS (Seddon et al. 1997) and biosurfactant activity (Seddon et al. 1997, 2000). Later Alenezi et al. (2017), compared the GS production potential of the Nagano and NCTC 7096 stains using liquid chromatography and high-resolution electrospray ionisation mass spectrometry (LC-HRESIMS) and arrived at the conclusion that the Nagano strain displayed higher biosurfactant activity compared to NCTC-7096 strain . Genome mining of both the strains indicated a difference of two amino acid units between GS genes of both the strains. The Nagano strain produced two additional GS-like molecules i.e. GS-1155 (molecular weight 1155) and GS-1169 (molecular weight 1169). Another interesting observation was that the NCTC 7096 strain is unable to form biofilms unlike the Nagano strain. The above findings were reinforced by the fact that *A. migulanus* Nagano was found to be more potent than NCTC 7096 in reducing the germination of *B. cinerea* spores.

Several members of the genus *Aneurinibacillus* promote growth and suppress plant diseases. Chauhan et al. (2017) described the plant growth promotion potential of the multi-trait *A. aneurinilyticus* strain CKMV1 from the rhizosphere of the Himalayan plant *Valeriana jatamansi*. This strain efficiently solubilized phosphates through production of various organic acids such as gluconic and oxalic acids. The presence of *pqq* gene encoding the pyrroloquinoline quinone synthase a cofactor of the enzyme glucose dehydrogenase that has been majorly attributed to phosphate solubilization was also been reported for the first time from the bacterium. Apart from solubilizing phosphorous, this strain was also able to fix the atmospheric nitrogen, and the presence of the *nif* H gene was reported from *A. aneurinilyticus* for the first time. This strain was also able to produce plant growth-promoting hormones such as indole-3-acetic acid (IAA) and siderophores, besides producing HCN. Inoculation of this strain in tomato resulted in significant improvement of the plant growth promotion parameters and nutrient uptake. The strain CKMV1 also possesses antifungal activity against number of pathogenic fungal species such as *Sclerotium rolfsii*, *Fusarium oxysporum*, *Dematophora necatrix*, *Rhizoctonia solani*, *Alternaria* sp. and *Phytophthora* species. The instances of plant growth promotion and other related environmentally useful traits of *Aneurinibacillus* strains are listed in Table 7.1.

Table 7.1 Plant growth-promoting and environmentally useful traits of *Aneurinibacillus* strains

Name of bacterium	PGPR/environmentally useful trait	Reference
Aneurinibacillus migulanus	Multiple antimicrobial peptides with varying plant growth-promoting traits – (i) biosurfactant property, (ii) presence of chitinase gene, (iii) the presence of single siderophore cluster and the *fur* gene, and (iv) useful properties of bioplastic production and biodegradation of pesticides	Alenezi et al. (2016), Jackson et al. (2000), Stern et al. (1968), Laiken et al. (1969), Seddon et al. (1997, 2000), Maget-Dana and Ptak (1995), Epand and Vogel (1999), Tsukagoshi et al. (1970), Kondejewski et al. (1996), Edwards and Seddon (2001), Belbahri et al. (2015)
A. aneurinilyticus strain SBP-11	Production of a lipopeptide biosurfactant	Desai and Banat (1997), Balan et al. (2017)
A. thermoaerophilus WBS2	Cellulase activity	Acharya and Chaudhary (2012)
Aneurinibacillus sp. XH2 strain	Bioplastics production	Xiao et al. (2015)

7.3 Environment-Friendly Applications of *Aneurinibacillus*

The lipopeptide biosurfactant aneurinifactin produced by *Aneurinibacillus* strains has broad spectrum antimicrobial activity and can replace the use of synthetic preservatives in many product formulations and bioremediation applications (Desai and Banat 1997). *A. aneurinilyticus* strain SBP-11 is known for the production of aneurinifactin with emulsifying capabilities at broader temperature and pH ranges (Balan et al. 2017). *A. migulanus* strain Nagano also has the ability to produce biosurfactants which tend to enhance the rate of evaporation from the plant surface, which leads to reduced wetness of the plant surface and repressed pathogen spore germination (Seddon et al. 1997, 2000).

Biofilm formation has been proposed as a mean of providing efficient attachment of bacteria to plant surfaces (Yaron and Römling 2014). Genome mining of *A. migulanus* strain Nagano and NCTC 7096 showed substantial similarity with bacterial genes responsible for biofilm formation and exopolysaccharide biosynthesis. The gene polyprenyl glycosylphosphotransferase, which encodes the first step of exopolysaccharide biosynthesis, was found in the genome of the Nagano strain but was absent in genome of the NCTC 7096 strain (Alenezi et al. 2016). Strain Nagano showed strong attachment ability to pine needles even after infection with the fungal pathogen *Dothistroma septosporum*, whereas strain NCTC 7096 showed weak attachment ability. The presence of the exopolysaccharide cluster genes suggested that biofilm formation could be an important mechanism used by *A. migulanus* Nagano to interact with plant roots and therefore provide biocontrol ability. To confirm the possibility of this mechanism, knock-out mutant studies are needed (Alenezi et al. 2017).

The *Aneurinibacillus* XH2 strain is characterized by its ability to produce bioplastics polyhydroxyalkanoates (PHAs) at thermophilic temperatures. Generally

PHAs were biosynthesized by mesophilic strains, which have their own shortcomings in terms of bacterial contamination during the fermentation processes (Xiao et al. 2015). This novel strain isolated from the Gudao Oilfield of China holds much promise. The annotated genomic sequence of this strain proved its function as a producer of thermophilic PHAs. *A. migulanus* has also been studied for the degradation of natural and synthetic plastics (Chaisu 2016). The outcome of the study was that *A. migulanus* can degrade plastic packaging in liquid medium within 20 days after inoculation at 30 °C. Though further detailed investigation requires to generalize the technique for broader capacity of organisms in plastic degradation, another relevant study was done on degradation of pesticides by *A. migulanus* (Palanimanickam and Sepperumal 2017). The organism was able to degrade the organophosphate pesticide Profenofos and the residues produced in the mineralization process have been documented to decrease the pollution load in paddy soil. *A. thermoaerophilus* WBS2 isolated from Bakreshwar hot springs of India produces an alkaline thermostable cellulase (Acharya and Chaudhary 2012). This represents an opportunity of meeting the huge requirements of thermophilic enzymes for industrial operations at high temperatures.

7.4 Conclusion

This chapter aimed to review the potential applications of *Aneurinibacillus* strains in agriculture and environmental applications that are very much at their infancy compared to the other well-known AEFB. Various strains of *Aneurinibacillus* possess plant growth-promoting and biological control traits. Therefore, these strains have high potential to commercial application in agriculture. Apart from this, the environmental applications of *Aneurinibacillus* strains are gaining importance especially in the production of bioplastics and the degradation of xenobiotic compounds. But this has to be done after due diligence of their biosafety status, though no adverse reports of their opportunistic pathogenic potential are available at present. If this can be established scrupulously, mankind will be blessed with another bacterial genus for environmentally safe crop production and protection.

References

Acharya S, Chaudhary A (2012) Bioprospecting thermophiles for cellulase production: a review. Braz J Microbiol 43:844–856

Alenezi FN, Rekik I, Bełka M, Ibrahim AF, Luptakova L, Jaspars M, Woodward S, Belbahri L (2016) Strain-level diversity of secondary metabolism in the biocontrol species *Aneurinibacillus migulanus*. Microbiol Res 182:116–124

Alenezi FN, Rekik I, Chenari Bouket A, Luptakova L et al (2017) Increased biological activity of *Aneurinibacillus migulanus* strains correlates with the production of new gramicidin secondary metabolites. Front Microbiol 8:517

Alenezi FN, Weitz HJ, Belbahri L, Nidhal J, Luptakova L, Jaspars M, Woodward S (2015) Draft genome sequence of *Aneurinibacillus migulanus* NCTC 7096. Genome Announc 30:3–15

Allan RN, Lebbe L, Heyrman J, De Vos P, Buchanan CJ, Logan NA (2005) *Brevibacillus levickii* sp. nov. and *Aneurinibacillus terranovensis* sp. nov., two novel thermoacidophiles isolated from geothermal soils of northern Victoria Land, Antarctica. Int J Syst Evol Microbiol 55:1039–1050

Aoyama S (1952) Studies on the thiamin-decomposing bacterium. I. Bacteriological researches of a new thiamin decomposing bacillus *Bacillus aneurinolyticus* Kimura et Aoyama. Acta Sch Med Univ Kioto 30:127

Balan SS, Kumar CG, Jayalakshmi S (2017) Aneurinifactin, a new lipopeptide biosurfactant produced by a marine *Aneurinibacillus aneurinilyticus* SBP-11 isolated from Gulf of Mannar: purification, characterization and its biological evaluation. Microbiol Res 194:1–9

Belbahri L, Alenezi FN, Luptakova L, Rateb ME, Woodward S (2015) Complete genome sequence of *Aneurinibacillus migulanus* E1, a gramicidin S-and D-phenylalanyl-L-propyl diketopiperazine-deficient mutant. Genome Announc 3:441–415

Chaisu K (2016) Study of plastic degrading bacteria by *Aneurinibacillus migulanus*. In: Management and Innovation Technology International Conference (MITicon), IEEE, p MIT-209

Chauhan A, Guleria S, Balgir PP, Walia A, Mahajan R, Mehta P, Shirkot CK (2017) Tricalcium phosphate solubilization and nitrogen fixation by newly isolated *Aneurinibacillus aneurinilyticus* CKMV1 from rhizosphere of *Valeriana jatamansi* and its growth promotional effect. Braz J Microbiol 48:294–304

Desai JD, Banat IM (1997) Microbial production of surfactants and their commercial potential. Microbiol Mol Biol Rev 61:47–64

Edwards SG, Seddon B (2001) Mode of antagonism of *Brevibacillus brevis* against *Botrytis cinerea* in vitro. J Appl Microbiol 91:652–659

Epand RM, Vogel HJ (1999) Diversity of antimicrobial peptides and their mechanisms of action. Biochim Biophys Acta 1462:11–28

Goto K, Fujita R, Kato Y, Asahara M, Yokota A (2004) Reclassification of *Brevibacillus brevis* strains NCIMB 13288 and DSM 6472 (= NRRL NRS-887) as *Aneurinibacillus danicus* sp. nov. and *Brevibacillus limnophilus* sp. nov. Int J Syst Evol Microbiol 54:419–427

Heyndrickx M, Lebbe L, Vancanneyt M, Kersters K et al (1997) A polyphasic reassessment of the genus *Aneurinibacillus*, reclassification of *Bacillus thermoaerophilus* (Meier-Stauffer et al. 1996) as *Aneurinibacillus thermoaerophilus* comb, nov., and emended descriptions of *A. aneurinilyticus* corrig., *A. migulanus*, and *A. thermoaerophilus*. Int J Syst Evol Microbiol 47:808–817

Jackson TJ, Burgess T, Colquhoun I, Hardy GS (2000) Action of the fungicide phosphite on *Eucalyptus marginata* inoculated with *Phytophthora cinnamomi*. Plant Pathol 49:147–154

Kondejewski LH, Farmer SW, Wishart DS, Hancock RE, Hodges RS (1996) Gramicidin S is active against both Gram-positive and Gram-negative bacteria. Int J Pept Protein Res 47:460–466

Laiken S, Printz M, Craig LC (1969) Circular dichroism of the tyrocidines and gramicidin SA. J Biol Chem 244:4454–4457

Lee KC, Kim KK, Eom MK, Kim JS, Kim DS, Ko SH, Lee JS (2014) *Aneurinibacillus soli* sp. nov., isolated from mountain soil. Int J Syst Evol Microbiol 64:3792–3797

Maget-Dana R, Ptak M (1995) Interactions of surfactin with membrane models. Biophys J 68:1937–1943

Meier-Stauffer K, Bisse HJ, Rainey FA, Burghardt J et al (1996) Description of *Bacillus thermoaerophilus* sp. nov., to include sugar beet isolates and *Bacillus brevis* ATCC 12990. Int J Syst Evol Microbiol 46:532–541

Palanimanickam A, Sepperumal U (2017) Profenofos degradation potential of *Bacillus cereus* and *Aneurinibacillus migulanus* isolated from Paddy crop field soil. J Pure Appl Microbiol 11:1

Schäffer C, Müller N, Christian R, Graninger M, Wugeditsch T, Scheber A, Messner P (1999) Complete glycan structure of the S-layer glycoprotein of *Aneurinibacillus thermoaerophilus* GS4-97 cg. Glycobiology 9:407–414

Schmitt A, Seddon B (2004) Biocontrol of plant pathogens with microbial BCAs and plant extracts–advantages and disadvantages of single and combined use. In: Proceedings of the 14th International Reinhardsbrunn symposium

Seddon B, Edwards SG, Markellou E, Malathrakis NE (1997) Bacterial antagonist fungal pathogen interactions on the plant aerial surface. In: Multitrophic interactions in terrestrial systems. The 36th symposium of The British Ecological Society, pp 5–25

Seddon B, McHugh RC, Schmitt A (2000) *Brevibacillus brevis*-a novel candidate biocontrol agent with broad-spectrum antifungal activity. In: The BCPC conference: pests and diseases, vol. 2. Proceedings of an international conference held at the Brighton Hilton Metropole Hotel, Brighton, UK, British Crop Protection Council, pp 563–570

Shida O, Takagi H, Kadowaki K, Komagata K (1996) Proposal for two new genera, *Brevibacillus* gen. nov. and *Aneurinibacillus* gen. nov. Int J Syst Evol Microbiol 46:939–946

Shida O, Takagi H, Kadowaki K, Yano H, Abe M, Udaka S, Komagata K (1994) *Bacillus aneurinolyticus* sp. nov., nom. rev. Int J Syst Evol Microbiol 44:143–150

Stern A, Gibbons WA, Craig LC (1968) A conformational analysis of gramicidin SA by nuclear magnetic resonance. Proc Natl Acad Sci 61:734–741

Subhash Y, Kim HJ, Lee SS (2017) *Aneurinibacillus sediminis* sp. nov., isolated from lagoon sediments. Int J Syst Evol Microbiol 67:2544–2548

Takagi H, Shida O, Kadowaki K, Komagata K, Udaka S (1993) Characterization of *Bacillus brevis* with descriptions of *Bacillus migulanus* sp. nov., *Bacillus choshinensis* sp. nov., *Bacillus parabrevis* sp. nov., and *Bacillus galactophilus* sp. nov. Int J Syst Evol Microbiol 43:221–231

Tsubouchi T, Mori K, Miyamoto N, Fujiwara Y et al (2015) *Aneurinibacillus tyrosinisolvens* sp. nov., a tyrosine-dissolving bacterium isolated from organics-and methane-rich seafloor sediment. Int J Syst Evol Microbiol 65:1999–2005

Tsukagoshi N, Tamura G, Arima K (1970) A novel protoplast-bursting factor (surfactin) obtained from *Bacillus subtilis* IAM 1213: II. The interaction of surfactin with bacterial membranes and lipids. Biochim Biophys Acta 196:211–214

Xiao Z, Zhang Y, Xi L, Huo F, Zhao JY, Li J (2015) Thermophilic production of polyhydroxyalkanoates by a novel *Aneurinibacillus* strain isolated from Gudao oilfield, China. J Basic Microbiol 55:1125–1133

Yaron S, Römling U (2014) Biofilm formation by enteric pathogens and its role in plant colonization and persistence. Microb Biotechnol 7:496–516

Chapter 8
Phylogeny and Taxonomy of Agriculturally Important *Bacillus* Species

Christopher Dunlap

8.1 Taxonomy of Prokaryotes

The taxonomy and nomenclature of prokaryotes are governed by the International Committee on Systematics of Prokaryotes (ICSP). The committee was organized by the International Union of Microbial Societies to develop rules and procedures for naming prokaryotes. The rules of naming prokaryotes are described in the most recent edition of the International Code of Nomenclature of Bacteria, 1990 revision (Lapage et al. 1992). This code only addresses how prokaryotes are named, while the criteria for describing new species are defined by taxonomic subcommittees that specialize in different bacterial taxa. In the case of *Bacillus* species, the most recent minimal standards for describing new species were defined in 2009 (Logan et al. 2009). These standards are often informally revised by journal editors and reviewers in the field as new technology becomes available.

For many decades, the "gold standard" of describing new prokaryotic species has been DNA-DNA hybridizations (DDH) with the type strains of the nearest species. A strain was considered a new species if the pairwise DDH was below 70% (Wayne et al. 1987). In addition to the DDH requirement, a new species was required to show a distinguishing phenotype. These were often based on differences in carbohydrate utilization or growth physiology. While in the last decade, the focus of taxonomists has increasingly been on a genome-based taxonomy for prokaryotes (Auch et al. 2010; Meier-Kolthoff et al. 2013; Richter and Rosselló-Móra 2009; Yoon et al. 2017). This has led to a much more stable and predictable system for the classification of prokaryotes.

C. Dunlap (✉)
Crop Bioprotection Research Unit, Agricultural Research Service,
U.S. Department of Agriculture, Peoria, IL, USA
e-mail: Christopher.dunlap@ars.usda.gov

M. T. Islam et al. (eds.), *Bacilli and Agrobiotechnology: Phytostimulation and Biocontrol*, Bacilli in Climate Resilient Agriculture and Bioprospecting,
https://doi.org/10.1007/978-3-030-15175-1_8

8.2 *Bacillus subtilis* Group Taxonomy

Bacillus is an important bacterial genus that is often isolated from soil. The genus currently is comprised of nearly 300 validly described species. The taxonomic heterogeneity of the large genus has long been known and reinforced with the early 16S rRNA studies of the genus (Ash et al. 1991b; Stackebrandt et al. 1987). It was also soon realized that the 16S rRNA gene was insufficient for the absolute resolution of some species in the genus (Fox et al. 1992). One of the important lineages in the genus is the *Bacillus subtilis* group, which was initially comprised of *Bacillus subtilis*, *Bacillus licheniformis*, and *Bacillus pumilus* (Gibson 1944). At the time of writing, the *Bacillus subtilis* group encompasses 22 validly described species and three subspecies. Figure 8.1 shows the phylogenetic relationship of all current members of the *Bacillus subtilis* group.

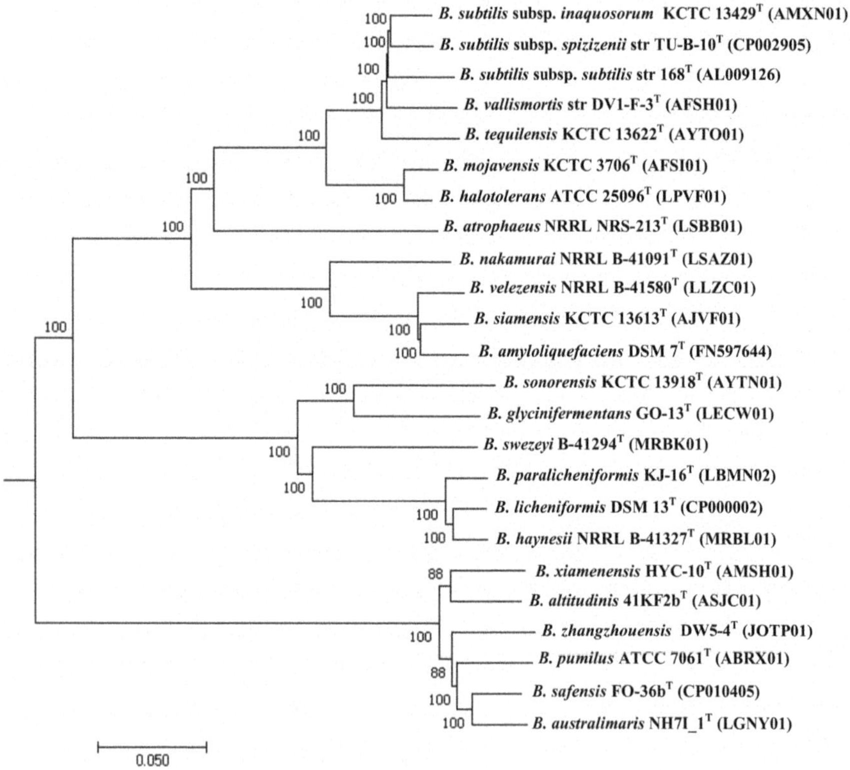

Fig. 8.1 Phylogenetic tree reconstructed from the core genomes of type strains of species from the *Bacillus subtilis* group (350 genes). Bootstrap values >50%, based on 1500 pseudoreplicates are indicated on branch points. *Bacillus cereus* was used as an outgroup, and only the relevant part of the tree is presented. The scale bar corresponds to 0.05 nucleotide substitutions per site

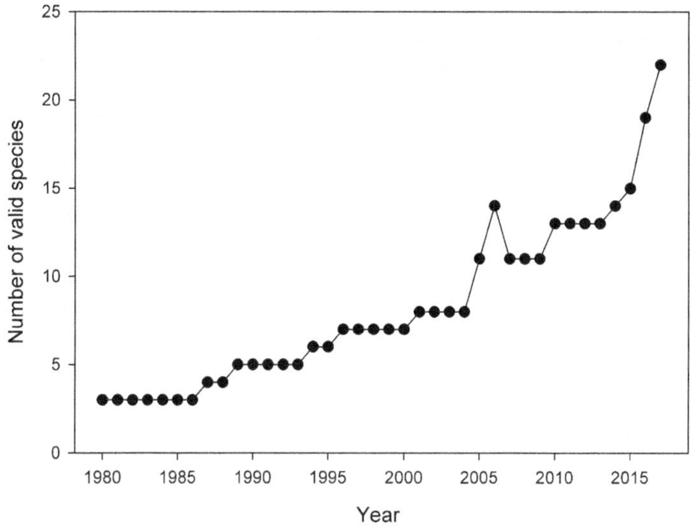

Fig. 8.2 Total number of validly described species in the *Bacillus subtilis* group

Species of the *B. subtilis* group have always been difficult to delineate due to the close phenotypes of the group members. To overcome these hurdles, a variety of approaches have been established such as fatty acid analysis (Kämpfer 1994) and metabolic profiles (Logan and Berkeley 1984). Advances in genotyping these species came with the analysis of single protein-coding genes, such as *gyrA* (Chun and Bae 2000) and *gyrB* (Wang et al. 2007). Later, multilocus sequence alignments (MLSA) were applied to the *B. subtilis* group (Rooney et al. 2009) and to strains in the *B. pumilus* clade (Liu et al. 2013) to successfully delineate these species. More recently, genomes have been used to support and correct the taxonomy of the *B. subtilis* group.

The number of species in the *B. subtilis* group has risen rapidly in recent years (Fig. 8.2), with 10 of the 22 species having been described in the past 4 years (Dunlap et al. 2015, 2016c, 2017; Kim et al. 2015; Lai et al. 2014; Liu et al. 2016). It is anticipated that this high rate of discovery will continue with genomics making it easier to identify novel species.

8.3 Recent Corrections in *Bacillus subtilis* Group Taxonomy

In addition to the discovery of new species, the genomics technology has led to the correction of many taxonomic mistakes in this area. The most common mistake in this area is describing the same species more than once. By naming rules, a species may have only one valid name, and the naming priority is given to the first validly described species. The worst example of this error was associated with *Bacillus velezensis*, some very important bacteria sold commercially around the world as antifungal/antibacterial plant pathogen antagonists and as plant growth promoters. This species was described four separate times with four different names. It is still sold around the world under at least six species names. Once the genomes of these strains were sequenced, it became clear these species were all the same taxa (Dunlap et al. 2016b) which resulted in *B. methylotrophicus*, "*B. oryzicola*," and *B. amyloliquefaciens* subsp. *plantarum* becoming junior synonyms of *B. velezensis*. There were numerous other synonyms identified in this area using genomics. "*Bacillus vanillea*" was found to be a junior synonym of *Bacillus siamensis* (Dunlap 2015). *B. malacitensis* and *B. axarquiensis* are junior synonyms of *B. halotolerans* (Dunlap et al. 2016a). *B. invictae* was found to be a junior synonym of *B. altitudinis* (Liu et al. 2015a). In addition to identifying synonyms, genomics data has been used to correct earlier studies that ruled two species to be synonyms when they were, in fact, separate species (Dunlap et al. 2016a, 2016b). Currently, genome data exists from all type strains in the *B. subtilis* group, which should limit the taxonomic mistakes going forward.

8.4 *Bacillus cereus* Group Taxonomy

The second important group of *Bacillus* species is known as the *Bacillus cereus* group. This group contains many species that have important agricultural implications. The group namesake *B. cereus* is an important species due to its prevalence as a food poisoning agent. The group also contains *B. anthracis*, which is a well-known animal/human pathogen and potential biological weapon. The group also provides a very important insect - pathogenic strain, *B. thuringiensis* (Schnepf et al. 1998). This strain is the most used prokaryotic species as an agricultural biopesticide (Gould 1998). Its insecticidal protein has been transgenetically introduced into plants and an important class of genetically modified plants (Bravo et al. 2011).

8.5 Controversies in *Bacillus cereus* Group Taxonomy

The taxonomy of the *B. cereus* group has had many controversies over the past few decades (Ash et al. 1991a; Bavykin et al. 2004; Helgason et al. 2000; Liu et al. 2015b; Nakamura and Jackson 1995; Radnedge et al. 2003). Most of the controversies were based on whether *B. anthracis*, *B. cereus*, and *B. thuringiensis* were separate species or should be considered conspecific (Ash et al. 1991a; Helgason et al. 2000; Nakamura and Jackson 1995). These species had distinct phenotypes based on their pathogenicity, but their 16S rRNA sequences were closely related. The problem is compounded since many of the distinctive phenotypes are based on virulence plasmids (Liu et al. 2015b). These plasmids can be subject to horizontal transfer or loss from the parent strain, creating taxonomic confusion. The release and analysis

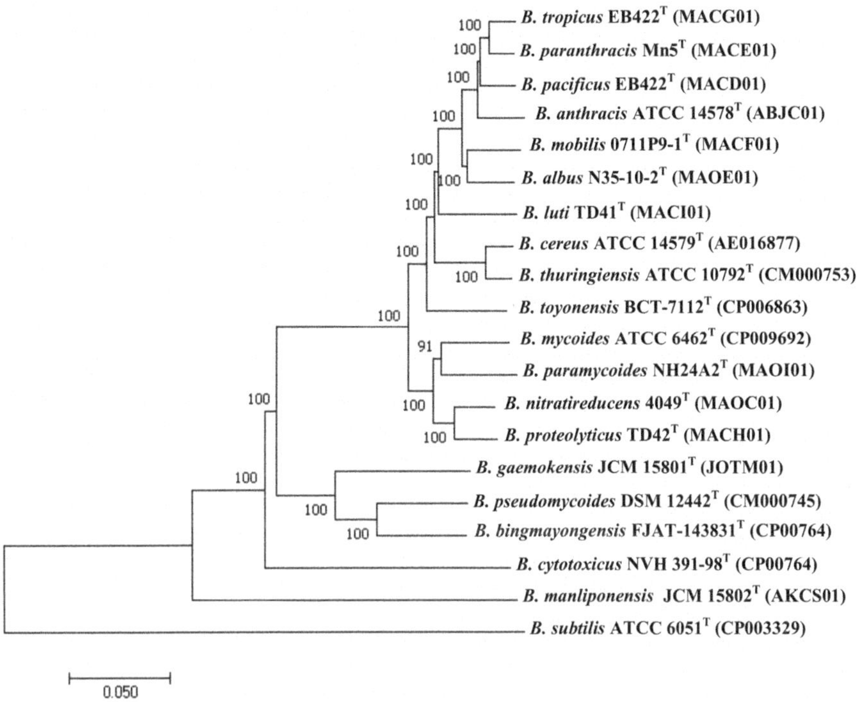

Fig. 8.3 Phylogenetic tree reconstructed from the core genomes of type strains of species from the *Bacillus cereus* group (2205 genes). Bootstrap values >50%, based on 1500 pseudoreplicates are indicated on branch points. *B. subtilis* was used as an outgroup. The scale bar corresponds to 0.05 nucleotide substitutions per site

of hundreds of genomes in this area have provided much needed clarity. Liu et al. (2017) provided genomic analysis-based average nucleotide identity (ANI) comparisons to delineate the species in the *B. cereus* group. This analysis identified many new putative species in this group which were subsequently described as nine new species (Liu et al. 2017). This new framework based on standard genome comparisons should allow for a more stable taxonomy in field. Based on these new species additions and a recent report of synonymy (Liu et al. 2018), the *B. cereus* group stands at 19 species, and an updated genome-based phylogeny is reported in Fig. 8.3.

8.6 Recommendations on Identifying Unknown *Bacillus* Strains

Now that the taxonomy of these species is stable and well-defined, it is much easier to recommend approaches to identify strains to the species level. Typically, unknown strains will first be tentatively assigned to the *B. subtilis* group or the *B. cereus* group using a standard high-throughput method of microbial identification. This is most commonly done with 16S rRNA gene sequencing and comparing the sequence with a reference database, such as EzBioCloud (Yoon et al. 2017). But there are many other techniques that can identify the bacteria, such as MALDI (Seng et al. 2009), Biolog (Morgan et al. 2009), FAME (Tighe et al. 2000), and many other methods. Many of these methods will identify the strain to within a few closely related species, if done correctly. But if a definitive identification is required, it is recommended that an additional protein-coding gene is sequenced. In the case of strains from the *B. subtilis* group, it is recommended that *gyrA* (Chun and Bae 2000) or *gyrB* (Wang et al. 2007) is sequenced and compared to sequence data from the type strains of the *B. subtilis* group. In the case of strains from the *B. cereus* group, it is recommended that *pycA* (Liu et al. 2015b) is sequenced and compared to sequence data from the type strains of the *B. cereus* group. In these cases, nucleotide reference data of the protein-coding gene can be extracted from the genomes of all type strains in either group. Genomes are available for all type strains in these two groups.

8.7 Conclusions

The past 5 years have seen incredible activity in the taxonomy of agriculturally important *Bacillus* strains. This activity has been driven by the introduction of low-cost high-throughput DNA sequencing. New genome-based phylogenies have identified many problems and inconsistencies in the previous 16S rRNA-based and chemotaxonomic-based species circumscriptions. These discoveries have led to many corrections and additions to the taxonomy of *Bacillus* strains. While these changes created short-term confusion in the field as nomenclatures were updated, the long-term benefits of an accurate and consistent taxonomy far outweigh any negatives. It is anticipated that the modern genome-based phylogeny, coupled with better methods to identify strains, will lead to improved communication of research results in the field.

References

Ash C, Farrow JAE, Dorsch M, Stackebrandt E, Collins MD (1991a) Comparative analysis of *Bacillus anthracis, Bacillus cereus,* and related species on the basis of reverse transcriptase sequencing of 16S rRNA. Int J Syst Bacteriol 41:343–346

Ash C, Farrow JAE, Wallbanks S, Collins MD (1991b) Phylogenetic heterogeneity of the genus *Bacillus* revealed by comparative analysis of small-subunit-ribosomal RNA sequences. Lett Appl Microbiol 13:202–206

Auch AF, von Jan M, Klenk HP, Göker M (2010) Digital DNA-DNA hybridization for microbial species delineation by means of genome-to-genome sequence comparison. Stand Genomic Sci 2:117–134

Bavykin SG, Lysov YP, Zakhariev V, Kelly JJ, Jackman J, Stahl DA, Cherni A (2004) Use of 16S rRNA, 23S rRNA, and gyrB gene sequence analysis to determine phylogenetic relationships of *Bacillus cereus* group microorganisms. J Clin Microbiol 42:3711–3730

Bravo A, Likitvivatanavong S, Gill SS, Soberón M (2011) *Bacillus thuringiensis*: a story of a successful bioinsecticide. Insect Biochem Mol Biol 41:423–431

Chun J, Bae KS (2000) Phylogenetic analysis of *Bacillus subtilis* and related taxa based on partial *gyrA* gene sequences. Anton Leeuw Int J Gen Mol Microbiol 78:123–127

Dunlap CA (2015) Phylogenomic analysis shows that '*Bacillus vanillea*' is a later heterotypic synonym of *Bacillus siamensis*. Int J Syst Evol Microbiol 65:3507–3510

Dunlap CA, Kwon S-W, Rooney AP, Kim S-J (2015) *Bacillus paralicheniformis* sp. *nov*., isolated from fermented soybean paste. Int J Syst Evol Microbiol 65:3487–3492

Dunlap CA, Bowman MJ, Schisler DA, Rooney AP (2016a) Genome analysis shows *Bacillus axarquiensis* is not a later heterotypic synonym of *Bacillus mojavensis;* reclassification of *Bacillus malacitensis* and *Brevibacterium halotolerans* as heterotypic synonyms of *Bacillus axarquiensis*. Int J Syst Evol Microbiol 66:2438–2443

Dunlap CA, Kim S-J, Kwon S-W, Rooney AP (2016b) *Bacillus velezensis* is not a later heterotypic synonym of *Bacillus amyloliquefaciens*; *Bacillus methylotrophicus, Bacillus amyloliquefaciens* subsp *plantarum* and '*Bacillus oryzicola*' are later heterotypic synonyms of *Bacillus velezensis* based on phylogenomics. Int J Syst Evol Microbiol:1212–1217

Dunlap CA, Saunders LP, Schisler DA, Leathers TD, Naeem N, Cohan F, Rooney AP (2016c) *Bacillus nakamurai* sp. nov., a black pigment producing strain. Int J Syst Evol Microbiol 66:2987–2991

Dunlap CA, Schisler DA, Perry EB, Connor N, Cohan F, Rooney AP (2017) *Bacillus swezeyi* sp. nov. and *Bacillus haynesii* sp. nov., isolated from desert soil. Int J Syst Evol Microbiol 67:2720–2725

Fox GE, Wisotzkey JD, Jurtshuk P Jr (1992) How close is close: 16S rRNA sequence identity may not be sufficient to guarantee species identity. Int J Syst Bacteriol 42:166–170

Gibson T (1944) A study of *Bacillus subtilis* and related organisms. J Dairy Res 13:248–260

Gould F (1998) Sustainability of transgenic insecticidal cultivars: integrating pest genetics and ecology. Annu Rev Entomol 43:701–726

Helgason E, Økstad OA, Caugant DA, Johansen HA, Fouet A, Mock M, Hegna I, Kolstø AB (2000) *Bacillus anthracis, Bacillus cereus,* and *Bacillus thuringiensis* – one species on the basis of genetic evidence. Appl Environ Microbiol 66:2627–2630

Kämpfer P (1994) Limits and possibilities of total fatty acid analysis for classification and identification of *Bacillus* species. Syst Appl Microbiol 17:86–98

Kim S-J, Dunlap CA, Kwon S-W, Rooney AP (2015) *Bacillus glycinifermentans* sp. *nov*., isolated from fermented soybean paste. Int J Syst Evol Microbiol 65:3586–3590

Lai Q, Liu Y, Shao Z (2014) *Bacillus xiamenensis* sp. *nov*., isolated from intestinal tract contents of a flathead mullet (*Mugil cephalus*). Anton Leeuw Int J Gen Mol Microbiol 105:99–107

Lapage SP, Sneath PHA, Lessel EF, Skerman VBD, Seeliger HPR, Clark WA (1992) International code of nomenclature of bacteria: bacteriological code, 1990 revision. ASM Press International Union of Microbiological Societies, Washington, DC

Liu Y, Lai Q, Dong C, Sun F, Wang L, Li G, Shao Z (2013) Phylogenetic diversity of the *Bacillus pumilus* group and the marine ecotype revealed by multilocus sequence analysis. PLoS One 8:e80097

Liu Y, Lai Q, Du J, Shao Z (2015a) Reclassification of *Bacillus invictae* as a later heterotypic synonym of *Bacillus altitudinis*. Int J Syst Evol Microbiol 65:2769–2773

Liu Y, Lai Q, Göker M, Meier-Kolthoff JP, Wang M, Sun Y, Wang L, Shao Z (2015b) Genomic insights into the taxonomic status of the *Bacillus cereus* group. Sci Rep 5:14082

Liu Y, Lai Q, Du J, Shao Z (2016) *Bacillus zhangzhouensis* sp. *nov*. and *Bacillus australimaris* sp. *nov*. Int J Syst Evol Microbiol 66:1193–1199

Liu Y, Du J, Lai Q, Zeng R, Ye D, Xu J, Shao Z (2017) Proposal of nine novel species of the *Bacillus cereus* group. Int J Syst Evol Microbiol 67:2499–2508

Liu Y, Lai Q, Shao Z (2018) Genome analysis-based reclassification of *Bacillus weihenstephanensis* as a later heterotypic synonym of *Bacillus mycoides*. Int J Syst Evol Microbiol 68:106–112

Logan NA, Berkeley RCW (1984) Identification of *Bacillus* strains using the API system. Microbiology 130:1871–1882

Logan NA, Berge O, Bishop AH, Busse HJ, De Vos P, Fritze D, Heyndrickx M, Kampfer P, Rabinovitch L, Salkinoja-Salonen MS, Seldin L, Ventosa A (2009) Proposed minimal standards for describing new taxa of aerobic, endospore-forming bacteria. Int J Syst Evol Microbiol 59:2114–2121

Meier-Kolthoff JP, Auch AF, Klenk HP, Göker M (2013) Genome sequence-based species delimitation with confidence intervals and improved distance functions. BMC Bioinf 14:60

Morgan MC, Boyette M, Goforth C, Sperry KV, Greene SR (2009) Comparison of the Biolog OmniLog Identification System and 16S ribosomal RNA gene sequencing for accuracy in identification of atypical bacteria of clinical origin. J Microbiol Methods 79:336–343

Nakamura LK, Jackson MA (1995) Clarification of the taxonomy of *Bacillus mycoides*. Int J Syst Bacteriol 45:46–49

Radnedge L, Agron PG, Hill KK, Jackson PJ, Ticknor LO, Keim P, Andersen GL (2003) Genome differences that distinguish *Bacillus anthracis* from *Bacillus cereus* and *Bacillus thuringiensis*. Appl Environ Microbiol 69:2755–2764

Richter M, Rosselló-Móra R (2009) Shifting the genomic gold standard for the prokaryotic species definition. Proc Natl Acad Sci U S A 106:19126–19131

Rooney AP, Price NPJ, Ehrhardt C, Sewzey JL, Bannan JD (2009) Phylogeny and molecular taxonomy of the *Bacillus subtilis* species complex and description of *Bacillus subtilis* subsp. *inaquosorum* subsp. nov. Int J Syst Evol Microbiol 59:2429–2436

Schnepf E, Crickmore N, Van Rie J, Lereclus D, Baum J, Feitelson J, Zeigler DR, Dean DH (1998) *Bacillus thuringiensis* and its pesticidal crystal proteins. Microbiol Mol Biol Rev 62:775–806

Seng P, Drancourt M, Gouriet F, Scola BL, Fournier PE, Rolain JM, Raoult D (2009) Ongoing revolution in bacteriology: routine identification of bacteria by matrix-assisted laser desorption ionization time-of-flight mass spectrometry. Clin Infect Dis 49:543–551

Stackebrandt E, Ludwig W, Weizenegger M, Dorn S, McGill TJ, Fox GE, Woese CR, Schubert W, Schleifer KH (1987) Comparative 16S rRNA oligonucleotide analyses and murein types of round-spore-forming bacilli and non-spore-forming relatives. J Gen Microbiol 133:2523–2529

Tighe SW, De Lajudie P, Dipietro K, Lindström K, Nick G, Jarvis BDW (2000) Analysis of cellular fatty acids and phenotypic relationships of *Agrobacterium, Bradyrhizobium, Mesorhizobium, Rhizobium* and *Sinorhizobium* species using the Sherlock Microbial Identification System. Int J Syst Evol Microbiol 50:787–801

Wang LT, Lee FL, Tai CJ, Kasai H (2007) Comparison of *gyrB* gene sequences, 16S rRNA gene sequences and DNA-DNA hybridization in the *Bacillus subtilis* group. Int J Syst Evol Microbiol 57:1846–1850

Wayne LG, Brenner DJ, Colwell RR, Grimont PAD, Kandler O, Krichevsky MI, Moore LH, Moore WEC, Murray RGE, Stackebrandt E, Starr MP, Truper HG (1987) Report of the Ad Hoc Committee on reconciliation of approaches to bacterial systematics. Int J Syst Bacteriol 37:463–464

Yoon SH, Ha SM, Kwon S, Lim J, Kim Y, Seo H, Chun J (2017) Introducing EzBioCloud: a taxonomically united database of 16S rRNA gene sequences and whole-genome assemblies. Int J Syst Evol Microbiol 67:1613–1617

Chapter 9
Endophytic *Bacillus* Species Induce Systemic Resistance to Plant Diseases

Mohammad Tofajjal Hossain and Young Ryun Chung

9.1 Introduction

Plants have evolved a myriad of types of defense mechanisms against pathogens depending on the respective pathogen and their interaction with the microbes in the host plant. These kinds of plant-microbe interactions rely on the different pathways of resistance induction in the host cells. Some endophytic rhizobacteria are directly involved in inducing the systemic resistance during their interactions with host plants and pathogens. This systemic resistance induced in plants by nonpathogenic-antagonistic rhizobacteria is known as ISR (induced systemic resistance) against pathogens (Ryu et al. 2004a, b; Walters et al. 2005). It is one of the resistance mechanisms by the rhizobacteria or other root-colonizing nonpathogenic endophytic bacteria. Thus, these bacteria trigger the resistance induction to the plants against pathogens. But, the interaction between the pathogen and any other root-colonizing nonpathogenic microbe would be an indirect counterpart, i.e., pathogenic rhizobacteria are not directly involved to pathogen (Pieterse et al. 2009). Detailed studies of the immune-related mechanisms through the plant-microbe interaction have been executed in *Arabidopsis* and rice plants (Jones and Dangl 2006). In the monocot model plant rice, a devastating fungal pathogen, *Fusarium fujikuroi* Nirenberg (anamorph), which causes bakanae disease has been controlled successfully by the *Bacillus oryzicola* YC7007 (Hossain et al. 2016). Some other endophytic bacteria were reported to suppress rice diseases by inducing the "ISR" against bacterial and fungal pathogens and also to promoted rice growth. Endophytic

M. T. Hossain (✉)
Plant Pathology Division, Regional Agricultural Research Station, BARI, Chittagong, Bangladesh

Y. R. Chung
Plant Molecular Biology and Biotechnology Research Center, Gyeongsang National University, Jinju, Republic of Korea

M. T. Islam et al. (eds.), *Bacilli and Agrobiotechnology: Phytostimulation and Biocontrol*, Bacilli in Climate Resilient Agriculture and Bioprospecting,
https://doi.org/10.1007/978-3-030-15175-1_9

Gluconacetobacter diazotrophicus strain PAL5 also controls the pathogen by enhancing the resistance through the JA signaling pathway (Alqueres et al. 2013; Hossain et al. 2016). Of the many endophytic bacteria, several *Bacillus* species stimulate the plant immune system and utilize the JA signaling pathway in the ISR to control plant diseases (Kloepper et al. 2004; McSpadden Gardener 2010). *Bacillus oryzicola* YC7007 and YC7010^T, which are two novel endophytic strains isolated from rice roots, were reported to induce systemic resistance against *F. fujikuroi*, *Burkholderia glumae*, and *Xanthomonas oryzae* pv. *oryzae* in rice (Chung et al. 2015). By inducing the expression of *OsLOS-L2* and *OsAOC* genes via the JA signaling pathway, strain *B. oryzicola* YC7007 successfully controlled rice bakanae disease (Hossain et al. 2016). In the dicot model plant *Arabidopsis*, endophytic *Bacillus* species such as *B. subtilis* GB03 and *B. amyloliquefaciens* IN937 against *Erwinia carotovora* subsp. *carotovora* and *B. cereus* AR156 and *B. subtilis* against *Pseudomonas syringae* pv. *tomato* DC3000 were also reported to switch on the defensive signaling network for the induction of systemic resistance in *Arabidopsis* (Kloepper et al. 2004; Kumar et al. 2012; Niu et al. 2011).

For controlling the diseases, chemical fungicides have been widely used during the past decades, but the efficacy of chemical pesticides has decreased recently due to the occurrence of resistance (Yang et al. 2012). Furthermore, application of some chemical fungicides encourages the fungus to produce more mycotoxins (D'Mello et al. 1998). So, the approach of alternative control measures like ISR using antagonistic microorganisms with the underlying defensive mechanisms would be a breakthrough for controlling plant diseases.

9.2 Biological Control Agents

According to Pal and McSpadden Gardener (2006), biological control can be defined as the use of antagonist microbes to suppress the pathogens and finally control diseases. According to the US National Research Council, biological control refers to "the use of natural or modified organisms, formulated product, genes, or gene products, to reduce the effects of undesirable organisms and to favor desirable organisms such as crops, beneficial insects, and microorganisms" (Anon. 1996). Biological agents, therefore, must be environmentally sound as a trigger bio-agent for controlling plant disease. It has a vast network of living organisms interacting in their natural environment. The presence of an organism is determined by favorable environment; presence of associated organisms (symbionts) for its development, or of organisms required for its survival (e.g., hosts for parasites); and the inhibition or absence of organisms (disease organisms, antagonistic, predators) to cause the extinction of pathogen. Thus, interaction is the essence of a population, and this continued existence would be evidence of biological balance. Mutualism, protocooperation, commensalism, antagonisms, competition, and neutralism to the nature are the principle for biological agents. Many mutualistic rhizobacteria, fungus, and yeasts are well reported as biological agents. Fungi, *Piriformospora indica* against the *Fusarium culmorum* (Harrach et al. 2013), *Talaromyces* sp. KNB-422, and *Trichoderma* isolates against bakanae (Bhramaramba and Nagamani 2013; Kato et al. 2012), and yeasts, *Metschnikowia pulcherrima* and *Pichia guilliermondii* against bakanae (Matic et al. 2014), have been reported as biological agents. Additionally, bacterial genera such as *Bacillus*, *Burkholderia*, *Lysobacter*, *Pantoea*, *Pseudomonas*, and *Streptomyces* have been used as biocontrol agents for controlling diseases of many crops. In recent years, much research has been done on biological agents for different crops using the *Bacillus* species. However, biological control of diseases of agricultural crops, especially rice, is still in its infancy compared with chemical pesticides. Meanwhile, there is a public demand for healthier foods free of contamination from chemical residues.

9.3 The Genus *Bacillus* Is the Good Source for Biological Control

The genus *Bacillus* was first described by Cohn in 1872 (Claus and Berkeley 1986). Numerous *Bacillus* strains have been reported as biocontrol agents for plant pathogens. They can lead to suppression of plant diseases as well as to stimulate plant growth directly (Niu et al. 2011). Many *Bacillus* species produce different types of antibiotic compounds, such as phenazines, pyrrolnitrin, and pyoluteorins, as well as lipopeptides, such as fengycin, iturin, or surfactin, which inhibit the growth of plant pathogens. Some of these species also produce phytohormones, including auxin indole acetic acid (IAA), cytokinin, and gibberellins that actively promote the plant growth (Arkhipova et al. 2005; Bais et al. 2004). A greater understanding of this genus with their many uses will help to accelerate the development and improvement of crop quality and yields. Recently, one endophytic *Bacillus oryzicola* YC7007 has been reported as a novel species that successfully controlled the rice bakanae and bacterial blast diseases (Chung et al. 2015; Hossain et al. 2016). The *Bacillus* species which are widely used for biological control of many plant diseases in different hosts include *B. amyloliquefaciens*, *B. subtilis*, *B. pasteurii*, *B. cereus*, *B. pumilus*, *B. mycoides*, and *B. sphaericus* (Kloepper et al. 2004; McSpadden Gardener 2010; Niu et al. 2011). *B. subtilis* GB03 and *B. amyloliquefaciens* IN937 were demonstrated to control the bacterial pathogen, *Erwinia carotovora* subsp. *carotovora* in *Arabidopsis* (Ryu et al. 2004b). *B. cereus* AR156 and *B. subtilis* were also demonstrated to control *Pseudomonas syringae* pv. *tomato* DC3000 successfully in *Arabidopsis* by inducing resistance (Niu et al. 2011). Some of these *Bacillus* species have been well characterized in terms of their anti-fungal, antibacterial, plant growth-promoting, and resistance-inducing activities in host plants (Park et al. 2009; Ryu 2013). Among the diverse antagonistic bacteria, several *Bacillus* species have been developed as commercial biopesticides because they can produce endospores and persist successfully in natural environments for a long period after treatment (Hu et al. 2011). *Bacillus* species have been reported as strong biological agents showing the dramatic action against the rice pathogens. Diverse species of

Table 9.1 *Bacillus* species used for controlling the major rice diseases

Bacillus species	Pathogen (disease)	Mechanism	References
Bacillus oryzicola YC7007	*B. glumae* (panicle blight), *F. fujikuroi* (bakanae)	ISR	Chung et al. (2015)
B. polymyxa	*Magnaporthe oryzae* (blast of rice)	ISR	Gnanamanickam and Mew (1992) and Kavitha (2002)
B. pumilus	*Magnaporthe oryzae* (blast of rice)	ISR	Gnanamanickam and Mew (1992) and Kavitha (2002)
B. coagulans	*Magnaporthe oryzae* (blast of rice)	ISR	Gnanamanickam and Mew (1992) and Kavitha (2002)
B. polymyxa	*Rhizoctonia solani* (sheath blight)	ISR	Gnanamanickam and Mew (1992)
B. cereus	*X. oryzae* pv. *oryzae* (bacterial blight)	ISR	Velusamy and Gnanamanickam (2003)

Bacillus have been isolated from various terrestrial and halophytic plants, and some of them have been shown to be endophytic (Bibi et al. 2012; Bibi et al. 2011) (Table 9.1).

9.4 History of Resistance Induction

History of resistance induction is not very precisely denoted in the literature. Many scientists have documented their opinions about the history of resistance induction. Biological control that encompasses resistance induction is a more interesting research topic compared with only biological control measures. From the beginning, in the 1970s to the 1980s, biological control research consisted screening of antagonistic microorganisms for their biological activity (Ryu 2013). However, mechanisms were not elucidated in many cases. Through the study of plant-microbe interactions it is revealed how microbes work in the defense signaling pathways such as induced systemic resistance (ISR), systemic acquired resistance (SAR), and primed induced resistance (PIS) against phytopathogens is an interesting subject in the biocontrol measures. Resistance induction, therefore, is called a safe fungicide (Walters et al. 2005). Resistance induction was first proposed as the "acquired physiological immunity" by Chester (1933). Systemic acquired resistance was first proposed by Ross (1961). Pathogen-related gene *PR* was discovered by Van loon (1982). Since then, the plant-microbe interaction was implemented in agriculture. When ISR was first proposed by Van Peer and Schippers (1992), resistance induction by the plant growth-promoting rhizobacteria (PGPR) was shown to be more protective against phytopathogens with indirect interaction. Since then, many scientists were involved in the ISR mechanisms for controlling plant diseases.

9.5 Resistance Induction by the Microbe-Associated Molecular Patterns

Plants have well-organized varieties of physical cell wall (cellular) and hormonal defense mechanisms to defend themselves against microbial pathogens. Cellular defense, innate immunity of plants, can be regulated through phytoalexin, camalexin, callose deposition, cell wall reinforcements, and hydrogen peroxide (H_2O_2) accumulation (Ahn et al. 2007; De Vleesschauwer et al. 2008). These types of innate immunity lead to pattern recognition receptors (PRRs) that are indicators for the receptor response of the molecules from the beneficiary microbes. Plants recognize chemically diverse molecules patenting from microbes (pathogen-/microbe-associated molecular patterns, PAMPs/MAMPs) through pattern recognition receptors (PRRs), inducing a set of defense responses known as pattern-triggered immunity (PTI) (Jones and Dangl 2006). This PTI also encodes PAMP-triggered immunity. In plants, pattern recognition receptors (PRRs) are all membrane-associated receptor-like kinases or receptor-like proteins. PRRs confer robustness to the whole PTI system in plant in which different PRRs are simultaneously involved with microbial attacks (Saijo et al. 2018). The functional significance of PRR-mediated microbe recognition with beneficial microbes is an important era for the plant immune system. MAMPs, viz., fungal chitin, glycans, and, their glycoconjugates, lipopolysaccharides (LPSs), flagellin, and peptidoglycan, are molecules derived from microbes that must be detected by receptors of the host cell to suppress the pathogen. MAMPs of root-associated microbiota can trigger defenses and promote the expense of plant growth. However, beneficial rhizobacteria, such as *Pseudomonas simiae* WCS417, *Martellela endophytica* YC6887, and *B. oryzicola* YC7007, promote plant growth and induce systemic resistance (Hossain et al. 2016; Khan et al. 2016; Stringlis et al. 2018). Recent studies point to a role for host PTI in the selection and management of plant-associated microbial communities that actually enhance the resistance induction and promotion (Hossain et al. 2019; Hacquard et al. 2017). These findings are consistent with the idea that PTI plays a central role in the establishment and maintenance of plant-associated microbiomes for resistance induction. Recently, elicitors of plant defenses such as bacterial flagellin have emerged as a novel generation of plant protection products. Expression of a number of defensive genes has been associated with plant defense transcriptomes and can be induced by MAMPs, ethylene (ET) and jasmonic acid (JA) or SA signaling pathways (Huffaker et al. 2013). Hormonal defense mechanisms are fulfilled by plant hormones salicylic acid (SA), jasmonic acid (JA), ethylene (ET), abscisic acid (ABA), cytokinins (CKs), and brassinosteroids (BR) in molding plant-pathogen interaction in the plant immune system. These types of hormonal inductions, which are regulated by hormonal networks of cross talks or interconnected by transductional signals, depend on the lifestyles of pathogens. Hormonal defense to control the diseases with SA or JA/ET, which is mainly involved with biotrophic or necrotrophic pathogens, respectively, is predominantly associated with those respective signaling molecules (Pieterse et al. 2009; Robert-Seilaniantz et al. 2011). Biotrophic pathogen is mainly associated with SA-dependent defense which leads to systemic acquired resistance (SAR) having a long-lasting plant immunity. The role of SA in plant immunity was maintained by the exogenous application of SA or endogenous accumulation of transcription levels and expressed in the network signaling against the biotrophic pathogen. The transcriptional levels of *PATHOGENESIS-RELATED* (*PR*) genes such as *PR1*, *PR2*, and *PR5* under the SA pathways protect the *Pst* DC3000 (Niu et al. 2011). Consistent with these *PR* genes in the SA pathway, three upregulating biosynthesis genes such as *Enhanced Disease Susceptibility* (*EDS1*, *EDS5*), *Phytoalexin Deficient* (*PAD4*), and *Salicylic Acid Induction Deficient* (*SID2*) are also essential against biotrophic pathogen responses (Brodersen et al. 2006; Wang et al. 2010). On the contrary, some receptors and signal molecules, which are required for defense responses against necrotrophic pathogens such as *Alternaria brassicicola* and *Botrytis cinerea*, are regulated through the JA and ET pathways. The transcription levels of *Plant Defensin1.2* (*PDF1.2*) and *PR* genes such as *PR3* and *PR4* were elevated in *Arabidopsis* against the necrotrophic pathogen infection (Thomma et al. 1998). Moreover, interactions between these two types of hormonal defenses based on SA or JA/ET are mostly antagonistic to one another. This multitude of defenses is performed or inducible through cellular reinforcement and hormonal defenses of SA or JA/ET signaling pathways that can be enhanced by biological agents or an abiotic inducer locally or systemically through subsequent pathogen infection or without attack. These hormonal inductions led by the endophytic *Bacillus* species are important for resistance induction against pathogens through PTI machineries.

9.6 Plant-Microbe Interaction by Resistance Induction

The beneficial rhizobacteria as biotic inducers play prominent roles in the defense system of the plant. These bacterial species produce phytohormones or convert the fixed nutrients to the available form for plant development and inhibit the phytopathogens by secreting various metabolites (Walters et al. 2005; Ryu et al. 2004b). These bacterial metabolites can assist in inducing hormonal and cellular defenses, and thus, some rhizobacteria can elicit the plant resistance induction by induced systemic resistance (ISR) or priming induced resistance depending on the lifestyle of pathogens (Ahn et al. 2007; De Vleesschauwer et al. 2012; Niu et al. 2011). Some nonpathogenic rhizobacteria elicited an ISR response through JA or ET pathways or JA and SA simultaneously via *NPR1* dependent and suppress the disease by expressing the specific defense genes (Niu et al. 2011; Ryu et al. 2004a, Thomma et al. 1998). Some plant growth-promoting rhizobacteria (PGPR)-mediated ISR was also switched on by the lipopolysaccharides, siderophores, and SA (Pieterse et al. 1996). On the contrary, some *Bacillus* species can activate the plant's defense system by enhancing the different hormonal pathways of either salicylic acid (SA) or ethylene/jasmonic (ET/JA) acid or, simultaneously, both pathways (Niu et al. 2011). Therefore, it is really interesting to induce the signaling molecules in the plant defense system by the PGPR strains. Several signaling molecules, such as SA, JA, ET, abscisic acid (ABA), cytokinins (CKs), brassinosteroids (BRs), and reactive oxygen species, have been implicated in inducible defense systems involving rhizobacterial interaction (Koornneef and Pieterse 2008). Most of these defense-related hormonal pathways are activated by rhizobacteria, *Bacillus*, and *Pseudomonas* species, which can elicit an induced systemic resistance (ISR) response through the JA or ET pathway or both pathways in a *NPR1*-dependent process (Niu et al. 2011). This phenomenon is well-defined in the *Arabidopsis*. ISR triggered by rhizobacteria suppresses the diseases by expressing the specific defense-related genes during the interaction (Bakker et al. 2007; Doornbos et al. 2011; Niu et al. 2011; Ton et al. 1999).

9.7 Conclusion and Future Trends

The genus *Bacillus* could be more effective in controlling rice disease than current chemical pesticides. The PGPR strains, especially *Bacillus* species, could turn on different signaling pathways against pathogens. Endophytic *Bacillus* species are superior bioactive agents against pathogens, induce systemic resistance, and make a good symbiotic relationship with the plant host. Their MAMP-mediated defense enhances the PTI and ultimately controls the plant disease with resistance induction.

References

Ahn IP, Lee SW, Suh SC (2007) Rhizobacteria-induced priming in *Arabidopsis* is dependent on ethylene, jasmonic acid, and *NPRI*. Mol Plant Microbe Interact 20:759–768

Alqueres S, Meneses C, Rouws L, Rothballer M, Baldani I, Schmid M, Hartmann A (2013) The bacterial superoxide dismutase and glutathione reductase are crucial for endophytic colonization of rice roots by *Gluconacetobacter diazotrophicus* strain PAL5. Mol Plant Microbe Interact 26:937–945

Anonymous (1996) U.S. Environmental Protection Agency. Microbial pesticide test guidelines. OPPTS 885.0001. Overview for microbial pest control agents. EPA 712-C-96-280

Arkhipova TN, Veselov SU, Melentiev AI, Martynenko EV, Kudyarova GR (2005) Ability of bacterium *Bacillus subtilis* produce cytokinins and to influence the growth and endogenous hormone content of lettuce plants. Plant Soil 272:201–209

Bais HP, Fall R, Vivanco JM (2004) Biocontrol of *Bacillus subtilis* against infection of *Arabidopsis* roots by *Pseudomonas syringae* is facilitated by biofilm formation and surfactin production. Plant Physiol 134:307–319

Bakker PAHM, Pieterse CMJ, Van Loon LC (2007) Induced systemic resistance by fluorescent *Pseudomonas* spp. Phytopathology 97:239–243

Bhramaramba S, Nagamani A (2013) Antagonistic *Trichoderma* isolates to control bakanae pathogen of rice. Agric Sci Dig 33:104–108

Bibi F, Chung EJ, Jeon CO, Chung YR (2011) *Bacillus graminis* sp. nov., an endophytic bacterium isolated from a coastal dune plant. Int J Syst Evol Microbiol 61:1567–1571

Bibi F, Yasir M, Song GC, Lee SY, Chung YR (2012) Diversity and characterization of endophytic bacteria associated with tidal flat plants and their antagonistic effects on oomycetous plant pathogens. Plant Pathol J 28:20–31

Brodersen P, Petersen M, Nielsen HB, Zhu S, Newman MA, Shokat KM, Rietz S, Parker J, Mundy J (2006) *Arabidopsis* MAP kinase 4 regulates salicylic acid-and jasmonic acid/ethylene-dependent responses via *EDS1* and *PAD4*. Plant J 47:532–846

Chester K (1933) The problem of acquired physiological immunity in plants. Q Rev Biol 8:129–151

Chung EJ, Hossain MT, Khan A, Kim KH, Jeon CO, Chung YR (2015) *Bacillus oryzicola* sp. nov., an endophytic bacterium isolated from the roots of rice with anti-microbial, plant-growth-promoting, and systemic resistance-inducing activities in rice. Plant Pathol J 31:152–164

Claus D, Berkeley RCW (1986) Genus *Bacillus* Cohn 1872. In: Sneath PHA, Mair NS, Sharpe ME, Holt JG (eds) Bergey's manual of systematic bacteriology, vol 2. Williams and Wilkins, Baltimore, pp 1105–1139

D'Mello JPF, Macdonald AMC, Postel D, Dijksma WTP, Dujardin A, Placinta CM (1998) Pesticide use and mycotoxin production in *Fusarium* and *Aspergillus* phytopathogens. Eur J Plant Pathol 104:741–751

De Vleesschauwer D, Djavaheri M, Bakker PA, Höfte M (2008) *Pseudomonas fluorescens* WCS374r-induced systemic resistance in rice against *Magnaporthe oryzae* is based on pseudobactin-mediated priming for a salicylic acid-repressible multifaceted defense response. Plant Physiol 148:1996–2012

De Vleesschauwer D, Van Buyten E, Satoh K, Balidion J, Mauleon R, Choi IR, Vera-Cruz C, Kikuchi S, Höfte M (2012) Brassinosteroids antagonize gibberellin and salicylate mediated root immunity in rice. Plant Physiol 158:1833–1846

Doornbos RF, Geraats BPJ, Kuramae EE, Van Loon LC, Bakker PAHM (2011) Effects of jasmonic acid, ethylene, and salicylic acid signaling on the rhizosphere bacterial community of *Arabidopsis thaliana*. Mol Plant Microbe Interact 24:395–407

Gnanamanickam SS, Mew TW (1992) Biological control of blast disease of rice (*Oryza sativa*, L.,) with antagonistic bacteria and its mediation by *Pseudomonas* antibiotics. Ann Phytopathol Soc Jpn 58:380–385

Hacquard S, Spaepen S, Garrido-Oter R, Schulze-Lefert P (2017) Interplay between innate immunity and the plant microbiota. Annu Rev Phytopathol 55:565–589

Harrach BD, Baltruschat H, Barna B, Fodor J, Kogel KH (2013) The mutualistic fungus *Piriformospora indica* protects barley roots from a loss of antioxidant capacity caused by the necrotrophic pathogen *Fusarium culmorum*. Mol Plant Microbe Interact 26:599–605

Hossain MT, Khan A, Rashid HO, Chung YR (2019) A volatile producing endophytic Bacillus siamensis YC7012 promotes root development independent on auxin or ethylene/jasmonic acid pathway. Plant Soil:1–6. https://doi.org/10.1007/s11104-019-04015-y

Hossain MT, Khan A, Chung EJ, Rashid HO, Chung YR (2016) Biological control of rice bakanae by an endophytic *Bacillus oryzicola* YC7007. Plant Pathol J 32:228–242

Hu X, Roberts DP, Maul JE, Emche SE, Liao X, Guo X, Li Y, McKenna LF, Buyer JS, Liu S (2011) Formulations of the endophytic bacterium *Bacillus subtilis* Tu-100 suppress *Sclerotinia sclerotiorum* on oilseed rape and improve plant vigor in field trials conducted at separate locations. Can J Microbiol 57:539–546

Huffaker A, Pearce G, Veyrat N et al (2013) Plant elicitor peptides are conserved signals regulating direct and indirect antiherbivore defense. Proc Natl Acad Sci U S A 110:5707–5712

Jones JD, Dangl JL (2006) The plant immune system. Nature 444:323–329

Kato A, Miyake T, Nishigata K, Tateishi H, Teraoka T, Arie T (2012) Use of fluorescent proteins to visualize interactions between the bakanae disease pathogen *Gibberella fujikuroi* and the biocontrol agent *Talaromyces* sp. KNB-422. J Gen Plant Pathol 78:54–61

Kavitha S (2002) Strategies for management of rice blast and sheath blight with bacterial biocontrol agents in combination with major genes for disease resistance. PhD dissertation, University of Madras

Khan A, Hossain MT, Park HC, Yun DJ, Shim SH, Chung YR (2016) Development of root system architecture of *Arabidopsis thaliana* in response to colonization by *Martelella endophytica* YC6887 depends on auxin signaling. Plant Soil 405:81–96

Kloepper JW, Ryu CM, Zhang S (2004) Induced systemic resistance and promotion of plant growth by *Bacillus* spp. Phytopathology 94:1259–1266

Koornneef A, Pieterse CMJ (2008) Cross talk in defense signaling. Plant Physiol 46:839–844

Kumar AS, Lakshmanan V, Caplan JL, Powell D, Czymmek KJ, Levia DF, Bais HP (2012) Rhizobacteria *Bacillus subtilis* restricts foliar pathogen entry through stomata. Plant J 72:694–706

Matic S, Spadaro D, Garibaldi A, Gullino ML (2014) Antagonistic yeasts and thermotherapy as seed treatments to control *Fusarium fujikuroi* on rice. Biol Control 73:59–67

McSpadden Gardener B (2010) Biocontrol of plant pathogens and plant growth promotion by *Bacillus*. In: Gisi U, Chet I, Gullino ML (eds) Recent developments in management of plant diseases, Plant pathology in the 21st century. Springer, Amsterdam, pp 71–79

Niu DD, Liu HX, Jiang CH, Wang YP, Wang QY, Jin HL, Guo JH (2011) The plant growth-promoting rhizobacterium *Bacillus cereus* AR156 induces systemic resistance in *Arabidopsis thaliana* by simultaneously activating salicylate-and jasmonate/ethylene-dependent signaling pathways. Mol Plant Microbe Interact 24:533–542

Pal KK, McSpadden Gardener B (2006) Biological control of plant pathogens. Plant Health Instruct. https://doi.org/10.1094/PHI-A-2006-1117-02

Park WS, Choi HW, Han SS, Shin DB, Shim HK, Jung ES, Lee SW, Lim CK, Lee YH (2009) Control of bakanae disease of rice by seed soaking into the mixed solution of prochloraz and fludioxonil. Res Plant Dis 15:94–100

Pieterse CMJ, Van Wees SCM, Hoffland E, Van Pelt JA, Van Loon LC (1996) Systemic resistance in *Arabidopsis* induced by biocontrol bacteria is independent of salicylic acid accumulation and pathogenesis-related gene expression. Plant Cell 8:1225–1237

Pieterse CMJ, Leon-Reyes A, Van der Ent S, Van Wees SC (2009) Networking by small-molecule hormones in plant immunity. Nat Chem Biol 5:308–316

Robert-Seilaniantz A, Grant M, Jones JD (2011) Hormone crosstalk in plant disease and defense: more than just jasmonate-salicylate antagonism. Annu Rev Phytopathol 49:317–343

Ross AF (1961) Systemic acquired resistance by localized virus infection in plants. Virology 14:340–358

Ryu CM (2013) Promoting plant protection by root-associated microbes. Plant Pathol J 29:123–124

Ryu CM, Murphy JF, Mysore KS, Kloepper JW (2004a) Plant growth promoting rhizobacteria systemically protect *Arabidopsis thaliana* against cucumber mosaic virus by a salicylic acid and *NPR1* independent and jasmonic acid dependent signaling pathway. Plant J 38:381–392

Ryu CM, Farag MA, Hu CH, Reddy MS, Kloepper JW, Pare PW (2004b) Bacterial volatiles induce systemic resistance in *Arabidopsis*. Plant Physiol 134:1017–1026

Saijo Y, Loo EP, Yasuda S (2018) Pattern recognition receptors and signaling in plant–microbe interactions. Plant J 93:592–613

Stringlis IA, Proietti S, Hickman R, Van Verk MC, Zamioudis C, Pieterse CM (2018) Root transcriptional dynamics induced by beneficial rhizobacteria and microbial immune elicitors reveal signatures of adaptation to mutualists. Plant J 93:166–180

Thomma BPHJ, Eggermont K, Penninckx IAMA, Mauch-Mani B, Vogelsang R, Cammue BPA, Broekaert WF (1998) Separate jasmonate-dependent and salicylate-dependent defense-response pathways in *Arabidopsis* are essential for resistance to distinct microbial pathogens. Proc Natl Acad Sci 95:15107–15111

Ton J, Pieterse CMJ, Van Loon LC (1999) Identification of a locus in *Arabidopsis* controlling both the expression of rhizobacteria-mediated induced systemic resistance (ISR) and basal resistance against *Pseudomonas syringae* pv. *tomato*. Mol Plant Microbe Interact 12:911–918

Van Loon LC (1982) Regulation of changes in proteins and enzymes associated with active defense against virus infection. In: Wood RKS (ed) Active defense mechanisms in plants, vol 1. Plenum Press, New York, pp 247–273

Van Peer R, Schippers B (1992) Lipopolysaccharides of plant growth promoting Pseudomonas sp. strain WCS417r induces resistance in carnation to Fusarium wilt. Neth J Pl Path 98:129–139

Velusamy P, Gnanamanickam SS (2003) Identification of 2,4-deacetylphloglucinol (DAPG) production by plant-associated bacteria and its role in suppression of rice bacterial blight in. India Curr Sci 85:1270–1273

Walters D, Walsh D, Newton A, Lyon G (2005) Induced resistance for plant disease control: maximizing the efficacy of resistance elicitors. Phytopathology 95:1368–1373

Wang C, Gao F, Wu J, Dai J, Wei C, Li Y (2010) *Arabidopsis* putative deacetylase *AtSRT2* regulates basal defense by suppressing *PAD4*, *EDS5* and *SID2* expression. Plant Cell Physiol 51:1291–1299

Yang YR, Kim YC, Lee SW, Lee SW, An GG, Kim IS (2012) Involvement of an efflux transporter in prochloraz resistance of *Fusarium fujikuroi* CF245 causing rice bakanae disease. J Korean Soc Appl Biol Chem 55:571–574

Chapter 10
Genomics and Post-genomics Approaches for Elucidating Molecular Mechanisms of Plant Growth-Promoting *Bacilli*

Musrat Zahan Surovy, Dipali Rani Gupta, Nur Uddin Mahmud, Zerihun T. Dame, Prodip Kumar Roy, and Md Tofazzal Islam

10.1 Introduction

Plant-associated bacteria from diverse taxonomic groups colonize plant tissues (root, shoot, and leaf) and provide beneficial effects on plant growth and development (Kloepper and Beauchamp 1992; Islam et al. 2005; Islam and Hossain 2012, 2013). These naturally - occurring bacteria play significant roles in sustainability in productivity of the ecosystems. They are generally called plant growth-promoting bacteria (PGPB). Bacterial genera such as *Bacillus*, *Pseudomonas*, *Alcaligenes*, *Arthrobacter*, *Azoarcus*, *Azospirillum*, *Azotobacter*, *Clostridium*, *Enterobacter*, *Delftia*, *Gluconacetobacter*, *Klebsiella*, *Paraburkholderia*, and *Serratia* are among the most reported PGPB (Podile and Kishore 2007; Saharan 2011; Kumar et al. 2011; Hurek and Reinhold-Hurek 2003; Hayat et al. 2010; Khan et al. 2016; Rahman et al. 2018). The PGPB promote plant growth and productivity through direct or indirect effects. Direct effects of the PGPB are as follows: (i) nitrogen fixation from the atmosphere and symbiotically support N nutrition to host plants; (ii) production of phytohormones (e.g., indole-3-acetic acid) and promote plant growth; (iii) solubilization of essential nutrient elements (e.g., phosphate, zinc, etc.) and promote nutrient uptake by plant roots; and (iv) control or inhibition of the activity of plant pathogens by secretion of antimicrobial compounds (Glick 1995; Babu et al. 2015). Some of the PGPB also promote plant growth indirectly by eliciting induced sys-

M. Z. Surovy · D. R. Gupta · N. U. Mahmud · M. T. Islam (✉)
Institute of Biotechnology and Genetic Engineering (IBGE), Bangabandhu Sheikh Mujibur Rahman Agricultural University, Gazipur, Bangladesh

Z. T. Dame
Department of Biology, College of Natural and Computational Science, Ambo University, Ambo, Ethiopia

P. K. Roy
Chemistry and Chemical Bioengineering Unit, OIST Graduate University, Okinawa, Japan

M. T. Islam et al. (eds.), *Bacilli and Agrobiotechnology: Phytostimulation and Biocontrol*, Bacilli in Climate Resilient Agriculture and Bioprospecting,
https://doi.org/10.1007/978-3-030-15175-1_10

temic resistance (ISR) in the hosts (Abdelrahman et al. 2016; Jogaiah et al. 2010). Due to an increased public awareness on the use of synthetic chemicals (e.g., pesticides and fertilizers) in agriculture, there have been a growing demand for application of the PGPB as alternatives to environmentally hazardous synthetic fertilizers and chemical pesticides to promote plant growth and development (Arrebola et al. 2010; Chen et al. 2008). Among them, the members of the genus *Bacillus* are the most important PGPB that have unique features for commercialization as biopesticides, plant growth stimulants, and biologics. The *Bacillus* is a Gram positive, heat and desiccation-resistant spore-forming bacterium that are rich in biosynthesis of antimicrobial substances, and thus, is an important genus for agricultural biotechnology (Chowdhury et al. 2015). *Bacillus thuringiensis*, *B. amyloliquefaciens*, *B. licheniformis*, *B. pumilus*, *B. velezensis*, and *B. subtilis* are some notable species that have been industrialized as biocontrol and plant growth-promoting agents for promoting sustainable agriculture (Mendis et al. 2018). Genes of *Bacillus* spp. such as *B. thuringiensis* have been well characterized and also been introduced to plants for the development of genetically modified plants (e.g., Bt-cotton, Bt-brinjal, etc.).

Members of the *Bacillus* spp. are well-known for their ability to produce diverse classes of secondary metabolites, including polyketides, and ribosomally and non-ribosomally synthesized peptides (Mondol et al. 2013; Tareq et al. 2014a, b). In the past decades, discovery of bioactive secondary metabolites and antimicrobial compounds from various isolates of *Bacillus* spp. was primarily based on fermentation, extraction, bioassay-guided chromatographic separation, and structure elucidation of pure compounds by spectroscopic methods (Tareq et al. 2014a, b). Mutant studies has been used for identification and characterization of gene(s) involved in biosynthesis of particular metabolites (Palazzotto and Weber 2018). Recent advances in genomics and post-genomics approaches including bioinformatics analyses of robust omics data are replacing the tedious and time-consuming processes of classical natural products chemistry (Palazzotto and Weber 2018). Bioinformatics tools are now used to predict important genes or gene clusters in the genomes of *Bacillus* spp. involved in secondary metabolite synthesis and to characterize them through analyses of transcriptomics, proteomics, or metabolomics databases (Harwood et al. 2018). *Bacillus* spp. harbor diverse classes of secondary metabolites that exhibit a wide range of biological activity (Mondol et al. 2013). Bacillibactin, bacitracin, plipastatin, surfactin, bacillaene, fengycin, bacilysin, iturin, macrolactin, and difficidin are some most commonly known bioactive metabolites produced by *Bacillus* spp.

Although a good number of reviews have been published on roles of *Bacillus* spp. on plant growth and productivity, no comprehensive review has so far been published on genomics and post-genomics studies that remarkably elucidated the underlying molecular mechanisms of bacilli on plants. This review aimed to update our understanding on molecular mechanisms of plant growth promotion by *Bacillus* spp. as known through genomics and post-genomics studies in recent years (Fig. 10.1).

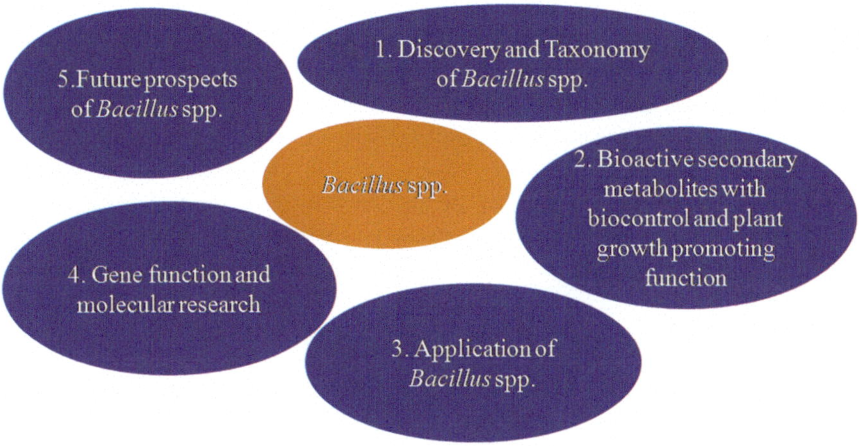

Fig. 10.1 Schematic representation of the discovery, beneficial effect, molecular mechanism of their functions, and application of *Bacillus* spp. in agrobiotechnology

10.2 Plant Growth Promotion

10.2.1 Nitrogen Fixation

Nitrogen is the most essential nutrient that limits plant growth. It is an essential component of nucleic acids, proteins, amino acids, many nitrogen-containing metabolites (e.g., alkaloids), and polysaccharides. The earth's atmosphere is composed of about 78% nitrogen gas, but eukaryotic organisms such as plants and animals cannot directly use this element from the atmosphere due to the absence of the enzymes required for their conversion to ammonium (e.g., nitrogenase) (Sa et al. 2012). Many plant-associated bacteria are also known to promote plant growth by fixing atmospheric nitrogen (Jha and Kumar 2007; Rodrigues et al. 2008; Mia et al. 2013), and *Bacillus* is one of them. Camacho et al. (2001) demonstrated that *Bacillus* spp. CECT 450 increased nodulation in bean (*Phaseolus vulgaris* L.) when co-inoculated with *Rhizobium tropici* CIAT 899 by the process of fixing atmospheric nitrogen. They also found that nodulation in soybean was reduced when co-inoculated with *Bradyrhizobium japonicum* USDA 110. Regulation of nitrogen fixation in *Bacillus* sp. is mediated by two transcriptional factor proteins, *gln*R (transcriptional factor/regulator) and *tnr*A (nitrogen-sensing transcriptional regulator), that belong to the merR protein family (Wray and Fisher 2008). *B. paralicheniformis* strain KMS 80 contains *glnR*, *tnrA*, and *nifH* (encodes nitrogenase iron protein also called component II of nitrogenase enzyme/protein) genes, which regulate the expression of N fixation genes in response to the changes in nitrogen availability (Annapurna et al. 2018). *Bacillus megaterium* NMp082 harbored *nifH* and *nod* gene and symbiosis with *E. medicae*. It expresses *nodABC* gene and forms nodules in alfalfa plants and helps in nitrogen fixation (Chinnaswamy et al. 2018). Li et al. (2008) reported that horizontal transfer of *nifH gene* from *Bradyrhizobium* *japonicum* to endophytic *Bacillus* spp. helps fix atmospheric nitrogen as well as promote plant growth by *Bacillus* spp. *B. amyloliquefaciens* could have an enhanced ability to import nitrate, transform nitrate to nitrite and nitrite to ammonium, and subsequently form glutamine (Irizarry and White 2018). This bacterium upregulated asparagine synthetase, asparaginase, and helps in the storage and transport of nitrogen in plants (Lam et al. 1996; Sieciechowicz et al. 1988; Lea et al. 2006). *B. subtilis* strain 330-2 expressed genes like *narI*, nitrate reductase production, and pckA gene, which were involved in nitrogen fixation in rice and maize plant (Ahmad et al. 2017).

10.2.2 Indole-3-Acetic Acid Production

Indole-3-acetic acid (IAA) is a phytohormone which plays critical roles in plant growth and development (Khan et al. 2016). *Bacillus* spp. have the potential to produce varying levels of IAA, which accelerates plant growth and development (Vacheron et al. 2014). *B. cereus* (Çakmakçi et al. 2007), *B. licheniformis* (Çakmakçi et al. 2007; Souza et al. 2015), *Bacillus circulans* (Thakuria et al. 2004), *B. megaterium*, *B.* altitudinis (Ambreetha et al. 2018), *B. amyloliquefaciens* (Irizary and White 2018), and *B. subtilis* (Ahmad et al. 2017) are the most common IAA producing species of *Bacillus*. In 2018, Ambreetha and his co-workers (2018) found that *B. altitudinis* FD48 modifies rice root architecture. They assessed the expression of auxin-responsive genes (AUX/IAA) by a reverse transcriptase quantitative PCR method. They found that the expression pattern of the *OsIAA1* and *OsIAA4* genes was responsible for shortening of primary roots. Originally, *OsIAA11* and *OsIAA13* genes were responsible for lateral root formation, and these genes were modulated by the inoculation of *B. altitudinis* FD48 and control endogenous IAA levels as well as change the root architecture of rice. Several studies have reported on modulating the root system traits of rice so as to promote its yield under unfavorable conditions (Steele et al. 2007; Uga et al. 2013; Rogers and Benfey 2015; Zou et al. 2015; Nestler et al. 2016). *B. megaterium* also modulates the primary and lateral root structure of *Arabidopsis thaliana* (Lopez-Bucio et al. 2007). Co-inoculation of *E. medicae* with *B. megaterium* NMp082 significantly enhanced growth of *Medicago* spp. plants (Chinnaswamy et al. 2018). Production of IAA by *B. megaterium* increased root surface area and length, secondary root formation, and number of root hairs and root tip number which increased plant nutrient uptake and growth (Ribaudo et al. 2006; Glick, 2012). *B. amyloliquefaciens* FZB42 containing nitrilase, which helps to catalyze the direct conversion of indole-3-acetonitrile to IAA (Hillebrand et al. 1998) and putative IAA acetyl transferase gene, which regulates the tryptophan biosynthesis (Idris et al. 2007), helped to promote plant growth. *B. amyloliquefaciens* encoded for WAT1-related protein which upregulated the root to transport auxin (Ranocha et al. 2013) and IAA synthetase *GH3.1* gene expression occurs in *B. amyloliquefaciens* inoculated cotton seedlings which is related to auxin homeostasis in cotton seedlings (Irizary et al. 2018). The *pabA*, the precursor of tryptophan (IAA synthesis), and bioA, responsible for biotin biosynthesis, were found to be differentially expressed in *B. subtilis* strain 330-2 which regulates the IAA production and enhances growth of maize and rice plant (Ahmad et al. 2017).

10.2.3 Phosphate Solubilization

The *Bacillus* spp. can solubilize insoluble organic and inorganic phosphates from soils and promote plant phosphorus nutrition (Islam and Hossain 2012). For example, *B. amyloliquefaciens* FZB strain produces phytase and phytate enzymes (Idris et al. 2007) under phosphate-limiting and nonphosphate-limiting conditions, contributing to plant growth-promoting activity (Idriss et al. 2002; Makarewicz et al. 2006).

10.2.4 Biofilm Formation

Cells in biofilms experience stringent growth conditions. Rhizosphere competence is linked to the capability to form sessile, multicellular communities (biofilms) (Islam et al. 2005). Biofilm matrices are composed of multiple components like proteins, exopolysaccharides, and nucleic acids. These play various roles in biofilm formation, such as in cell attachment to surfaces, biofilm stabilization via promoting interactions between components, and the development of three-dimensional biofilm architectures. The formation of biofilm matrices is controlled by several extracellular polysaccharides and genes of protein synthesis, such as *epsD* and *tasA* (Fong and Yildiz 2015). In addition, the formation of the extracellular matrix of the biofilm is also enhanced by root-secreted signals (Zhang et al. 2015). In *Bacillus*, biofilm formation is genetically controlled by the regulatory repressor and anti-repressor proteins, SinI and SinR (Branda et al. 2001; Terra et al. 2012; Mielich-Süss and Lopez 2015). In *B. subtilis*, the SinR controlled operon contains the gene triad tapA-sipW-tasA (Kearns et al. 2005; Chu et al. 2006). The membrane-bound peptidase SipW cleaves the signal peptides of TapA and TasA before secretion (Terra et al. 2012). Intriguingly, in pathogenic *B. cereus* or *B. anthracis* strains, the tasA operon contains a SipW homolog together with tasA-like genes named calY1 and calY2. The respective proteins are members of the zinc-dependent M73 metalloproteinase family, usually termed camelysins (Gao et al. 2015), and are considered important for the pathogenicity of *B. cereus* and *B. anthracis* (Rawlings et al. 2016; Pflughoeft et al. 2011). Diehl et al. (2018) found that structural change of TasA protein from globular state to a fibrillar form occurs on the molecular level and regulates biofilm formation of *B. subtilis*. The ability of *Bacillus* spp. to efficiently colonize surfaces of plant roots is a prerequisite for phytostimulation. The unique genes RBAM00750, RBAM00751, and RBAM00754 of *B. amyloliquefaciens* FZB42 encode proteins with a collagen-related GXT structural motif and are probably involved in surface adhesion or biofilm formation. ClpYQ (also known as HslVU) is highly conserved in ATP-dependent proteases in bacteria. Yu et al. (2018)

found that in *B. subtilis*, the Δ*clpYQ* deletion mutant formed early and robust biofilms, while swarming motility was severely impaired and ClpYQ protease is primarily involved in the multicellular development in *B. subtilis*. Hence, ClpYQ does not play an essential role in heat-shock response in both *B. subtilis* and *Bacillus cereus*. The gene cluster encoding iturin A synthase under the control of the pitu promoter is expressed in *B. velezensis* CC09 in wheat tissues; formed biofilm on the root surface, epidermis, cortex, and xylem vessels; and even migrated to stems and leaves. This reveals that iturin A contributes to in vivo antifungal activity and disease control (Kang et al. 2018a, b). High doses of nitric oxide are deleterious for bacteria, and nitric oxide (NO) is primarily synthesized by nitric oxide synthase (NOS) encoded *yflM* gene which positively regulates biofilm formation in *B. amyloliquefaciens* SQR9. Knockdown of yflM gene from *B. amyloliquefaciens* SQR9 mutant, biofilm formation was reduced fivefold compared to the wild type. The *Hmp1* gene plays an important role in the detoxification of high nitric oxide doses in *B. amyloliquefaciens* SQR9 (Dong et al. 2018).

10.3 Protection of Plants from Biotic Stresses

The regulator *PhoP* and its sensor kinase *PhoR* compose a two-component system in *B. subtilis* for production of fengycin (a product to control microbes). The *phoR*- and *phoP*-knockout mutants of *B. subtilis* NCD-2 dramatically reduced its inhibition ability against *Botrytis cinerea* growth in vitro compared to the strain *B. subtilis* NCD-2 wild type. The regulation of fengycin production by the PhoR/PhoP two-component system occurred in a low phosphate condition but not in a high phosphate condition because fengycin synthetase gene *fenC* was positively regulated by *phoP* in the low phosphate condition (Qing-gang et al. 2018). Previously, Dong et al. (2014) demonstrated that the inactivation of *phoP* in *B. subtilis* decreased its antifungal ability. Wang et al. (2015) reported that DegQ also regulated fengycin production in the *B. subtilis* strain NCD-2. DegS is a sensor histidine kinase and DegU is a response regulator. DegQ stimulates the phosphor transfer from DegS-P to DegU and regulates fengycin production. However, the relationship between PhoR/PhoP and DegU/DegS has not been clarified yet. Iturin A is an antibiotic that is useful in fighting plant pathogens and fungi (Meena and Kanwar 2015; Kalai-Grami et al. 2016). The inhibition mechanism of iturin A functions by interacting with the sterol compounds in fungal cell membranes, changing the membrane permeability to allow potassium ion leakage, and result in cell damage (Maget-Dana and Peypoux 1994; Phister et al. 2004). The *ituC* gene of *B. amyloliquefaciens* PBD1 incorporates serine as well as completes the cyclization step during iturin A synthesis (Wu et al. 2018). *B. subtilis* BSD-2 produces small molecule antibiotics that effectively control gray mold. The pHV1249 plasmid with a mini-Tn10 transposon utilized for contraction of *B. subtilis* BSD-2 mutant library. Mutants containing the *yodF* gene encode a Na+/metabolite permease and control cucumber gray mold. But, knockdown of the *yodF* gene in *B. subtilis* mutant lost its inhibitory effect on gray mold (Liu et al. 2018).

Enzymatic properties of the acetolactate synthase (*AlsS*) gene of *Bacillus* spp. are responsible for acetoin production (Huo et al. 2018). *AlsS* from *B. subtilis* was overexpressed to improve the 2,3-BD production in *Saccharomyces cerevisiae* (Ng et al. 2012), and the ethanol titer significantly improved by the knockout of *alsS* in *Pyrococcus furiosus* (Nguyen et al. 2016). *B. licheniformis* generally contains the acetolactate synthase (*AlsS*) gene, but mutants containing knockout of *AlsS* gene eliminate the biosynthetic pathway of acetoin and valine. Acetoin and 2,3-BD biosynthetic pathways can prevent acidification through conversion of excess pyruvate into neutral compounds (acetoin and 2,3-BD) (Tsau et al. 1992). Huo et al. (2018) confirmed that overexpression of the AlsS could enhance the titers of acetoin/2,3-BD and L-valine in *B. licheniformis* and enhanced antibiotics production by *B. licheniformis*. Zimmerman et al. (1986) identified difficidin and oxydifficidin in *B. subtilis*. 3-Hydroxy-2-butanone (acetoin) and 2,3-butanediol, released by *B. subtilis* and *B. amyloliquefaciens*, trigger enhanced plant growth (Ryu et al. 2003). Ryu et al. (2003) observed that both *B. subtilis* GB03 and *B. amyloliquefaciens* IN937a produced volatiles that significantly reduced the severity of disease on *Arabidopsis thaliana* caused by *Pectobacterium carotovorum* subsp. *carotovorum*. The volatile organic compounds (VOCs) produced by *Bacillus* spp. identified as acetoin as elicitors of induced systemic resistance (ISR) in *A. thaliana* against *P. carotovorum* subsp. *carotovorum*. Acetoin elicit ISR in several *Arabidopsis* mutants, including a jasmonic acid-insensitive line (*coi1* mutant), an ethylene-insensitive line (*ein2* mutant), a salicylic acid (SA)-degrading transgenic line (*nahG* mutant), and a non-expressor of PR proteins (*npr1* mutant). The VOCs from strain *B. amyloliquefaciens* IN937 elicited ISR on all of these lines, whereas VOCs from *B. subtilis* GB03 failed to elicit ISR on *ein2* mutant plants. The ethylene-dependent signaling pathway for elicitation of ISR by VOCs from *B. subtilis* GB03 was confirmed by using GUS fusions to the PDF1.2 gene, the indicator for ethylene response. Ethylene signaling and bacterial VOCs-elicited ISR expressed by upregulation of three major ethylene biosynthesis proteins (i.e., aspartate aminotransferase, *S*-adenosylmethionine synthetase 2, and methionine adenosyltransferase 3) following bacterial VOCs treatment (Kwon et al. 2010). Induction of ethylene (ET)-responsive PDF1.2 gene by bacterial VOCs requires a salicylic acid (SA)-dependent resistance response (Naznin et al. 2014). In addition, 2,3-butanediol and acetoin produced by *Bacillus subtilis* strain FB17 can induce ISR against *P. syringae* pv. *tomato* DC3000 (Rudrappa et al. 2010). Production of two major volatiles, i.e., 2,3-butanediol and acetoin from *Bacillus* spp., orchestrated SA and ET signaling to protect plants against two different types of pathogens, i.e., necrotrophs and biotrophs (Kim et al. 2017). *B. subtilis* and *B. cereus* enhanced pathogen-related protein 1 (PR1) gene expression in maize, melon, pepper, and *A. thaliana* (Yang et al. 2009; Niu et al. 2011; Garcia-Gutierrez et al. 2013; Gond et al. 2015). Peroxidase, ascorbate oxidase, and monodehydroascorbate peroxidase were upregulated in cotton seedling roots inoculated with *B. amyloliquefaciens* and stimulate the production of a variety of antioxidant enzymes in cotton (Irizzary and White 2018), which relates to host tolerance to oxidative stress and suppresses the induction of cell death in the host's tissues (White and Torres 2010; Hamilton et al. 2012).

B. subtilis strain 330-2 triggered the production of secondary metabolites, including antibiotics such as macrolactin (*mln*), difficidin (*dfn*), iturin A (*ituA*), bacillibactin (*dhbF*), penicillin-binding protein 2B (*pbpB*), and beta-lactamase (*penP*). These are important for the plant's defense mechanism, so it can cope with competing microorganisms and inhibit the growth of phytopathogenic fungi or bacteria (Ahmad et al. 2017). Interestingly, Munoz-moreno et al. (2018) found CheA and CheC genes in the genome of *B. subtilis* 2C-9B. These proteins are usually involved in chemotaxis and adaptation. Non-ribosomal peptide synthetases and a beta-glucanase gene play a part in antifungal activity (Tapi et al. 2010). The butanediol dehydrogenase gene and a spermidine synthase gene of this bacterium regulate to induce systemic resistance in plants (Yi et al. 2016). The spermidine gene is associated with plant growth promotion (Xie et al. 2014).

B. thuringiensis contains ZwittermicinA gene cluster exhibited an inhibitory or interfering effect on the growth of the phytopathogens carried putative mycolitic-chitinases, which might contribute to antifungal activities against *Verticillium* species (Hollensteiner et al. 2017). Bacteriocin gene clusters (e.g., thuricine-like, lichenicidins-like, amylolysin-like, paeninodin-like, subtilosin A, amylocyclicin, lactococcin 972 family bacteriocin) are present in higher frequency in *B. toyonensis* BAC3151, which regulates *N*-acyl homoserine lactonase and chitinases, revealing a genetic repertoire for antimicrobial synthesis in plants (Lopes et al. 2017). *B. atrophaeus* GQJK17 contains eight candidate gene clusters for producing antimicrobial secondary metabolites, including surfactin, bacillaene, fengycin, and bacillibactin, and controls phytopathogens (Ma et al. 2018). The *B. amyloliquefaciens* FZB42 contained alkaline serine protease *AprE* gene encoding proteins that eventually kill plant pathogenic nematodes (Chen et al. 2007). Insecticidal protein gene (*Sip1Ab*) expression of *B. thuringiensis* involves controlling coleopteran insect *Colaphellus bowringi* Baly (Sha et al. 2018).

Various *Bacillus* spp. have been shown to be nematicidal. *B. cereus* produces sphingosine, a nematicidal toxin, as a secondary metabolite (Engelbrecht et al. 2018). The production of kanosamine, zwittermicin A, C16 sphingosine, and phytosphingosine inhibits the growth of phytopathogens as well as nematodes (Gao et al. 2016; Ramezani Moghaddam et al. 2014). *B. firmus* act as plant growth promoters by paralysis and inhibition of egg hatching of plant pathogenic nematodes by production of nematicidal serine protease Sep1 (Jansen-Girgan et al. 2016; Xiong et al. 2015). *B. subtilis* produces various antibiotics which are largely lipopeptides that can be divided into surfactin, iturin, or fengycin families. Presence of the purL gene also helps in production of enzymes such as protease, chitinase, and gelatinase linked to its nematicidal activity (Abbasi et al. 2017; Roy et al. 2017). *B. nematocida* produces two extracellular alkaline serine proteases, Bace16 and a neutral protease Bae16, which cause degradation of the nematode cuticle (Niu et al. 2007). *B. velezensis* (Xiang et al. 2017), *B. mojavensis* (Xiang et al. 2017), *B. circulans* (El-Hadad et al. 2010), *B. megaterium* (Huang et al. 2010; Padgham and Sikora 2007), *B. pumilus* (Lee and Kim 2016; Ramezani Moghaddam et al. 2014), *B. coag-*

ulans (Abbasi et al. 2017), *B. methylotrophicus* (Zhou et al. 2016), *B. amyloliquefaciens* (Jamal et al. 2017; Sarangi et al. 2014), *B. licheniformis* (Colagiero et al. 2017), and *B. weihenstephanensis* (Sarangi et al. 2014) also control plant pathogenic nematodes.

B. thuringiensis is well-known for its broad spectrum of insects (e.g., Coleoptera, Diptera, Hemiptera, Hymenoptera, and Lepidoptera) and larvicidal and anti-cancer activities leading to an original toxin combination (Palma et al. 2014). It is postulated that one species of *B. thuringiensis* is capable of inhibiting growth of another sensitive strain in a mixed infection, as well as inhibiting growth of the normal gut flora (Borchert et al. 2010; Dieterich et al. 2008). *B. thuringiensis* BLB406 is completely divergent from known *B. thuringiensis* strains, containing five cry genes (*cry11A*, *cry22A*, *cry2A*, *cry60B*, and *cry64A*) responsible for quinolone, glyoxalase/bleomycin, tetracycline, fosmidomycin, and methicillin antibiotic resistance and two *vip* genes (VanR-F and VanZ-F) responsible for vancomycin resistance, which encode for novel toxins in the biological control (Zghal et al. 2018). The genome also harbors genes encoding bacitracin and corresponds to a broad-spectrum antibiotic which targets both Gram-positive and Gram-negative bacteria. The resistance to bacitracin could be very important in the production of *B. thuringiensis* BLB406 bioinsecticides against *Aedes aegypti* (Zghal et al. 2018). Chemistry and biological diversity of secondary metabolites from marine *Bacillus* spp. have already been reviewed (Mondol et al. 2013) (Table 10.1, Fig. 10.2).

10.4 Enhancement of Abiotic Stress Tolerance in Plants

The major abiotic stresses such as salinity, alkalinity, heat stress, metal-induced stress, water stress, drought, and cold stress significantly impact on agricultural productivity. *Bacillus* spp. have huge potential to enhance abiotic tolerance to stress in plants. *B. megaterium* NMp082, containing 1-Aminocyclopropane-1-carboxylate deaminase activity (Chinnaswamy et al. 2018), which helps to reduce stress-induced ethylene levels, inducing tolerance, root elongation and growth promotion in alphalpha plants subjected to salt, drought and heavy metal-induced stresses (Ali et al. 2014; Glick 2014). *Bacillus* spp. help to uptake more water and essential nutrients (N, P, K^+, Ca^{2+}, Mg^{2+}, Zn^{2+}, Mn^{2+}, Fe^{2+}, and Cu^{2+}) and thus inhibit the formation of reactive oxygen species as well as mitigate cellular disruption of plants during water stress conditions (Lopes et al. 2018). Gagne-Bourque et al. (2016) observed that *Bacillus* spp. stimulate the synthesis of chlorophylls a and b and carotenoids, triggering the production of metabolites, and increase photosynthesis. Drought stress-related genes are differentially expressed by *Bacillus* spp. by production of stress-related hormones (salicylic acid, jasmonic acid, and abscisic acid) (Radhakrishnan et al. 2017). Passive flow of Na^+ into plant cells was also reduced by *Bacillus* spp. under saline conditions (Ashraf et al. 2004). It reduces heavy metal toxicity of plants by stimulating pigment synthesis and modulating endogenous antioxidants (Lopes et al. 2018; Ahmad et al. 2014; Jamil et al. 2014). *B. subtilis*

Table 10.1 Important bioactive compounds discovered from *Bacillus* spp.

Name	Function	Molecular mechanisms	References
Bacillibactin	Plant growth promotion through siderophore production	Exported via YmfD and Mta gene, taken up when Fe^{3+}loaded by the FeuABC transporter, and cleaved intracellularly by the BesA esterase to release its iron	Miethke et al. (2006, 2008)
Bacitracin	Inhibit pathogenic bacterial growth	Inhibit pathogenic bacterial growth by preventing the dephosphorylation of C55-undecaprenyl pyrophosphate (bactoprenol) and subsequent recycling of the lipid carrier	Harwood et al. (2018)
Plipastatin	Biosurfactant and antifungal activity	Involve the inhibition of phospholipase A2 and the formation of pores in fungal membranes	Honma et al. (2012), Gong et al. (2015)
Surfactin	Surfactant activity	Non-specific cytolytic activity, composition of the target phospholipid bilayer influences its penetration into host cells	Gong et al. (2015), Okada et al. (2015)
Bacillaene	Antibacterial agent	Unknown	Chen et al. (2018)
Fengycin	Biosurfactant and antifungal activity	Reduce the mitochondrial membrane potential (MMP), induce bursts of reactive oxygen species (ROS), and downregulate the expression level of ROS-scavenging enzymes resulting in the upregulation of DNA repair-related proteins expression and the cleavage of poly (ADP-ribose) polymerase (PARP)	Zhang and Sun (2018)
Bacilysin	Antimicrobial activities	Inhibit the activity of glucosamine-6-phosphate synthase through the release of anticapsin	Wang et al. (2018)
Iturin	Antifungal activity against important fungal pathogens (e.g., *Rhizoctonia*, *Penicillium*, *Aspergillus*, *Fusarium*, and *Pyricularia*)	The antifungal activity of iturins is related to its interaction with the cytoplasmic membrane of target cells, resulting in the formation of ion-conducting pores and increased K+ permeability	Gong et al. (2015)
Macrolactin	Antimicrobial activities		Ahmad et al. (2017), Belbahri et al. (2017)
Difficidin	Antimicrobial activities		Ahmad et al. (2017), Belbahri et al. (2017)

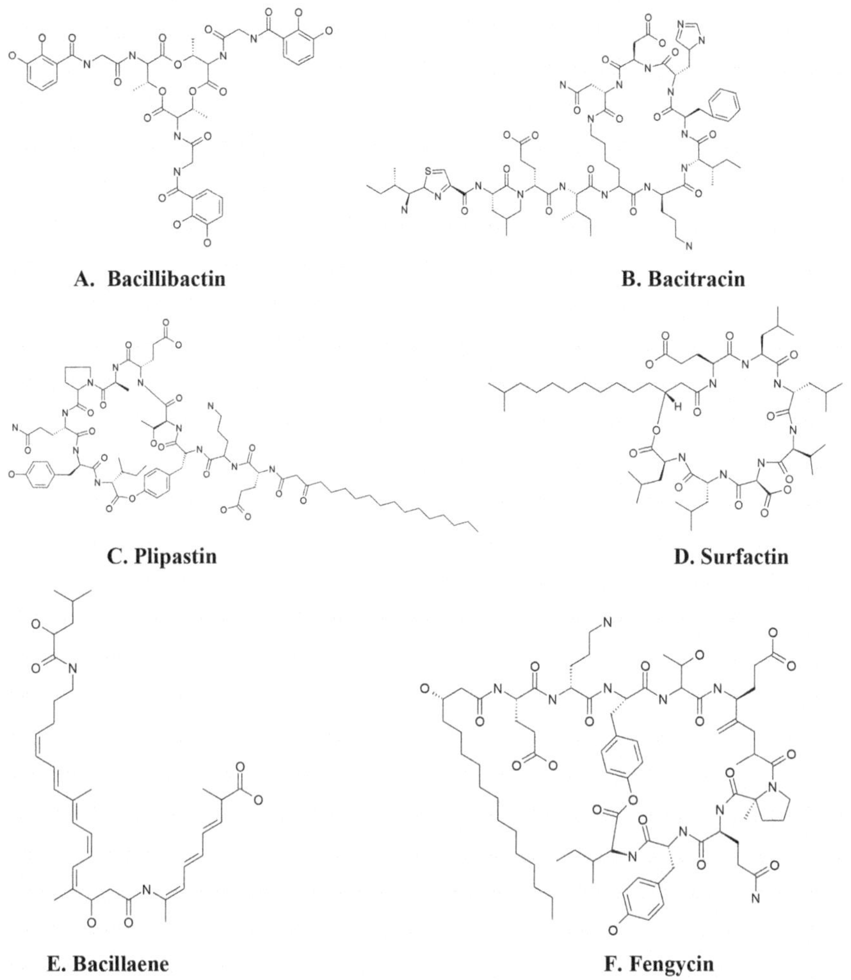

Fig. 10.2 Chemical structures (**A**-**J**) of some major secondary metabolites discovered from *Bacillus* spp.

strain 330-2 contains genes associated with tolerance to heat (elongation factor Tu; aspartokinase II; and dihydroorotase, *pyrC*), salinity [cardiolipin synthase, *ywiE*; glutaminase-1, *ybgJ*; phosphor glyceratemutase, *gpml*; 4-hydroxy-3-methylbut-2-enyl diphosphate reductase, *ispH* and glutamate synthase (NADPH/NADH), *gltAI*], drought (cystathionine beta-lyase and cysteine desulfhydrase) and cold stress (sporulation cortex protein, *coxA*), and differentially expressed in the *B. subtilis* strain 330-2. NADH-dependent butanol dehydrogenase A (*yugJ*) and formate dehydrogenase (*yjgC*) are commonly involved in tolerance to heat, salinity, and oxidative stresses. Also, expressed glutamate symporter (*gltT*) and *S*-adenosylmethionine synthase (*metK*) genes are commonly involved in enhancing the plant's tolerance to

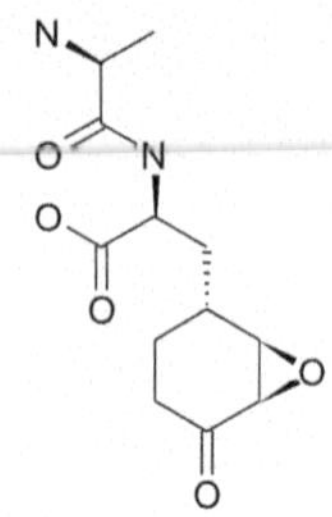

G. Bacilysin

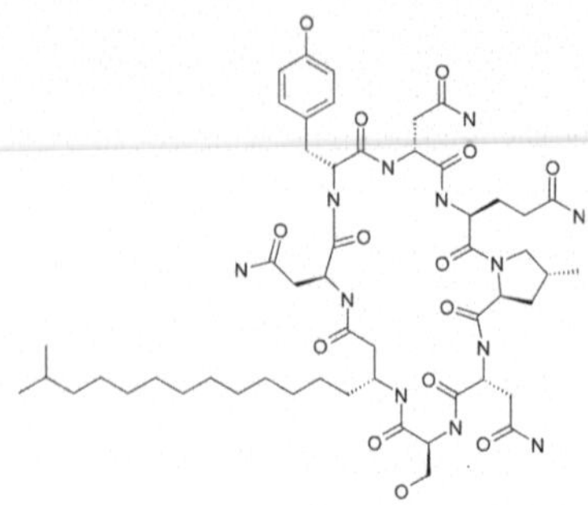

H. Iturin

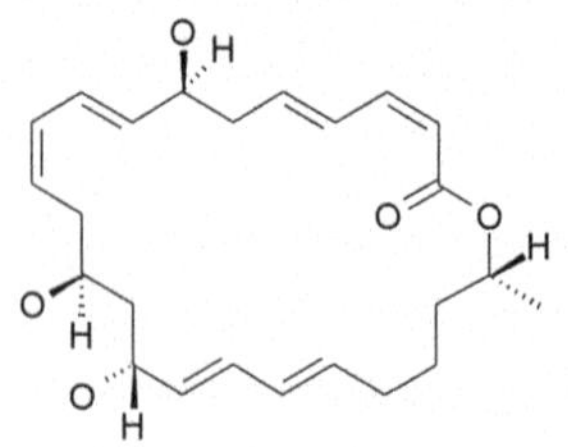

I. Macrolactin

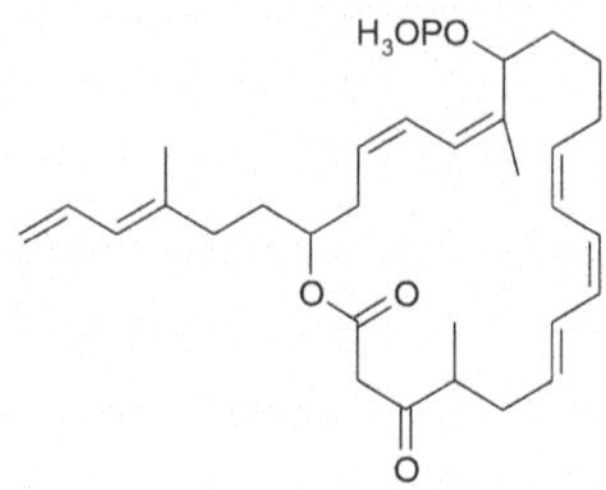

J. Difficidin

Fig. 10.2 (continued)

drought and salinity (Ahmad et al. 2017). *B. licheniformis* K11 upregulated the differentially expressed genes, viz., Cadhn (dehydrin-like protein), VA (vacuolar H^+-ATPase), sHSP (small heat-shock protein), and CaPR-10 (pathogenesis-related protein 10) in pepper root and induced tolerance to both biotic and abiotic stresses and suppressed phytophthora blight of pepper (Lim and Kim 2013). Secretion of 1-aminocyclopropane-1-carboxylate (ACC) deaminase by *Bacillus* sp. inhibits ethylene synthesis and thus plant senescence (Yaish et al. 2015). Antifungal volatile compound, 5-methyl-2-phenyl-1H-indole, produced by *B. megaterium* KU143 strain, controls *Aspergillus* and *Penicillium* spp. during rice grain storage (Mannaa and Kim 2018).

10.5 Genomic Analysis of Bacilli

The advent of next-generation sequencing (NGS) technologies and the possibility of obtaining high-quality whole-genome sequences have allowed scientists to show the potential metabolic landscape of *Bacilli*. Thus, this has created greater potential to produce specialized metabolites now compared to what has been discovered by classic screening-based methods in the past five decades. The genetic potential to produce approximately a tenfold higher number of secondary metabolites, antibiotics, enzymes, and volatiles has been detected during screening and chemical analytics (Bentley et al. 2002; Challis, 2014). These observations have strongly contributed to a renewed interest in studying the molecular mechanisms of *Bacilli* in recent years. Bioinformatics tools helped to uncover the untapped metabolic potential of *Bacilli*, triggering a clear understanding of its molecular mechanisms. Genomic analyses of *Bacilli* revealed that silent gene clusters are present even within its genome. Genome mining has been treated as powerful tool to estimate the genetic potential of *Bacilli* by scanning the genome of interest and identifying mechanisms of its growth-promoting attributes (Ziemert et al. 2016). Many tools were extensively used to determine the growth-promoting traits of *Bacilli* (Tables 10.2 and 10.3).

Table 10.2 Bioinformatics tools used for genome mining of *Bacillus* spp.

Function	Tools	References
Computational mining for secondary metabolite biosynthetic gene clusters	Shell (antiSMASH) and connected antiSMASH database	Blin et al. (2017), Medema et al. (2011), Palazzotto and Weber (2018)
Prediction of secondary metabolomes	Prediction Informatics for Secondary Metabolomes (PRISM)	Skinnider et al. (2015), Belbahri et al. (2017)
Allows automatic matching of polyketide and NRPS structural data to predicted biosynthetic gene clusters	Global alignment for natural products cheminformatics (GARLIC)	Dejong et al. (2016)
Allows automatic matching of polyketide and NRPS structural data to predicted biosynthetic gene clusters	Generalized Retrobiosynthetic Assembly Prediction Engine (GRAPE) platform	Dejong et al. (2016)
Largest database of automatically mined biosynthetic gene clusters identified in microbial genomes	IMG/ABC	Hadjithomas et al. (2017)
Growth-promoting and antagonistic genes identification by multilocus phylogeny	Multilocus phylogenetic analysis (MLPA)	Douriet-Gamez et al. (2018)
Bacteriocin-specific software	BAGEL4	van Heel et al. (2018)
Prediction of secondary metabolomes	NaPDos	Ziemert et al. (2016), Belbahri et al. (2017)
Prediction of secondary metabolomes	NP.searcher	Li et al. (2009), Belbahri et al. (2017)
Secondary metabolomes	Bacteriocin-specific software BAGEL3	van Heel et al. (2013), Belbahri et al. (2017)

Table 10.3 Examples of gene or gene clusters predicted for plant growth promotion by *Bacilli*

Gene clusters/ gene	Natural products	Growth-promoting traits	References
NifDHK	Nitrogenase	Biological nitrogen fixation	Seldin et al. (1989), Oliveira et al. (1993)
Nas	Nitrate uptake	Nitrogen assimilation	Luque-Almagro et al. (2011)
phyABCPR	3-Phytase	Phosphorus mineralization	Kim et al. (1998)
phoAB	Phosphatase	Phosphorus mineralization	Hulett et al. (1990)
Gdh	Glucose 1dehydrogenase	Gluconic acid biosynthesis	Goa (2001), Camon et al. (2004)
krtABCD	Ktr system potassium uptake protein	Potassium uptake	Holtmann et al. (2003)
dhbABCF	Bacillibactin	Siderophore biosynthesis	May et al. (2001)
yvrE	Gluconolactonase	Ascorbate and aldarate metabolism and caprolactam degradation	Reshma et al. (2017)
kdpABCD	K uptake ATPase	Potassium uptake	Vieira-Pires et al. (2013)
ypqP, CapD	Polysaccharide biosynthesis enzyme	Biofilm formation	Niazi et al. (2014a, b)
Che	Chemotaxis	Quorum sensing	Niazi et al. (2014a, b)
spsABCDEFG	Subtilis synthesis enzymes	Sporulation and biofilm formation	Charnock and Davies (1999)
yhcX, ysnE, ipdC	Nitrilase, acetyltransferase, indole-3-pyruvate decarboxylase	Tryptophan-dependent IAA biosynthesis	Shao et al. (2015), Read et al. (2003)
metK, acdS	Adenosylmethionine and (ACC) 1-aminocyclopropane-1-carboxylate	Ethylene biosynthesis	McDaniel et al. (2006)
Cyt	Cytokinin	Cytokinin presence found	Ortíz-Castro et al. (2008)
srf ABCD	Surfactin	Biosurfactant activity	Niazi et al. (2014a, b), Jasim et al. (2016), Govindasamy et al. (2017)
lipAB	Lipase	Lysis, stress resistance	Shi et al. (2010), Eggert et al. (2003), Eppinger et al. (2011)

(continued)

Table 10.3 (continued)

Gene clusters/ gene	Natural products	Growth-promoting traits	References
abnA	Endo-arabinase	Cell lysis	Niazi et al. (2014a, b)
xynA	Xylanase	Cell lysis	
bglS	Lichenase	Cell lysis	
ydbD	Mn catalase	Cell lysis	
nrsABCDEF	Putative peptide	Lysis, biocontrol	
bacABCDEGH	Bacilysin	Antibiosis, biocontrol	
arsABC	Arsenite-transporting ATPase	Arsenate resistance	
KatA, *KatE*, *KatX*	Catalases	Oxidative stress reduction	
SodACF	Superoxide dismutases	Oxidative stress reduction	
alsSD	Acetoin	VOC, biocontrol	
bdhA	R,R-butanediol dehydrogenase	VOC, biocontrol	
swrA	Flagellar motility	Movement	
htrABC	Serine protease	Lysis, biocontrol	Noone et al. (2001)
Hsp	Heat-shock protein	IST	Todd et al. (1985)
eglS	Glucanase	Lysis, biocontrol	Han et al. (1995)
pelB	Pectin lyase	Lysis, biocontrol	Kunst et al. (1997)
pelC	Pectate lyase	Lysis, biocontrol	Soriano et al. (2006)
chiACHW	Chitinase	Lysis, biocontrol	Pleban et al. (1997), Huang and Chen (2005), Driss et al. (2005), Earl et al. (2012), Tang et al. (2017)
fenABCDE	Fengycin	Antibiosis, biocontrol	Ahmad et al. (2017), Lopes et al. (2017), Blacutt et al. (2016)
mlnABCDEFG	Macrolactin	Antibiosis, biocontrol	Niazi et al. (2014a, b), Ahmad et al. (2017)
lchABCTE	Lipoheptapeptide biosurfactants	Biosurfactants production	Anuradha (2010), Shaligram et al. (2016)
glgXYZ	Trehalose biosynthesis enzyme	Abiotic tress tolerance	
treABPR	Trehalohydrolase synthesis	Abiotic tress tolerance	Schöck and Dahl (1996)
baeJLMNR	Bacillaene	Antibiosis, biocontrol	Niazi et al. (2014a, b), Ahmad et al. (2017)
Hmp	Nitric oxide dioxygenase	Adaptive detoxification	Nakano (2002), Han et al. (2006)

(continued)

Table 10.3 (continued)

Gene clusters/ gene	Natural products	Growth-promoting traits	References
ituAB	Iturin	Secondary metabolites production	Ahmad et al. (2017)
Pen P	Beta-lactamase	Secondary metabolites production	Ahmad et al. (2017)
Pbp B	Penicillin-binding protein 2B	Secondary metabolites production to control phytopathogen	Ahmad et al. (2017)
Cry	Cry gene	Entomotoxic effect	Djenane et al. (2017)
cshABCD	Cold RNA helicases	IST cold	Hunger et al. (2006)
bmy ABCD	Bacillomycin D	Antibiosis, biocontrol	Jasim et al. (2016)
dfnDEFGHIJ	Difficidin	Antibiosis, biocontrol	Niazi et al. (2014a, b)

10.6 Comparative Genomics for Genetic Engineering

Genome mining and comparative genomic analysis have initiated the prospect of prioritizing *Bacilli* for genetic engineering. The complete genome sequencing and comparative analysis allowed for the reconstruction of both primary and secondary biosynthetic pathways suggesting key growth-promoting genes for genetic engineering. The genomic analysis of *Bacillus* sp. underlined the genetic potential to synthesize novel specialized metabolites of this genus. By genomics approaches, biosynthetic gene clusters were detected, including NRPS, PKS, terpenes, and siderophore clusters (Carro et al. 2018). The analysis of whole-genome sequence data of the genus *Bacillus* provided insight into its metabolic and biotechnological potential (Table 10.4).

There is a correlation between genome size and the number of gene clusters of different *B. amyloliquefaciens* strains known to be involved in secondary metabolite biosynthesis and mined by antiSMASH and PRISM. About 65% of the variance in the number of secondary metabolite clusters can be explained by genome size for antiSMASH, for instance. Dynamics of evolution of the clusters of different genes were also investigated using comparative genomics across all known core and accessory genomes of *B. amyloliquefaciens* strains (Belbahri et al. 2017).

Through comparative genomics, Douriet-Gamez et al. (2018) compared the *B. cereus* group and biocontrol agents of phytopathogenic fungi. They unveiled that *B. cereus* contains the genes necessary for controlling the growth of maize pathogen *Fusarium verticillioides*, including competing for nutrients, mycoparasitism, antibiosis, and biofilm formation (Fig. 10.3).

Table 10.4 Comparative genomic features of different *Bacillus* spp.

Strain	Host plant	Size (Mb)	G+C%	Total gene	Features	Accession number (GenBank)	References
B. subtilis BSn5	*Amorphophallus konjac*	4.09	43.80	4237	Antibacterial activity	CP002468.1	Deng et al. (2011)
B. velezensis CC09	*Cinnamomum camphora*	4.17	46.10	4128	Antifungal activity	CP015443.1	Cai et al. (2016), Pylro et al. (2018)
B. flexus KLBMP 4941	*Limonium sinense*	4.10	37.97	4244	Plant growth promotion	CP016790.1-CP016792.1	Wang et al. (2017)
B. pumilus GB34 (*B. pumilus* INR7)	*Cucumis sativus*	3.68	41.30	3820	Elicitation of ISR and plant growth promotion	AYTK00000000.1	Jeong et al. (2014)
B. amyloliquefaciens XK-4-1	*Gossypium* spp.	3.94	46.40	3915	Antifungal activity and plant growth promotion	LJDI00000000.1	Sun et al. (2015)
B. mojavensis RRC101	*Zea mays*	4.03	43.70	4227	Antifungal activity and plant growth promotion	ASJT00000000.1	Gold et al. (2014)
B. toyonensis BAC3151	*Phaseolus vulgaris*	5.74	34.90	6043	Antibacterial activity	LDKD00000000.2	Lopes et al. (2017)
B. aryabhattai SQU-R12	*Phoenix dactylifera*	5.58	37.70	5718	Plant growth promotion	NHZZ00000000.1	Yaish et al. (2015)
B. thuringiensis KB1	*Arabidopsis thaliana*	5.75	35.00	6175	Antifungal and antibacterial activity	LSNJ00000000.1	Jeong et al. (2016)
B. velezensis strain AGVL-005	*Soybean seed*	4.14	45.98%	6261	Phytopathogen biocontrol	CP024922	Pylro et al. (2018)
B. amyloliquefaciens IT-45		3.93		3945	Plant growth promotion	CP004065.1	Belbahri et al. (2017)

(continued)

Table 10.4 (continued)

Strain	Host plant	Size (Mb)	G+C%	Total gene	Features	Accession number (GenBank)	References
B. amyloliquefaciens Y2	wheat rhizosphere	4.23			Suppresses a broad spectrum of pathogenic fungi, such as *Phytophthora capsici*, *Colletotrichum orbiculare*, *Fusarium moniliforme*, and *Magnaporthe griseus*	CP003332.1	Belbahri et al. (2017)
Bacillus sp. B25		5.11	35.60%	5360	Maize pathogen *Fusarium verticillioides*		Douriet-Gamez et al. (2018)
B. subtilis 2C-9B	Rhizosphere of wild grass	4.21		4823	Biocontrol of chili pepper rot pathogen	MOXE01000001	Munoz-Moreno et al. (2018)
B. toyonensis BAC3151	Leaves of the common bean	5.74	34.90%,	6055	Biocontrol of plant pathogens	LDKD00000000.1	Lopes et al. (2017)
B. atrophaeus GQJK17	Rhizosphere of *Lycium barbarum* L.	4.32	43.30%.	4294	Biocontrol of plant pathogens	NZ_CP022653.1	Ma et al. (2018)
*B. velezensis*OSY-S3	Silage	3.90	46.50%		Antimicrobial against *L. innocua*, *Escherichia coli*, *Penicillium* sp., *Cladosporium* sp., and *S. aureus*	CP024706	Gerst et al. (2018)
B. paralicheniformis strain KMS 80	Rice	4.58	45.50	4598	Biological nitrogen fixation and plant growth-promoting abilities	MUEI00000000	Annapurna et al. (2018)
B. altitudinis Lc5,	Black rice	3.74	41.89	3950	Plant growth-promoting abilities and biocontrol of phytopathogens	QCWN00000000.	Potshangbam et al. (2018)

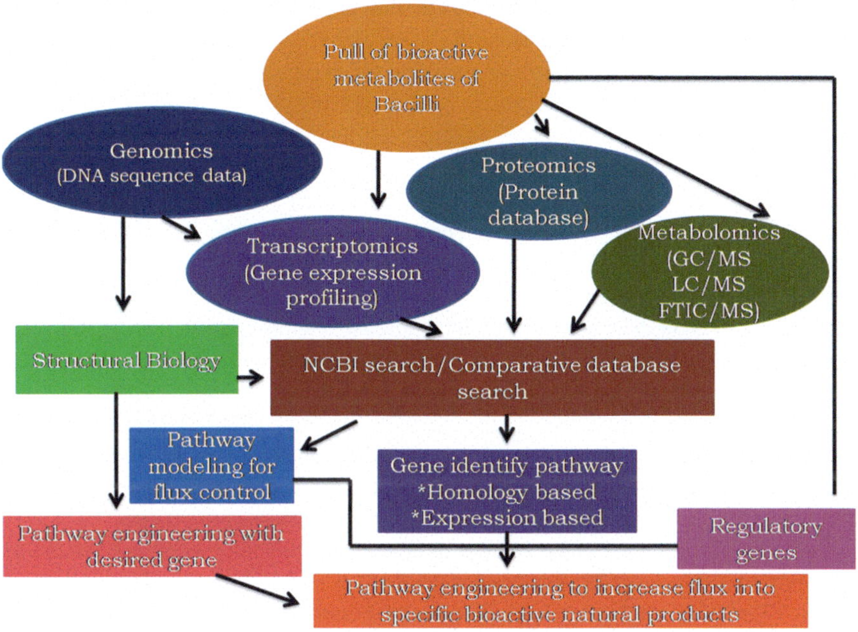

Fig. 10.3 Genomics and post-genomics approaches for dissecting the roles of bioactive natural products in *Bacilli*

10.7 Transcriptomics Analysis of Bacilli

The DNA sequencing has revealed the high abundance of biosynthetic gene clusters in the *Bacilli* genome, but a majority of its actual products have not been characterized properly. These clusters often remain silent under laboratory conditions due to the complex regulatory mechanisms at transcriptional, translational, and posttranslational levels. To unveil the mechanisms controlling the metabolic regulation, development differentiation, and bioactive natural products biosynthesis, transcriptional changes in gene expression levels have been extensively studied (Liu et al. 2013; Palazzotto et al. 2015; Vockenhuber et al. 2011).

A widely used strategy to trigger the expression of gene clusters and production of bioactive natural products is the use of biological, chemical, and molecular agents/elicitors (Abdelmohsen et al. 2015). Different studies highlight the transcriptomic analysis of *Bacillus* spp. which trigger the growth promotion of plants. The different expression profiles suggest that endophytic *B. mycoides* showed different genetic adaptation strategies to recognize or respond to plant-released signals. The upregulation of genes related to amino acids metabolism, several proteolytic enzymes, and *O*-glycosyl hydrolases helped in specific adaptation to the ecological niche and a good rhizosphere fitness of the endophytic *B. mycoides* (Yi et al. 2017). The endophytic *Bacilli* is able to change its gene expression pattern and thus adapt the metabolism toward a physiological state that enables an optimal nutrient acquisition, competition with species from the same niche, and colonization of the plant. Its IclR gene is related to multidrug resistance (Song et al. 2004; Molina-Henares et al. 2006) and a stress induced protein (BG05_RS01935), a stress response protein (BG05_RS28400), and a general stress protein (BG05_RS26065) were specifically upregulated and genes encoding a methanol dehydrogenase (BG05_RS02305) and a streptomycin biosynthesis protein (BG05_RS25055) were also upregulated (Yi et al. 2017). The sigma-28 factor is reported to transcribe the flagellin gene and control the transcription of a regulon specifying flagellar, chemotaxis, and motility functions in *B. subtilis* (Helmann and Chamberlin 1987; Mirel and Chamberlin 1989). *B. subtilis* S499 activates a systemic defense response against *Colletotrichum lagenarium* (Ongena et al. 2005). In *Arabidopsis*, seed treatment with *B. subtilis* GB03 activates the signaling pathway of ethylene, independently of the salicylic acid or jasmonic acid signaling pathways (Kang et al. 2018a, b), triggering defense responses against *Erwinia carotovora* subsp. (Ryu et al. 2004). In soybean, seed treatment with *Bacillus simplex* strain Sneb545 induces resistance against soybean cyst nematode (Xiang et al. 2013).

Two non-ribosomal peptide synthetases and a beta-glucanase gene were found in the genome of *B. subtilis* 2C-9B (Muñoz-Moreno et al. 2018), suggesting antifungal activity (Tapi et al. 2010). Also, a butanediol dehydrogenase gene and a spermidine synthase gene were found, with butanediol being a potential inducer of systemic resistance in chili pepper plants against chili pepper root pathogens (Yi et al. 2016).

10.8 Proteomics Analyses of Bacilli

In addition to genomics and transcriptomics, proteomics also has been used as powerful process to shed light on the relationship among metabolic pathways and natural products synthesis, like antibiotics, enzymes, volatiles, etc. Comparing protein expression levels, proteomics provides information on differential pathway regulation highlighting key players in biosynthesis of bioactive products that can be used as target-oriented rational engineering.

An extensive analysis of proteomic investigations found that certain functionally important defense-related proteins, viz., putative late blight resistance protein homolog, toll-interleukin-resistance domain-containing protein, translation initiation factor IF1, disease resistance protein putative Kalata-B1, and β-1,3-glucanase, were induced by *B. subtilis* which are involved in the induction of defense response of host against the pathogen, *M. incognita* (Govindasamy et al. 2017). Kim et al. (2017) demonstrated that *Bacillus subtilis* GB03 and *B. amyloliquefaciens* IN937a were reported to promote *Arabidopsis* growth via the production of bacterial volatile 2,3-butanediol (Ryu et al. 2003). Moreover, *B. megaterium* XTBG34 promoted the growth in *Arabidopsis* by emitting 2-pentylfuran (Zou et al. 2010). Proteomic analysis revealed transcriptional regulator (*PhcA*) of virulence of tomato wilt pathogen *Ralstonia solani* downregulated by *B. amyloliquefaciens* SQR-9 (Raza et al. 2016).

As a result, *B. amyloliquefaciens* SQR-9 reduces the metabolic or respiratory activity of *Ralstonia solani* and decreased the production of antioxidant enzymes of *R. solani*. The ABC transporter system of *B. amyloliquefaciens* SQR-9 has phosphate transporter and binding protein which caused structural changes on the cell envelop or damaged the cell wall of *R. solani*. Overexpression of ABC transporter proteins helps the osmo-adaptation and maintenance of cell integrity, which facilitates them in survival of undesirable changes of the host cell. Iborta et al. (2018) found that *B. megaterium* inoculation increased SlPT1 phosphate transporter expression in non-ripening tomato plants. It induced an oxidative stress-related proteins such as glutathione *S*-transferase (GST) (Edwards and Dixon 2010; Edwards and Dixon 2005), heat-shock protein 70 (Hsp70) (Al-Whaibi 2011; Scarpeci et al. 2008), and neutral ceramidase, which improve resistance to oxidative stress responses (Li et al. 2015) of non-ripening tomato. Wang et al. (2017) demonstrated the proteomic alternations induced by Cry6A toxin produced by *B. thuringiensis* in *Caenorhabditis elegans* altering the protein profile of *C. elegans*. Cry6Aa2 protein significantly altered the immune defense, insulin-like receptor (ILR) signaling pathway, energy metabolism, and muscle assembly of *C. elegans*. The differentially expressed proteins like DIM-1 and galectin LEC-6 of *B. thuringiensis* functioning in diverse biological processes participate in the defense and stress responses of *C. elegans*.

10.9 Metabolomics Data on *Bacillus* spp.

The metabolomics is based on the comprehensive analysis of the low-molecular-weight metabolite complement of a biological sample. In fact, the metabolites are downstream of both gene transcription and enzyme activities; metabolomics has the potential to give a more accurate picture or snapshot of the actual physiological state of a cell or cells including bacteria (Griffin 2003). Metabolomic analysis has been successfully applied, providing valuable information on plant-pathogen interactions and system-wide variations in plant metabolism under pathogen infection and allowing the identification of compounds that play a pivotal role in plant innate immunity (Hagemeier et al. 2001; Desbrosses et al. 2005; Colebatch et al. 2004; Kang et al. 2018a, b). Metabolomics data allow for an in-depth study of the molecular mechanisms of *Bacillus* spp. in suppression of microorganisms (Kang et al. 2018a, b). A large body of literature is available on metabolomics studies of *Bacillus* species using major metabolomics analytical tools such as capillary electrophoresis-mass spectrometry, GC-MS, LC-MS, LS-MS-MS, and ^{1}H-NMR analyses (Bundy et al. 2005; Meyer et al. 2013; Kang et al. 2018a, b). By using a comprehensive metabolomics study, Vinci et al. (2018) concluded that *B. amyloliquefaciens* inoculation in combination with composts led to an increased content of sugar monomers (glucose and fructose), amino acids (GABA, alanine, and leucine), and membrane phospholipids (phosphatidylcholine), which results in improved photosynthetic activity, growth, and nutrition (Vinci et al. 2018). Gas chromatography-mass spectroscopy (GC-MS) provides a high-throughput platform to investigate bacterial strains. By this method, four nematicidal metabolites (4-vinylphenol, L-methionine, piperine, and palmitic acid) were identified in *Bacillus simplex* Sneb545 which inhibited the soybean cyst nematode of soybean (Kang et al. 2018a). *B. subtilis* RR4 induced malate synthase and reduced malate transporter *OsALMT* in rice roots, resulting higher malic acid synthesis in rice roots but limits the exudation of malic acid through roots. It suggests that the metabolism of rice plant is enhanced and could ultimately lead to the growth promotion of rice (Rekhaa et al. 2017). *Bacillus* sp. produces volatile organic compounds (VOCs), such as acetoin, 2,3-butanediol, albuterol, and 1, 3-propanediol, which regulate the metabolism of phytohormones (Tahir et al. 2017; Rath et al. 2018).

10.10 Metagenomics

Metagenomics is one of the most popular approaches to study the chemistry of uncultivated bacteria. Through metagenomic analysis of ginseng plant roots, Dong et al. (2018) confirmed that *B. subtilis* 50-1 acts as a biocontrol agent against the root rot pathogen *Fusarium oxysporum* (Palazzotto and Weber 2018). In 2017, Borris group worked on a large-scale lysine malonylation analysis in Gram-positive *B. amyloliquefaciens* FZB42. Their bioinformatic analysis revealed that the expression of various enzymes involved in the plant-bacteria interaction and that bacteria produce various antibiotics which are beneficial for plants. Using the PRISM method from unsequenced environmental *Bacillus* spp., the cyclic Imine koranimine shed a light on the mechanisms used by *Bacillus* spp. to regulate physiological processes and natural product production as well as helps in plant growth (Palazzotto and Weber 2018) (Table 10.5).

10.11 Genome Editing of *Bacilli* Through CRISPR-Cas9 System

Clustered regularly interspaced short palindromic repeats (CRISPR) with RNA-guided DNA endonuclease Cas9 is a system which not only eliminated the use of selection markers but also dramatically increased the genome editing efficiency (Jinek et al. 2012; Haque et al. 2018; Bhowmik et al. 2018). This system is widely adopted for genome editing in many organisms including *Bacillus* spp. (Hong et al. 2018). Single plasmid CRISPR-Cas9 method decreased the barrier to the host species and results in high rate of survival of *B. subtilis*. Previously several scientists reported that CRISPR-Cas9 system generated a precise gene deletion (Altenbuchner 2016), point mutation and gene insertion (So et al. 2017), and chromosomal maintenance (Westbrook et al. 2016) in *B. subtilis*. To overcome the instability of plasmid, transformation efficiency, and metabolic burden on the host cell, chromosomally

Table 10.5 Some examples of the omics approaches used in *Bacilli* strains

Bacilli strains	Omics approaches	Major findings	References
B. pumilus Jo2	Proteomics and metabolomics	Functional secretion system including the components of the Sec- and Tat-secretion machinery	Handtke et al. (2014)
B. megaterium	Proteomics	Modifying levels of stress-related and interaction proteins and showing bacterial inoculation effects on antioxidant content and phosphorus acquisition capacity	Iborta et al. (2018)
B. altitudinis FD48	Transcriptomics	*OsIAA1* and *OsIAA4* genes responsible for shortening of primary roots and *OsIAA11* and *OsIAA13* genes for lateral root formation	Ambreetha et al. (2018)
Bacillus simplex Sneb545	Transcriptomics and metabolomics	Nematicidal metabolites, viz., 4vinylphenol, methionine, piperine, and palmitic acid	Kang et al. (2018a)
B. amyloliquefaciens	Metabolomics	Glucose, fructose, alanine, and GABA metabolites increased in maize leaves	Vinci et al. (2018)
B. thuringiensis	Metagenomics	Cry proteins are predisposed shifts in the bacterial endophytes' community associated with maize shoots	Mashiane et al. (2017)
Bacillus spp.	Metagenomics	Major endophytic genera in rice roots and playing a key role in nitrogen fixation	Sengupta et al. (2017)
B. subtilis	Proteomics	Defense-related proteins, viz., putative late blight resistance protein homolog, Toll-interleukin-resistance domain-containing protein, translation initiation factor IF1, disease resistance protein putative Kalata-B1, and β-1,3-glucanase, were induced by *B. subtilis*	Govindasamy et al. (2017)
Bacillus sp. JS	Proteomics	Chlorophyll a-/b-binding proteins were significantly upregulated and increased total chlorophyll content	Kim et al. (2017)
B. amyloliquefaciens SQR-9	Proteomics	Downregulated antioxidant activity, virulence, carbohydrate and amino acid metabolism, protein folding, and translation of *Ralstonia solani*. Upregulated proteins involved in the ABC transporter system, amino acid synthesis, detoxification of aldehydes and ketones, methylation, protein translation, folding, and energy transfer	Raza et al. (2016)
B. thuringiensis	Proteomics	Defense pattern of *Caenorhabditis elegans* against *Cry6Aa2*	Wang et al. (2017)

(continued)

Table 10.5 (continued)

Bacilli strains	Omics approaches	Major findings	References
B. mycoides	Comparative transcriptomics	Different genetic adaptations of plant-associated endophytes and soil isolates of *B. mycoides* in response of potato root exudates	Yi et al. (2017)
Bacillus sp. B25	Comparative Genomics	Identified 24 genomic islands (chitinases, glycoside hydrolases, siderophores, antibiotics, biofilm-producing genes, etc.) and 3 CRISPR sequences	Douriet-Gamez et al. (2018)
B. thuringiensis	Genomics	Gene cluster for zwittermicin A, inhibit growth of *Verticillium longisporum*	Hollensteiner et al. (2017)
B. licheniformis K11	Proteomics	Stress protein specific genes of Cadhn, VA, sHSP, and CaPR-10	Lim and Kim (2013)
B. toyonensis BAC3151	Genomics	Biosynthetic genes of non-ribosomal peptides, *N*-acyl homoserine lactonase, and chitinases	Lopes et al. (2017)
B. subtilis	Proteomics	Biofilm-producing TasA protein changes its structure globular state to a fibrillar form at molecular level	Diehl et al. (2018)

encoded CRISPR-Cas9 cassette performed effectively to edit the genes of *B. subtilis* (Westbrook et al. 2016). Cas9 protein increased the availability of target space, but gRNA:FnCas9 are prone to target CG or TG (Hirano et al. 2016). Cpf 1 protein requires additional PAM on the 5′ region of the protospacer as well as tracerRNA during genome editing process (Dong et al. 2016; Jinek et al. 2012). Therefore, CRISPR-Cas9 system improves the efficiency and shortcoming of existing gene manipulation methods. This technique revolutionizes the editing of desirable genes or nucleotides in the genomes of Bacilli for the development commercially useful bioinoculants for agricultural use.

10.12 Is *Bacillus* Useful or Harmful for Humans?

B. cereus, *B. clausii*, and *B. pumilus* were widely used as human probiotics due to their high colonization, immune stimulation, and antimicrobial activity (Duc et al. 2004). Orally ingested nonindigenous *Bacillus* spp. have a probiotic effect in a host by (i) immune modulation (i.e., stimulation of the gut-associated lymphoid tissue by induction of cytokines), (ii) competitive exclusion of gastrointestinal pathogens (e.g., competition for adhesion sites), and (iii) secretion of antimicrobial compounds which suppress the growth of harmful bacteria (Fuller 1991). It is reported that *B. coagulans* is a spore-forming bacterium which exhibits characteristics of both *Bacillus* and *Lactobacillus* genera and is highly resistant to high temperatures with its probiotic activity (Konuray and Erginkaya 2018). Researchers reported that if *B. clausii*, *B. coagulans*, and *B. subtilis* containing probiotics are consumed on a regular basis, gastrointestinal disorders can be prevented (Hong et al. 2005; Jurenka 2012; Fijan 2014), and respiratory tract infections in children are reduced (Marseglia et al. 2007). Endres et al. (2009) reported *B. coagulans* for human consumption by bacterial reverse mutation test, chromosomal abnormality test, micronucleus test, acute and 90-day sub-chronic recurrent toxicity test, and an acute eye and skin irritation test in order to determine toxicological properties of *B. coagulans* (Konuray and Erginkaya 2018).

Bacillus anthracis possesses two plasmids, pX01 and pX02, which encode the lethal toxin and the poly-D-glutamic acid capsule, respectively, which would normally distinguish this bacterium from the closely related *B. cereus*. Although these two species differ in phenotypes and disease spectra produced, reports of pulmonary infections mimicking anthrax have been attributed to *B. cereus* strains harboring *B. anthracis* toxin genes (Hoffmaster et al. 2004). *B. cereus* strains, which produce the Hbl and Nhe enterotoxins, are unsafe for human use (Duc et al. 2004). On the other hand, plant-associated or plant probiotic and fish-associated or fish probiotics are commercially being used as safe growth-promoting and immune-enhancing agents in agriculture.

10.13 Conclusion

Plant growth-promoting *Bacillus* spp. produce a large spectrum of bioactive molecules with various antagonistic activities. *Bacillus* spp. have shown to be promising candidates for the biocontrol of phytopathogens and plant growth promotion. They are ubiquitous within the rhizosphere, promote plant growth, produce highly resistant endospores, and produce a plethora of secondary metabolites, including lipopeptides, polyketides, bacteriocins, lantibiotics, and dipeptides. Modern genomics and post-genomics approaches are able to profile and quantify the secondary metabolites of *Bacillus* spp. responsible for plant growth-promoting activities. The metabolomics and proteomics data of *Bacillus* spp. could be helped in synthetic engineering of beneficial bioactive secondary metabolites and used to combat phytopathogens as well as biotic and abiotic stresses. Metabolomics provides a valuable tool which can be used to screen a multitude of biologically active molecules with the aim of discovering features which are highly selective for agricultural pests. This emerging field of "genomics" studies will enable scientists to understand the complex chemistry of *Bacillus* spp. (Tyc et al. 2017). The field of omics can thus be applied to agricultural research that focuses on developing safe, environmentally friendly alternatives to agrochemicals like fertilizer and pesticides. This approach directly addresses national and international concerns regarding sustainable food production for the ever-growing global population (Aliferis and Chrysayi-Tokousbalides 2011). Standardized and universal procedures for omics are required to comprehensively analyze bacterial metabolites and directly influence the final omics data. This should aid in the identification and understanding of the various mechanisms used by *Bacillus* spp. Further, the identification of *Bacillus* strains and the accurate characterization of their secondary metabolites will enable researchers to develop biocontrol agents which are a lot more effective and appropriate for use in specific climates. The current review work provides a valuable background for future in silico analyses (e.g., transcriptomic, metabolic network, reconstruction, and functional-structural protein modeling) as well as biotechnological applications of *Bacillus* spp. Further genomics and post-genomics analyses of a large number of both cultivated and uncultivated *Bacilli* from diverse ecosystem will shed light in their interactions with plants, fishes, animals, and humans that surely help to develop more effective *Bacillus*-based industrial products for the promotion of sustainable agriculture. Application of newly discovered and useful genes from *Bacillus* spp. would accelerate genetic engineering in commercial crop that would accelerate industrial production of crops with less or no synthetic chemicals, and also would facilitate biopharming for the production of complex medicinally important metabolites.

Acknowledgment The authors are thankful to the Bangladesh Academy of Sciences for funding this work under the BAS-USDA project to MTI. The authors are also thankful to the RMC of BSMRAU, University Grants Commission of Bangladesh, and Ministry of Science and Technology of Bangladesh for partial funding to this work.

References

Abbasi MW, Marium T, Khan MQ, Zaki MJ (2017) Assessment of extracellular metabolites from Bacillus species against root-knot nematodes and root-infecting fungi in *Abelmoschus esculentus* (L.) Moench. Pak J Bot 49(S1):289–294

Abdelmohsen UR, Grkovic T, Balasubramanian S, Kamel MS, Quinn RJ, Hentschel U (2015) Elicitation of secondary metabolism in actinomycetes. Biotechnol Adv 33:798–811

Abdelrahman M, Abdel-Motaal F, El-Sayed M, Jogaiah S, Shigyo M et al (2016) Dissection of *Trichoderma longibrachiatum*-induced defense in onion (Allium cepa L.) against *Fusarium oxysporum* f. sp. cepa by target metabolite profiling. Plant Sci 246:128–138. https://doi.org/10.1016/j.plantsci.2016.02.008. PMID: 26993243

Ahmad I, Akhtar MJ, Zahir ZA, Naveed M, Mitter B, Sessitscha A (2014) Cadmium-tolerant bacteria induce metal stress tolerance in cereals. Environ Sci Pollut Res 21:11054–11065

Ahmad Z, Wu J, Chen L, Dong W (2017) Isolated *Bacillus subtilis* strain330-2 and its antagonistic genes identified by the removing PCR. Sci Rep 7:1777. https://doi.org/10.1038/s41598-017-01940-9

Ali S, Charles TC, Glick BR (2014) Amelioration of high salinity stress damage by plant-growth promoting bacterial endophytes that contain ACC-deaminase. Plant Physiol Biochem 80:160–167

Aliferis KA, Chrysayi-Tokoubalides M (2011) Metabolomics in pesticide research and development: review and future perspectives. Metabolomics 7:35–53

Altenbuchner J (2016) Editing of the *Bacillus subtilis* genome by the CRISPR-Cas9 system. Appl Environ Microbiol 82(17):5421–5427. https://doi.org/10.1128/AEM.01453-16

Al-Whaibi MH (2011) Plant heat-shock proteins: a mini review. J King Saud Univ Sci 23:139–150. https://doi.org/10.1016/j.jksus.2010.06.022

Ambreetha S, Chinnadurai C, Marimuthu P, Balachandar D (2018) Plant-associated bacillus modulates the expression of auxin-responsive genes of rice and modifies the root architecture. Rhizosphere 5:57–66. https://doi.org/10.1016/j.rhisph.2017.12.001
Annapurna K, Govindasamy V, Sharma M, Ghosh A, Chikara SK (2018) Whole genome shotgun sequence of *Bacillus paralicheniformis* strain KMS 80, a rhizobacterial endophyte isolated from rice (*Oryza sativa* L.). 3 Biotech 8:223
Anuradha SN (2010) Structural and molecular characteristics of lichenysin and its relationship with surface activity. In: Biosurfactants, pp 304–315
Arrebola E, Jacobs R, Korsten L (2010) Iturin A is the principal inhibitor in the biocontrol activity of *Bacillus amyloliquefaciens* PPCB004 against postharvest fungal pathogens. J. Appl. Microbiol 108:386–395. https://doi.org/10.1111/j.1365-2672.2009.04438.x. PMID:19674188
Ashraf M, Hasnain S, Berge O, Mahmood T (2004) Inoculating wheat seedlings with exopolysaccharide-producing bacteria restricts sodium uptake and stimulates plant growth under salt stress. Biol Fertil Soils 40:157–162
Babu AN, Jogaiah S, Ito S, Nagaraj AK, Tran LSP (2015) Improvement of growth, fruit weight and early blight disease protection of tomato plants by rhizosphere bacteria is correlated with their beneficial traits and induced biosynthesis of antioxidant peroxidase and polyphenol oxidase. Plant Sci 231:62–73. https://doi.org/10.1016/j.plantsci.2014.11.006. PMID: 25575992
Belbahri L, Bouket AC, Rkik I, Alenezi FN, Vallat A, Luptakova L, Petrovova E, Oszako T, Cherrad S, Vacher S, Rateb ME (2017) Comparative genomics of B*acillus amyloliquefaciens* strains reveals a core genome with traits for habitat adaptation and a secondary metabolites rich accessory genome. Front Microbiol 8:1438. https://doi.org/10.3389/fmicb.2017.01438
Bentley SD, Chater KF, Cerdeño-Tarraga AM, Challis GL, Thomson NR, James KD, Harris DE, Quail MA, Kieser H, Harper D et al (2002) Complete genome sequence of the model actinomycete *Streptomyces coelicolor* A3(2). Nature 417:141–147
Bhowmik P, Ellison E, Polley B, Bollina V, Kulkarni M, Ghanbarnia K et al (2018) Targeted mutagenesis in wheat microspores using CRISPR/Cas9. Sci Rep 8:6502
Blacutt AA, Mitchell TR, Bacon CW, Gold SE (2016) *Bacillus mojavensis* RRC101 lipopeptides provoke physiological and metabolic changes during antagonism against *Fusarium verticillioides*. Mol Plant-Microbe Interac 29(9):713–723
Blin K, Medema MH, Kottmann R, Lee SY, Weber T (2017) The antiSMASH database, a comprehensive database of microbial secondary metabolite biosynthetic gene clusters. Nucleic Acids Res 45:D555–D559
Borchert N, Dieterich C, Krug K, Schütz W, Jung S, Nordheim A, Sommer RJ, Macek B (2010) Proteogenomics of *Pristionchus pacificus* reveals distinct proteome structure of nematode models. Genome Res 20:837–846
Branda SS, González-Pastor JE, Ben-Yehuda S, Losick R, Kolter R (2001) Fruiting body formation by *Bacillus subtilis*. Proc Natl Acad Sci USA 98:11621–11626
Bundy JG, Willey TL, Castell RS, Ellar DJ, Brindle KM, K. M. (2005) Discrimination of pathogenic clinical isolates and laboratory strains of Bacillus cereus by NMR-based metabolomic profiling. FEMS Microbiol Lett 242:127–136. https://doi.org/10.1016/j.femsle.2004.10.048
Cai X, Kang X, Xi H, Liu C, Xue Y (2016) Complete genome sequence of the endophytic biocontrol strain *Bacillus velezensis* CC09. Genome Announc 4:e01048–e01016
Cakmakci R, Dönmez MF, Erdoğan A (2007) The effect of plant growth promoting rhizo-bacteria on barley seedling growth, nutrient uptake, some soil properties, and bacterial counts. Turkish J Agril Forestry 31:189–199
Camacho M, Santamaria C, Temprano F, Rodriguez-Navarro DN, Daza A (2001) Co-inoculation with *Bacillus* sp. CECT 450 improves nodulation in *Phaseolus vulgaris* L. Can J Microbiol 47:1058–1062
Camon E, Magrane M, Barrell D, Lee V, Dimmer E, Maslen J, Apweiler R (2004) The gene ontology annotation (GOA) database: Sharing knowledge in UniProt with gene ontology. Nucleic Acids Res 32(1):262–266

Carro L, Nouioui I, Sangal V, Meier-Kolthoff JP, Trujillo ME, Montero-Calasanz MDC, Sahin N, Smith DL, Kim KE, Peluso P et al (2018) Genome-based classification of micromonosporae with a focus on their biotechnological and ecological potential. Sci Rep 8:525

Challis GL (2014) Exploitation of the *Streptomyces coelicolor* A3(2) genome sequence for discovery of new natural products and biosynthetic pathways. J Ind Microbiol Biotechnol 41:219–232

Charnock SJ, Davies GJ (1999) Structure of the nucleotide-diphospho-sugar transferase, SpsA from *Bacillus subtilis*, in native and nucleotide-complexed forms. Biochemistry 3820:6380–6385

Chen XH, Koumoutsi A, Scholz R, Eisenreich A, Schneider K, Heinemeyer I et al (2007) Comparative analysis of the complete genome sequence of the plant growth–promoting bacterium *Bacillus amyloliquefaciens* FZB42. Nat Biotechnol 25(9):1007–1014. https://doi.org/10.1038/nbt1325

Chen X-H, Koumoutsi A, Scholz R, Borriss R (2008) More than anticipated-production of antibiotics and other secondary metabolites by *Bacillus amyloliquefaciens* FZB42. J Mol Microbiol Biotechnol 16:14–24. https://doi.org/10.1159/000142891. PMID: 18957859

Chen K, Tian Z, Luo Y, Cheng Y, Long C (2018) Antagonistic activity and the mechanism of *Bacillus amyloliquefaciens* dh-4 against citrus green mold. Plant Pathol. https://doi.org/10.1094/PHYTO-01-17-0032-R

Chinnaswamy A, Coba de la Peña T, Stoll A, de la Peña Rojo D, Bravo J, Rincon A, Lucas MM, Pueyo JJ (2018) A nodule endophytic *Bacillus megaterium* strain isolated from *Medicago polymorpha* enhances growth, promotes nodulation by *Ensifer medicae* and alleviates salt stress in alfalfa plants. Ann Appl Biol 172:295–308. https://doi.org/10.1111/aab.12420

Chu F, Kearns DB, Branda SS, Kolter R, Losick R (2006) Targets of the master regulator of biofilm formation in *Bacillus subtilis*. Mol Microbiol 59:1216–1228

Colagiero M, Rosso LC, Ciancio A (2017) Diversity and biocontrol potential of bacterial consortia associated to root-knot nematodes. Biol Control. https://doi.org/10.1016/j.biocontrol.2017.07.010

Colebatch G, Desbrosses G, Ott T, Krusell L, Montanari O, Kloska S, Kopka J, Udvardi MK (2004) Global changes in transcription orchestrate metabolic differentiation during symbiotic nitrogen fixation in *Lotus japonicus*. Plant J 39(4):487–512

De Souza R, Ambrosini A, Passaglia LMP (2015) Plant growth-promoting bacteria as inoculants in agricultural soils. Gen Mol Biol 38:401–419

Dejong CA, Chen GM, Li H, Johnston CW, Edwards MR, Rees PN, Skinnider MA, Webster AL, Magarvey NA (2016) Polyketide and nonribosomal peptide retro-biosynthesis and global gene cluster matching. Nat Chem Biol 12:1007–1014

Deng Y, Zhu Y, Wang P, Zhu L, Zheng J, Li R et al (2011) Complete genome sequence of *Bacillus subtilis* BSn5, an endophytic bacterium of *Amorphophallus konjac* with antimicrobial activity for the plant pathogen *Erwinia carotovora* subsp. *carotovora*. J Bacteriol 193:2070–2071

Desbrosses GG, Kopka J, Udvardi MK (2005) Lotus japonicus metabolic profiling. Development of gas chromatography-mass spectrometry resources for the study of plant-microbe interactions. Plant Physiol 137(4):1302–1318

Diehl A, Roske Y, Ball L, Chowdhury A, Hiller M, Molière N, Kramer R, Stöppler D, Worth CL, Schlegel B, Leidert M, Cremer N, Erdmann N, Lopez D, Stephanowitz H, Krause E, Rossum BJ, Schmieder P, Heinemann U, Turgay K, Kbey U, Oschkinat H (2018) Structural changes of TasA in biofilm formation of *Bacillus subtilis*. PNAS 115(13):3237–3242

Dieterich C, Clifton SW, Schuster LN, Chinwalla A, Delehaunty K, Dinkelacker I, Fulton L, Fulton R, Godfrey J, Minx P, Mitreva M, Roeseler W, Tian H, Witte H, Yang SP, Wilson RK, Sommer RJ (2008) The *Pristionchus pacificus* genome provides a unique perspective on nematode lifestyle and parasitism. Nat Genet 40:1193–1198

Djenane Z, Nateche F, Amziane M, Gomis-Cebolla J, El-Aichar F, Khorf H, Ferré J (2017) Assessment of the antimicrobial activity and the entomocidal potential of *Bacillus thuringiensis* isolates from Algeria. Toxins 9:139

Dong W, Li S, Lu X, Zhang X, Wang P, Ma P, Guo Q (2014) Regulation of fengycin biosynthase by regulator PhoP in the *Bacillus subtilis* strain NCD-2. Acta Phytopathol Sin 44:180–187. (in Chinese)

Dong D, Ren K, Qiu X, Zheng J, Guo M, Guan X, Liu H, Li N, Zhang B, Yang D (2016) The crystal structure of Cpf1 in complex with CRISPR RNA. Nature 532(7600):522. https://doi.org/10.1038/nature17944

Dong X, Liu Y, Zhang G, Wang D, Zhou X, Shao J, Shen Q, Zhang R (2018) Synthesis and detoxification of nitric oxide in the plant beneficial rhizobacterium *Bacillus amyloliquefaciens* SQR9 and its effect on biofilm formation. Biochem Biophys Res Commun 503:784–790

Douriet-Gamez NR, Maldonado-Mendoza IE, Ibarra-Laclette E, Blom J, Calderon-Vazquez CL (2018) Genomic analysis of Bacillus sp. strain B25, a biocontrol agent of maize pathogen Fusarium verticillioides. Curr Microbiol 75:247–255. https://doi.org/10.1007/s00284-017-1372-1

Driss F, Kallassy Awad M, Zouari N, Jaoua S (2005) Molecular characterization of a novel chitinase from *Bacillus thuringiensis* subsp. *kurstaki*. J Appl Microbiol 99(4):945–953

Duc LH, Hong HA, Barbosa TM, Henriques AO, Cutting SM (2004) Characterization of Bacillus probiotics available for human use. Appl Environ Microbiol 70(4):2161

Earl AM, Eppinger M, Fricke WF, Rosovitz MJ, Rasko DA, Daugherty S, Ravel J (2012) Whole-genome sequences of *Bacillus subtilis* and close relatives. J Bacteriol 194(9):2378–2379

Edwards R, Dixon DP (2005) Plant glutathione transferases. Methods Enzymol 401:169–186. https://doi.org/10.1016/S0076-6879(05)01011-6

Edwards R, Dixon DP (2010) Glutathione transferases. Arab B 8:e0131. https://doi.org/10.1199/tab.0131

Eggert T, Brockmeier U, Dröge MJ, Quax WJ, Jaeger KE (2003) Extracellular lipases from *Bacillus subtilis*: regulation of gene expression and enzyme activity by amino acid supply and external pH. FEMS Microbiol Letters 225(2):319–324

El-Hadad ME, Mustafa MI, Selim SM, Mahgoob AEA, El-Tayeb TS, Abdel Aziz NH (2010) In vitro evaluation of some bacterial isolates as biofertilizers and biocontrol agents against the second stage juveniles of *Meloidogyne incognita*. World J Microbiol Biotechnol 26(12):2249–2256

Endres JR, Clewell A, Jade KA, Farber T, Hauswirth J, Schauss AG (2009) Safety assessment of a proprietary preparation of a novel probiotic, *Bacillus coagulans*, as a food ingredient. Food and Chemical Toxicol 47:1231–1238

Engelbrecht G, Horak I, van Rensburg PJJ, Claassens S (2018) *Bacillus*-based bionematicides: development, modes of action and commercialization. Biocontrol Sci Tech 28(7):629–653. https://doi.org/10.1080/09583157.2018.1469000

Eppinger M, Bunk B, Johns MA, Edirisinghe JN, Kutumbaka KK, Koenig SS, Martin M (2011) Genome sequences of the biotechnologically important *Bacillus megaterium* strains QM-B1551 and DSM319. J Bacteriol 193(16):4199–4213

Fijan S (2014) Microorganisms with claimed probiotic properties: an overview of recent literature. Int J Environ Res Public Health 11:4745–4767

Fong JNC, Yildiz FH (2015) Biofilm matrix proteins. Microbiol Spectr 3

Fuller R (1991) Probiotics in human medicine. Gut 32:439–442

Gagne-Bourque F, Bertrand A, Claessens A, Aliferis KA, Jabaji S (2016) Alleviation of drought stress and metabolic changes in timothy (*Phleumpratense* L.) colonized with *Bacillus subtilis* B26. Front Plant Sci 7:584

Gao T, Foulston L, Chai Y, Wang Q, Losick R (2015) Alternative modes of biofilm formation by plant-associated *Bacillus cereus*. Microbiol Open 4:452–464

Gao H, Qi G, Yin R, Zhang H, Li C, Zhao X (2016) *Bacillus cereus* strain S2 shows high nematicidal activity against *Meloidogyne incognita* by producing sphingosine. Sci Rep 6:605. https://doi.org/10.1038/srep28756

Gold SE, Blacutt AA, Meinersmann RJ, Bacon CW (2014) Whole genome shotgun sequence of *Bacillus mojavensis* strain RRC101, an endophytic bacterium antagonistic to the mycotoxigenic endophytic fungus *Fusarium verticillioides*. Genome Announc 2:e01090

Garcia-Gutierrez L, Zeriouh H, Romero D, Cubero J, deVicente A, Perez-Garcia A (2013) The antagonistic strain *Bacillus subtilis* UMAF6639 also confers protection to melon plants against cucurbit powdery mildew byactivation of jasmonate-and salicylic acid-dependent defense responses. Microb Biotechnol 6:264–274

Gerst MM, Yesil M, Yousef AE (2018) Draft genome sequence of *Bacillus velezensis* OSY-S3, a producer of potent antimicrobial agents active against bacteria and fungi. Genome Announc 6:e01465–e01417. https://doi.org/10.1128/genomeA.01465-17

Glick BR (1995) The enhancement of plant growth by free-living bacteria. Can J Microbiol 41:109–117

Glick BR (2012) Plant growth-promoting bacteria: mechanisms and applications. Scientifica:963401

Glick BR (2014) Bacteria with ACC deaminase can promote plant growth and help to feed the world. Microbiol Res 169:30–39

Goa M (2001) Gene Ontology annotation based on Enzyme Commission mapping. Genomics 74:121–128

Gond SK, Torres MS, Bergen MS, Helsel Z, White JF (2015) Induction of salt tolerance and up-regulation of aquaporin genes in tropical corn by rhizobacterium *Pantoea agglomerans*. Lett Appl Microbiol 60:392–399

Gong AD, Li H-P, Yuan Q-S, Song X-H, Yao W, He W-J, Zhang J-B, Liao Y-C (2015) Antagonistic mechanism of Iturin A and Plipastatin A from *Bacillus amyloliquefaciens* S76-3 from wheat spikes against *Fusarium graminearum*. PLoS ONE 10(2):e0116871. https://doi.org/10.1371/journal.pone.0116871

Govindasamy K, Sevugapperumal N, Devi A (2017) A proteomic approach to identify *Bacillus subtilis* induced defense related proteins in noni challenged with *Meloidogyne incognita*. Int J Plant Anim Environ Sci:110–116. https://doi.org/10.21276/Ijpaes

Griffin JL (2003) Metabonomics: NMR spectroscopy and pattern recognition analysis of body fluids and tissues for characterisation of xenobiotic toxicity and disease diagnosis. Curr Opin Chem Biol:7648–7654

Hadjithomas M, Chen IA, Chu K, Huang J, Ratner A, Palaniappan K, Andersen E, Markowitz V, Kyrpides NC, Ivanova NN (2017) IMG-ABC: new features for bacterial secondary metabolism analysis and targeted biosynthetic gene cluster discovery in thousands of microbial genomes. Nucleic Acids Res 45:D560–D565

Hagemeier J, Schneider B, Oldham NJ, Hahlbrock K (2001) Accumulation of soluble and wall-bound indolic metabolites in Arabidopsis thaliana leaves infected with virulent or avirulent Pseudomonas syringae pathovar tomato strains. Proc Natl Acad Sci 98(2):753–758

Hamilton CE, Gundel PE, Helander M, Saikkonen K (2012) Endophytic mediation of reactive oxygen species and antioxidant activity in plants: a review. Fungal Divers 54:1–10

Han SJ, Yoo YJ, Kang HS (1995) Characterization of a bifunctional cellulase and its structural gene The cell gene of *Bacillus* sp. D04 has exo-and endoglucanase activity. J Biol Chem 270(43):26012–26019

Han CS, Xie G, Challacombe JF, Altherr MR, Bhotika SS, Bruce D, Cleland C (2006) Pathogenomic sequence analysis of *Bacillus cereus* and *Bacillus thuringiensis* isolates closely related to *Bacillus anthracis*. Bacteriol 188(9):33823390

Handtke S, Volland S, Methling K, Albrecht D, Becher D, Nehls J, Bongaerts J, Maurer K-H, Lalk M, Liesegang H, Voigt B, Daniel R, Hecker M (2014) Cell physiology of the biotechnological relevant bacterium Bacilluspumilus—an omics-based approach. J Biotechnol 192:204–214

Haque E, Taniguchi H, Hassan MM, Bhowmik P, Karim MR, Smiech M, Zhao K, Rahman M, Islam T (2018) Application of CRISPR/Cas9 genome editing technology for the improvement of crops cultivated in tropical climates: recent progress, prospects, and challenges. Front Plant Sci 9:617. https://doi.org/10.3389/fpls.2018.00617

Harwood CR, Mouillon J-M, Pohl S, Arnau J (2018) Secondary metabolite production and the safety of industrially important members of the *Bacillus subtilis* group. FEMS Microbiol Rev 42(6):721–738

Hayat R, Ali S, Amara U, Khalid R, Ahmed I (2010) Soil beneficial bacteria and their role in plant growth promotion: a review. Ann Microbiol 60:579–598

Helmann JD, Chamberlin MJ (1987) DNA sequence analysis suggests that expression of flagellar and chemotaxis genes in *Escherichia coli* and *Salmonella typhimurium* is controlled by an alternative sigma factor. Proc Natl Acad Sci USA 84:6422–6424. https://doi.org/10.1073/pnas.84.18.6422

Hillebrand H, Bartling D, Weiler EW (1998) Structural analysis of the *nit2/nit1/nit3* gene cluster encoding nitrilases, enzymes catalyzing the terminal activation step in indole-acetic acid biosynthesis in *Arabidopsis thaliana*. Plant Mol Biol 36:89–99

Hirano H, Gootenberg JS, Horii T, Abudayyeh OO, Kimura M, Hsu PD, Nakane T, Ishitani R, Hatada I, Zhang F, Nishimasu H, Nureki O (2016) Structure and engineering of *Francisella novicida* Cas9. Cell 164(5):950–961. https://doi.org/10.1016/j.cell.2016.01.039

Hoffmaster AR, Ravel J, Rasko DA, Chapman GD, Chute MD, Mariston CK, De BK, Sacchi CT, Fitzgerald C, Mayer LW, Maiden MC, Priest FG, Barker M, Jiang L, Cer RZ, Rilstone J, Reed TD, Popovic T, Fraiser CM (2004) Identification of anthrax toxin genes in *Bacillus cereus* associated with an illness resembling inhalation anthrax. Proc Natl Acad Sci USA 101:8449–8454

Hollensteiner J, Wemheuer F, Harting R, Kolarzyk AM, Valerio SMD, Poehlein A, Brzuszkiewicz EB, Nesemann K, Braus-Stromeyer SA, Braus GH, Daniel R, Liesegang H (2017) *Bacillus thuringiensis* and *Bacillus weihenstephanensis* inhibit the growth of phytopathogenic Verticillium species. Front Microbiol 7:2171. https://doi.org/10.3389/fmicb.2016.02171

Holtmann G, Bakker EP, Uozumi N, Bremer E (2003) *KtrAB* and *KtrCD*: two K^+ uptake systems in *Bacillus subtilis* and their role in adaptation to hypertonicity. J Bacteriol 185(4):1289–1298

Hong HA, Duc LH, Cutting SM (2005) The use of bacterial spore formers as probiotics. FEMS Microbiol Rev 29:813–835

Hong K-Q, Liu D-Y, Hen T, Wang Z-W (2018) Recent advances in CRISPR/Cas9 mediated genome editing in *Bacillus subtilis*. World J Microbiol Biotechnol 34:153

Honma M, Tanaka K, Konno K, Tsuge K, Okuno T, Hashimoto M (2012) Termination of the structural confusion between plipastatin A1 and fengycin IX. Bioorg Med Chem 20:3793–3798

Huang CJ, Chen CY (2005) High-level expression and characterization of two chitinases, ChiCH and ChiCW, of *Bacillus cereus* 28-9 in *Escherichia coli*. Biochem Biophys Res Commun 327(1):8–17

Huang Y, Xu C, Ma L, Zhang K, Duan C, Mo M (2010) Characterisation of volatiles produced from *Bacillus megaterium* YFM3.25 and their nematicidal activity against *Meloidogyne incognita*. Eur J Plant Pathol 126:417–422. https://doi.org/10.1007/s10658-009-9550-z

Hulett FM, Bookstein CRESCENCE, Jensen K (1990) Evidence for two structural genes for alkaline phosphatase in *Bacillus subtilis*. J Bacteriol 172(2):735–740

Hunger K, Beckering CL, Wiegeshoff F, Graumann PL, Marahiel MA (2006) Cold induced putative DEAD box RNA helicases CshA and CshB are essential for cold adaptation and interact with cold shock protein B in *Bacillus subtilis*. J Bacteriol 188(1):240–248

Huo Y, Zhan Y, Wang Q, Li S, Yang S, Nomura CT, Wang C, Chen S (2018) Acetolactate synthase (AlsS) in *Bacillus licheniformis* WX-02:enzymatic properties and efficient functions for acetoin/butanediol and L-valine biosynthesis. Bioprocess Biosyst Eng 41:87–96. https://doi.org/10.1007/s00449-017-1847-2

Hurek T, Reinhold-Hurek B (2003) *Azoarcus sp.* strain BH72 as a model for nitrogen-fixing grass endophytes. J Biotechnol 106:169–178. PMID:14651859

Iborta P, Imaib H, Uemurab M, Arocaa R (2018) Proteomic analysis reveals that tomato interaction with plant growth promoting bacteria is highly determined by ethylene perception. J Plant Physiol 220:43–59

Idris EE, Iglesias DJ, Talon M, Borriss R (2007) Tryptophan-dependent production of indole-3-acetic acid (IAA) affects level of plant growth promotion by *Bacillus amyloliquefaciens* FZB42. Am Phytopathol Soc 20(6):619–626

Idriss EES, Makarewicz O, Farouk A, Rosner K, Greiner R, Bochow H, Richter T, Borriss R (2002) Extracellular phytase activity of *Bacillus amyloliquefaciens* FZB 45 contributes to its plant growth-promoting effect. Microbiology 148:2097–2109

Irizarry I, White JF (2018) *Bacillus amyloliquefaciens* alters gene expression, ROSproduction and lignin synthesis in cotton seedling roots. J Appl Microbiol 124:1589–1603. https://doi.org/10.1111/jam.13744

Islam MT, Hossain MM (2012) Plant probiotics in phosphorus nutrition in crops, with special reference to rice. In: Maheshwari DK (ed) Bacteria in agrobiology: plant probiotics. Springer, Berlin, pp 325–363

Islam MT, Hossain MM (2013) Biological control of peronosporomycete phytopathogens by antagonistic bacteria. In: Maheshwari DK (ed) Bacteria in agrobiology: plant disease management. Springer, Berlin, pp 167–218

Islam MT, Hashidoko Y, Deora A, Ito T, Tahara S (2005) Suppression of damping-off disease in host plants by the rhizoplane bacterium Lysobacter sp. strain SB-K88 is linked to plant colonization and antibiosis against soilborne Peronosporomycetes. Appl Environ Microbiol 71:3786–3796

Jamal Q, Cho J-Y, Moon J-H, Munir S, Anees M, Kim KY (2017) Identification for the first time of Cyclo(d-Pro-l-Leu) produced by *Bacillus amyloliquefaciens* Y1 as a nematocide for control of *Meloidogyne incognita*. Molecules (Basel, Switzerland) 22(11):1839. https://doi.org/10.3390/molecules22111839

Jamil M, Zeb S, Anees M, Roohi A, Ahmed I et al (2014) Role of *Bacillus licheniformis* in phytoremediation of nickel contaminated soil cultivated with rice. Int J Phytoremed 16:554–571

Jansen-Girgan C, Claassens S, Fourie H (2016) In vitro evaluations to determine the effect of *Bacillus firmus* strains on the motility of *Meloidogyne incognita* second-stage juveniles. Tropical Plant Pathol 41(5):320–324

Jasim B, Sreelakshmi KS, Mathew J, Radhakrishnan EK (2016) Surfactin, Iturin, and Fengycin Biosynthesis by Endophytic *Bacillus* sp. from *Bacopa monnieri*. Microb Ecol 72(1):106–119. https://doi.org/10.1007/s00248-016-0753-5

Jha PN, Kumar A (2007) Endophytic colonization of *Typha australis* by a plant growth-promoting bacterium *Klebsiella oxytoca* strain GR-3. J Appl Microbiol 103:1311–1320

Jinek M, Chylinski K, Fonfara I, Hauer M, Doudna JA, Charpentier E (2012) A programmable dual-RNA-guided DNA endonuclease in adaptive bacterial immunity. Science 337(6096):816–821. https://doi.org/10.1126/science.1225829

Jeong H, Choi SK, Kloepper JW, Ryu CM (2014) Genome sequence of the plant endophytic *Bacillus pumilus* triggering induced systemic resistance in field crops. Genome Announc 2(5):e01093–e01014. https://doi.org/10.1128/genomeA.01093-14

Jeong H, Jo SH, Hong CE, Park JM (2016) Genome sequence of the endophytic bacterium *Bacillus thuringiensis* strain KB1, a potential biocontrol agent against phytopathogens. Genome Announc 4:e00279–e00216

Jogaiah S, Shivanna RK, Gnanaprakash PH, Hunthrike SS (2010) Evaluation of plant growth-promoting rhizobacteria for their efficiency to promote growth and induce systemic resistance in pearl millet against downy mildew disease. Arch Phytopathol Plant Protect 43:368–378

Jurenka JS (2012) *Bacillus coagulans*. Altern Med Rev 17:76–81

Kalai-Grami L, Karkouch I, Naili O, Slimene IB, Elkahoui S, Zekri RB, Touati I, Mnari-Hattab M, Hajlaoui MR, Limam F (2016) Production and identification of iturin A lipopeptide from *Bacillus methyltrophicus* TEB1 for control of *Phoma tracheiphila*. J Basic Microbiol 56:864e871

Kang W, Zhu X, Wang Y, Chen L, Duan Y (2018a) Transcriptomic and metabolomic analyses reveal that bacteria promote plant defense during infection of soybean cyst nematodein soybean. BMC Plant Biol 18:86. https://doi.org/10.1186/s12870-018-1302-9

Kang X, Zhang W, Cai X, Zhu T, Xue Y, Liu C (2018b) *Bacillus velezensis* CC09: a potential 'vaccine' for controlling wheat diseases. Mol Plant-Microbe Interact 31(6):623–632. https://doi.org/10.1094/MPMI-09-17-0227-R

Kearns DB, Chu F, Branda SS, Kolter R, Losick R (2005) A master regulator for biofilm formation by *Bacillus subtilis*. Mol Microbiol 55:739–749

Khan MMA, Khatun A, Islam MT (2016) Promotion of plant growth by phytohormone producing bacteria. In: Garg N, Aeron A (eds) Microbes in action. Nova Science Publishers, New York, pp 45–76

Kim YO, Lee JK, Kim HK, Yu JH, Oh TK (1998) Cloning of the thermostable phytase gene (phy) from *Bacillus* sp. DS11 and its overexpression in *Escherichia coli*. FEMS Microbiol Lett 162(1):185–191

Kim JS, Lee JE, Nie H, Lee YJ, Kim ST, Kim S-H (2017) Physiological and proteomic analysis of plant growth enhancement by the rhizobacteria *Bacillus* sp. JS Genes Genom. https://doi.org/10.1007/s13258-017-0615-7

Kloepper JW, Beauchamp CJ (1992) A review of issues related to measuring colonization of plant roots by bacteria. Can J Microbiol 38:1219–1232

Konuray G, Erginkaya Z (2018) Potential use of Bacillus coagulans in the food industries. Foods 7:92. https://doi.org/10.3390/foods7060092

Kumar A, Prakash A, Johri B (2011) Bacillus as PGPR in crop ecosystem. In: Bacteria in agrobiology: crop ecosystems. Springer, pp 37–59

Kunst F, Ogasawara N, Moszer I, Albertini AM, Alloni GO, Azevedo V, Borriss R (1997) The complete genome sequence of the gram-positive bacterium *Bacillus subtilis*. Nature 390(6657):249–256

Kwon YS et al (2010) Proteome analysis of *Arabidopsis* seedlings exposed to bacterial volatiles. Planta 232:1355–1370

Lam HM, Coschigano KT, Oliveira IC, Melo-Oliveira R, Coruzzi GM (1996) The molecular-genetics of nitrogen assimilation into amino acids in higher plants. Annu Rev Plant Physiol Plant Mol Biol 47:569–593

Lea PJ, Sodek L, Parry MAJ, Shewry PR, Halford NG (2006) Asparagine in plants. Ann Appl Biol 150:1–26

Lee YS, Kim KY (2016) Antagonistic potential of *Bacillus pumilus* L1 against root-knot nematode, *Meloidogyne arenaria*. J Phytopathol 164(1):29–39. https://doi.org/10.1111/jph.12421

Li JH, Wang ET, Chen WF, Chen WX (2008) Genetic diversity and potential for promotion of plant growth detected in nodule endophytic bacteria of soybean grown in Heilongjiang province of China. Soil Biol Biochem 40:238–246

Li MHT, Ung PMU, Zajkowski J, Garneau-Tsodikova S, Sherman DH (2009) Automated genome mining for natural products. BMC Bioinf 10:185. https://doi.org/10.1186/1471-2105-10-185

Li J, Bi F-C, Yin J, Wu J-LJ-X, Rong C, Wu J-LJ-X, Yao N (2015) An Arabidopsis neutral ceramidase mutant ncer1 accumulates hydroxy ceramides and is sensitive to oxidative stress. Front Plant Sci 6:460. https://doi.org/10.3389/fpls.2015.00460

Lim JH, Kim SD (2013) Induction of drought stress resistance by multi-functional PGPR *Bacillus licheniformis* K11 in pepper. Plant Pathol J 29(2):201–208. https://doi.org/10.5423/PPJ.SI.02.2013.0021

Liu G, Chater KF, Chandra G, Niu G, Tan H (2013) Molecular regulation of antibiotic biosynthesis in Streptomyces. Microbiol Mol Biol Rev 77:112–143

Liu H, Fan X, Cheng H, Yin S, Zhang G, Xi Y, An L, Zhang L (2018) Characterization of a gene regulating antibiotic production in Bacillus subtilis BSD-2. Biotechnol Biotechnol Equip 32(2):332–338. https://doi.org/10.1080/13102818.2017.1329633

Lopes R, Cerdeira L, Tavares GS, Ruiz JC, Blom J, Horacio ECA et al (2017) Genome analysis reveals insights of the endophytic *Bacillus toyonensis* BAC3151 as a potentially novel agent for biocontrol of plant pathogens. World J Microbiol Biotechnol 33:185

Lopes R, Tsui S, Gonçalves PJRO, Vieira de Queiroz M (2018) A look into a multifunctional toolbox: endophytic *Bacillus* species provide broad and underexploited benefits for plants. World J Microbiol Biotechnol 34:94. https://doi.org/10.1007/s11274-018-2479-7

Lopez-Bucio J, Campos-Cuevas JC, Hernández-Calderón E et al (2007) *Bacillus megaterium* rhizobacteria promote growth and alter root-system architecture through an auxin-and ethylene-independent signaling mechanism in *Arabidopsis thaliana*. Mol Plant Microbe Interact 20:207–217. https://doi.org/10.1094/MPMI-20-2-0207

Luque-Almagro VM, Gates AJ, Moreno-Vivián C, Ferguson SJ, Richardson DJ, Roldán MD (2011) Bacterial nitrate assimilation: gene distribution and regulation. Biochem Soc Trans 39(6):1838–1843

Ma J, Wang C, Wang H, Liu K, Zhang T, Yao L, Zhao Z, DU B, Ding Y (2018) Analysis of the complete genome sequence of *Bacillus atrophaeus* GQJK17 reveals its biocontrol characteristics as a plant growth-promoting rhizobacterium. BioMed Res Internal 2018:9473542. https://doi.org/10.1155/2018/9473542

Maget-Dana R, Peypoux F (1994) Iturins, a special class of pore-forming lipopeptides:biological and physicochemical properties. Toxicology 87:151e174

Makarewicz O, Dubrac S, Msadek T, Borriss R (2006) Dual role of the PhoP~P response regulator: *Bacillus amyloliquefaciens* FZB45 phytase gene transcription is directed by positive and negative interaction with the *phyC* promoter. J Bacteriol 188:6953–6965

Mannaa M, Kim KD (2018) Biocontrol activity of volatile-producing *Bacillus megaterium* and *Pseudomonas protegens a*gainst *Aspergillus* and *Penicillium* spp. predominant in stored rice grains: study II. Mycobiology 46(1):52–63. https://doi.org/10.1080/12298093.2018.1454015

Marseglia GL, Tosca M, Cirillo I, Licari A, Leone M, Marseglia A, Castellazzi AM, Ciprandi G (2007) Efficacy of *Bacillus clausii* spores in theprevention of recurrent respiratory infections in children: a pilot study. Ther Clin Risk Manag 3:13e17. Available at: https://www.dovepress.com/articles.php?article_id-1150

Mashiane RA, Ezeokoli OT, Adeleke RA, Bezuidenhout CC (2017) Metagenomic analyses of bacterial endophytes associated with the phyllosphere of a Bt maize cultivar and its isogenic parental line from South Africa. World J Microbiol Biotechnol 33:80

May JJ, Wendrich TM, Marahiel MA (2001) The *dhb* operon of *Bacillus subtilis* encodes the biosynthetic template for the catecholic siderophore 2, 3dihydroxybenzoate-glycine-threonine trimeric ester bacillibactin. J Biol Chem 276(10):7209–7217

McDaniel BA, Grundy FJ, Kurlekar VP, Tomsic J, Henkin TM (2006) Identification of a mutation in the *Bacillus subtilis* S-adenosylmethionine synthetase gene that results in derepression of S-box gene expression. J Bacteriol 188(10):3674–3681

Medema MH, Breitling R, Bovenberg R, Takano E (2011) Exploiting plug-and-play synthetic biology for drug discovery and production in microorganisms. Nat Rev Microbiol 9:131–137

Meena KR, Kanwar SS (2015) Lipopeptides as the antifungal and antibacterial agents: applications in food safety and therapeutics. BioMed Res Int 9:473050

Mendis HC, Thomas VP, Schwientek P, Salamzade R, Chien J-T, Waidyarathne P et al (2018) Strain-specific quantification of root colonization by plant growth promoting rhizobacteria *Bacillus firmus*I-1582 and *Bacillus amyloliquefaciens*QST713 in non-sterile soil and field conditions. PLoS ONE 13(2):e0193119. https://doi.org/10.1371/journal.pone.0193119

Meyer H, Weidmann H, Lalk M (2013) Methodological approaches to help unravel the intercellular metabolome of *Bacilus subtilis*. Microb Cell Factories 12:69

Mia MAB, Hossain MM, Shamsuddin ZH, Islam MT (2013) Plant-associated bacteria in nitrogen in crops, with special reference to rice and banana. In: Maheshwari DK et al (eds) Bacteria in agrobiology: crop productivity. Springer-Verlag, Berlin/Heidelberg, pp 97–126

Mielich-Süss B, Lopez D (2015) Molecular mechanisms involved in *Bacillus subtilis* biofilm formation. Environ Microbiol 17:555–565

Miethke M, Bisseret P, Beckering CL et al (2006) Inhibition of aryl acid adenylation domains involved in bacterial siderophore synthesis. FEBS J 273:409–419. https://doi.org/10.1111/j.1742-4658.2005.05077.x

Miethke M, Schmidt S, Marahiel MA (2008) The major facilitator superfamily-type transporter YmfE and the multidrug-efflux activator Mta mediate bacillibactin secretion in Bacillus subtilis. J Bacteriol 190:5143–5152. https://doi.org/10.1128/JB.00464-08

Mirel DB, Chamberlin MJ (1989) The *Bacillus subtilis* flagellin gene(hag) is transcribed by the sigma 28 form of RNA polymerase. J Bacteriol 171:3095–3101. https://doi.org/10.1128/jb.171.6.3095-3101.1989

Molina-Henares AJ, Krell T, Eugenia Guazzaroni M, Segura A, Ramos JL (2006) Members of the IclR family of bacterial transcriptional regulators function as activators and/or repressors. FEMS Microbiol Rev 30:157–186. https://doi.org/10.1111/j.1574-6976.2005.00008.x

Mondol MAM, Shin HJ, Islam MT (2013) Diversity of secondary metabolites from marine *Bacillus* species: chemistry, biosynthesis and biological activity. Mar Drugs 11:2846–2872

Muñoz-Moreno CY, De La Cruz-Rodríguez L, Vega-Arreguín J, Alvarado-Rodríguez M, Gómez-Soto JM, Alvarado-Gutiérrez A, Fraire-Velázquez S (2018) Draft genome sequence of *Bacillus subtilis* 2C-9B, a strain with biocontrol potential against chili pepper root pathogens and tolerance to Pb and Zn. Genome Announc 6:e01502–e01517

Nakano MM (2002) Induction of ResDE-dependent gene expression in *Bacillus subtilis* in response to nitric oxide and nitrosative stress. J Bacteriol 184(6):1783–1787

Naznin HA, Kiyohara D, Kimura M, Miyazawa M, Shimizu M, Hyakumachi M (2014) Systemic Resistance Induced by Volatile Organic Compounds Emitted by Plant Growth-Promoting Fungi in *Arabidopsis thaliana*. PLoS ONE 9(1):e86882. https://doi.org/10.1371/journal.pone.0086882

Nestler J, Keyes SD, Wissuwa M (2016) Root hair formation in rice (*Oryza sativa* L.) differs between root types and is altered in artificial growth conditions. J Exp Bot 67:3699–3708. https://doi.org/10.1093/jxb/erw115

Ng CY, Jung MY, Lee J, Oh MK (2012) Production of 2,3-butanediol in Saccharomyces cerevisiae by in silico aided metabolic engineering. Microb Cell Fact 11:68

Nguyen DM, Lipscomb GL, Schut GJ, Vaccaro BJ, Basen M, Kelly RM, Adams MW (2016) Temperature-dependent acetoin production by *Pyrococcus furiosus* is catalyzed by a biosynthetic acetolactate synthase and its deletion improves ethanol production. Metab Eng 34:71–79

Niazi A, Manzoor S, Asari S, Bejai S, Meijer J, Bongcam-Rudloff E (2014a) Genome analysis of *Bacillus amyloliquefaciens* subsp. *plantarum* UCMB5113: A rhizobacterium that improves plant growth and stress management. PloS One 9(8):e104651

Niazi A, Manzoor S, Bejai S, Meijer J, Bongcam-Rudloff E (2014b) Complete genome sequence of a plant associated bacterium *Bacillus amyloliquefaciens* subsp. *plantarum* UCMB5033. Stand Genomic Sci 9(3):718

Niu Q, Huang X, Zhang L, Lian L, Li Y, Li J, Zhang K (2007) Functional identification of the gene bace16 from nematophagous bacterium Bacillus nematocida. Appl Microbiol Biotechnol 75(1):141–148. https://doi.org/10.1007/s00253-006-0794-7

Niu DD, Liu HX, Jiang CH, Wang YP, Wang QY, Jin HL, Guo JH (2011) The plant growth promoting rhizobacterium *Bacillus cereus* AR156 induces systemic resistance in *Arabidopsis thaliana* by simultaneously activating salicylate-and jasmonate/ethylene-dependent signaling pathways. Mol Plant Microbe Interact 24:533–542

Noone D, Howell A, Collery R, Devine KM (2001) YkdA and YvtA, HtrA-like serine proteases in *Bacillus subtilis*, engage in negative autoregulation and reciprocal cross regulation of *ykdA* and *yvtA* gene expression. J Bacteriol 183(2):654663

Okada M, Nakamura Y, Hayashi S et al (2015) Chemical structure and biological activity of a quorum sensing pheromone from Bacillus subtilis subsp natto. Bioorg Med Chem Lett 19:4293–4296

Oliveira SS, Seldin L, Bastos MCF (1993) Identification of structural nitrogen fixation (*nif*) genes in *Bacillus polymyxa* and *Bacillus macerans*. World J Microbiol Biotechnol 9(3):387–389

Ongena M, Duby F, Jourdan E, Beaudry T, Jadin V, Dommes J, Thonart P (2005) *Bacillus subtilis* M4 decreases plant susceptibility towards fungal pathogens by increasing host resistance associated with differential gene expression. Appl Microbiol Biotechnol 67(5):692–698

Ortíz-Castro R, Valencia-Cantero E, López-Bucio J (2008) Plant growth promotion by *Bacillus megaterium* involves cytokinin signaling. Plant Signal Behav 3(4):263–265

Padgham JL, Sikora RA (2007) Biological control potential and modes of action of *Bacillus megaterium* against *Meloidogyne graminicola* on rice. Crop Protect 26(7):971–977. https://doi.org/10.1016/j.cropro.2006.09.004

Palazzotto E, Weber T (2018) Omics and multi-omics approaches to study the biosynthesis of secondary metabolites in microorganisms. Curr Opin Microbiol 8(45):109–116. https://doi.org/10.1016/j.mib.2018.03.004

Palazzotto E, Renzone G, Fontana P, Botta L, Scaloni A, Puglia AM, Gallo G (2015) Tryptophan promotes morphological and physiological differentiation in Streptomyces coelicolor. Appl Microbiol Biotechnol 99:10177–10189

Palma L, Muñoz D, Berry C, Murillo J, Caballero P (2014) *Bacillus thuringiensis* toxins: an overview of their biocidal activity. Toxins 6:3296–3325

Pflughoeft KJ, Sumby P, Koehler TM (2011) *Bacillus anthracis* sin locus and regulation of secreted proteases. J Bacteriol 193:631–639

Phister TG, O'Sullivan DJ, McKay LL (2004) Identification of bacilysin, chlorotetaine, and iturin A produced by Bacillus sp. strain CS93 isolated from pozol, a mexican fermented maize dough. Appl Environ Microbiol 70:631e634

Pleban S, Cherin L, Chet I (1997) Chitinolic activity of an endophytic strain of Bacillus cereus. Lett Appl Microbiol 25:284–288

Podile AR, Kishore GK (2007) Plant growth-promoting rhizobacteria. In: Plant-associated bacteria. Springer, pp 195–230

Potshangbam M, Sahoo D, Verma P, Verma S, Kalita MC, Indira DS (2018) Draft genome sequence of *Bacillus altitudinis*Lc5, a biocontrol and plant growth-promoting endophyte strain isolated from indigenous black rice of Manipur. Genome Announc 6:e00601–e00618. https://doi.org/10.1128/genomeA.00601-18

Pylro VS, Dias ACF, Andreote FD, Morais DK, Varani AM, Andreote CCF, Bernardo ERA, Zucchie T (2018) Closed genome sequence of phytopathogen biocontrol agent *Bacillus velezensis* strain AGVL-005, isolated from soybean. Genome Announc 6:e00057–e00018. https://doi.org/10.1128/genomeA.00057-18

Qing-gang G, Li-hong D, Pei-pei W, She-zeng L, Wei-song Z, Xiu-yun L, Xiao-yun Z, Ping M (2018) The PhoR/PhoP two-component system regulates fengycin production in *Bacillus subtilis* NCD-2 under low-phosphate conditions. J Integr Agric 17(1):149–157. https://doi.org/10.1016/S2095-3119(17)61669-1

Radhakrishnan R, Hashem A, Abd Allah EF (2017) *Bacillus*: a biological tool for crop improvement through bio-molecular changes in adverse environments. Front Physiol 8:667

Rahman M, Sabir AA, Mukta JA, Khan MMA, Mohi-ud-Din M, Miah MG, Rahman M, Islam MT (2018) Plant probiotic bacteria Bacillus and Paraburkhulderia improve growth, yield and content of antioxidants in strawberry fruit. Sci Reports 8:2504

Ramezani Moghaddam M, Mahdikhani Moghaddam E, Baghaee Ravari S, Rouhani H (2014) The nematicidal potential of local Bacillus species against the root-knot nematode infecting greenhouse tomatoes. Biocontrol Sci Technol 24(3):279–290. https://doi.org/10.1080/09583157.2013.858100

Ranocha P, Dima O, Nagy R, Felten J, Corratge-Faillie C, Novak O, Morreel K, Lacombe B et al (2013) Arabidopsis WAT1 is a vacuolar auxin transport facilitator required for auxin homeostasis. Nat Commun 4:2625

Rath M, Mitchell TR, Gold SE (2018) Volatiles produced by *Bacillusmojavensis*RRC101 act as plant growth modulators and are strongly culture-dependent. Microbiol Res 208:76–84

Rawlings ND, Barrett AJ, Finn R (2016) Twenty years of the MEROPS database of proteolytic enzymes, their substrates and inhibitors. Nucleic Acids Res 44:D343–D350

Raza W, Ling N, Yang L, Huang Q, Shen Q (2016) Response of tomato wilt pathogen *Ralstonia solanacearum* to the volatile organic compounds produced by a biocontrol strain *Bacillus amyloliquefaciens* SQR-9. Sci Rep 6:24856. https://doi.org/10.1038/srep24856

Read TD, Peterson SN, Tourasse N, Baillie LW, Paulsen IT, Nelson KE, Holtzapple EK (2003) The genome sequence of *Bacillus anthracis* Ames and comparison to closely related bacteria. Nature 423(6935):81–86

Rekhaa K, Baskarb B, Srinathb S, Ushaa B (2017) Plant growth promoting rhizobacteria *Bacillus subtilis* RR4 1 isolated from rice 2 rhizosphere induces malic acid biosynthesis in rice roots. Can J Microbiol 64:20–27

Reshma SV, Sathyanarayanan N, Nagendra HG (2017) Characterization of a hypothetical protein YVRE from Bacillus subtilis indicates its key role as glucono-lactonase in pentose phosphate pathway and glucose metabolism. Bioinformation 13(12):430

Ribaudo CM, Krumpholz EM, Cassan FD, Botin R, Cantore ML, Cura JA (2006) *Azospirillum* sp. promotes root hair development in tomato plants through a mechanism that involves ethylene. J Plant Growth Regul 24:175–185

Rodrigues EP, Rodrigues LS, Oliveira ALM, Baldani VLD, Teixeira KRS, Urquiaga S, Reis VM (2008) Azospirillum amazonense inoculation: effects on growth, yield and N 2 fixation of rice (*Oryza sativa* L.). Plant Soil 302(1):249–261

Rogers ED, Benfey PN (2015) Regulation of plant root system architecture: implications for crop advancement. Curr Opin Biotechnol 32:93–98. https://doi.org/10.1016/j.copbio.2014.11.015

Roy S, Rathod A, Pramanik A (2017) Nematicidal activity of Bacillus subtilis against root knot nematode *Meloidogyne incognita* (kofoid and white) chitwood in tomato. Environ Ecol 35(1B):469–473

Rudrappa T, Biedrzycki ML, Kunjeti SG, Donofrio NM, Czymmek KJ, Pare PW, Bais HP (2010) The rhizobacterial elicitor acetoin induces systrmic resistance in *Arabidopsis thaliana*. Commun Integr Biol 3(2):130–138

Ryu C-M, Hu C-H, Reddy MS, Kloepper JW (2003) Different signaling pathways of induced resistance by rhizobacteria in *Arabidopsis thaliana* against two pathovars of *Pseudomonas syringae*. New Phytol 160:413–420

Ryu CM, Farag MA, Hu CH, Reddy MS, Kloepper JW, Paré PW (2004) Bacterial volatiles induce systemic resistance in Arabidopsis. Plant Physiol 134(3):1017–1026

Sa ALB, Dias ACF, Teixeira MA, Vieira RF (2012) Contribution of N_2 fixation for the world agriculture. In: Maheshwari DK (ed) Bacteria in agrobiology: plant probiotics. Springer, Berlin/Heidelberg, pp 315–324. https://doi.org/10.1007/978-3-642-27515-9_17

Saharan B (2011) Plant growth promoting Rhizobacteria: a critical review. Life Sci Med Res 21:1–30

Sarangi T, Ramakrishnan S, Nakkeeran S (2014) Isolation and evaluation of indigenous isolates of *Bacillus* spp. against *Meloidogyne incognita*. Int J Tropical Agric 32(1/2):15–18

Scarpeci TE, Zanor MI, Carrillo N, Mueller-Roeber B, Valle EM (2008) Generation of superoxide anion in chloroplasts of Arabidopsis thaliana during active photosynthesis: a focus on rapidly induced genes. Plant Mol Biol 66:361–378. https://doi.org/10.1007/s11103-007-9274-4

Schöck F, Dahl MK (1996) Analysis of DNA flanking the *treA* gene of *Bacillus subtilis* reveals genes encoding a putative specific enzyme II Tre and a potential regulator of the trehalose operon. Gene 175(1):59–63

Seldin L, Bastos MCF, Penido EGC (1989) Identification of *Bacillus azotofixans* nitrogen fixation genes using heterologous *nif* probes. In: Nitrogen fixation with nonlegumes. Springer, Dordrecht, pp 179–187

Sengupta S, Ganguli S, Singh PK (2017) Metagenome analysis of the root endophytic microbial community of Indian rice (*O. sativa* L.). Genomics Data 12:41–43

Sha J, Zhang J, Chi B, Liu R, Li H, Gao J (2018) Sip1Ab gene from a native Bacillus thuringiensis strain QZL38 and its insecticidal activity against *Colaphellus bowringi* Baly. Biocontol Sci Technol. https://doi.org/10.1080/09583157.2018.1460313

Shao J, Li S, Zhang N, Cui X, Zhou X, Zhang G, Zhang R (2015) Analysis and cloning of the synthetic pathway of the phytohormone indole-3-acetic acid in the plant-beneficial *Bacillus amyloliquefaciens* SQR9. Microb Cell Fact 14(1):2–13

Shaligram S, Kumbhare SV, Dhotre DP, Muddeshwar MG, Kapley A, Joseph N, Purohit HP, Shouche YS, Pawa SP (2016) Genomic and functional features of the biosurfactant producing *Bacillus* sp. AM13 Functional & Integrative Genomics 16(5):557–566

Shi B, Wu W, Wen J, Shi Q, Wu S (2010) Cloning and expression of a lipase gene from *Bacillus subtilis* FS1403 in *Escherichia coli*. Annals of Microbiol 60(3):399–404

Sieciechowicz KA, Joy KW, Ireland RJ (1988) The metabolism of asparagine in plants. Phytochemistry 27:663–671

Skinnider MA, Dejong CA, Rees PN, Johnston CW, Li H, Webster AL, Wyatt MA, Magarvey NA (2015) Genomes to natural products PRediction Informatics for Secondary Metabolomes (PRISM). Nucleic Acids Res 43:9645–9662

So Y, Park SY, Park EH, Park SH, Kim EJ, Pan JG, Choi SK (2017) A highly efficient CRISPR-Cas9-ediated large genomic deletion in *Bacillus subtilis*. Front Microbiol 8:1167. https://doi.org/10.3389/fmicb.2017.01167

Song Y, Tong Z, Wang J, Wang L, Guo Z, Han Y et al (2004) Complete genome sequence of Yersinia pestis strain 91001, an isolate a virulent to humans. DNA Res 11:179–197. https://doi.org/10.1093/dnares/11.3.179

Soriano M, Diaz P, Pastor FIJ (2006) Pectate lyase C from *Bacillus subtilis*: a novel endo-cleaving enzyme with activity on highly methylated pectin. Microbiology 152(3):617–625

Steele K, Virk D, Kumar R et al (2007) Field evaluation of upland rice lines selected for QTLs controlling root traits. Field Crops Res 101:180–186. https://doi.org/10.1016/j.fcr.2006.11.002

Sun Z, Hsiang T, Zhou Y, Zhou J (2015) Draft genome sequence of *Bacillus amyloliquefaciens* XK-4-1, a plant growth-promoting endophyte with antifungal activity. Genome Announc 3:e01306–e01315

Tahir HA, Gu Q, Wu H, Raza W, Hanif A, Wu L et al (2017) Plant growth promotion by volatile organic compounds produced by *Bacillus subtilis* SYST2. Front Microbiol 8:171

Tang M, Sun X, Zhang S, Wan J, Li L, Ni H (2017) Improved catalytic and antifungal activities of *Bacillus thuringiensis* cells with surface display of Chi9602DSP. J Appl Microbiol 122:106–118

Tapi A, Chollet-Imbert M, Scherens B, Jacques P (2010) New approach for the detection of non-ribosomal peptide synthetase genes in *Bacillus strains* by polymerase chain reaction. Appl Microbiol Biotechnol 85:1521–1531. https://doi.org/10.1007/s00253-009-2176-4

Tareq FS, Lee MA, Lee HS, Lee YJ, Lee JS, Hasan CM, Islam MT, Shin HJ (2014a) Gageotetrins A-C, noncytotoxic antimicrobial linear lipopeptides from a marine bacterium *Bacillus subtilis*. Org Lett *16*:928–931

Tareq FS, Lee MA, Lee HS, Lee YJ, Lee JS, Hasan CM, Islam MT, Shin HJ (2014b) Non-cytotoxic antifungal agents: isolation and structures of gageopeptides from a marine-derived *Bacillus subtilis* 109GGC020. J Agric Food Chem 62(24):5565–5572

Terra R, Stanley-Wall NR, Cao G, Lazazzera BA (2012) Identification of Bacillus subtilis SipW as a biofunctional signal peptidase that controls surface-adhered biofilm formation. J Bacteriol. https://doi.org/10.1128/JB.06780-11

Thakuria D, Talukdar NC, Goswami C, Hazarika S, Boro RC, Khan MR (2004) Characterization and screening of bacteria from the rhizosphere of rice grown in acidic soils of. Assam Curr Sci 86:978–985

Todd JA, Hubbard TJP, Travers AA, Ellar DJ (1985) Heat-shock proteins during growth and sporulation of *Bacillus subtilis*. FEBS Lett 188(2):209–214

Tsau JL, Guffanti AA, Montville TJ (1992) Conversion of pyruvate to acetoin helps to maintain pH homeostasis in *Lactobacillus* plantarum. Appl Environ Microbiol 58:891–894

Tyc O, Song C, Dickschat JS, Vos M, Garbeva P (2017) The ecological role of volatile and soluble secondary metabolites produced by soil bacteria. Trends in Microbiol. https://doi.org/10.1016/j.tim.2016.12.002

Uga Y, Sugimoto K, Ogawa S et al (2013) Control of root system architecture by DEEPER ROOTING 1 increases rice yield under drought conditions. Nat Genet 45:1097. https://doi.org/10.1038/ng.2725

Vacheron J, Desbrosses G, Bouffaud M-L et al (2014) Plant growth-promoting rhizobacteria and root system functioning. Front Plant Sci 4:1–19. https://doi.org/10.3389/fpls.2013.00356

van Heel AJ, de Jong A, Montalbán-López M, Kok J, Kuipers OP (2013) BAGEL3: automated identification of genes encoding bacteriocins and(non-)bactericidal post translationally modified peptides. Nucleic Acids Res 41:W448–W453. https://doi.org/10.1093/nar/gkt391

van Heel AJ, de Jong A, Song C, Viel JH, Kok J, Kuipers OP (2018) BAGEL4: a user-friendly web server to thoroughly mine RiPPs and bacteriocins. Nucleic Acids Res 46(1):278–281

Vieira-Pires RS, Szollosi A, Morais-Cabral JH (2013) The structure of the *KtrAB* potassium transporter. Nature 496(7445):323–328

Vinci G, Cozzolino V, Mazzei P, Monda H, Savy D, Drosos M, Piccolo A (2018) Effects of *Bacillus amyloliquefaciens* and different phosphorus sources on Maize plants as revealed by NMR and GC-MS based metabolomics. Plant Soil. https://doi.org/10.1007/s11104-018-3701-y

Vockenhuber MP, Sharma CM, Statt MG, Schmidt D, Xu Z, Dietrich S, Liesegang H, Mathews DH, Suess B (2011) Deep sequencing-based identification of small non-coding RNAs in *Streptomyces coelicolor*. RNA Biol 8:468–477

Wang P, Guo Q, Ma Y, Li S, Lu X, Zhang X, Ma P (2015) DegQ regulates the production of fengycins and biofilm formation of the biocontrol agent *Bacillus subtilis* NCD-2. Microbiol Res 178:42–50

Wang B, Wang H, Xiong J, Zhou Q, Wu H, Xia L, Li L, Yu Z (2017) A Proteomic analysis provides novel insights into the stress responses of *Caenorhabditis elegans* towards nematicidal Cry6AToxin from *Bacillus thuringiensis*. Sci Rep 7:14170. https://doi.org/10.1038/s41598-017-14428-3

Wang T, Liu X-H, Wu M-B, Ge S (2018) Molecular insights into the antifungal mechanism of bacilysin. J Mol Model 24:118. https://doi.org/10.1007/s00894-018-3645-4.

Westbrook AW, Moo-Young M, Chou CP (2016) Development of a CRISPR-Cas9 Tool Kit forcomprehensive engineering of *Bacillus subtilis*. Appl Environ Microbiol 82(16):4876–4895. https://doi.org/10.1128/AEM.01159-16

White JF, Torres MS (2010) Is plant endophyte mediated defensive mutualism the result of oxidative stress protection? Physiol Plant 138:440–446

Wray LV Jr, Fisher SH (2008) *Bacillus subtilis* GlnR contains an autoinhibitory C-terminal domain required for the interaction with glutamine synthetase. Mol Microbiol 68:277–285

Wu J-Y, Liao J-H, Shieh C-J, Hsieh F-C, Liu Y-C (2018) Kinetic analysis on precursors for iturin A production from *Bacillus amyloliquefaciensBPD1*. J Biosci Bioeng. In Press

Xiang P, Chen L, Zhu X, Wang Y, Duan Y (2013) Screening and identification of bacterium to induce resistance of soybean against *Heterodera glycines*. Chin J Biol Control 29(4):661–665

Xiang N, Lawrence KS, Kloepper JW, Donald PA, McInroy JA, Lawrence GW (2017) Biological control of *Meloidogyne incognita* by spore-forming plant growth-promoting rhizobacteria on cotton. Plant Dis 101(5):774–784

Xie SS, Wu HJ, Zang HY, Wu LM, Zhu QQ, Gao XW (2014) Plant growth promotion by spermidine-producing *Bacillus subtilis* OKB105. Mol Plant Microbe Interact 27:655–663. https://doi.org/10.1094/MPMI-01-14-0010-R

Xiong J, Zhou Q, Luo H, Xia L, Li L, Sun M, Yu Z (2015) Systemic nematicidal activity and biocontrol efficacy of *Bacillus firmus* against the root-knot nematode *Meloidogyne incognita*. World J Microbiol Biotechnol 31(4):661–667. https://doi.org/10.1007/s11274-015-1820-7

Yaish MW, Antony I, Glick BR (2015) Isolation and characterization of endophytic plant growth-promoting bacteria from date palm tree (*Phoenix dactylifera* L.) and their potential role in salinity tolerance. Antonie Van Leeuwenhoek 107:1519–1532

Yang JW, Yu SH, Ryu CM (2009) Priming of defense-related genes confers root-colonizing bacilli elicited induced systemic resistance in pepper. Plant Pathol J 25:389–399

Yi HS, Ahn YR, Song GC, Ghim SY, Lee S, Lee G, Ryu CM (2016) Impact of a bacterial volatile 2,3-butanediol on *Bacillus subtilis* rhizosphere robustness. Front Microbiol 7:993. https://doi.org/10.3389/fmicb.2016.00993

Yi Y, Jong A, Frenzel E, Kuipers OP (2017) Comparative transcriptomics of *Bacillus mycoides* strains in response to potato-root exudates reveals different genetic adaptation of endophytic and soil isolates. Front Microbiol 8:1487. https://doi.org/10.3389/fmicb.2017.01487

Yu Y, Yan F, He Y, Qin Y, Chen Y, Chai Y, Guo J (2018) The ClpY-ClpQ protease regulates multicellular development in *Bacillus subtilis*. Microbiology 164:848–862. https://doi.org/10.1099/mic.0.000658

Zghal RZ, Ghedira K, Elleuch J, Kharrat M, Tounsi S (2018) Genome sequence analysis of a novel *Bacillus thuringiensis* strain BLB406 active against *Aedes aegypti* larvae, a novel potential bioinsecticide. Int J Biol Macromol. https://doi.org/10.1016/j.ijbiomac.2018.05.119

Zhang L, Sun C (2018) Cyclic lipopeptides fengycins from marine bacterium *Bacillus subtilis* kill plant pathogenic fungus *Magnaporthe grisea* by inducing reactive oxygen species production and chromatin condensation. Appl Environ Microbiol. https://doi.org/10.1128/AEM.00445-18

Zhang H, Yang D, Wang D, Miao Y, Shao J, Zhou X, Xu Z, Li Q, Feng H, Li S (2015) Whole transcriptomic analysis of the plant-beneficial rhizobacterium *Bacillus amyloliquefaciens* SQR9 during enhanced biofilm formation regulated by maize root exudates. BMC Genom 16:685

Zhou L, Yuen G, Wang Y, Wei L, Ji G (2016) Evaluation of bacterial biological control agents for control of root-knot nematode disease on tomato. Crop Protect 84:8–13. https://doi.org/10.1016/j.cropro.2015.12.009

Ziemert N, Alanjary M, Weber T (2016) The evolution of genome mining in microbes – a review. Nat Prod Rep 33:988–1005

Zimmerman SB, Schwartz CD, Monaghan RL, Pelak BA, Weissberger B, Gilfillan EC (1986) Difficidin and oxydifficidin: novel broad spectrum antibacterial antibiotics produced by *Bacillus subtilis*: production, taxonomy and antibacterial activity. J Antibiot 12:1677–1681

Zou C, Li Z, Yu D (2010) *Bacillus megaterium* strain XTBG34 promotes plant growth by producing 2-pentylfuran. J Microbiol 48:460–466

Zou H, Wenwen Y, Zang G et al (2015) OsEXPB2, a β-expansin gene, is involved in rice root system architecture. Mol Breed 35:41. https://doi.org/10.1007/s11032-015-0203-y

Chapter 11
Tapping the Potential of Metabolomics in New Natural Products Discovery from *Bacillus* Species

Zerihun T. Dame and Md Tofazzal Islam

11.1 Introduction

The genus *Bacillus* exhibits enormous taxonomic and metabolic diversity within the species (Fu et al. 2014). They produce diverse classes of bioactive compounds with a potential application in agro-industry, pharmaceutical, and biotechnological sectors. While peptide antibiotics are the most common metabolites produced by *Bacillus* species, other classes of bioactive compounds have also been isolated from the group; macrolactins (Jaruchoktaweechai et al. 2000), polyene antibiotics (Patel et al. 1995), terpenoids (Kontnik et al. 2008), phospholipid antibiotics (Tamehiro et al. 2002), lipopeptides (Villegas-Escobar et al. 2013), siderophores (Heidarzadeh and Baghaee-Ravari 2015), enzymes (Gomaa 2012), and macrocyclic polyene (Ravu et al. 2015) are some of the classes of molecules they produce. However, the discovery of bioactive metabolites from microbial sources has shown a decline in the past, partly because of difficulties in finding high-quality natural products screening libraries, lack of research funding commitments from the stakeholders, difficulty in dereplication, lack of modern screening assays, and challenges in production scale-up (Koehn 2008). Some of these challenges can be addressed by employing emerging technologies such as metabolomics and genomics. In this chapter application of metabolomics in microbial natural product discovery is discussed.

Z. T. Dame (✉)
Department of Biology, College of Natural and Computational Science, Ambo University, Ambo, Ethiopia

M. T. Islam
Institute of Biotechnology and Genetic Engineering (IBGE), Bangabandhu Sheikh Mujibur Rahman Agricultural University, Gazipur, Bangladesh

M. T. Islam et al. (eds.), *Bacilli and Agrobiotechnology: Phytostimulation and Biocontrol*, Bacilli in Climate Resilient Agriculture and Bioprospecting, https://doi.org/10.1007/978-3-030-15175-1_11

11.2 *Bacillus* Chemistry

Bacillus species has continued to be a rich source of bioactive natural products. Examples of metabolites recently isolated from the group are given here (Fig. 11.1). From a marine sediment strain of *B. subtilis* two cyclic-lipotetrapeptides, bacilotetrins A (**1**) and B (**2**) with anti-MRSA (methicillin-resistant *Staphylococcus aureus*) activity were isolated using LC-MS and NMR spectroscopy guided metabolic profiling and dereplication of a crude extract of the strain (Tareq and Shin 2017).

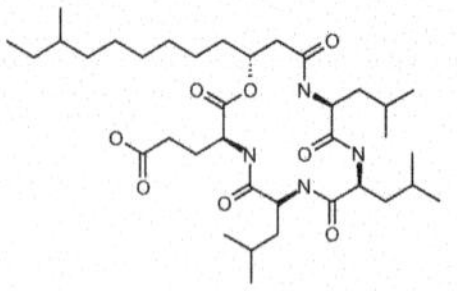
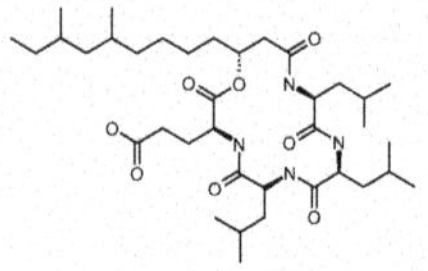
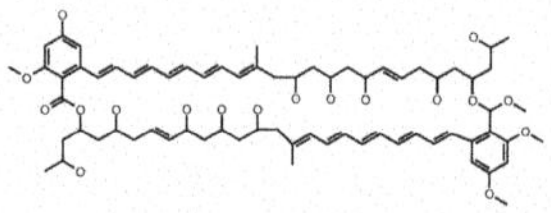

1 **2** **3**

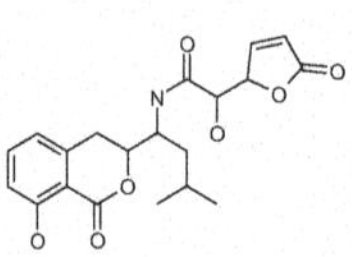
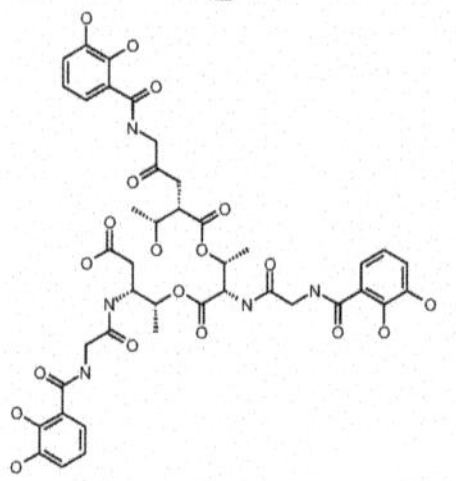
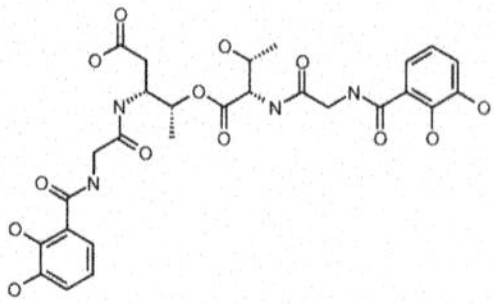

4 **5** **6**

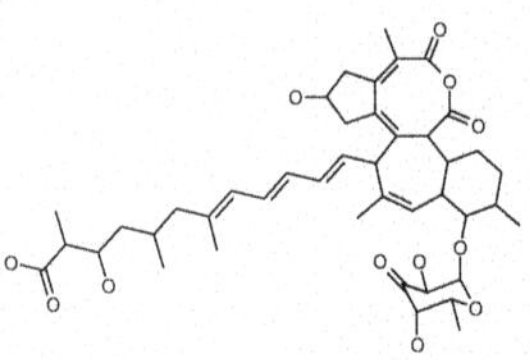
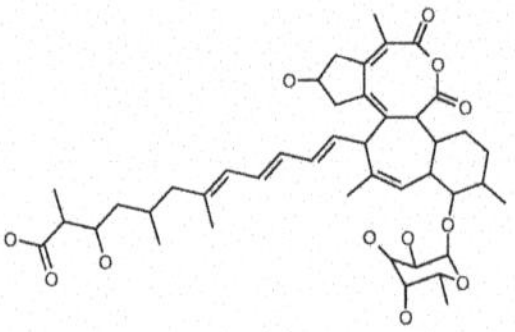
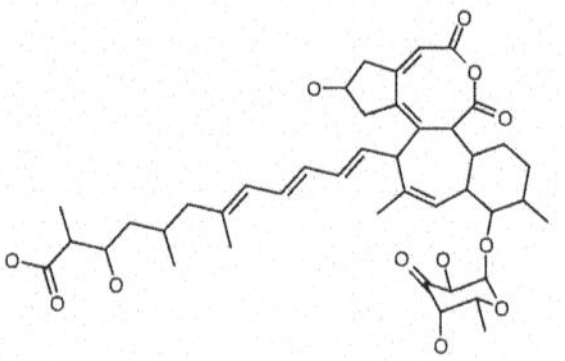

7 **8** **9**

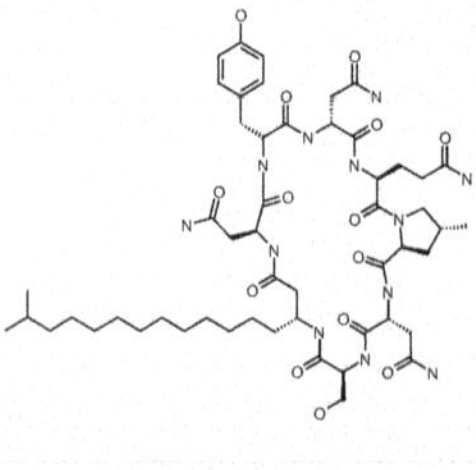
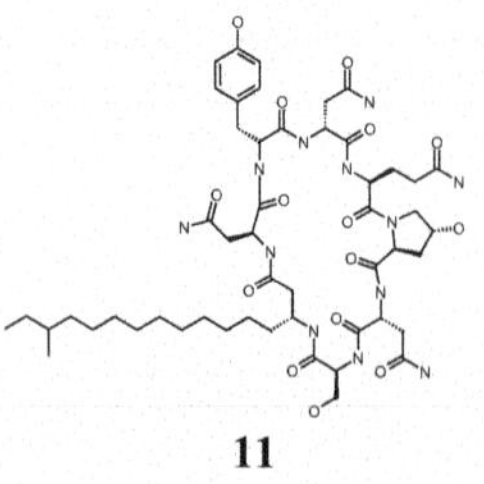
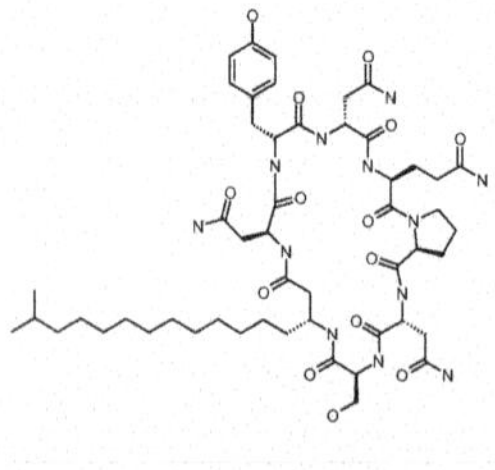

10 **11** **12**

Fig. 11.1 Structures (**1–24**) of secondary metabolites discovered from *Bacillus* species using various "omics" techniques

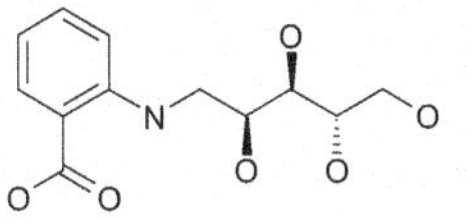

13

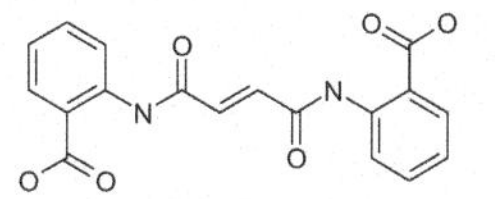

14

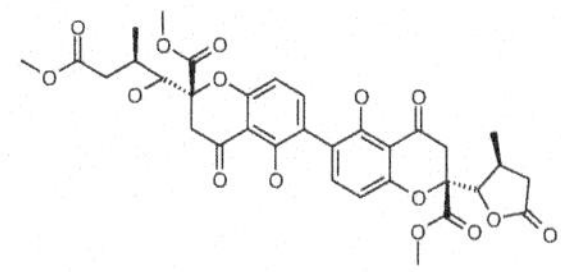

15

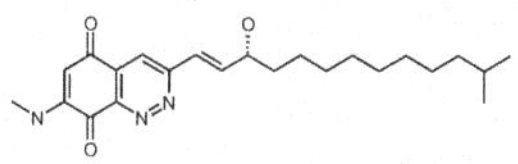

16

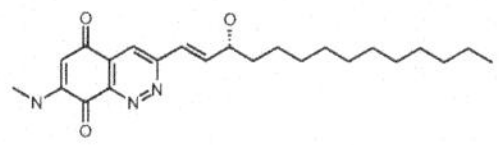

17

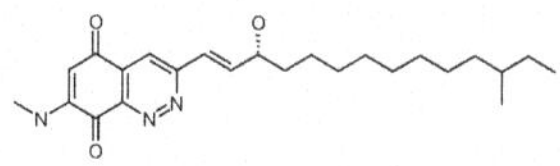

18

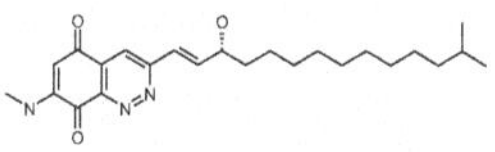

19

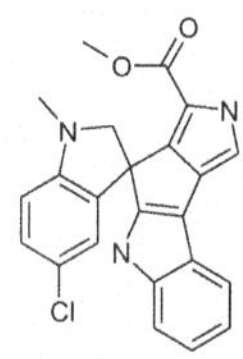

20

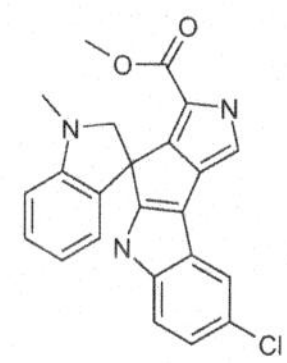

21

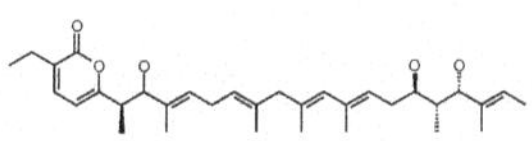

22

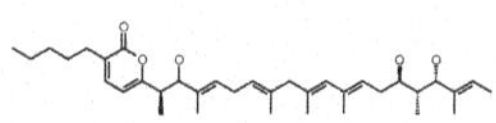

23

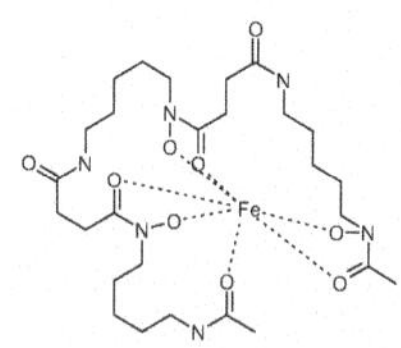

24

Fig. 11.1 (continued)

*Bacillus*in A (**3**) a rare macrocyclic polyene antibiotic, also with strong activity against MRSA and vancomycin-resistant *Enterococcus faecium*, was isolated by Ravu et al. (2015) from crude extract of a strain of *B. amyloliquefaciens*. The compound has a minimum inhibitory concentration ranges between 0.6 and 1.2 μg/mL.

A new metabolite, (S)-2-hydroxy-N-((S)-1-((S)-8-hydroxy-1-oxoisochroman-3-yl)-3-methylbutyl)-2-((S)-5-oxo-2,5-dihydrofuran-2-yl)acetamide) (**4**) with enzyme inhibitory activity, was isolated from *Oryza sativa*-associated endophytic *B. amyloliquefaciens* RWL-1 (Shahzad et al. 2018). The compound showed inhibitory activities against α-glucosidase (37 ± 0.09%) and urease (49.4 ± 0.53%), respectively.

From a marine *Bacillus* strain, two new non-ribosomal peptides (**5** and **6**) were isolated and characterized (Zhou et al. 2018). They are analogues of the bioactive metabolites, bacillibactin.

From fermentation of a *B. subtilis* strain that was isolated from compost, three new polyene antibiotics, aurantinins B (**7**), C (**8**), and D (**9**) were isolated. The compounds displayed activity against MRSA. Their antibacterial activity was shown to be related to their ability to disrupt the cell membrane (Yang et al. 2016). Likewise, a *Bacillus* sp. KCB14S006 isolated from a natural hypersaline environment was studied for their metabolomic competence leading to the isolation of three new lipopeptides: iturin F_1 (**10**), iturin F_2 (**11**), and iturin A_9 (**12**). These compounds showed strong antifungal activities against various pathogenic fungi and moderate cytotoxic activities toward HeLa and src(ts)-NRK cell lines (Son et al. 2016).

The potential of *Bacillus* species to produce structurally diverse bioactive metabolites has been widely reviewed. Metabolites from marine (Mondol et al. 2013) and soil (Sansinenea and Ortiz et al. 2011) *Bacillus* species have been discussed, respectively. Some of the isolated compounds and their activities are summarized in Table 11.1.

Besides being primary producers of various bioactive metabolites, *Bacillus* species are attracting investigator's attention in co-culturing experiments as a strategy to induce production of new natural products. The discovery of bionectriamines A (**13**) and B (**14**) through co-cultivation of a fungal strain with *B. subtilis* is a good example (Kamdem et al. 2018).

A related study by Arora et al. (2018) on a strain of *Setophoma terrestris* co-cultured with B. *amyloliquefaciens* led them to the discovery of a novel compound, blennolide K (**15**), which was active against PC-3 (prostate) and MCF-7 (breast) cell lines with IC_{50} values of $3{\cdot}7 \pm 0{\cdot}6$ and $4{\cdot}8 \pm 0{\cdot}4$ μmol l^{-1}, respectively.

11.3 Metabolomics

Metabolomics, an omic science in systems biology, is the global quantitative assessment of endogenous metabolites within a biological system (Fiehn et al. 2000). It targets a complete analysis of small molecules that participate in general metabolic reactions and produced by normal or abnormal cellular processes. Its main purposes are identification and quantification of endogenous metabolites under a given set of standard conditions and perturbations (Rochfort 2005). To this effect, metabolomics uses various analytical approaches to identify and quantify all metabolites thereby helping researchers to understand the complex molecular interactions in biological systems (Hall et al. 2002). Likewise, advances in analytical tools have created the opportunities to analyze metabolites in a biofluid, cell cultures, or a tissue (O'Connell 2012). Metabolomics follows a different approach in studying small molecules as opposed to the traditional natural product discovery method. It involves various databases, spectral libraries, extract collections, and bioinformatics tools with a potential to facilitate rapid identification of known metabolites. Moreover, it needs small material for the analysis (Macintyre et al. 2014). Krug and Müller (2014) in their review argued that traditional methods of natural product discovery have so far only scratched the surface of the real microbial "secondary metabolome landscape" showing the huge potential microbial metabolome carries, which could be put in to use through metabolomics-based technologies. Metabolomics has also been reported to play a role in strain prioritization during a microbial drug discovery program (Hou et al. 2012; Huang et al. 2014).

Table 11.1 Bioactive metabolites recently isolated from *Bacillus* spp.

Compounds	Producing Strain	Structure	Reported Activity	Reference
Aeritracin	*Aeribacillus pallidus*	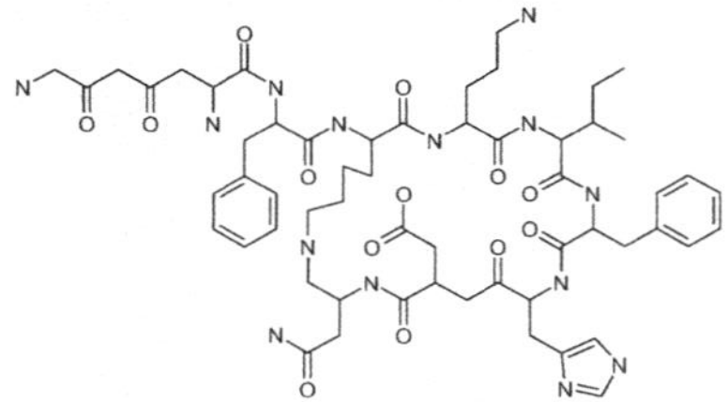	Antibacterial	Muhammad and Ahmed (2015)
Paenialvin A	*Paenibacillus alvei*	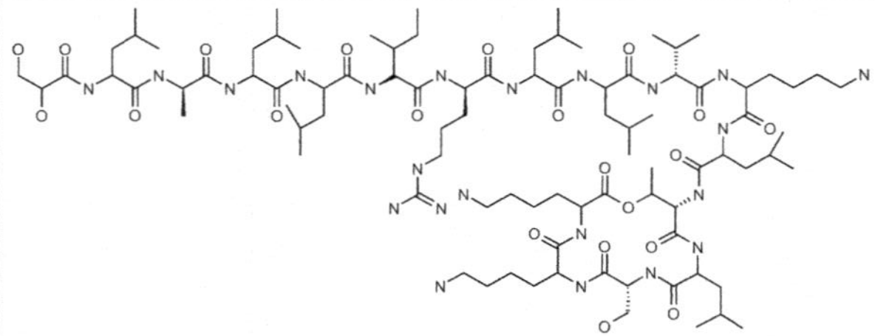	Antimicrobial	Meng et al. (2018)
Paenialvin B	*P. alvei*	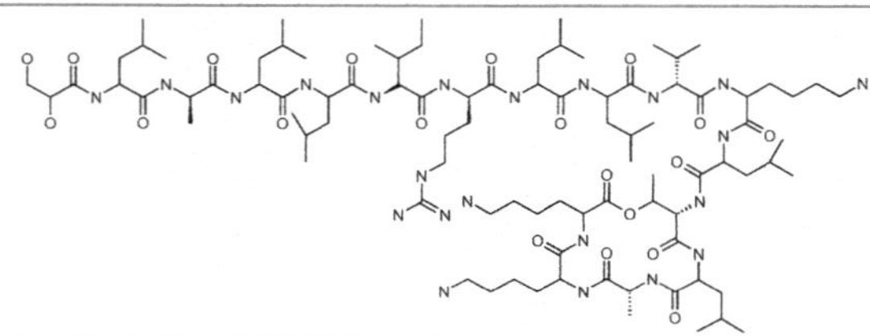	Antimicrobial	Meng et al. (2018)

(continued)

Table 11.1 (continued)

Compounds	Producing Strain	Structure	Reported Activity	Reference
Paenialvin C	*P. alvei*	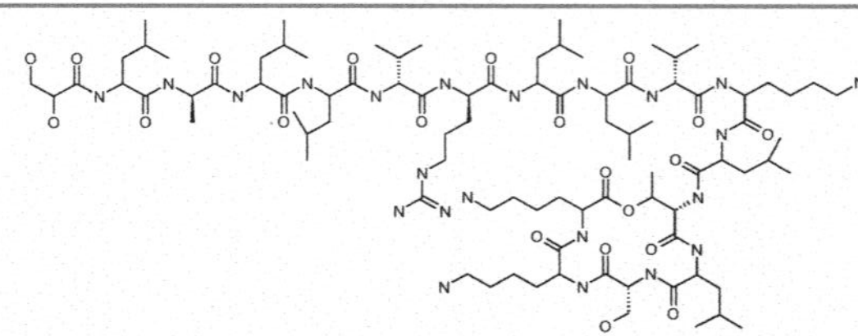	Antimicrobial	Meng et al. (2018)
Paenialvin D	*P. alvei*	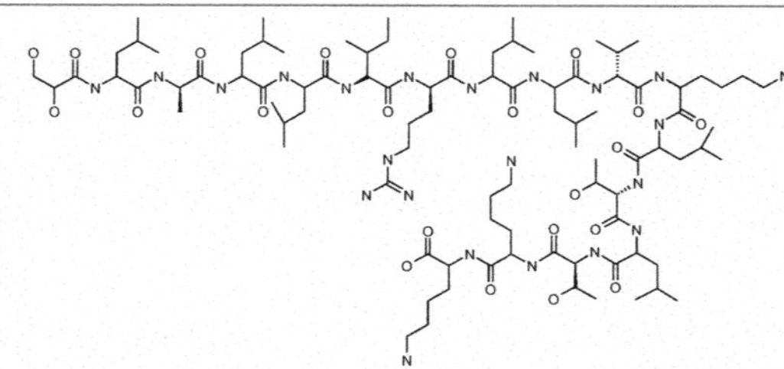	Antimicrobial	Meng et al. (2018)
Gageotetrins A	*B. subtilis*	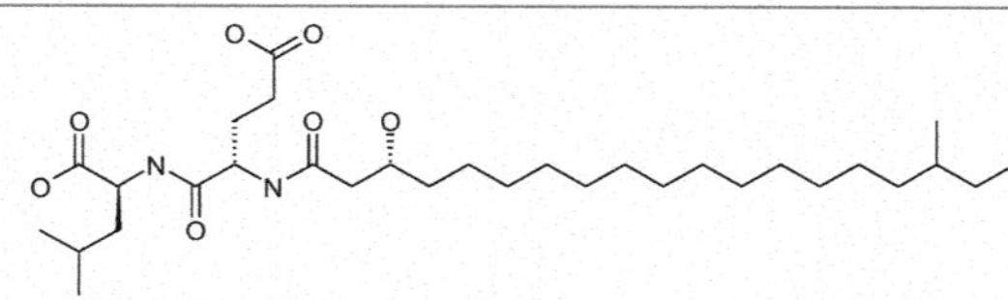	Antimicrobial	Tareq et al. (2014a)
Gageotetrins B	*B. subtilis*	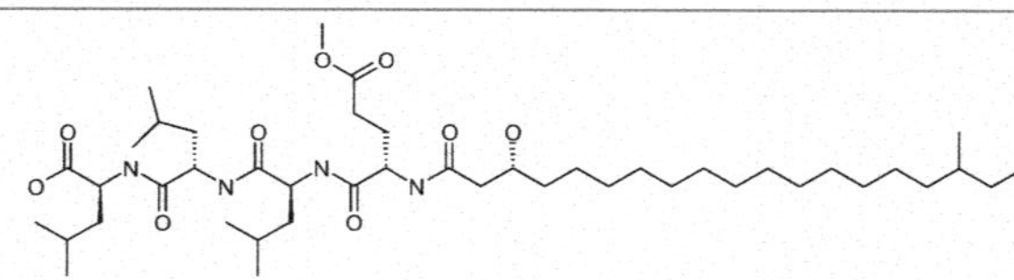	Antimicrobial	Tareq et al. (2014a)

Gageotetrins C	*B. subtilis*	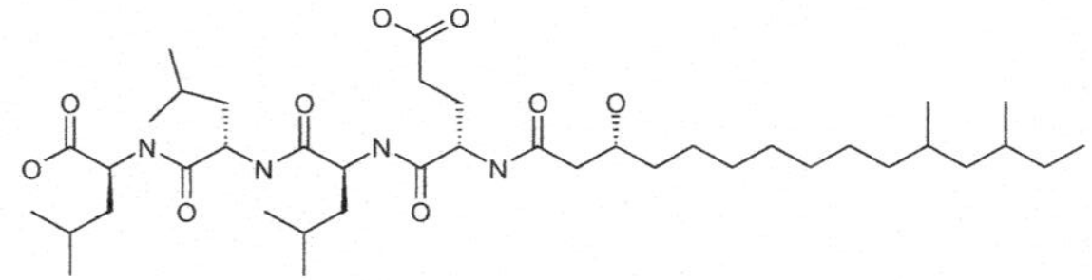	Antimicrobial	Tareq et al. (2014a)
Gageopeptide A	*B. subtilis*	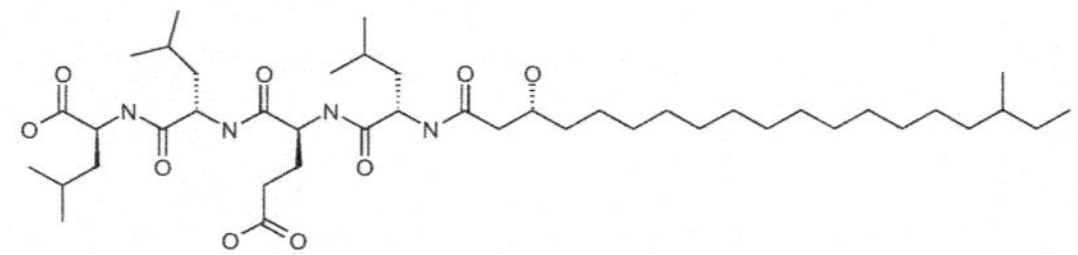	Antimicrobial	Tareq et al. (2014b)
Gageopeptide B	*B. subtilis*	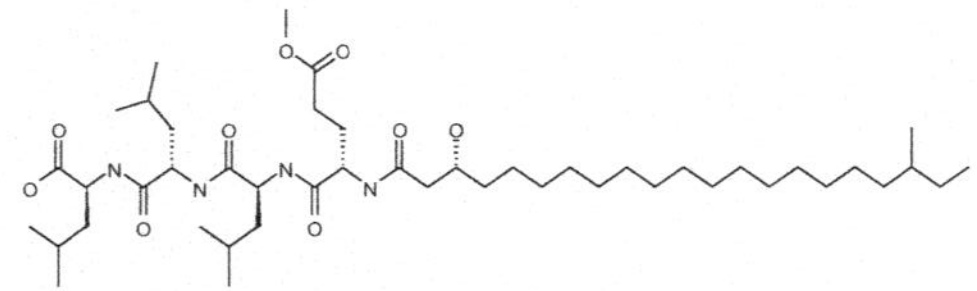	Antimicrobial	Tareq et al. (2014b)
Gageopeptide C	*B. subtilis*	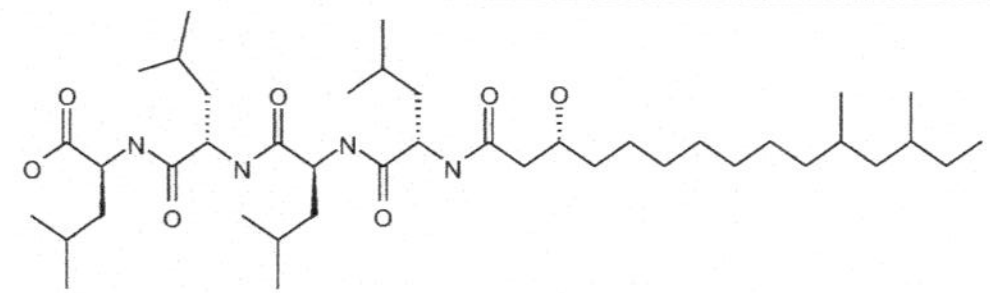	Antimicrobial	Tareq et al. (2014b)
Gageopeptide D	*B. subtilis*	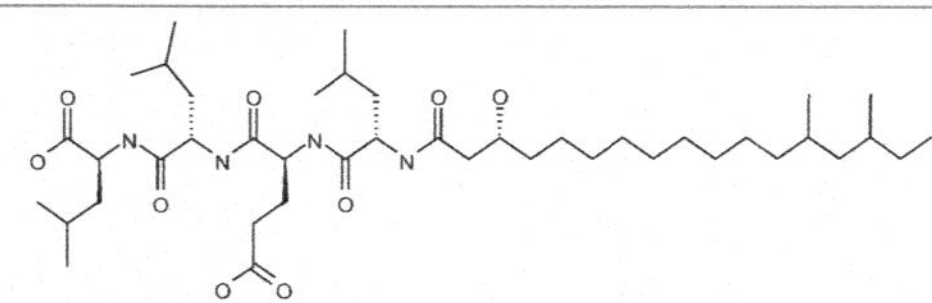	Antimicrobial	Tareq et al. (2014b)

(continued)

Table 11.1 (continued)

Compounds	Producing Strain	Structure	Reported Activity	Reference
Macrolactin	*B. subtilis*	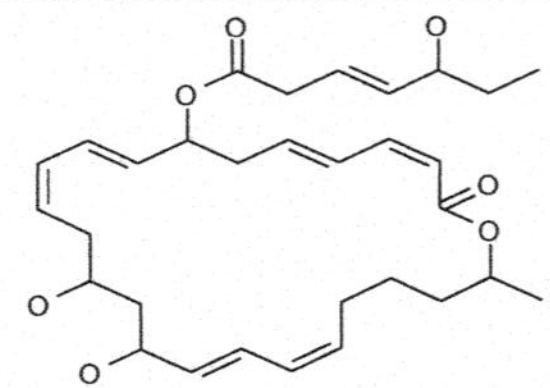	Antimicrobial	Chakraborty et al. (2014) Agri and food shem
Plipastatin A	*B. amyloliquefaciens*	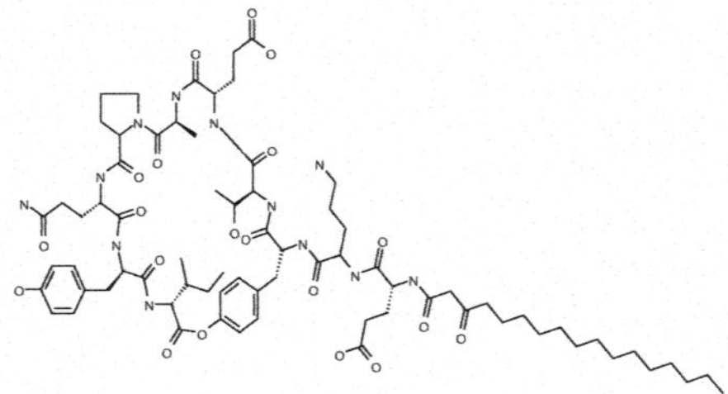	Antifungal	Ma and Hu (2018)
Furanoterpenoids	*B. subtilis*	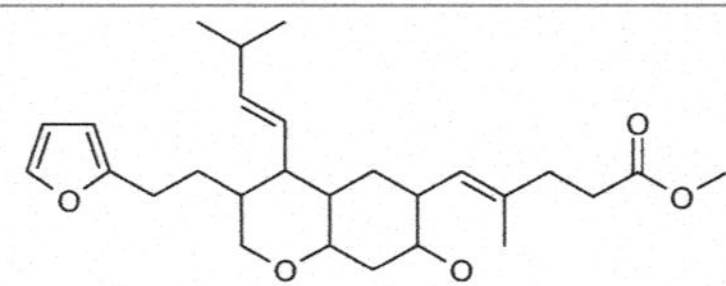	Antibacterial	Chakraborty et al. (2017a)
Furanoterpenoids	*B. subtilis*	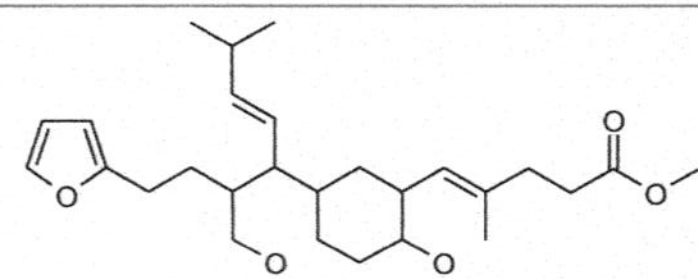	Antibacterial	Chakraborty et al. (2017a)

Turnagainolide A	*Bacillus* sp.	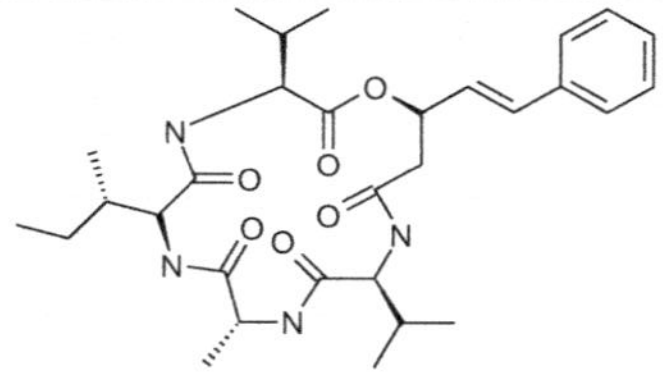	NA	Li et al. (2011)
Turnagainolide B	*Bacillus* sp.	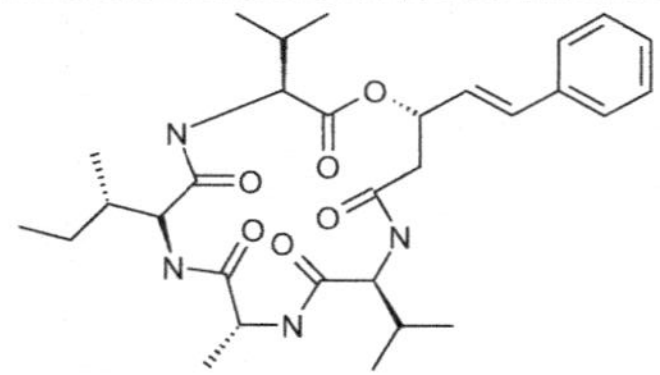	*Activation* of *SHIP1*	Li et al. (2011)
Bacilsubteramide A	*B. subterraneus*	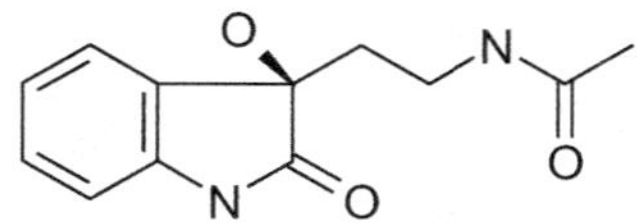	-	Xie et al. (2018)
2.3.1. 3-(Octahydro-9-isopropyl-2H-benzo [h] chromen-4-yl)-2-methylpropyl benzoate	*B. amyloliquefaciens*	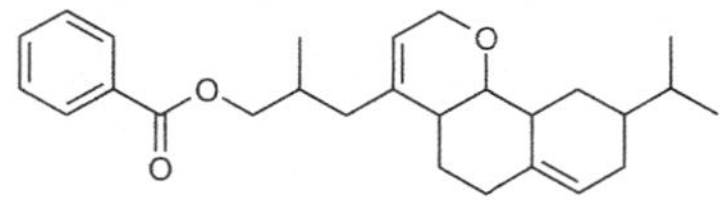	Antibacterial	Chakraborty et al. (2017a)
2.3.2. Methyl 8-(2-(benzoyloxy) ethyl)–hexahydro-4-((E)–pent-2-enyl)-2H-chromene-6-carboxylate	*B. amyloliquefaciens*	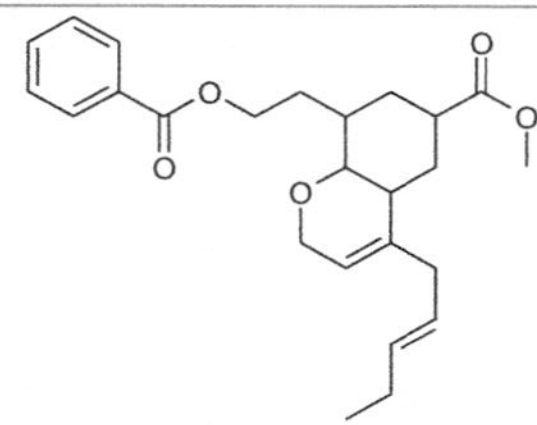	Antibacterial	Chakraborty et al. (2017a)

(continued)

Table 11.1 (continued)

Compounds	Producing Strain	Structure	Reported Activity	Reference
Pseurotin A	Marine *Bacillus* sp.	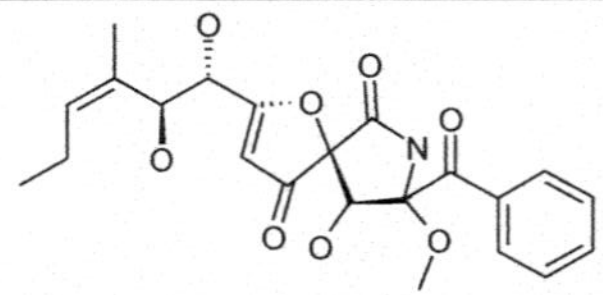	Anticancer	Anjum et al. (2018)
11-(15-Butyl-13-ethyl-tetrahydro-12-oxo-2H-pyran-13-yl) propyl-2-methylbenzoate	*B. amyloliquefaciens*	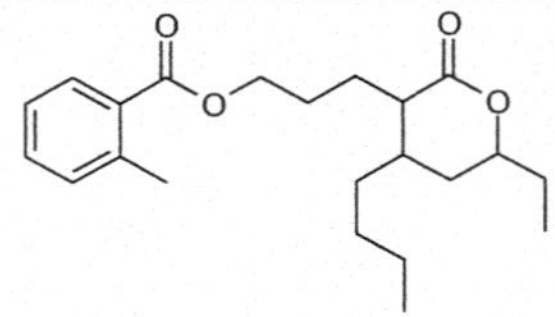	Antibacterial	Chakraborty et al. (2018)
9-(Tetrahydro-12-isopropyl-11-oxofuran-10-yl)-ethyl-4-ethoxy-2-hydroxybenzoate (**2**)	*B. amyloliquefaciens*	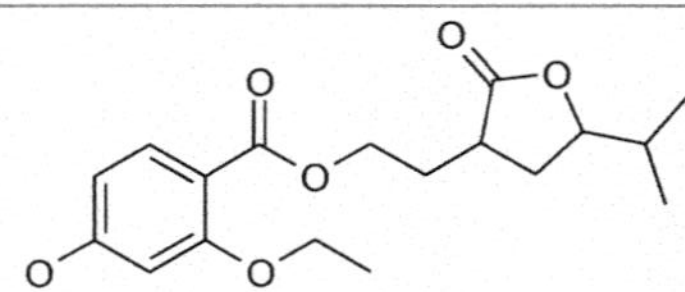	–	Chakraborty et al. (2018)
Cereusitin A	*B. cereus*	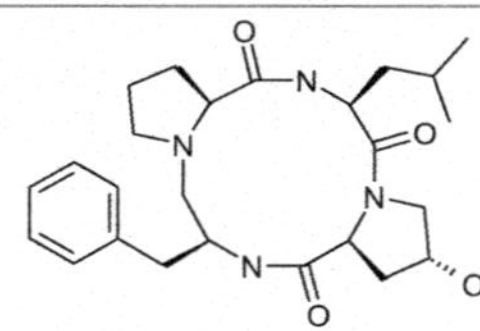	Antifungal	Pinzón-Espinosa et al. (2017)

Bacillomycin DC	*B. amyloliquefaciens*	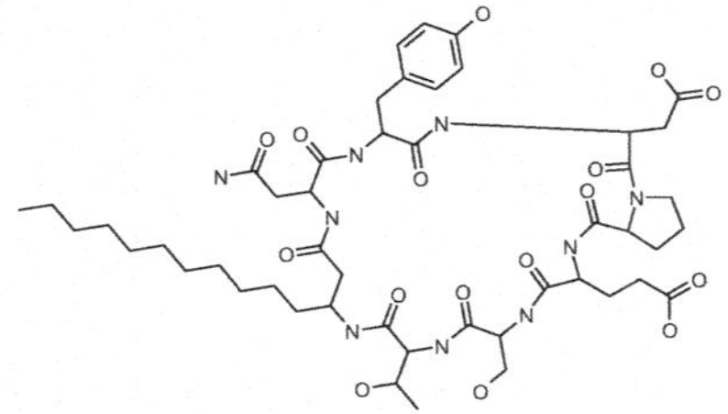	Antifungal	Jin et al. (2018)
Isotryptophan	*B. amyloliquefaciens*	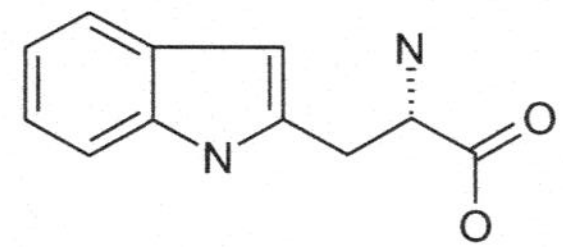	Antifungal	Cui et al. (2017)
2-(7-(2-Ethylbutyl)-2,3,4,4a,6,7-Hexahydro-2-oxopyrano-[3,2b]-pyran-3-yl)-Ethyl benzoate	*B. subtilis*	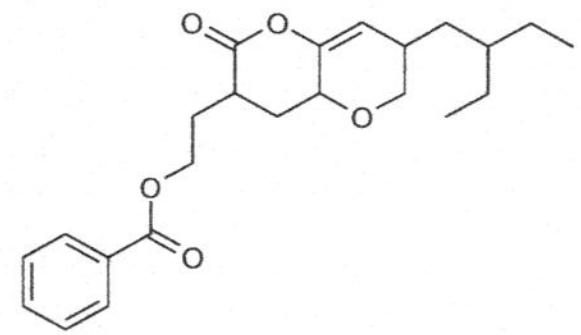	Antibacterial	Chakraborty et al. (2017b)
2-((4Z)-2-ethyl-octahydro-6-oxo-3-((E)-pent-3-enylidene)-Pyrano-[3,2b]-pyran-7-yl)-ethyl benzoate	*B. subtilis*	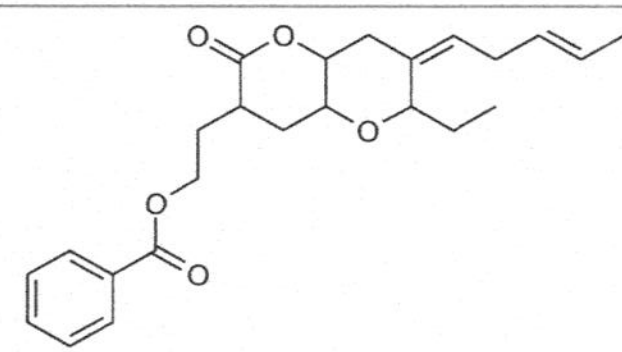	Antibacterial	Chakraborty et al. (2017c)

11.3.1 Metabolomics in Microbial Natural Product Discovery

In recent years, various research groups have begun employing metabolomic techniques for natural product discovery, and some interesting results are being reported. Wu et al. (2015) used NMR-based metabolomics to discover a new antibiotic through tracking the target molecule proton signal during the isolation process to effectively separate it from the crude extract. Likewise, the use of NMR-based metabolomics coupled with bioinformatics tools has been reported to assist in chemical profiling of crude extract without further fractionation (Bakiri et al. 2017). Yang et al. (2014) developed a natural product library of column fractions using a high-throughput, automated fractionation system coupled with UPLC-MS-ELSD-PDA information that made it possible to implement an improved dereplication and facilitated the discovery of new metabolites. A natural product library of this type with detailed information on fractions and/or pure compounds, spectroscopic data, and taxonomy of the producing strains facilitate natural product discovery with a minimum cost and time. The AntiBase, a microbial natural product database, developed by the group of Professor Laatsch at Gottingen University, Germany, is a good example.

Kurita et al. (2015) demonstrated that untargeted metabolomics based novel natural product discovery through their isolation of the unusual compounds with cinnoline core, quinocinnolinomycins A–D (**16**–**19**), from a library of natural product extracts. They used a compound activity mapping platform coupled with untargeted metabolomics, a method that facilitates rapid identification of bioactivity and compound's identity in complex natural product libraries.

On a related study, the use of metabolomics techniques with genomic study of marine bacteria led Paulus et al. (2017) to the discovery of new metabolites, spiroindimicins E (**20**) and F (**21**) and lagunapyrones D (**22**) and E (**23**), respectively.

Sidebottom et al. (2013) were able to discover a group of siderophores including ferrioxamine D1 (**24**) from a metabolome of an actinomycete strain using integrated metabolomics. They employed an approach that combines bacterial growth perturbation, accurate mass spectrometry, comparative mass spectral data analysis, and fragmentation spectra clustering.

Moreover, recent advancement in metabolomics-based databases have facilitated compound identification and dereplication at early stage of separation. Databases such as NAPROC-13 (López-Pérez et al. 2007), the universal natural products database (UNPD) with information on more than 200,000 natural products (Gu et al. 2013), the KNApSAcK database (Afendi et al. 2012), and the SuperNatural database (Banerjee et al. 2015) with detailed information of more than 300,000 metabolites could be used in microbial natural product research.

11.4 Conclusions

Bacilli are continuing to be a source of bioactive metabolites that could be used as a lead compound for agriculture and health application. Tapping this potential needs the use of emerging technologies such as metabolomics. Metabolomics facilitate compounds identification and characterization with a minimal cost and lesser time. Therefore, adopting this technology in studying metabolites of bacilli presents a great opportunity in the discovery of new natural products.

Acknowledgments The authors are thankful to the Bangladesh Academy of Sciences for funding this work under the BAS-USDA project to MTI. They are also thankful to the RMC of BSMRAU, University Grants Commission of Bangladesh, and Ministry of Science and Technology of Bangladesh for partial funding to this work.

References

Afendi FM, Okada T, Yamazaki M, Hirai-Morita A, Nakamura Y, Nakamura K et al (2012) KNApSAcK family databases: integrated metabolite-plant species databases for multifaceted plant research. Plant Cell Physiol 53:e1. https://doi.org/10.1093/pcp/pcr165

Anjum K, Bi H, Chai W, Lian XY, Zhang Z (2018) Antiglioma pseurotin A from marine *Bacillus* sp. FS8D regulating tumour metabolic enzymes. Nat Prod Res 32(11):1353–1356

Arora D, Chashoo G, Singamaneni V, Sharma N, Gupta P, Jaglan S (2018) *Bacillus amyloliquefaciens* induces production of a novel blennolide K in coculture of *Setophoma terrestris*. J Appl Microbiol 124(3):730–739

Bakiri A, Hubert J, Reynaud R, Lanthony S, Harakat D, Renault JH, Nuzillard JM (2017) Computer-aided 13C NMR chemical profiling of crude natural extracts without fractionation. J Nat Prod 80(5):1387–1396

Banerjee P, Erehman J, Gohlke BO, Wilhelm T, Preissner R, Dunkel M (2015) Super natural II – a database of natural products. Nucleic Acids Res 43:D935–D939. https://doi.org/10.1093/nar/gku886

Chakraborty K, Thilakan B, Raola VK (2014) Polyketide family of novel antibacterial 7-O-Methyl-5′-hydroxy-3′-heptenoate-macrolactin from seaweed-associated *Bacillus subtilis* MTCC 10403. J Agric Food Chem 62(50):12194–12208

Chakraborty K, Thilakan B, Chakraborty RD, Raola VK, Joy M (2017a) O-heterocyclic derivatives with antibacterial properties from marine bacterium *Bacillus subtilis* associated with seaweed, *Sargassum myriocystum*. Appl Microbiol Biotechnol 101(2):569–583

Chakraborty K, Thilakan B, Raola VK (2017b) Antimicrobial polyketide furanoterpenoids from seaweed-associated heterotrophic bacterium *Bacillus subtilis* MTCC 10403. Phytochemistry 142:112–125

Chakraborty K, Thilakan B, Raola VK, Joy M (2017c) Antibacterial polyketides from *Bacillus amyloliquefaciens* associated with edible red seaweed *Laurenciae papillosa*. Food Chem 218:427–434

Chakraborty K, Thilakan B, Raola VK (2018) Previously undescribed antibacterial polyketides from heterotrophic *Bacillus amyloliquefaciens* associated with seaweed padina gymnospora. Appl Biochem Biotechnol 184(2):716–732

Cui P, Guo W, Chen X (2017) Isotryptophan from Antarctic *Bacillus amyloliquefaciens* Pc3: purification, identification, characterization, and antifungal activity. Nat Prod Res 31(18):2153–2157

Fiehn O, Kopka J, Dörmann P, Altmann T, Trethewey RN, Willmitzer L (2000) Metabolite profiling for plant functional genomics. Nat Biotechnol 18(11):1157
Fu P, Johnson M, Chen H, Posner BA, MacMillan JB (2014) Carpatamides A–C, cytotoxic arylamine derivatives from a marine-derived Streptomyces sp. J Nat Prod 77(5):1245–1248
Gomaa EZ (2012) Chitinase production by *Bacillus thuringiensis* and *Bacillus licheniformis*: their potential in antifungal biocontrol. J Microbiol 50(1):103–111
Gu J, Gui Y, Chen L, Yuan G, Lu H-Z, Xu X (2013) Use of natural products as chemical library for drug discovery and network pharmacology. PLoS ONE 8:e62839. https://doi.org/10.1371/journal.pone.0062839
Hall R, Beale M, Fiehn O, Hardy N, Sumner L, Bino R (2002) Plant metabolomics: the missing link in functional genomics strategies. Plant Cell 14:1437–1440
Heidarzadeh N, Baghaee-Ravari S (2015) Application of *Bacillus pumilus* as a potential biocontrol agent of Fusarium wilt of tomato. Arch Phytopathol Plant Protect 48(13–16):841–849
Hou Y, Braun DR, Michel CR, Klassen JL, Adnani N, Wyche TP, Bugni TS (2012) Microbial strain prioritization using metabolomics tools for the discovery of natural products. Anal Chem 84(10):4277–4283
Huang T, Yang D, Rudolf JD, Xie P, Xie G, Teng Q, Lohman JR, Zhu X, Huang Y, Zhao LX, Jiang Y (2014) Strain prioritization for natural product discovery by a high-throughput real-time PCR method. J Nat Prod 77(10):2296–2303
Jaruchoktaweechai C, Suwanborirux K, Tanasupawatt S, Kittakoop P, Menasveta P (2000) New macrolactins from a marine *Bacillus* sp. Sc026. J Nat Prod 63(7):984–986
Jin P, Wang H, Liu W, Fan Y, Miao W (2018) A new cyclic lipopeptide isolated from *Bacillus amyloliquefaciens* HAB-2 and safety evaluation. Pestic Biochem Physiol 147:40–45
Kamdem RS, Wang H, Wafo P, Ebrahim W, Özkaya FC, Makhloufi G et al (2018) Induction of new metabolites from the endophytic fungus Bionectria sp. through bacterial co-culture. Fitoterapia 124:132–136
Koehn FE (2008) High impact technologies for natural products screening. In: Natural compounds as drugs, vol I. Birkhäuser Verlag, Basel, pp 175–210
Kontnik R, Bosak T, Butcher RA, Brocks JJ, Losick R, Clardy J, Pearson A (2008) Sporulenes, heptaprenyl metabolites from *Bacillus subtilis* spores. Org Lett 10(16):3551–3554
Krug D, Müller R (2014) Secondary metabolomics: the impact of mass spectrometry-based approaches on the discovery and characterization of microbial natural products. Nat Prod Rep 31(6):768–783
Kurita KL, Glassey E, Linington RG (2015) Proc Natl Acad Sci 112(39):11999–12004
Li D, Carr G, Zhang Y, Williams DE, Amlani A, Bottriell H, Mui AL, Andersen RJ (2011) Turnagainolides A and B, cyclic depsipeptides produced in culture by a *Bacillus* sp.: isolation, structure elucidation, and synthesis. J Nat Prod 74(5):1093–1099
López-Pérez JL, Theron R, del Olmo E, Diaz D (2007) NAPROC-13: a database for the dereplication of natural product mixtures in bioassay-guided protocols. Bioinformatics 23:3256–3257
Ma Z, Hu J (2018) Plipastatin A1 produced by a marine sediment-derived *Bacillus amyloliquefaciens* SH-B74 contributes to the control of gray mold disease in tomato. 3 Biotech 8(2):125
Macintyre L, Zhang T, Viegelmann C, Martinez IJ, Cheng C, Dowdells C et al (2014) Metabolomic tools for secondary metabolite discovery from marine microbial symbionts. Mar Drugs 12(6):3416–3448
Meng J, Zhong Z, Qian PY (2018) Paenialvin A-D, four peptide antibiotics produced by *PaeniBacillus alvei* DSM 29. J Antibiot 71:769–777
Mondol MAM, Shin HJ, Islam MT (2013) Diversity of secondary metabolites from marine *Bacillus* species: chemistry and biological activity. Mar Drugs 11(8):2846–2872
Muhammad SA, Ahmed S (2015) Production and characterization of a new antibacterial peptide obtained from *AeriBacillus pallidus* SAT4. Biotechnol Rep 8:72–80
O'Connell TM (2012) Recent advances in metabolomics in oncology. Bioanalysis 4(4):431–451
Patel PS, HuANG S, Fisher S, Pirnik D, Aklonis C, Dean L, Meyers E, Fernandes P, Mayerl F (1995) Bacillaene, a novel inhibitor of procaryotic protein synthesis produced by *Bacillus subtilis*. J Antibiot 48(9):997–1003

Paulus C, Rebets Y, Tokovenko B, Nadmid S, Terekhova LP, Myronovskyi M, Zotchev SB, Rückert C, Braig S, Zahler S, Kalinowski J (2017) New natural products identified by combined genomics-metabolomics profiling of marine *streptomyces* sp. MP131-18. Sci Rep 7:42382

Pinzón-Espinosa A, Martinez-Matamoros D, Castellanos L, Duque C, Rodríguez J, Jiménez C, Ramos FA (2017) Cereusitin A, a cyclic tetrapeptide from a *Bacillus cereus* strain isolated from the soft coral Antillogorgia (syn. Pseudopterogorgia) elisabethae. Tetrahedron Lett 58(7):634–637

Ravu RR, Jacob MR, Chen X, Wang M, Nasrin S, Kloepper JW, Liles MR, Mead DA, Khan IA, Li XC (2015) *Bacillus*in A, an antibacterial macrodiolide from *Bacillus amyloliquefaciens* AP183. J Nat Prod 78(4):924–928

Rochfort S (2005) Metabolomics reviewed: a new "omics" platform technology for systems biology and implications for natural products research. J Nat Prod 68(12):1813–1820

Sansinenea E, Ortiz A (2011) Secondary metabolites of soil *Bacillus* spp. Biotechnol Lett 33(8):1523–1538

Shahzad R, Latif Khan A, Ali L, Bilal S, Imran M, Choi KS, Al-Harrasi A, Al-Harrasi A, Lee IJ (2018) Characterization of new bioactive enzyme inhibitors from endophytic *Bacillus amyloliquefaciens* RWL-1. Molecules 23(1):114

Sidebottom AM, Johnson AR, Karty JA, Trader DJ, Carlson EE (2013) Integrated metabolomics approach facilitates discovery of an unpredicted natural product suite from *Streptomyces coelicolor* M145. ACS Chem Biol 8(9):2009–2016

Son S, Ko SK, Jang M, Kim JW, Kim GS, Lee JK, Jeon ES, Futamura Y, Ryoo I-J, Lee J-S, Oh H, Hong Y-S, Kim BY, Takahashi S, Osada H, Jang JH, Oh H (2016) New cyclic lipopeptides of the iturin class produced by saltern-derived *Bacillus* sp. KCB14S006. Mar Drugs 14(4):72

Tamehiro N, Okamoto-Hosoya Y, Okamoto S, Ubukata M, Hamada M, Naganawa H, Ochi K (2002) Bacilysocin, a novel phospholipid antibiotic produced by *Bacillus subtilis* 168. Antimicrob Agents Chemother 46(2):315–320

Tareq FS, Shin HJ (2017) Bacilotetrins A and B, Anti-Staphylococcal Cyclic-Lipotetrapeptides from a marine-derived *Bacillus subtilis*. J Nat Prod 80(11):2889–2892

Tareq FS, Lee MA, Lee HS, Lee YJ, Lee JS, Hasan CM, Islam MT, Shin HJ (2014a) Gageotetrins A–C, noncytotoxic antimicrobial linear lipopeptides from a marine bacterium *Bacillus subtilis*. Org Lett 16(3):928–931

Tareq FS, Lee MA, Lee HS, Lee YJ, Lee JS, Hasan CM, Islam MT, Shin HJ (2014b) Non-cytotoxic antifungal agents: isolation and structures of gageopeptides A–D from a *Bacillus* strain 109GGC020. J Agric Food Chem 62(24):5565–5572

Villegas-Escobar V, Ceballos I, Mira JJ, Argel LE, Orduz Peralta S, Romero-Tabarez M (2013) Fengycin C produced by *Bacillus subtilis* EA-CB0015. J Nat Prod 76(4):503–509

Wu C, Du C, Gubbens J, Choi YH, van Wezel GP (2015) Metabolomics-driven discovery of a prenylated isatin antibiotic produced by streptomyces species MBT28. J Nat Prod 78(10):2355–2363

Xie CL, Xia JM, Su RQ, Li J, Liu Y, Yang XW, Yang Q (2018) Bacilsubteramide A, a new indole alkaloid, from the deep-sea-derived *Bacillus subterraneus* 11593. Nat Prod Res:1–5

Yang J, Liang Q, Wang M, Jeffries C, Smithson D, Tu Y, Boulos N, Jacob MR, Shelat AA, Wu Y, Ravu RR (2014) UPLC-MS-ELSD-PDA as a powerful dereplication tool to facilitate compound identification from small-molecule natural product libraries. J Nat Prod 77(4):902–909

Yang J, Zhu X, Cao M, Wang C, Zhang C, Lu Z, Lu F (2016) Genomics-inspired discovery of three antibacterial active metabolites, aurantinins B, C, and D from compost-associated *Bacillus subtilis* fmb60. J Agric Food Chem 64(46):8811–8820

Zhou M, Liu F, Yang X, Jin J, Dong X, Zeng KW, Liu D, Zhang Y, Ma M et al (2018) Bacillibactin and bacillomycin analogues with cytotoxicities against human cancer cell lines from marine *Bacillus* sp. PKU-MA00093 and PKU-MA00092. Mar Drugs 16(1):22

Chapter 12
Genomic Insights and Comparative Genomics of *Bacillus* Species Having Diverse Mechanisms of Biocontrol Against Fungal Phytopathogens

Jina Rajkumari and Piyush Pandey

12.1 Introduction

Emerging developments in next-generation sequencing (NGS) technologies allow exploration of whole genome sequence of bacteria and other organisms (Schuster 2008). Systemic analysis of whole genome data and identification of genes that contribute to the plant growth promotion and biocontrol activity has aided in the understanding of the molecular mechanism of many bacterial species (MacLean et al. 2009). NGS have been employed to study genomes of several PGPRs such as *Pseudomonas* spp. and *Bacillus* spp. (Song et al. 2012; Duan et al. 2013). Previously, insufficient knowledge underlying the mechanisms of interaction between plants and *Bacilli* (Qiao et al. 2014) with the lack of genetic data had hindered their application in agriculture and biotechnology, although certain progress had been made in the last decade (Borriss et al. 2011). However, we assume that competence of rhizosphere and function of the biocontrol in *Bacilli* are partly caused by nonribosomally produced cyclic lipopeptides acting against phytopathogenic microorganisms (Stein et al. 1996). Over the past few years, the draft or complete genome sequencing of bacteria isolated from the rhizosphere has been carried out providing the genetic attributes involved in plant growth promotion and biocontrol. The genomic overview of the *Bacillus* isolates had provided genetic mechanisms underlying plant growth promotion and antagonism against pathogens. The data allow us to use bioinformatics to mine the genome for potential secondary metabolites effective against plant pathogenic fungi for developing agro-biotechnological agents with predictable features. Also, it is necessary to reveal and elucidate the genetic mechanisms involved in plant-associated lifestyle and whole biocontrol process achieved by PGPR (Palazzini et al. 2016). Genome mining has been used to predict

J. Rajkumari · P. Pandey (✉)
Soil and Environment Microbiology Laboratory, Department of Microbiology, Assam University, Silchar, Assam, India

M. T. Islam et al. (eds.), *Bacilli and Agrobiotechnology: Phytostimulation and Biocontrol*, Bacilli in Climate Resilient Agriculture and Bioprospecting,
https://doi.org/10.1007/978-3-030-15175-1_12

uncharacterized genes clusters, although few known secondary metabolites have been identified; hundreds of secondary metabolites are yet to be identified accurately. The sequencing of a number of prokaryotic genomes had allowed comparing the genomes, and, accordingly, evolution of the clusters was explored using comparative genomics across all known core and accessory genomes of *Bacillus* strains (Belbahri et al. 2017). Core genome is the shared genetic material among nearly all the strains of the species. It contains the housekeeping genes and is interspersed with accessory genomic parts. An accessory genome is assumed to be present in some strains while absent in the rest of the species strains (Ozer et al. 2014).

Phytopathogenic fungi are microbial agents that cause major biotic stress which contribute significantly to the global loss in yield of crop plants (Ribera and Zuniga 2012). The fungi of the divisions *Ascomycetes*, *Basidiomycetes*, or *Oomycetes* are most prevalent as plant pathogens. The main fungal diseases of plants comprise mildew, powdery mildew, rusts, coals, guts and deformation, necrosis, cankers, leaf wilts, vascular, root rots, flower rots, fruit rots, etc. For instance, *Phytophthora* spp. cause black pod, frosty pod, and witches' broom in cacao that leads to yield loss. *Cladosporium herbarum* cause passion fruit rust; *Mycosphaerella fijiensis* cause black Sigatoka of banana which results in reduction of yield and quality of fruit (Wood and Lass 2001; Phillips-Mora et al. 2006; Cuervo-Parra et al. 2011). Phytopathogenic *Fusarium* fungi cause several diseases of grain cereals. *F. graminearum*, *F. culmorum*, *F. poae*, *F. avenaceum*, and *Microdochium nivale* are common pathogens of wheat and barley (Parry et al. 1995; Sutton 1982; Miedaner 1997; Tekauz et al. 2000; Brennan et al. 2003; Leslie et al. 1996), and *F. graminearum*, *F. moniliforme*, and *F. subglutinans* on maize (Sutton 1982; Leslie et al. 1996; Velluti et al. 2000; Torres et al. 2001) cause significant yield losses worldwide (Popovski and Celar 2013). The phytopathogenic plant disease *Fusarium* head blight (FHB) (head scab) is one of the most destructive diseases for cereals, and it is caused by a group of *Fusarium* species including *F. graminearum* and *F. culmorum* (Nasraoui et al. 2007). *Fusarium* foot rot and *Fusarium* root rot on cereals are also caused by soilborne fungi during wet seasons (Scherm et al. 2013).

Earlier, farmers have become more dependent on agrochemicals as a consistent method of crop production as agricultural production has been intensified (Santos et al. 2012). However, wide usage of chemicals has negative effect on the environment as well as human health (Tournas 2005). The plant-associated nonpathogenic microorganisms have the potential to increase soil fertility and plant health, and it has been used to replace chemical fertilizers. Due to development of resistance, various chemicals had become ineffective to control diseases, and therefore, biological control through the use of beneficial natural microflora of soil popularly known as plant growth-promoting rhizobacteria (PGPR) offers the opportunity for sustainable agricultural systems protecting the plants from pathogens while simultaneously lowering chemical residues (Vessey 2003). Biocontrol agents are able to protect plants from phytopathogenic organism infection (Bloemberg and Lugtenberg 2001). It is environmentally safe, and it employs natural antagonists of pests and pathogens (Cook 1993). Therefore, the use of biological control is considered as a safer and more sustainable strategy for profitable agricultural production.

Bacterial endophytes benefit plants by imparting biotic and abiotic stress to hosts, colonizing the same niche similar to that of phytopathogens (Rai et al. 2007). Some bacterial endophytes have been shown to produce anti-pest compounds to protect plants from various pathogens as well as to promote growth of the host plants (Gheler et al. 2013). *Bacillus* species comprise a physiologically versatile group of bacteria isolated from diverse habitats. *Bacillus*-based biocontrol agents have proved to be effective against a broad range of plant pathogens. Mostly, *Bacillus* species used as biocontrol agents are mobile with peritrichous flagella (Driks 2004). The species has versatile metabolic activities, and it has been regarded as a safe biological agent (Kim et al. 2003). Various *Bacillus* species were shown to have antifungal activity against phytopathogenic fungi that mark them as a good biocontrol agent (Li et al. 2014). The main mechanisms by which biocontrol agents suppress pathogens include antibiosis, competition, plant growth promotion, and induction of systemic resistance. Antibiotic production also plays a key role in biocontrol activities (Szczech and Shoda 2006). *Bacillus* strains are usually isolated as biological control agents or plant growth promoters, due to their capacities to produce a wide range of antifungal compounds, including volatiles, enzymes, lipopeptides, and several small peptides (Moyne et al. 2004; Romero et al. 2007).

Many genome sequences of PGPR isolates from the genus *Bacillus* have been published. To extend the understanding of the potential antifungal capacities, 286 genome assemblies possessing the biocontrol properties have been reported within the species of *B. velenzensis*, *B. subtilis*, *B. amyloliquefaciens*, *B. methylotrophicus*, and *B. thuringiensis*. Forty-seven strains within *B. velenzensis* have been sequenced, and 20 of them have been assembled completely (Chen et al. 2007; Cai et al. 2013; Yang et al. 2014).

This chapter highlights the genomic analysis and comparative analysis of genomes of different biocontrol *Bacillus* spp., with the aim to study the genetic traits and secondary metabolites production to reveal the potential of *Bacillus* spp. for biocontrol of phytopathogens and plant growth promotion.

12.2 Biological Control of Phytopathogens

Various *Bacillus* species including *B. subtilis*, *B. licheniformis*, *B. pumilus*, *B. cereus*, *B. mycoides*, *B. amyloliquefaciens*, *B. amyloliquefaciens* subsp. *plantarum*, *B. oryzicola*, and *B. thuringiensis* have been reported to control fungal diseases of plants because of their ability to produce antibiotics and/or suppress growth of several fungal pathogens such as *Rhizoctonia* spp., *Fusarium* spp., *Sclerotinia* spp., *Gaeumannomyces* spp., *Nectria* spp., *Pythium* spp., *Phytophthora* spp., and *Verticillium* spp. (Romero et al. 2004; Leclere et al. 2005; Kotan et al. 2009; Zhang et al. 2009). Enzymatic and endosporic products of *B. subtilis* were found to be active against many fungal pathogens (Denner and Gillanders 1996). Several strains of *B. amyloliquefaciens*, a close relative of *B. subtilis*, have been reported to be effective in control of plant pathogens (Yu et al. 2002). Generally, the mode of action of *Bacillus* is (i) antibiosis (ii) production of extracellular enzymes (iii) competition to inhabit an ecological niche by utilizing root exudates and thus affecting growth of pathogens (Suarez-López 2010; Doornbos et al. 2012). *B. subtilis* strains have been reported to have an antagonistic effect against *Fusarium* spp. (Baysal et al. 2013).

12.3 Genomes of *Bacillus*, Which Are Antagonist of Fungal Phytopathogens

The strain *B. subtilis* RC 218 isolated from wheat anthers was suggested as a potential antagonist of *Fusarium graminearum*. It was demonstrated to reduce disease and the associated mycotoxin (deoxynivalenol) accumulation under field conditions. The genome of *B. subtilis* RC 218 was observed to encode five non-ribosomal peptide synthetase clusters – surfactin, iturin A, fengycin, the siderophore bacillibactin, and the antibacterial bacilysin along with three gene clusters for polyketide synthetase (PKS) – bacillaene, difficidin, and macrolactin (Palazzini et al. 2016).

Biocontrol strain *B. subtilis* XF-1 isolated from the rhizosphere soil of cabbage infected by *Plasmodium brassicae* had been proposed to show suppression effect on 21 fungal pathogens belonging to oomycetes, ascomycetes, basidiomycetes, and deuteromycetes (Xiong et al. 2009). The genome of XF-1 contains gene clusters for antifungal lipopeptides, i.e., surfactin and fengycin, and the siderophore bacillibactin (Guo et al. 2015).

In agriculture, several *B. velezensis* have been used as plant growth promoters and antagonist of plant pathogens (Yao et al. 2006; Chen et al. 2007; Dunlap et al. 2015). *B. velezensis* had been used as biocontrol agent as it has the ability to suppress pathogenic fungi. It was reclassified as a synonym of *B. amyloliquefaciens* subsp. *plantarum*, *B. methylotrophicus*, and *B. oryzicola* (Dunlap et al. 2015). A number of *B. velezensis* species had been described to produce lipopeptides, which have antifungal properties synthesized by non-ribosomal peptide synthetases. The whole genome sequence of endophytic biocontrol *B. velezensis* strain CC09, isolated from healthy leaves of *Cinnamomum camphora*, was reported to harbor 13 clusters for secondary metabolite production that showed a broad spectrum of antifungal activities against numerous phytopathogens such as *Glomerella glycines*, *Rhizoctonia solani*, *Alternaria alternata*, and *Blumeria graminis*. The metabolites detected are surfactin, bacillaene, macrolactin, bacitracin, difficidin, iturins, plipstatins (fengycins), bacillibactin, and bacilysin. The strain showed inhibition of growth of fungi including *Geotrichum candidum*, *Aspergillus niger*, *Phytophthora capsici*, and *Mycosphaerella pomi*, when the enzyme activity assay was performed in presence of iturin A (0.66 mg/mg) (Yang et al. 2014; Cai et al. 2017). The whole genome of *B. velezensis* M75, which is an isolate from cotton waste used for mushroom cultivation in Korea, have been reported to have antagonistic activity against fungal plant pathogens. The genome comprises operons encoding various non-ribosomal peptide synthetases and polyketide synthases that are responsible for the biosynthesis of secondary metabolites (Kim et al. 2017).

B. velezensis strain LM2303 was reported to be effective in inhibiting the growth of *F. graminearum*, *F. culmorum*, *A. flavus*, *F. moniliforme*, *Coniothyrium olivaceum*, *Rhizomorpha roth*, and *Alternaria tenuissima* (Chen et al. 2017). The genome of LM2303 was proposed to detect a total of 29 biosynthetic gene clusters (BGCs), out of which 13 (18% of the whole genome) were identified including 4 NRPSs, 3 PKSs, and 2 PKS-NRPS hybrid synthetases. The three annotated BGCs encoding antifungal metabolites include fengycin, iturin, and surfactin. Iturin was found to be encoded by PKS-NRPS hybrid cluster with a length of 37.3 kb in LM2303 (Chen et al. 2018). Also, *srf* gene cluster spans for 26.2 kb in LM2303. The four *srf* genes (*srfAA-AD*) were found to be 78% similar with the strain FZB42. Further, the *sfp* gene was detected together with the regulatory gene *yczE*. The synthetase genes for the metabolites were shown to be 100% similar with *B. velenzensis* strains (Chen et al. 2018).

B. amyloliquefaciens FZB42 isolated from the rhizosphere of *Beta vulgaris* is a representative of a group of endophytic *Bacillus* strains. The strain had been proposed to consider this specific group of rhizobacteria as an ectomorph, which was distinct from the *B. amyloliquefaciens* type strain F (Reva et al. 2004). The strain was developed as biocontrol agent (BCA) to control diseases of tomato, cucumber, cotton, tobacco, and lettuce (Grosch et al. 1999; Yao et al. 2006, Wang et al. 2009a; Chowdhury et al. 2013, 2015), and it showed extraordinary colonizing ability on *Lemna minor*, *Arabidopsis thaliana*, *Zea mays*, *Lycopersicon esculentum*, and *Lactuca sativa* (Fan et al. 2012). The genome of *B. amyloliquefaciens* FZB42 was reported to dedicate approximately 9% (340 kb) of its genetic capacity to the synthesis of antimicrobial metabolites. It has several gene clusters directing to the synthesis of a huge range of secondary metabolites such as bacillaene, bacillibactin, bacillomycin D, bacilysin, difficidin, fengycin, macrolactin, and surfactin (Chen et al. 2009).

The plant-associated *B. amyloliquefaciens* GA1 isolated from strawberry had been observed to exhibit inhibitory activity toward the growth of fungal and oomycete plant pathogens. Incidence of post-harvest infection caused by *Botrytis cinerea*, which is the causative agent of gray mold disease in apple, was reported to be reduced by the strain GA1 (Toure et al. 2004). From the genome analysis, it has been investigated that the partial ORFs of eight gene clusters directing the synthesis of non-ribosomal peptide and polyketide synthetase have been identified in GA1. Fourteen gene clusters were observed to direct the synthesis of cyclic lipopeptides surfactin, iturin A, fengycin, and iron-siderophore bacillibactin. Of these, 11 gene clusters had a high amino acid identity with *srf* (80–96%) or *fen* (41–92%) operon that directs the synthesis of surfactin and fengycin, respectively, in *B. amyloliquefaciens* FZB42 (Chen et al. 2007).

The antagonist *B. amyloliquefaciens* AS 43.3 isolated from a wheat head was reported to reduce the incidence of *Fusarium* head blight (FHB) which is caused by *Gibberella zeae* (*Fusarium graminearum*) (Khan et al. 2001; Schisler et al. 2002). The genome sequence of the strain AS 43.3 was reported to encode nine biosynthetic clusters associated with biocontrol activity including three lipopeptides (surfactin, iturin, and fengycin) and a siderophore (bacillibactin) (Dunlap et al. 2013).

The genome of *B. amyloliquefaciens* subsp. *plantarum* CAU B946, an isolate from rice rhizosphere, was described to commit about 8.5% (nine gene clusters) of the whole genome to be involved in non-ribosomal synthesis of antimicrobial compounds and siderophores. The strain has been reported to show great potential as a biofungicide for the control of various plant disease like tobacco black shank, rice sheath blight, cotton *Fusarium* wilt, cotton *Verticillium* wilt, and wheat scab (Blom et al. 2012; Hao et al. 2012).

The complete genome sequence of *B. amyloliquefaciens* subsp. *plantarum* YAU B9601-Y2 strain isolated from wheat rhizosphere had been reported to produce NRPS and PKS (9 gene clusters, 8.5% of the whole genome) such as bacillomycin D, fengycin, and difficidin. The strain was suggested to suppress a broad spectrum of pathogenic fungi, promote growth and rooting of crops and vegetables, and develop drought resistance of wheat, corn, and broad bean, and it was also found to reduce the number of nematodes in tomato and tobacco rhizosphere. Presence of diverse antifungal activities in these strains was unique to their genomic features (Hao et al. 2012).

The genome of *B. amyloliquefaciens* subsp. *plantarum* UCMB5113 isolated from soil has been described to colonize plant roots and stimulate growth of the plants and limit the growth of several fungal pathogens on oilseed rape such as *Alternaria brassicae*, *Botrytis cinerea*, *Leptosphaeria maculans*, and *Verticillium longisporum* (Danielsson et al. 2007; Sarosh et al. 2009). The genome harbors gene clusters for the synthesis of lipopeptides including surfactin (*srf*), bacillomycin D (*bmy*), and fengycin (*fen*). The organization of gene clusters in UCMB5113 was found to be similar with genomic segments in the strain FZB42 (Niazi et al. 2014).

12.4 Comparative Genomes of *Bacillus* Which Are Antagonist of Fungal Phytopathogens

The genome sequencing allowed us to accurately determine the taxonomy of the strain using a phylogenomic approach. Comparative genome studies of four strains, *B. amyloliquefaciens* FZB42, *B. subtilis* 168, *B. licheniformis*, and *B. pumilus*, suggested that they belong to a closely related taxonomic unit called *B. subtilis* group (Fritze 2004). The four members of *B. subtilis* group consisted of 2139 genes sharing more than 50% identity to each other by core genome analysis. The majority of the genes of FZB42 were found to be conserved in *B. subtilis* (3271), whereas *B. amyloliquefaciens* and *B. pumilus* share 2378 genes in common. A significant portion of the FZB42 genome was found to be conserved in *B. cereus* ZK (2342), *B. anthracis* (2338), *B. thuringiensis* (2335), *B. clausii* (2162), *B. halodurans* (2105), and *Geobacillus kaustophilus* (1995) (Chen et al. 2009).

The strain GA1 which was initially identified as *B. subtilis* was proposed to assign as *B. amyloliquefaciens* which resulted from phylogenetic tree of *recA* and *recN* genes. The two genes, encode DNA repair and recombination proteins, respectively, and had been suggested to exhibit in resolving closely related taxa (Zeigler 2003). The protein-coding sequence detected in the genome of the strain GA1 was investigated to be highly similar with *B. amyloliquefaciens* FZB42 (Arguelles-Arias et al. 2009). Five *dhb* gene fragments of GAI directing the synthesis of the siderophore bacillibactin was described to be similar with in *B. amyloliquefaciens* FZB42, located between *CDS304* and *yuilV* ORFs. The *comS* gene in GA1, which encodes a competence signal molecule, was described to be embedded within *srfAB* that is similar to *B. amyloliquefaciens* FZB42 and *B. subtilis* 168. This proposed that an interspecies horizontal gene transfer might follow between *B. subtilis* and *B. amyloliquefaciens* (Arguelles-Arias et al. 2009).

Borriss et al. (2011) proposed that *B. amyloliquefaciens* comprises two subspecies: the plant-associated *B. amyloliquefaciens* subsp. *plantarum* and the non-plant-associated *B. amyloliquefaciens* subsp. *amyloliquefaciens* based on phylogenic analysis, physiological characteristics, and production of antibiotics. A comparative genome analysis of *B. amyloliquefaciens* and *B. plantarum* subsp. strains showed a high level of genomic homogeneity sharing 72–84% coding genes among the genomes which specified that these two groups have recently diverged. In addition, two gene clusters encoding difficidin and macrolactin were not found in the genome sequence of *B. amyloliquefaciens*. Genes which are unique to plant-associated strains differentiate biocontrol strains from non-plant-associated species (Niazi et al. 2014).

Dunlap et al. (2015) stated that it was difficult to differentiate *B. amyloliquefaciens*, *B. subtilis*, and *B. velezensis* on the basis of 16S rRNA sequence similarity, morphological characteristics, and physiological and chemical reactions. Several heterotypic synonyms such as *B. subtilis*, *B. methylotrophicus*, *B. amyloliquefaciens* subsp. *plantarum*, and *B. oryzicola* have been regrouped as *B. velezensis* based on the genome sequencing, comparative genomics analysis, and DNA-DNA relatedness calculations (Dunlap et al. 2015). Earlier reports regrouped *B. subtilis* RC 218 and *B. amyloliquefaciens* IT45, Y2, and LFB112 into *B. velezensis* (Palazzini et al. 2016). Wang et al. (2008) described *B. velezensis* as a heterotypic synonym of *B. amyloliquefaciens*. The unique characteristics of *B. velezensis* consist of plant growth promotion, antifungal metabolite production, and competent colonization on plants (Madhaiyan et al. 2010; Borriss et al. 2011; Dunlap et al. 2015).

The core genome of 35 strains of *B. velezensis*, *B. amyloliquefaciens*, *Bacillus* sp. Pc3, and *B. subtilis* 168 was reported to share 257,259 bp in length including 2574 CDSs. Phylogenomic tree constructed from these core genome sequences grouped the strain CC09 as *B. velezensis* which was previously identified as *B. amyloliquefaciens* (Cai et al. 2017).

Based on a comparative genomic analysis of *Bacillus* strains, three clusters are conserved in strains of *B. velezensis*, *B. amyloliquefaciens*, and *B. subtilis* 168, and nine clusters were conserved in *B. velezensis* and *B. amyloliquefaciens* and two in all strains of *B. velezensis*. Until now, there are few reports on comparative genome data of *B. velezensis* strains regarding the production of secondary metabolites. Earlier reports suggested two clusters encoding NRPS and NRPS-Trans ATPKS, respectively, were found only in some strains of *B. velezensis* that might lead to the synthesis of new bioactive compounds (Ongena and Jacques 2008; Chen et al. 2009; Arrebola et al. 2010; Alvarez et al. 2012).

A comparative analysis of four *B. velezensis* genomes M75, FZB42, CAU B946, and YAU B9601 had shown that the genome sequence of M75 strain (4,007,450 bp) was found to be larger than other genome of FZB42 strain (3,918,589 bp) and smaller than other *B. velezensis* strains (4,242,774–4,019,861 bp) (Kim et al. 2017). In another study, He et al. (2012) suggested that the genomes of *B. amyloliquefaciens* B9601 and the closely related type strain have 3426 CDS in common. The M75 strain was observed to have 145 unique genes that were not found in other *B. velezensis* strains. The strain M75 possesses 22–23 extra genes associated with secondary metabolites biosynthesis as compared to the other *B. velezensis* strains. It has been investigated that M75 strains has the highest number of genes (87 genes and 2.6% of the whole genome) associated with the secondary metabolite biosynthesis, transport, and catabolism. The core genome analysis of four strains M75, CAU B946, FZB42, and YAU B9601 consisted of 3414 CDSs with average identity of 95%, and pan genome consisted of 4558 CDSs. All the four strains have been suggested to have strong antagonistic activities toward plant pathogens as the operons that encode the biosynthetic enzymes for the secondary metabolites were found to be highly homologous. However, the operon encoding biosynthetic enzymes in the four *B. velezensis* strains showed a difference in gene clusters (Kim et al. 2017).

The whole genome of the strain LM2303 has been reported to harbor the largest number of gene clusters associated with secondary metabolite biosynthesis, transport, and catabolism (119 genes, 3.08% of the whole genome) when compared with three biocontrol strains FZB42, M75, and CAU B946, which specify that LM2303 has higher potential as biocontrol agent than other *B. velezensis* strains. Using phylogenomic analysis of single-copy core genes, *Bacillus* strain LM2303 was described as a member of *B. velezensis* clade as it was found to cluster closely with *B. velezensis* strains CAU B946 and M75 (Chen et al. 2018).

Another comparative analysis of eight genomes of *B. amyloliquefaciens* was reported to identify 3,316,600 bp core genome and 5,529,004 bp pan genomes for the species. It was observed that the alignment of four biocontrol genomes were similar, with *B. amyloliquefaciens* AS 43.3 most similar to *B. amyloliquefaciens* FZB42. The strain AS 43.3 comprised 263,690 bp genes unique to the biocontrol strains. The nine metabolite clusters present in the strains were investigated to be found in the same relative physical location in the genome with same gene order (Dunlap et al. 2013). The core genome of biocontrol strains identified three large synthetic cluster (macrolactin, difficidin, and fengycin) that are specific for plant-associated *B. amyloliquefaciens* subsp. *plantarum*. Although a part of fengycin cluster (*fenDE*) was found in strain *B. amyloliquefaciens* DSM-7, it is not functional due to lack of other needed enzymes (*fenABC*) (Borriss et al. 2011).

Genome comparison of *B. amyloliquefaciens* subsp. *amyloliquefaciens* $DSM7_T$, UCMB5113, FZB42 identified 112 unique coding sequence in the genome

UCMB5113. The core genome of *Bacillus* species comprised of 2391 orthologs. The genome of *B. subtilis* 168, *B. amyloliquefaciens* FZB42, and *B. amyloliquefaciens* UCMB5113 was found to share 3421 orthologs, whereas UCMB5113 and DSM7$_T$ were found to share 3345 orthologs (Niazi et al. 2014). The comparative genomes of *Bacillus* strains with the gene clusters encoding antifungal metabolites are given in Table 12.1.

The core genomes of all representatives of *B. amyloliquefaciens* subsp. *plantarum* was detected to contain 3347 CDS, on the other hand, representatives of the two *B. amyloliquefaciens* subsp. "*plantarum*" and "*amyloliquefaciens*" contained 3153 CDS. Also, the core genome analysis of *B. amyloliquefaciens* B9601 with other representatives of *B. amyloliquefaciens*, *B. atrophaeus*, and *B. subtilis* was found to contain 2459 CDS. The strain B9601-Y2 was identified to harbor 81 unique genes. Moreover, it revealed that 130 genes were found in the representatives of *B. amyloliquefaciens* subsp. *plantarum*; however, it was absent in *B. amyloliquefaciens* subsp. *amyloliquefaciens* (He et al. 2012).

Furthermore, phylogenomic analysis of the sequenced genome allows us to check taxonomic validity of the isolates and define the extent of interspecies genome variability within the strains. Phylogenomic analysis based on the core genomes (1906 genes) of *B. methylotrophicus*, *B. amyloliquefaciens* subsp. *plantarum*, and several closely related strain showed that *B. methylotrophicus* KACC 130105^T and *B. amyloliquefaciens* subsp. *plantarum* FZB42^T were closely related and they were found to share 95% of the same genes. This approach confirmed that the strain *B. amyloliquefaciens* subsp. *plantarum* FZB42^T was not clustered with other *B. amyloliquefaciens* and proposed that the subspecies *B. amyloliquefaciens* subsp. *plantarum* should be regrouped as *B. methylotrophicus* (Dunlap et al. 2015).

12.5 Plant-Microbe Interaction

The microorganisms that exist within the tissues of plants without producing negative effects to the host are termed as endophytes. They produce anti-pest compounds to protect plants from several pathogens and insects (Wilson 1995). Plant-associated bacteria are described to play an important role in stimulating plant growth and protecting them from phytopathogens with the production of secondary metabolites. Efficient colonization on plant roots is essential for the biocontrol strains for disease suppression and for plant growth promotion (Timmusk et al. 2005).

The process for colonization in the plant roots relies on the passive movement in the water fluxes or by surface motility and capable of biofilm formation by bacterial cells (Guo et al. 2015; Zeriouh et al. 2014; Arguelles-Arias et al. 2009). Establishment of plant-microbe interaction is the next efficient step for colonization by non-specific binding or by forming of bacterial biofilm formed by various extracellular matrix (Guo et al. 2015). In many *Bacillus* strains, the expression of genes for flagellar movement is important to confirm cell motility (*fli*D, *fli*P, *flg*M) and chemotaxis (*che*C, *che*D, *che*V) in response to root exudates (Fredrick and Helmann 1994).

Table 12.1 Comparative genomes of *Bacillus* strains with gene clusters encoding antifungal metabolites

Features	*B. velezensis*			*B. amyloliquefaciens*			*B. amyloliquefaciens subsp. plantarum*		*B. subtilis*
Strains	CC09	M75	LM2303	FZB42	As 43.3	GA1	YAU B9601	UCMB5113	RC218
Genome size (mb)	4,167,153	4,007,450	3,989,393	3,918,589	3,961,291	461.5 kbp	4,242,774	3,889,530	3,873,821
GC content (%)	46.1	46.6	46.68	46.5	46.6	46.6	45.85	46.71	46.5
CDS	4141	3803	3866	3695	3919	358	3991	3106	3603
Surfactin	*srfABCD*	*srfABCD*	–	*srfABCD*	*srfABCD*	*srfABCD*	*srfABCD*	*srfABCD*	*srfABCD*
Iturin	*ituDABC*	–	*ituDABC*	–	*ituDABC*	*ituDABC*	*ituDABC*	–	*ituDABC*
Bacilomycin	–	–	–	*bmyCBAD*	–	–	*bmyD*	*bmyCBAD*	–
Fengycin	*fenABCDE*	*fenABCDE*	*fenABCDE*	*fenABCDE*	*fenABCDE*	*fenABCDE*	*fenABCDE*	*fenABCDE*	*fenABCDE*
Bacillibactin	*dhbABCDEF*	*dhbACEBF*	–	*dhbABCDEF*	*besA*, *dhbACEBF*	*dhbABCDEF*	*dhbABCDEF*	*dhbACEBF*	*besA*, *dhbACEBF*
References	Cai et al. (2017)	Kim et al. (2017)	Chen et al. (2018)	Chen et al. (2009)	Dunlap et al. (2013)	Arguelles-Arias et al. (2009)	Hao et al. (2012)	Niazi et al. (2014)	Palazzini et al. (2016)

The presence of gene clusters (*flgBCDEGKLMN*, *flhABFOP*) and *swrABC* have been investigated in the genomes of different *B. amyloliquefaciens* (Ghelardi et al. 2012). *B. subtilis* XF-1 has flagellar biosynthesis operon (*fli/che*) and two stator elements, *mot*AB and *mot*PS (Werhane et al. 2004). These gene clusters which are responsible for cell envelope and cellular processes motility and chemotaxis have been identified in different genomes of *Bacillus* strains (Niazi et al. 2014). *Bacillus velezensis* LM2303 has gene clusters for flagellar assembly (*flg* and *fli* cluster) and bacterial chemotaxis (*che* cluster). LM2303 has operons encoding Tas*A* protein for biofilm (*yqxM-sipW-tasA*), regulator gene (*sinR* and *sinI*), and exopolysaccharide (*eps* cluster) for colonization (Chen et al. 2018). Presence of gene clusters (*flg*, *flh*, *fli*) for production of functional flagellar components along with *swrA* gene for regulation of expression of flagellar genes in *B. amyloliquefaciens* subsp. *plantarum* UCMB5113 made it to exhibit swarming motility. Also operons *epsA-O* and *yqxM-sipW-tasA* for making up biofilm were found in strain UCMB5113 (Niazi et al. 2014). Homology-based searches and annotation of genes have been conducted in *Bacillus* genomes encoding exopolysaccharide using *B. subtilis epsA-O* operon genes *tapA*, *tasA*, *sipW*, *pgsB*, and *bslA* (Vlamakis et al. 2013). Moreover, a positive correlation between surfactin production and biofilm formation was observed, insufficiency in production of surfactin directed to a defect in biofilm formation and ultimate reduction of suppression of disease (Zeriouh et al. 2014). Further, surfactin was described to be able to inhibit formation of biofilm by pathogenic bacteria, thus providing colonization of biocontrol strains with antagonistic advantage (Chen et al. 2009). In the biocontrol system, certain interactions of plant, pathogen, and biocontrol strains can be generated by biocontrol strains, for example, direct antagonism toward pathogens, plant growth promotion, and ISR. Biocontrol strains promote plant growth by making it easier for plants to get nutrients or by producing plant growth-promoting hormones (Niazi et al. 2014). There are many *Bacillus* species that have been reported as endophytic bacteria in higher plants which promote plant growth and secrete antimicrobial compounds (Wang et al. 2009b; White et al. 2014). Surfactin and iturin favor plant root colonization by stimulating cell spreading, swarming, and biofilm formation (Kinsinger et al. 2003; Hofemeister et al. 2004). Also, iturins and fengycin exhibit strong activity against fungus and are inhibitory for the growth of various plant pathogens (Maget-Dana et al. 1992).

12.6 Protection Against Phytopathogens

Antagonism or mycotrophic ability of any organism includes special strategies to expand the pathogens and pests. It could be through a direct mycotrophic interaction or indirectly by inducing a defense response in plants to combat against the attackers. The effectiveness of biological control agents are determined by its diversity of mechanisms through which biological control agents are able to attack antagonistically on plant pathogens and benefit plants by suppressing disease (Van-Loon et al. 1998; Pozo and Azcón-Aguilar 2007; Jamalizadeh et al. 2011). About 2428 antimicrobial peptides identified from various organisms, 756 peptides have several degrees of antifungal properties (Microbiology UDoPa 2016). The mechanisms by which fungi lead to death include blockage, distraction, and holes formation in the cell wall and cell membranes of the fungi induced by antifungal metabolites produced by biocontrol organisms. Furthermore, some peptides disintegrate fungal intracellular organs such as nucleic acid and mitochondria (Zhao et al. 2013).

Chen et al. (2018) proposed that *B. velezensis* LM2303 can control fungal disease in wheat by four different mechanisms: (i) direct antagonistic action against *F. graminearum* and other pathogens mediated by *Bacillus* lipopeptides and antibacterial metabolites, (ii) stimulation of ISR in wheat by surfactin and volatiles, and (iii) plant growth promotion by producing plant growth hormones, and (iv) competition for space and nutrients through efficient colonization and persistence. A similar mechanism was found in *Bacillus* species which produce other metabolites like chitinases and other cell wall-degrading enzymes and volatile compounds that kill the fungus, and they also provoke plant resistance mechanisms (Pelletier and Sygusch 1990). Chitosanase and proteases have been reported to play an important role in dissolving and penetrating the cell walls of *Rhizoctonia solani* (Mcquilken and Gemmell 2004). Besides β-1,3-glucanase synthesized by *Paenibacillus*, *B. cepacia* has the capability to control the growth of *F. oxysporum*, *Rhizoctonia solai*, *S. rolfsii*, and *Pythium ultimum* cell walls.

Siderophore production had also been suggested to be one of the mechanisms of biocontrol actions of various rhizobacteria (Loper and Buyer 1991). Siderophores are ferric iron chelators that enable bacterial cells to accumulate and acquire iron in environments where bioavailability of iron is limited. Biocontrol antagonists have been recognized for competing for essential micronutrients like iron in the rhizosphere whose availability is dependent on soil. Concentration of iron in ferric form in aerated soil may be as low as 10^{-8} mol/L which is insufficient for growth and development of microorganisms. Presence of microorganisms which are able to produce siderophore in the rhizosphere adds plant health by completing iron and making it less available to phytopathogens (Kloepper et al. 2004). A functional gene cluster (*dhb*) had shown to direct the synthesis of catecholate iron-siderophore bacillibactin. It is the cyclic trimeric lactone of the 2, 3-dihydroxybenzoyl-Gly-Thr (Chen et al. 2007). Synthesis of bacillibactin is dependent on a functional Ppant transferase (*sfp*). Purified bacillibactin from *B. subtilis* AH18 has exhibited suppression against *Phytophthora capsici* which demonstrates the role of siderophore as biocontrol systems (Woo and Kim 2008). It had been proposed that bacillibactin was not produced in *B. subtilis* 168, although the respective gene cluster was present (May et al. 2001). Also, the gene cluster (*dhbACEBF)* in *B. amyloliquefaciens* FZB42 was examined to be 87–93% similar with *B. amyloliquefaciens* GA1 (Chen et al. 2009).

Studies on mechanisms of biocontrol by PGPR have revealed that various PGPR strains have the ability to protect plants from pathogenic organisms by colonizing the roots. The resulting subsequent resistance due to an inducing agent when infected by a pathogen is called induced systemic resistance (ISR) (Hammerschmidt and Kuc 1995). Concentration of antifungal lipopeptides determined in plants have been reported to be low, and other antimicrobial activity were not detected so far in surrounding area of plant roots colonized by PGPR bacilli. Therefore, ISR is assumed as one of the main factors for suppressing phytopathogens by PGPR bacilli. It depends on the capability of some strains to actuate defense systems in the host plants, thereby reducing the susceptibility of host to successive infection in plant tissues, without provoking any symptoms themselves (Stein 2005). ISR acts through two signaling pathways, i.e., salicylic acid (SA) and jasmonic acid (JA) pathway. ISR is induced when the plant is challenged by pathogenic organisms, and it stimulated physical and mechanical strength of the cell wall and physiological and biochemical reaction of the host and thus invades pathogen on its ability to build a defense chemicals against them by the host (Benhamou et al. 1996). It has been investigated that PGPR induces structural modification of the cell wall in response to pathogenic attack (M'Piga et al. 1997). *B. pumilus* strain SE 34 used in seed treatment have shown to induce strengthening of cell wall in tomato against *F. oxysporum* f.sp. *radicislycopersici* (Benhamou et al. 1998).

It depends on the capability of some strains to actuate defense systems in the host plants, thereby reducing the susceptibility of host to successive infection in plant tissues, without provoking any symptoms themselves (Stein 2005). Various strains of *Bacillus* like *B. amyloliquefaciens*, *B. subtilis*, *B. pasteurii*, *B. cereus*, *B. pumilus*, *B. mycoides*, and *B. sphaericus* have been reported to produce induced systemic disease resistance which caused reduction of disease incidence of host (Ryu et al. 2003). Figure 12.1 shows the different mechanisms of *Bacillus* species in controlling plant disease. Five *Bacillus* species have been evaluated to suppress bacterial wilt

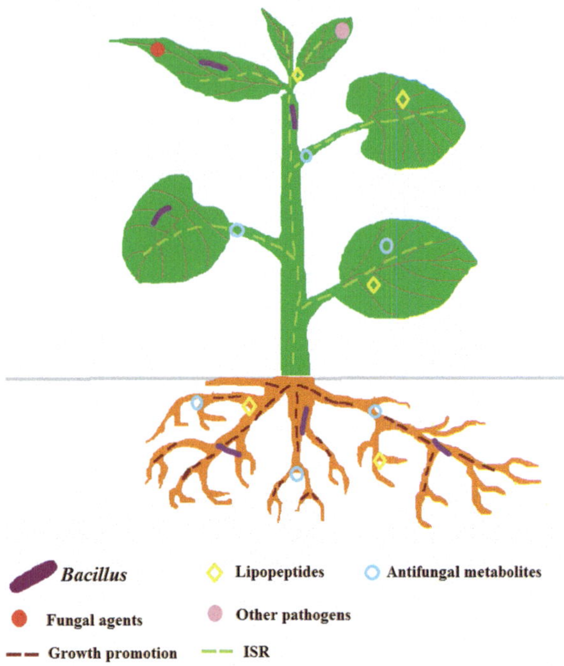

Fig. 12.1 Mode of action of *Bacillus* in rhizosphere. The figure illustrates the interaction of biocontrol *Bacillus* species (purple rod), fungal pathogens (red filled circle), and plant. The bacteria produce variety of secondary metabolites (light blue) and form a protective zone. Lipopeptides (yellow rhombus) act against fungal agents, inhibiting the growth of pathogenic fungi. Other metabolites stimulate ISR-mediated protection against the pathogen. The growth of the plant (red lines) is enhanced, by plant growth-promoting nutrients and growth regulators

caused by *R. solanacearum*. The bacterial strains *B. vallismortis* EXTN-1 have shown to reduce disease in plant which resulted from production of host plant resistance genes. The main constituents of systemic-induced resistance are phenolic compounds, genetic and structural modifications, plant resistance activators, and activation of enzymatic weapons (Park et al. 2007). PGPR strains can induce systemic resistance in plants through activation of various defense-related enzymes like chitinases, β-1,3-glucanase, peroxidase (PO), phenylalanine ammonia-lyase (PAL), and polyphenol oxidase (PPO) (Bharathi 2004).

Biocontrol *Bacillus* PGPR strains secrete specific metabolites that elicit plant defenses. The two well-known plant growth promoters 3-hydroxy-2-butanone (acetoin) and 2,3-butendiol act as volatile elicitors for ISR, and exposure to such volatile organic compounds results in reduced disease incidence in *Arabidopsis* (Ryu et al. 2004). Acetoin production in *Bacillus subtilis* GB03 and *B. amyloliquefaciens* IN937 involves two enzyme-encoding genes *alsS* and *alsD* that encode acetolactate synthase and acetolactate decarboxylase, respectively, and *alsR* regulates these genes (Ryu et al. 2004). Rate et al. (1999) stated that some common plant metabolites like organic acids especially oxaloacetate activate the *alsSD* operon which is required for acetoin production. Acetolactate synthase (*alsS*) catalyzes the condensation of two pyruvate molecules into acetolactate, and acetolactate decarboxylase (*alsD*) converts it into acetoin. The *als* gene cluster (*alsR*, *alsS*, *alsD*) which encodes 3-hydroxy-2-butanone and the gene *bdhA* which encodes the enzyme (R-R)-butanediol dehydrogenase for catalyzing 3-hydroxy-2-butanone to 2,3-butanediol were identified in LM2303 (Chen et al. 2018). The UCMB5113 chromosome also harbors genes that ferment pyruvate to acetoin and 2,3-butanediol (Niazi et al. 2014). Two synthetic pathways (*alsSDR* and *ilvHC*) for plant growth-promoting volatile compounds were identified in the genome of XF-1. The later pathways are the parts of leucine, valine, and isoleucine biosynthesis. The XF-1 genome had observed to possess six alcohol dehydrogenase whose function is thought to catalyze the reduction of acetoin to 2,3-butanediol (Guo et al. 2015). Cassan et al. (2009) suggested that the biosynthesis systems of acetoin and 2,3-butanediol in PGPR can change from acidic products, lactate, and toxic compounds to neutral volatile compounds which are plant growth promoters to carbohydrate catabolism. This shift may be beneficial for strain colonization and plant symbiotic relationship in the rhizosphere.

12.7 Conclusion

The *Bacillus* genus includes species that have been isolated from a wide range of sources, and they are used as biocontrol agents because the spores produced by them have the ability to withstand extreme environmental conditions. The effective application of PGPR in agriculture is interesting, and appropriate information is required to understand the biological processes mediating the modes of action that enhance plant growth. They synthesize several kinds of lipopeptides with specific activity against pathogens, and it gives a unique importance in agriculture. The mechanisms of biological control of plant diseases are diverse and complex, and the biocontrol ability of microbial strains is the outcome of complex interaction with each other. *Bacillus* biocontrol agents use a variety of mechanisms, and the most reported ones are attributed to non-ribosomally produced lipopeptides (Ongena and Jacques 2008). There are numerous reports where different species of *Bacillus* have been reported for their plant growth enhancement and biocontrol activities. The genomic data of the isolates have facilitated a better understanding of the *Bacillus* strains as biocontrol agents against phytopathogenic fungi. Nevertheless, detailed studies are needed to investigate the genetic components associated with plant pathogens' suppression. Various genomes of *Bacillus* have been reported to encode gene clusters for synthesis of various NRPS (bacillibactin, fengycin, surfactin, iturin) which regulate broad antifungal activities as well as their roles in disease control. Genes dedicated for plant colonization, motility, plant growth promotion, and ISR have been identified by genomic approach. In addition, comparative genome studies allowed us to identify a subset of genes shared in the species and unique genes of the isolate characterizing the potential metabolite for biocontrol application. An overview of comparative genomics of plant growth-promoting *Bacillus* strains may perhaps aid in studying the evolution of plant growth promotion and mechanisms of biocontrol species. These data, together with experiments performed in the laboratory, will help to clarify the potential *Bacillus* bacteria have for effective biocontrol agents to be used for plant growth and plant disease control.

References

Alvarez F, Castro M, Principe A, Borioli G, Fischer S, Mori G, Jofre E (2012) The plant-associated *Bacillus amyloliquefaciens* strains MEP2 18 and ARP23 capable of producing the cyclic lipopeptides iturin or surfactin and fengycin are effective in biocontrol of sclerotinia stem rot disease. J Appl Microbiol 112(1):159–174

Arguelles-Arias A, Ongena M, Halimi B, Lara Y, Brans A, Joris B et al (2009) *Bacillus amyloliquefaciens* GA1 as a source of potent antibiotics and other secondary metabolites for biocontrol of plant pathogens. Microb Cell Factories 8(1):63. https://doi.org/10.1186/1475-2859-8-63. PMID: 19941639

Arrebola E, Jacobs R, Korsten L (2010) Iturin A is the principal inhibitor in the biocontrol activity of *Bacillus amyloliquefaciens* PPCB004 against postharvest fungal pathogens. J Appl Microbiol 108(2):386–395

Baysal O, Lai D, Xu H, Siragusa M, Casiskan M, Carimi F, Silva JAT, Tor M (2013) A proteomic approach provides new insights into the control of soil-borne plant pathogens by *Bacillus* species. PLoS One 8:e53182

Belbahri L, Bouket AC, Rekik I, Alenezi FN, Vallat A, Luptakova L, Petrovova E, Oszako T, Cherrad S, Vacher S, Rateb ME (2017) Comparative genomics of *Bacillus amyloliquefaciens* strains reveals a core genome with traits for habitat adaptation and a secondary metabolites rich accessory genome. Front Microbiol 8:1438. https://doi.org/10.3389/fmicb.2017.01438

Benhamou N, Kloepper JW, Quadt-Hallmann A, Tuzun S (1996) Induction of defence-related ultrastructural modifications in pea root tissues inoculated with endophytic bacteria. Plant Physiol 112:919–929

Benhamou N, Kloepper JW, Tuzun S (1998) Induction of resistance against *Fusarium* wilt of tomato by combination of chitosan with an endophytic bacterial strain: ultra structure and cytochemistry of the host response. Planta 204:153–168

Bharathi S (2004) Development of botanical formulations for the management of major fungal diseases of tomato and onion. PhD thesis, Tamil Nadu Agricultural University, Coimbatore, India, p 152

Bloemberg GV, Lugtenberg BJ (2001) Molecular basis of plant growth promotion and biocontrol by Rhizobacteria. Curr Opin Plant Biol 4:343–350

Blom J, Rueckert C, Niu B, Wang Q, Borriss R (2012) The complete genome of *Bacillus amyloliquefaciens subsp. plantarum* CAU B946 contains a gene cluster for nonribosomal synthesis of iturin A. J Bacteriol 194(7):1845–1846

Borriss R, Chen XH, Rueckert C, Blom J, Becker A, Baumgarth B, Fan B, Pukall R, Schumann P, Sproer C, Junge H, Vater J, Puhler A, Klenk HP (2011) Relationship of *Bacillus amyloliquefaciens* clades associated with strains DSM7^{T} and FZB42^{T}: a proposal for *Bacillus amyloliquefaciens subsp. amyloliquefaciens* subsp. nov. and *Bacillus amyloliquefaciens subsp. plantarum* subsp. nov. based on complete genome sequence comparisons. Int J Syst Evol Microbiol 61(8):1786–1801

Brennan JM, Fagan B, Van Maanen A, Cooke BM, Doohan FM et al (2003) Studies on *in vitro* growth and pathogenicity of *Fusarium* fungi. Eur J Plant Pathol 109:577–587

Cai X, Li H, Xue Y, Liu C (2013) Study of endophytic *Bacillus amyloliquefaciens* CC09 and its antifungal cyclic lipopeptides. J Appl Biol Biotechnol 1(1):1–5

Cai XC, Liu CH, Wang BT, Xue YR (2017) Genomic and metabolic traits endow *Bacillus velezensis* CC09 with a potential biocontrol agent in control of wheat powdery mildew disease. Microbiol Res 196:89–94

Cassan F, Maiale S, Masciarelli O, Vidal A, Luna V, Ruiz O (2009) Cadaverine production by *Azospirillum brasilense* and its possible role in plant growth promotion and osmotic stress mitigation. Eur J Soil Biol 45(1):12–19

Chen X, Koumoutsi A, Scholz R, Eisenreich A, Schneider K, Heinemeyer I, Morgenstern B, Voss B, Hess W, Reva O, Junge H, Voigt B, Jungblut P, Vater J, Sussmuth R, Liesegang H, Strittmatter A, Gottschalk G, Borriss R (2007) Comparative analysis of the complete genome sequence of the plant growth-promoting bacterium *Bacillus amyloliquefaciens* FZB42. Nat Biotechnol 25:1007–1014

Chen X, Koumoutsi A, Scholz R, Schneider K, Vater J, Süssmuth R, Piel J, Borriss R (2009) Genome analysis of *Bacillus amyloliquefaciens* FZB42 revealsits potential for biocontrol of plant pathogens. J Biotechnol 140(1–2):27–37

Chen L, Li RJ, Qin SY, Huang L, Wei LC, Bian K (2017) Screening and inhibition effect of antagonistic strain against *Fusarium* head blight by *Fusarium graminearum*. China Plant Prot 37(5):12–17

Chen L, Heng J, Qin S, Bian K (2018) A comprehensive understanding of the biocontrol potential of *Bacillus velezensis* LM2303 against *Fusarium* head blight. PLoS One 13(6):e0198560. https://doi.org/10.1371/journal.pone.0198560

Chowdhury SP, Dietel K, Randler M, Schmid M, Junge H, Borriss R, Hartmann A, Grosch R (2013) Effects of *Bacillus amyloliquefaciens* FZB42 on lettuce growth and health under pathogen pressure and its impact on therhizosphere bacterial community. PLoS One 8(7):e68818

Chowdhury SP, Hartmann A, Gao X, Borriss R (2015) Biocontrol mechanism byroot-associated *Bacillus amyloliquefaciens* FZB42 – a review. Front Microbiol 6:1–11

Cook RJ (1993) Making greater use of microbial inoculants in agriculture. Annu Rev Phytopathol 31:53–80

Cuervo-Parra JA, Sánchez-López V, Ramirez-Suero M, Ramírez-Lepe M (2011) Morphological and molecular characterization of *Moniliophthora roreri* causal agent of frosty pod rot of cocoa tree in Tabasco. Mexico Plant Pathol J 10:122–127

Danielsson J, Reva O, Meijer J (2007) Protection of oilseed rape (*Brassica napus*) toward fungal pathogens by strain of plant-associated *Bacillus amyloliquefaciens*. Microb Ecol 54(1):134–140

Denner W, Gillanders T (1996) The legislative aspects of the use of industrial enzymes in the manufacture of food and food ingredients. In: Godfrey T, Reichelt J (eds) Industrial enzymology. Stockton Press, New York, pp 397–412

Doornbos RF, Loon LC, Bakker PAHM (2012) Impact of root exudates and plant defense signaling on bacterial communities in the rhizosphere. Agron Sustain Dev 32:227–243

Driks A (2004) The *Bacillus* spore coat. Phytopathology 94(11):1249–1251

Duan J, Jiang W, Cheng Z, Heikkila JJ, Glick BR (2013) The complete genome sequence of the plant growth-promoting bacterium *Pseudomonas* sp. UW4. PLoS One 8:462–469

Dunlap CA, Bowman MJ, Schisler DA (2013) Genomic analysis and secondary metabolite production in *Bacillus amyloliquefaciens* AS 43.3: a biocontrol antagonist of Fusarium head blight. Biol Control 64:166–175

Dunlap CA, Kim SJ, Kwon SW, Rooney AP (2015) Phylogenomic analysis shows that *Bacillus amyloliquefaciens subsp. plantarum* is a later heterotypic synonym of *Bacillus methylotrophicus*. Int J Syst Evol Microbiol 65:2104–2109

Fan B, Rainer B, Wilfrid B, Wu X (2012) Gram-positive rhizobacterium *Bacillus amyloliquefaciens* FZB42 colonizes three types of plants in different patterns. J Microbiol 50(1):38–44

Fredrick KL, Helmann JD (1994) Dual chemotaxis signaling pathways in *Bacillus subtilis*: a sigma D-dependent gene encodes a novel protein with both *CheW* and *CheY* homologous domains. J Bacteriol 176(9):2727–2735

Fritze D (2004) Taxonomy of the genus *Bacillus* and related genera: the aerobic endospore-forming bacteria. Phytopathology 94:1245–1248

Ghelardi E, Salvetti S, Ceragioli M, Gueye SA, Celandroni F, Senesi S (2012) Contribution of surfactin and SwrA to flagellin expression, swimming, and surface motility in *Bacillus subtilis*. Appl Environ Microbiol 78:6540–6544. https://doi.org/10.1128/AEM.01341-12

Gheler CF, Domingues ZT, de Soares MI (2013) Biological control of phytopathogenic fungi by endophytic actinomycetes isolated from maize (*Zea mays* L.). Braz Arch Biol Technol 56:948–955

Grosch R, Junge H, Krebs B, Bochow H (1999) Use of *Bacillus subtilis* as abiocontrol agent. III: influence of *Bacillus subtilis* on fungal root diseases Andon yield in soilless culture. J Plant Dis Protect 106(6):568–580

Guo S, Li X, He P, Ho H, Wu Y, He Y (2015) Whole-genome sequencing of *Bacillus subtilis* XF-1 reveals mechanisms for biological control and multiple beneficial properties in plants. J Ind Microbiol Biotechnol 42(6):925–937. https://doi.org/10.1007/s10295-015-1612-y. PMID: 25860123

Hammerschmidt R, Kuc J (1995) Induced resistance to disease in plants. Kluwer Academic, Dordrecht, p 182

Hao K, He P, Blom J, Rueckert C, Mao Z, Wu Y, He Y, Borriss R (2012) The genome of plant growth-promoting *Bacillus amyloliquefaciens subsp. plantarum* strain YAU B9601-Y2 contains a gene cluster for mersacidin synthesis. J Bacteriol 194:3264–3265

He P, Hao K, Blom J, Rückert C, Vater J, Mao Z, Wu Y, Hou M, He P, He Y, Borriss R (2012) Genome sequence of the plant growth promoting strain *Bacillus amyloliquefaciens subsp. plantarum* B9601-Y2 and expression of mersacidin and other secondary metabolites. J Biotechnol 164:281–291

Hofemeister J, Conrad B, Adler B, Hofemeister B, Feesche J, Kucheryava N, Steinborn G, Franke P, Grammel N, Zwintscher A, Leenders F, Hitzeroth G, Vater J (2004) Genetic analysis of the biosynthesis of non-ribosomal peptide- and polyketide-like antibiotics, iron uptake and biofilm formation by *Bacillus subtilis* A1/3. Mol Genet Genomics 272:363–378

Jamalizadeh M, Etebarian HR, Aminian H, Alizadeh A (2011) A review of mechanisms of action of biological control organisms against post-harvest fruit spoilage. EPPO Bull 41:65–71

Khan NI, Schisler DA, Boehm MJ, Slininger PJ, Bothast RJ (2001) Selection and evaluation of microorganisms for biocontrol of *Fusarium* head blight of wheat incited by *Gibberella zeae*. Plant Dis 85:1253–1258

Kim HS, Park J, Choi SW, Choi KH, Lee GL, Ban SJ, Lee CH, Kim CS (2003) Isolation and characterization of *Bacillus* strains for biological control. J Microbiol 41(3):196

Kim SY, Lee SY, Weon HY, Sang MK, Song J (2017) Complete genome sequence of *Bacillus velezensis* M75, a biocontrol agent against fungal plant pathogens, isolated from cotton waste. J Biotechnol 241:112–115

Kinsinger R, Shirk M, Fall R (2003) Rapid surface motility in *Bacillus subtilis* is dependent on extracellular surfactin and potassium ion. J Bacteriol 185:5627–5631

Kloepper JW, Ryu CM, Zhang S (2004) Induced systemic resistance and promotion of plant growth by *Bacillus* spp. Phytopathology 94(11):1259–1266. https://doi.org/10.1094/PHYTO.2004.94.11.1259

Kotan R, Dikbas N, Bostan H (2009) Biological control of post-harvest disease caused by *Aspergillus flavus* on stored lemon fruits. Afr J Biotechnol 8:209–214

Leclere V, Bechet M, Adam A, Guez JS, Wathelet B, Ongena M, Thonart P, Gancel F, Chollet-Imbert M, Jacques P (2005) Mycosubtilin overproduction by *Bacillus subtilis* BBG100 enhances the organism's antagonistic and biocontrol activities. Appl Environ Microbiol 71:4577–4584

Leslie JF, Marasas WF, Shephard GS, Sydenham EW, Stockenstrom S et al (1996) Duckling toxicity and the production of fumonisin and moniliformin by isolates in the A and E mating populations of *Gibberella fujikuroi* (*Fusarium moniliforme*). Appl Environ Microbiol 62:1182–1187

Li XY, Yang JJ, Mao ZC, Ho HH, Wu YX, He YQ (2014) Enhancement of biocontrol activities and cyclic lipopeptides production by chemical mutagenesis of *Bacillus subtilis* XF-1, a biocontrol agent of *Plasmodiophora brassicae* and *Fusarium solani*. Indian J Microbiol 54:476–479

Loper JE, Buyer JS (1991) Siderophores in microbial interactions on plant-surfaces. Mol Plant Microbe Interact 4:5–13

M'Piga P, Belanger RR, Paulitz TC, Benhamou N (1997) Increased resistance to *Fusarium oxysporum* f. sp. *radicis-lycopersici* in tomato plants treated with the endophytic bacterium *Pseudomonas fluorescens* strain 63-28. Physiol Mol Plant Pathol 50:301–320

MacLean D, Jones JD, Studholme DJ (2009) Application of 'next-generation' sequencing technologies to microbial genetics. Nat Rev Microbiol 7:287–296

Madhaiyan M, Poonguzhali S, Kwon SW, Sa TM (2010) *Bacillus methylotrophicus* sp. nov, a methanol-utilizing, plant-growth-promoting bacterium isolated from rice rhizosphere soil. Int J Syst Evol Microbiol 60(10):2490–2495

Maget-Dana R, Thimon L, Peypoux F, Ptak M (1992) Surfactin/iturin A interactions may explain the synergistic effect of surfactin on the biological properties of iturin A. Biochimie 74:1047–1051

May JT, Wendrich TM, Marahiel MA (2001) The *dhb* operon of *Bacillus subtilis* encodes the biosynthetic template for the catecholic siderophore 2,3- dihydroxybenzoate-glycine-threonine trimeric ester bacillibactin. J Biol Chem 276:7209–7217

McQuilken MP, Gemmell J (2004) Enzyme production by the mycoparasite *Verticillium biguttatum* against *Rhizoctonia solani*. Mycopathologia 157:201–205

Microbiology UDoPa (2016) Antimicrob Pept database; 18 April. Available from: http://aps.unmc.edu/AP/main.php

Miedaner T (1997) Breeding wheat and rye for resistance to *Fusarium* diseases. Plant Breed 116:201–220

Moyne AL, Cleveland TE, Tuzun S (2004) Molecular characterization and analysis of the operon encoding the antifungal lipopeptide bacillomycin D. FEMS Microbiol Lett 234:43–49

Nasraoui B, Hajlaoui MR, Aïssa AD, Kremer RJ (2007) Biological control of wheat take-all disease: I-characterization of antagonistic bacteria from diverse soils toward *Gaeumannomyces graminis* var. *tritici*. Tunisian J Plant Prot 2:23–34

Niazi A, Manzoor S, Asari S, Bejai S, Meijer J, Bongcam-Rudloff E (2014) Genome analysis of *Bacillus amyloliquefaciens subsp. plantarum* UCMB5113: a rhizobacterium that improves plant growth and stress management. PLoS One 9(8):e104651. https://doi.org/10.1371/journal.pone.0104651. PMID: 25119988

Ongena M, Jacques P (2008) *Bacillus* lipopeptides: versatile weapons for plant disease biocontrol. Trends Microbiol 16:115–125. https://doi.org/10.1016/j.tim.2007.12.009

Ozer EA, Allen JP, Hauser AR (2014) Characterization of the core and accessory genomes of *Pseudomonas aeruginosa* using bioinformatic tools Spine and AGEnt. BMC Genomics 15:737. https://doi.org/10.1186/1471-2164-15-737

Palazzini JM, Dunlap CA, Bowman MJ, Chulze SN (2016) *Bacillus velezensis* RC 218 as a biocontrol agent to reduce *Fusarium* head blight and deoxynivalenol accumulation: genome sequencing and secondary metabolite cluster profiles. Microbiol Res 192:30–36

Park K, Paul D, Kim YK et al (2007) Induced systemic resistance by *Bacillus vallismortis* EXTN-1 suppressed bacterial wilt in tomato caused by *Ralstonia solanacearum*. Plant Pathol J 23(1):22

Parry DW, Jenkinson P, McLeod L (1995) *Fusarium* ear blight (scab) in small-grain cereals – a review. Plant Pathol 44:207–238

Pelletier A, Sygusch J (1990) Purification and characterization of three chitosanase activities from *Bacillus megaterium* P1. Appl Environ Microbiol 56(4):844–848

Phillips-Mora W, Coutino A, Ortiz CF, Lopez AP, Hernandez J et al (2006) First report of *Moniliophthora roreri* causing frosty pod rot (moniliasis disease) of cacao in Mexico. Plant Pathol 55:584

Popovski S, Celar FA (2013) The impact of environmental factors on the infection of cereals with *Fusarium* species and mycotoxin production – a review. Acta Agric Slov 101:105–116

Pozo MJ, Azcón-Aguilar C (2007) Unravelling mycorrhiza-induced resistance. Curr Opin Plant Biol 10:393–398

Qiao JQ, Wu HJ, Huo R, Gao XW, Borriss R (2014) Stimulation of plant growth and biocontrol by *Bacillus amyloliquefaciens subsp. plantarum* FZB42 engineered for improved action. Chem Biol Technol Agric 1:1–14

Rai R, Dash PK, Prasanna BM, Singh A (2007) Endophytic bacterial flora in the stem tissue of a tropical maize (*Zea mays* L.) genotype: isolation, identification and enumeration. World J Microbiol Biotechnol 23:853–858

Rate DN, Cuenca JV, Bowman GR, Guttman DS, Greenberg JT (1999) The gain-of-function *Arabidopsis thaliana* acd6 mutant reveals novel regulation and function of the salicylic acid signaling pathway in controlling cell death, defenses and cell growth. Plant Cell 11:1695–1708

Reva ON, Dixelius C, Meijer J, Priest FG (2004) Taxonomic characterization and plant colonizing abilities of some bacteria related to *Bacillus amyloliquefaciens* and *Bacillus subtilis*. FEMS Microbiol Ecol 48:249–259

Ribera AE, Zuniga G (2012) Induced plant secondary metabolites for phytopatogenic fungi control: a review. J Soil Sci Plant Nutr 12:893–911

Romero D, Perez-Garcia A, Rivera ME, Cazorla FM, de Vicente A (2004) Isolation and evaluation of antagonistic bacteria towards the cucurbit powdery mildew fungus *Podosphaera fusca*. Appl Microbiol Biotechnol 64(2):263–269

Romero D, de Vicente A, Rakotoaly RV, Dufour SE, Veening JW, Arrebola E et al (2007) The iturin and fengycin families of lipopeptides are key factors in antagonism of *Bacillus subtilis* toward *Podosphaera fusca*. Mol Plant-Microbe Interact 20:430–440

Ryu CM, Hu CH, Reddy MS, Kloepper W (2003) Different signaling pathways of induced resistance by rhizobacteria in *Arabidopsis thaliana* against two pathovars of *Pseudomonas syringae*. New Phytol 160:413–420

Ryu CM, Farag MA, Hu CH, Reddy MS, Kloepper JW et al (2004) Bacterial volatiles induce systemic resistance in *Arabidopsis*. Plant Physiol 134:1017–1026

Sarosh BR, Danielsson J, Meijer J (2009) Transcript profiling of oilseed rape (*Brassica napus*) primed for biocontrol differentiate genes involved in microbial interactions with beneficial *Bacillus amyloliquefaciens* from pathogenic Botrytis cinerea. Plant Mol Biol 70(1–2):31–45

Santos VB, Araujo SF, Leite LF, Nunes LA, Melo JW (2012) Soil microbial biomass and organic matter fractions during transition from conventional to organic farming systems. Geoderma 170:227–231

Scherm B, Balmas V, Spanu F, Pani G, Delogu G, Pasquali M, Migheli Q (2013) *Fusarium culmorum*: causal agent of foot and root rot and head blight on wheat. Mol Plant Pathol 14:323–341

Schisler DA, Khan NI, Boehm MJ, Slininger PJ (2002) Greenhouse and field evaluation of biological control of *Fusarium* head blight on durum wheat. Plant Dis 86:1350–1356

Schuster SC (2008) Next-generation sequencing transforms today's biology. Nat Methods 5:16–18

Song JY et al (2012) Genome sequence of the plant growth-promoting rhizobacterium *Bacillus* sp. strain JS. J Bacteriol 194:3760–3761

Stein T (2005) *Bacillus subtilis* antibiotics: structure syntheses and specific functions. Mol Microbiol 56:845–857

Stein T, Vater J, Krufts V, Otto A, Wittmann-Liebold B, Franke P, Panico M, McDowell R, Morriss HR (1996) The multiple carrier model of nonribosomal peptide biosynthesis at modular multienzymatic templates. J Biol Chem 271:15428–15435

Suarez-López F (2010) Evaluación de microorganismos promotores de crecimiento en jitomate (*L. esculentum* L.) bajo condiciones de invernadero. Tesis de Licenciatura, Universidad Autónoma Agraria Antonio Narro, Buenavista, Saltillo, Coahuila

Sutton JC (1982) Epidemiology of wheat head blight and maize ear rot caused by *Fusarium graminearum*. Can J Plant Pathol 4:195–209

Szczech M, Shoda M (2006) The effect of mode of application of *Bacillus subtilis* RB14-C on its efficacy as a biocontrol agent against *Rhizoctonia solani*. Phytopathology 154:370–377

Tekauz A, McCallum B, Gilbert J (2000) Review: *Fusarium* head blight of barley in western Canada. Can J Plant Pathol 22:9–16

Timmusk S, Grantcharova N, Wagner EGH (2005) *Paenibacillus polymyxa* invades plant roots and forms biofilms. Appl Environ Microbiol 71(11):7292

Torres AM, Reynoso MM, Rojo FG, Ramirez ML, Chulze SN (2001) *Fusarium* species (section Liseola) and its mycotoxins in maize harvested in northern Argentina. Food Addit Contam 18:836–843

Toure Y, Ongena M, Jacques P, Guiro A, Thonart P (2004) Role of lipopeptides produced by *Bacillus subtilis* GA1 in the reduction of grey mould disease caused by *Botrytis cinerea* on apple. J Appl Microbiol 96:1151–1160. https://doi.org/10.1111/j.1365-2672.2004.02252.x

Tournas V (2005) Spoilage of vegetable crops by bacteria and fungi and related health hazards. Crit Rev Microbiol 31:33–44

Van-Loon LC, Bakker PAHM, Pieterse CMJ (1998) Systemic resistance induced by rhizosphere bacteria. Annu Rev Phytopathol 36:453–483

Velluti A, Marín S, Bettucci L, Ramos AJ, Sanchis V et al (2000) The effect of fungal competition on colonisation of maize grain by *Fusarium moniliforme*, *F. proliferatum* and *F. graminearum* and on fumonisin B1 and zearalenone formation. Int J Food Microbiol 59:59–66

Vessey JK (2003) Plant growth promoting rhizobacteria as biofertilizers. Plant Soil 255(2):571–586

Vlamakis H, Chai Y, Beauregard P, Losick R, Kolter R (2013) Sticking together: building a biofilm the Bacillus subtilis way. Nat Rev Microbiol 11:157–168. https://doi.org/10.1038/nrmicro2960

Wang LT, Lee FL, Tai CJ, Kuo HP (2008) *Bacillus velezensis* is a later heterotypic synonym of *Bacillus amyloliquefaciens*. Int J Syst Evol Microbiol 58:671–675. https://doi.org/10.1099/ijs.0.65191-0

Wang S, Wu H, Qiao J, Ma L, Liu J, Xia Y, Gao X (2009a) Molecular mechanism of plant growth promotion and induced systemic resistance to tobacco mosaic virus by *Bacillus* spp. J Microbiol Biotechnol 19(10):1250–1258

Wang H, Wen K, Zhao X, Wang X, Li A, Hong H (2009b) The inhibitory activity of endophytic *Bacillus* sp. strain CHM1 against plant pathogenic fungi and its plant growth-promoting effect. Crop Prot 28:634–639

Werhane H, Lopez P, Mendel M, Zimmer M, Ordal G, Márquez-Magaña L (2004) The last gene of the *fla/che* operon in *Bacillus subtilis*, *ylxL*, is required for maximal σ^D function. J Bacteriol 186(12):4025–4029

White JF, Torres MS, Sullivan RF, Jabbour RE, Chen Q, Tadych M et al (2014) Microscopy research and technique: occurrence of *Bacillus amyloliquefaciens* as a systemic endophyte of vanilla orchids. Microsc Res Tech. https://doi.org/10.1002/jemt.22410

Wilson D (1995) Endophyte: the evolution of a term, and clarification of its use and definition. Oikos 73(2):274–276

Woo SM, Kim SD (2008) Structural identification of siderophore from *Bacillus subtilis* AH18, a biocontrol agent of phytophthora blight disease in red pepper. Korean J Microbiol Biotechnol 36:326–335

Wood GAR, Lass RA (2001) Cacao, 4th edn. Blackwell Science, Oxford

Xiong G, Zhao G, Fan C, He Y (2009) Identification and fungistatic effect of a biocontrol strain. J Yunnan Agric Univ 24:190–194

Yang H, Xue Y, Yu X, Liu C (2014) Colonization of *Bacillus amyloliquefaciens* CC09 in wheat leaf and its biocontrol effect on powdery mildew disease. Microbiol China 30(4):481–488

Yao A, Bochow H, Karimov S, Boturov U, Sanginboy S, Sharipov AK (2006) Effect of FZB 24 *Bacillus subtilis* as a biofertilizer on cotton yields in field tests. Arch Phytopathol Plant Protect 39(4):323–328

Yu GY, Sinclair JB, Hartman GL, Bertagnolli BL (2002) Production of iturin A by *Bacillus amyloliquefaciens* suppressing *Rhizoctonia solani*. Soil Biol Biochem 34:955–963

Zeigler D (2003) Gene sequences useful for predicting relatedness of whole genomes in bacteria. Int J Syst Evol Microbiol 53:1893–1900

Zeriouh H, de Vicente A, Perez-Garcia A, Romero D (2014) Surfactin triggers biofilm formation of *Bacillus subtilis* in melon phylloplane and contributes to the biocontrol activity. Environ Microbiol 16(7):2196–2211. https://doi.org/10.1111/1462-2920.12271. PMID: 24308294

Zhang JX, Xue AG, Tambong JT (2009) Evaluation of seed and soil treatments with novel *Bacillus subtilis* strains for control of soybean root rot caused by *Fusarium oxysporum* and *F. graminearum*. Plant Dis 93(12):1317–1323

Zhao X, Zhou ZJ, Han Y et al (2013) Isolation and identification of antifungal peptides from *Bacillus* BH072, a novel bacterium isolated from honey. Microbiol Res 168(9):598–606

Chapter 13
Bacillus Species as Biocontrol Agents for Fungal Plant Pathogens

Çisem Bulut Albayrak

13.1 Introduction

Various organisms like insects, bacteria, and fungi can affect plant health by chewing damage or infections and diseases (Cawoy et al. 2011). Among them, pathogenic fungi not only cause diseases but also some of them produce mycotoxins and threat animal and human health in different ways in the food chain (Nguyen et al. 2017). These diseases and toxins lead to considerable economic losses in crop production and storage worldwide (Patriarca and Pinto 2017). In general, many of the filamentous fungi are phytopathogenic and/or mycotoxigenic. Mycotoxins are poisonous chemical compounds produced as secondary metabolites by toxigenic fungi. These toxins have great significance in the food safety with respect to human and animal health issues (Marroquín-Cardona et al. 2014). Most of them are immunosuppressive agents, and some are carcinogens, hepatotoxins, nephrotoxins, and neurotoxins (Marroquín-Cardona et al. 2014; Zain 2011). Mycotoxicosis is the name of the illness caused by mycotoxin-contaminated foods and feeds. Mycotoxicosis can be classified as acute (exposed to high dose and result in severe illness very quickly) and chronic (exposed to low dose of toxins and result in long-term adverse effects). Especially soilborne filamentous fungi belonging to *Penicillium*, *Aspergillus*, and *Fusarium* genera produce these toxic compounds as secondary metabolites (Palumbo et al. 2008; Patriarca and Pinto 2017). Among these toxic compounds, aflatoxins, ochratoxin A, deoxynivalenol, and fumonisins are the most important and dangerous mycotoxins in agriculture and food industry (Nguyen et al. 2017). Occurrence of mycotoxigenic fungi not only leads to crop quality and yield losses but also their toxins can resist common food processing steps, and

Ç. B. Albayrak (✉)
Engineering Faculty, Food Engineering Department, Adnan Menderes University, Aydın, Turkey
e-mail: cisem.bulut@adu.edu.tr

M. T. Islam et al. (eds.), *Bacilli and Agrobiotechnology: Phytostimulation and Biocontrol*, Bacilli in Climate Resilient Agriculture and Bioprospecting,
https://doi.org/10.1007/978-3-030-15175-1_13

mycotoxin-contaminated food consumption threats public health. Therefore, development of strategies for inhibition of mycotoxin-producing fungi is one of the important issues for scientist and producers. In this chapter, potential of *Bacillus* genus to control these fungal strains was reviewed and discussed with action mechanisms.

13.2 Main Mycotoxins

By considering frequency of occurrence in the food chain and risks for human and animal health, aflatoxins, ochratoxin, and fusarium toxins (fumonisins, trichothecenes, and zearalenone) are of great importance (Zain 2011; Tola and Kebede 2016; Hussein and Brasel 2001). Mycotoxin-producing fungi can be classified mainly into two groups as field fungi (invade before harvest) and storage fungi (exist after harvest). Toxicogenic field fungi can be categorized mainly into three groups as the first group includes plant pathogens such as *Fusarium graminearum*; the second group is composed of fungi that grow on stressed plants such as *Fusarium moniliforme* and sometimes *Aspergillus flavus*; and the third group is consisted of fungi that initially colonize plant before harvesting, and mycotoxin contamination occurs after harvest, such as *Penicillium verrucosum* and *A. flavus*. It is important to develop strategies to prevent or control toxin-producing fungal strains during pre-harvest, harvest, post-harvest, and storage periods. There are many reviews on harmful effect of mycotoxins for human and animal health (Hussein and Brasel 2001; Zain 2011; da Rocha et al. 2014). Fungal growth and mycotoxin productions are influenced by various factors. Both physical parameters such as water activity, moisture, temperature, and physical integrity of the plant chemical parameters such as substrate type and amount, pH, and presence of oxygen may determine the level of fungal growth and mycotoxin accumulation. To control mycotoxin-producing fungi, these parameters should be also evaluated. Due to the safety and health concerns, regulatory actions do exist to limit contamination of agricultural products.

13.2.1 Aflatoxins

Aflatoxins (AF) are natural chemicals in difuranocoumarin derivatives synthesized by a polyketide pathway. The chemical structures of main aflatoxins are presented in Fig. 13.1. *Aspergillus flavus*, *A. parasiticus*, *A. nomius*, *A. bombycis*, *A. ochraceoroseus*, *A. pseudotamarii*, and *A. tamarii* are the major producers of aflatoxins (Pitt 2000). AF were discovered in the early 1960s as the causative agent of the turkey X disease and recognized as the first identified mycotoxins. Feeding contaminated peanut meal results in the death of thousands of turkey pouts, ducklings, and chicks (Bradburn et al. 1994). Due to the toxicity and occurrence in wide range

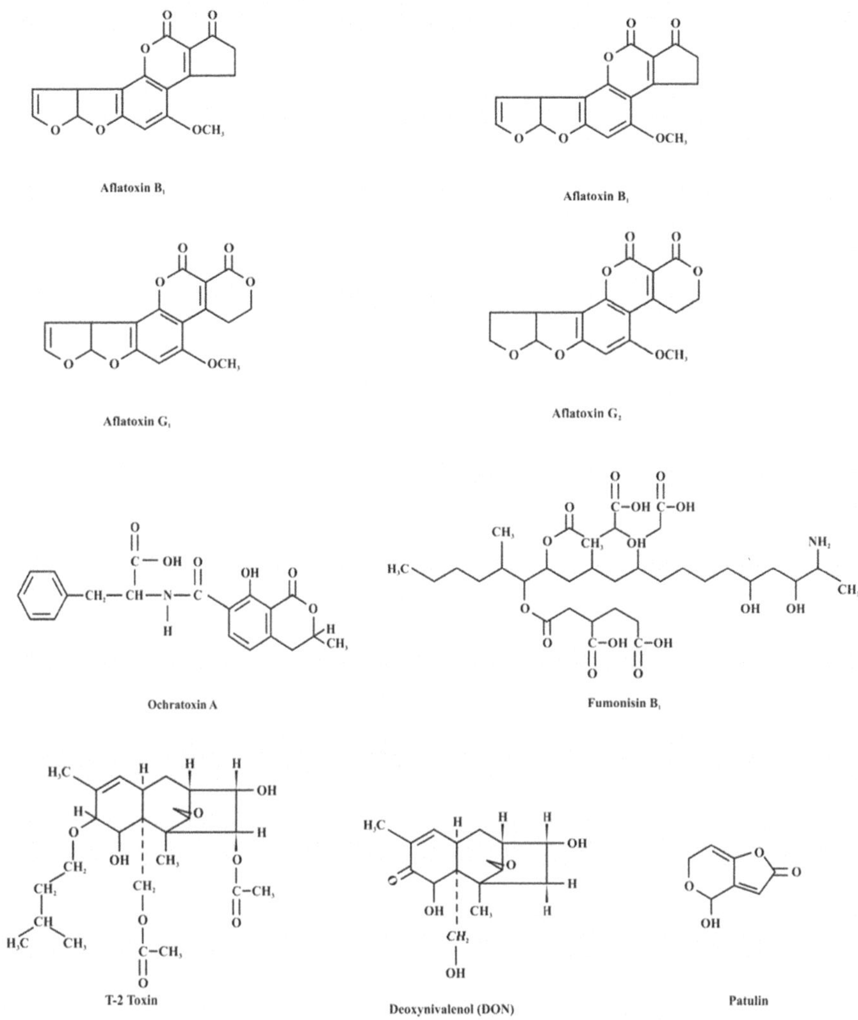

Fig. 13.1 Chemical structure of some major mycotoxins

of agricultural products, AF can be considered the most important mycotoxins in agriculture and food industry. They were found in a wide range of agricultural products such as corn (Majeed et al. 2013); peanut (Kamika and Takoy 2011); cotton seed (Ashworth and McMeans 1966); rice (Reddy et al. 2009); cereals like wheat, barley, and oats (Rahmani et al. 2010); dried fruits as dried fig (Bircan and Koç 2012) and raisins (Juan et al. 2008); spices (Bircan 2005); nuts (Baquião et al. 2012); milk; cheese (Tekinşen and Eken 2008); etc. Additionally, other genera such as some *Penicillium* and *Rhizopus* species can produce AF. Although there are more than 20 chemical structures, the most common forms are aflatoxin B1 (AFB1), aflatoxin B2 (AFB2), aflatoxin G1 (AFG1), aflatoxin G2 (AFG2), aflatoxin M1

(AFM1), and aflatoxin M2 (AFM2). The AFB1, B2, G1, and G2 are synthesized in natural way, but AFM1 is the main metabolite of AFB1 and it is bio-transformed during lactation. Therefore, it can be present in milk. AF have some carcinogen, mutagen, teratogen, hepatotoxic, and immune system suppressive effects, and toxicity can be in the order of AFB1 > AFM1 > AFG1 > AFB2 > AFM2 ≠ AFG2. AFB1 is of particular importance because it is the most toxic and potent hepatocarcinogenic compound. The International Agency for Research on Cancer classified it as group 1 human carcinogen (IARC 1993). Regulatory limits have been set in different countries, and they range from 0 to 30 mg/kg for AFB1 in foodstuffs and from 0 to 50 mg/kg for total AF (Patriarca and Pinto 2017).

13.2.2 Ochratoxins

Ochratoxin A (OTA) is the most toxic compound among ochratoxins. Several *Aspergillus* and *Penicillium* species produce ochratoxins. OTA is a pentaketide, derived from the dihydrocoumarins family coupled with b-phenylalanine, and chemical structure is given in Fig. 13.1. The first discovery of OTA was in 1965, and it is identified as a metabolite of *A. ochraceus* (Van der Merwe et al. 1965). The most common and important adverse health effects are observed due to the nephrotoxic properties of OTA. This toxin is nephrotoxic to all animal species, and it is related to the Balkan endemic nephropathy (da Rocha et al. 2014). OTA has been found in different cereals, such as wheat, barley, and rye, and other commodities such as in coffee and cocoa beans, rice, dried fruits, nuts, spices, and other plant-originated fermented beverages such as wine and beer (Zain 2011). Due to the relevance of this toxin to human and animal health, international regulations have limits for OTA in a wide range of foods such as cereals, coffee beans, soluble coffee, wine, grape juice, beer, cocoa and cocoa products, etc., which ranges from 0.5 to 10 mg/kg (Marin et al. 2013; Patriarca and Pinto 2017).

13.2.3 Fusarium *Toxins*

Fumonisins (FUM), trichothecenes, zearalenone (ZEN), and emerging mycotoxins that include beauvericin, enniatins, fusaproliferin, and moniliforme are the most common *Fusarium* toxins in agricultural products and food chains (Marin et al. 2013). The FUM have structures composed of a long-chain hydrocarbon backbone similar to that of sphinganine, and the chemical form of fumonisin B1 (FB1) is presented in Fig. 13.1. This toxin group consists of four main classes A, B, C, and P. Among different kinds of FUM (FA1, FA2, FB1, FB2, FB3, FB4, etc.), FB1 is the most toxic and prevalent, and it is found in cereals commonly in corn and maize. Especially *Fusarium* species, such as *F. proliferatum* and *F. verticillioides*, produce FUM. International legal limits for maize range from 200 to 3000 mg/kg in different countries (Marin et al. 2013; Patriarca and Pinto 2017; Tola and Kebede 2016). Trichothecenes are a complex group of chemically related sesquiterpenoids. They share a tricyclic nucleus called trichodiene characterized by a double bond at the 9 and 10 position and an epoxide group between carbons 12 and 13, responsible for its toxicity (Marin et al. 2013). All trichothecenes have tricyclic structure. The variations among trichothecenes exist due to the occurrence, number, and position of hydroxyl, ester, and carboxyl groups. On the basis of chemical structure, they can be divided into four subgroups (A, B, C, and D). Groups A (T-2 and HT-2 toxins, diacetoxyscirpenol) and B (deoxynivalenol, nivalenol) can be considered as the most toxic and prevalent ones. Several *Fusarium* species, including *F. graminearum*, *F. sporotrichioides*, *F. poae*, and *F. culmorum*, produce trichothecene.

Zearalenone (ZEA) is a mycooestrogenic toxin known as 6-(10-hydroxy-6-oxo-trans-1-undecenyl)-b-resorcylic acid l-lactone. Actually, it works similar to sexual hormone. It is an estrogenic lactone synthesized by some *Fusarium* species, *F. graminearum* being the main one. Other *Fusarium* spp. such as *F. crookwellense*, *F. culmorum*, *F. moniliforme*, *F. equiseti*, and *F. oxysporum* can also produce ZEA (Tola and Kebede 2016). Commonly observed health problems were associated with reproductive problems (Zain 2011; da Rocha et al. 2014). Besides estrogenic effects, it leads to problems in hepatic function (Zinedine et al. 2007). Although most prevalent contamination was observed in corn, other plant products such as wheat, barley, oat, and rye may also get contaminated with this toxin (Patriarca and Pinto 2017). Limits for ZEA in maize and other cereals vary from 20 to 1000 mg/kg around the world (Marin et al. 2013; Patriarca and Pinto 2017).

13.2.4 Patulin

Patulin is a mycotoxin with polyketide lactone structure (Fig. 13.1). Although it has antimicrobial properties, it is also toxic to humans and animals. It has genotoxic and teratogenic effects and shows immune suppressive activity. *Penicillium* spp. in particular *P. expansum* produced patulin. Also, *Aspergillus* spp. and *Byssochlamys* can synthesize it. Besides mycotoxin production, they can cause some post-harvest diseases. For instance, *Penicillium expansum* causes blue mold, and it is most commonly observed as a post-harvest disease of fruits such as apples and pears. In recent works, production of patulin in apples, pears, and their by-products were also reported (da Rocha et al. 2014; Wang et al. 2015). It can be stable after common heat process applied in food industry.

13.3 Main Phytogenic and Mycotoxigenic Fungi

Besides mycotoxin production, occurrence of *Aspergillus*, *Fusarium*, and *Penicillium* species can cause soilborne diseases and lead to post-harvest problems in plant products. The genus *Aspergillus* is grouped in the ascomycete class and reproduces by both asexual and sexual means. They produce large amount of spores and can spread through air. When the conditions become favorable, they germinate. Although many species of this genus can be defined as opportunistic pathogens, some of the species can play a role in plant diseases such as ear rot in corn, aflaroot in peanuts, and other post-harvest diseases (Klich 2007) and leads to mycotoxin productions. For example, *Aspergillus* section Nigri is associated with many diseases and mycotoxin production in grapes and wine. Growth of this fungus causes problems especially during post-harvest period. The potential of *Bacillus* species to control this fungus in the post-harvest period was investigated for grapes (Jiang et al. 2014). Grape fruits were treated with either *B. subtilis* liquid culture or cell-free culture supernatant, and both applications inhibited growth of *A. carbonarius* and other fungal contamination (Jiang et al. 2014).

Fusarium is one of the most common fungal pathogens infecting various plants. Occurrence of some *Fusarium* strains in plant products may cause plant diseases and also produce mycotoxin. This large fungi group attacks mostly cereals such as wheat, corn, barley, and oat (Logrieco et al. 2007; Luo et al. 1990; Muthomi et al. 2008; Schöneberg et al. 2018), but also other agricultural plants such as tomato, cucumber, alfalfa, soybeans, and lentil can be effected by this fungus (Cawoy et al. 2011). The most common *Fusarium* species were reported as *Fusarium graminearum*, *F. sporotrichioides*, *F. poae*, *F. avenaceum*, *F. culmorum*, *F. accuminatum*, *F. langsethiae*, *F. verticillioides*, *F. proliferatum*, *F. oxysporum*, *F. anthophilum*, and *F. paranaense* (Nguyen et al. 2017). *Fusarium* can cause problems in a variety of plant organs, such as heads, roots, or stem base, and causes mainly cereal diseases like Fusarium head blight (FHB), foot and root rot, and crown rot. Particularly, FHB has economic importance in worldwide cereal production, and *Fusarium graminearum* sensu stricto is known as the main causal agent of FHB in wheat. The main problem is not only significant yield losses, but also it is associated with dangerous mycotoxins produced by *Fusarium* species. Furthermore, soilborne *F. oxysporum* pathogenic strains can result in *Fusarium* wilt disease. Diverse species under genus *Fusarium*, especially pathogenic ones, lead to vascular wilt disease (Leslie and Summerell 2008). The pathogenic strains are particularly host specific, and they cause yield losses in various crops like tomato, cucumber, melons, flax, lettuce, strawberry, and banana. For instance, *Fusarium* wilt is one of the important diseases for strawberry, and the responsible fungus is *Fusarium oxysporum* f. sp. *fragariae* (Fof); it was first found in Australia in 1962 (Winks and Williams 1965). In addition, *Fusarium* cf. *incarnatum* is a known causing agent for ginseng root rot (Durairaj et al. 2018). Again, *Fusarium* species can lead to fig endosepsis, and these fungi both cause rot and mycotoxin production (Moretti et al. 2010).

Penicillium is grouped in ascomycetous class and can be considered as ubiquitous, opportunistic pathogens and contaminate various products such as cereals, nuts, vegetables, and fruits. *Penicillium* spp. produce several mycotoxins such as citreoviridin, cyclopiazonic acid, patulin, citrinin, and OTA. *Penicillium* spp. such as *P. expansum*, *P. digitatum*, *P. italicum*, *P. solitum*, and *P. ulaiense* cause rots in various vegetables and fruits (Nguyen et al. 2017). Especially, *P. expansum* and *P. digitatum* are known as the main post-harvest pathogens of pome and citrus fruit, respectively. During post-harvest handling and storage, occurrence of these species causes quality loss. Besides these losses, mycotoxin productions of patulin and OTA seem to be another health concern for these agricultural products.

13.4 *Bacillus* as Biocontrol Agent for Mycotoxigenic and Pathogenic Fungi

In order to combat plant pathogenic and mycotoxigenic fungi, common approaches include plant breeding to develop resistant varieties, production of genetically modified resistant plants, and utilization of chemical compounds such as fungicides. Chemical fungicides have common applications in practice. Despite high efficacy and ease of use of these products, increasing concerns exist for agrochemicals such as development of resistance in organisms against these chemicals, negative effects on human health, wide spectrum of toxicity, and environmental pollution. One of the new environmentally friendly solutions for safer and more effective plant disease control is utilization of biological control agents (BCAs). Biological control can be described as the use of natural antagonistic organisms to combat pests or suppress plant diseases, which offers an alternative to chemicals. In this context, the term biopesticide can be defined as living organisms or natural products derived from these organisms that are used to suppress pathogen population. Biopesticides can be classified into four main groups. These are microbial pesticides (Anderson and Kim 2018), other organisms (nematodes, insects, etc.), natural products obtained from living organisms (biochemical pesticides), and plant-incorporated protectants (Cawoy et al. 2011). The use of biopesticides has several advantages over synthetic chemicals (Mishra et al. 2015). For instance, they can be degraded in a short period of time in the environment and are known as less toxic due to the targeted action. Additionally, resistance phenomenon is not expected due to the multiplicity of mode of action. Only a low amount can be effective because microorganisms can colonize, multiply, and produce bioactive compounds in the phytosphere continuously. These microorganism-based products can be fungi, bacteria, and viruses and often isolated from suppressive environments such as aerial or underground parts of plants. Moreover, these organisms not only have potential for reducing disease incidences, but they also have other beneficial roles such as promoting plant growth and nutrition.

The majority of BCA is composed of bacterial products, and the main bacterial species are included in *Bacillus*, *Streptomyces*, and *Pseudomonas* genera (Mnif and Ghribi 2015; Shoda 2000). Spore-forming bacteria, like the *Bacillus* genus, have large genetic biodiversity and can exist in a wide range of environments such as seawater, soil, plants, and even extreme environments such as hot springs. It possesses some advantages and can be good candidates for use as biological controls. They have a unique ability to replicate rapidly and they have broad spectrum for biocontrol property. They produce different types of antimicrobial substances. They induce growth and defense responses by producing some compounds such as volatiles. Additionally, due to their spore-forming properties, they show resistance to adverse environmental conditions such as high or low temperatures, unsuitable pH, and scarcity in nutrients or water. When environmental conditions are unfavorable, spores help them in the phytosphere. This property also provides easy formulation and storage of commercial products because bacterial suspensions are converted to easy-to-handle powder formulation. Moreover longer shelf life and less precaution requirements are other advantages. Bacillus-based commercial producers such as Avogreen®, Bacillus SPP®, Biosubtilin®, Biosafe®, Cease®, Ballad®, Companion®, and Kodiak®a are available for control of plant pathogens (Cawoy et al. 2011). Particularly, *B. subtilis* is defined as GRAS (generally recognized as safe) organisms for food industries by the US Food and Drug Administration (US FDA) (Denner and Gillanders 1996). HiSticka, Kodiak, Larminar, Rhapsody, and Subtilex manufacture BCA by using *B. subtilis* as an active agent for controlling fungal growth in the agricultural products. Other *Bacillus* species such as *B. amyloliquefaciens*, *B. licheniformis*, and *B. pumilus* have also been commercialized to control fungal growth (Pérez-García et al. 2011; Bisutti et al. 2017).

The beneficial roles of *Bacillus* species against fungal diseases have been described in many reports (Abd-Allah et al. 2007; Chulze et al. 2015; Pérez-García et al. 2011; Cawoy et al. 2011; Shafi et al. 2017; Ab Rahman et al. 2017; Dunlap et al. 2013; Gutierrez-Monsalve et al. 2015; Heydari and Pessarakli 2010; Islam et al. 2012; Kildea et al. 2008; Lucon et al. 2010; Moreira et al. 2014; Pan et al. 2017; Palazzini et al. 2016; Senthil et al. 2011) and recent works tabulated (Table 13.1). These diseases are categorized mainly in three groups as root diseases (such as avocado root rot, tomato damping-off, and wheat take-all), foliar diseases (such as cucurbit and strawberry powdery mildews), and post-harvest diseases (such as green, gray, and blue molds).

Fusarium oxysporum f. sp. *lycopersici* is one of the fungal plant pathogens, and it is responsible for vascular wilt for tomato (Heidarzadeh and Baghaee-Ravari 2015). Two *Bacillus pumilus* strains were tested to control *Fusarium* wilt, and some biocontrol properties were determined by biofilm assay, root colonization, and trials under pot conditions. Both *B. pumilus* strains (ToIrMA and ToIrFT strains) displayed biocontrol traits in addition to plant growth factor. In vivo experiments indicated that 73% reduction was observed in disease incidence (Heidarzadeh and Baghaee-Ravari 2015).

Baffoni et al. (2015) studied the inhibition potential of bacterial species belonging to *Lactobacillus* and *Bacillus* genera against *Fusarium* spp. in in vitro and

Table 13.1 Recent studies for antifungal properties of *Bacillus* species

Bacteria	Mode of action	Target organisms	Crop	References
Bacillus stratosphericus	Antagonistic	*Ilyonectria* sp., *Neurospora* sp., *Cladosporium* sp., *Eutypella* sp., *Aschersonia* sp., and *Fusarium* sp.	Root rot infected ginseng root samples	Durairaj et al. (2018)
B. amyloliquefaciens subsp. *plantarum*	Antagonistic	*Fusarium* cf. *incarnatum*	Ginseng root rot	Song et al. (2014)
Bacillus mojavensis	Growth inhibition	*Aspergillus flavus* and *A. parasiticus*		Pereyra et al. (2018)
B. subtilis				
Bacillus sp.		*Fusarium oxysporum* f. sp. *lycopersici*		Jangir et al. (2018)
B. megaterium (B5), *B. amyloliquefaciens*		*Cochliobolus sativus*, *Alternaria alternata*, and *Fusarium graminearum*	Wheat plants	El-Gremi et al. (2017)
B. pumilus		*Xanthomonas campestris* and *F. oxysporum* f. sp. *melonis*		Suárez-Estrella et al. (2013)
B. atrophaeus		*Colletotrichum* species	Soursop, avocado	Guardado-Valdivia et al. (2018)
B. subtilis		*Aspergillus parasiticus*		Siahmoshteh et al. (2018)
B. amyloliquefaciens				
Bacillus isolates from soil and rhizosphere		*Fusarium oxysporum*	Lentil seeds	El-Bendary et al. (2016)
		Rhizoctonia solani		
		Macrophomina phaseolina		
B. toyonensis, *B. cereus*, *B. aryabhattai*, *B. megaterium*, *B. aerius*, *B. stratosphericus*, and *Paenibacillus barcinonensis*	Antagonistic with proteolytic and cellulolytic activity	*F. oxysporum* f. sp. *lycopersici*	Tomato/ fusarium wilt	Rocha et al. (2017)
B. amyloliquefaciens		*Fusarium graminearum*	Wheat fusarium head blight	Crane and Bergstrom (2014)
B. atrophaeus strain B5		*Colletotrichum gloeosporioides*	Anthracnose/ soursop and avocado	Guardado-Valdivia et al. (2018)

(continued)

Table 13.1 (continued)

Bacteria	Mode of action	Target organisms	Crop	References
Bacillus spp. strains isolated from the phyllosphere of lupin pods		*Colletotrichum acutatum*	Anthracnose/ Andean lupin seeds	Yánez-Mendizábal and Falconí (2018)
B. megaterium (B5), *B. amyloliquefaciens* (B28)		*Cochliobolus sativus*, *Alternaria alternata*, and *Fusarium graminearum*	Kernel black point disease/ wheat	El-Gremi et al. (2017)
Bacillus spp.		*F. oxysporum* f. sp. lycopersici	Tomato	Jangir et al. (2018)
B. stratosphericus (FW3)		*Ilyonectria* sp., *Neurospora* sp., *Cladosporium* sp., *Eutypella* sp., *Aschersonia* sp., and *Fusarium* sp.	Ginseng/root rot disease	Durairaj et al. (2018)
B. subtilis CICC 10034		*Penicillium expansum*	Apples	Wang et al. (2016a, b)
B. cereus AR156		*P. expansum*	Sweet cherry fruit blue rot	Wang et al. (2015)
B. subtilis ABS-S14		*P. digitatum*	Citrus fruit	Waewthongrak et al. (2015)
B. amyloliquefaciens		*F. oxysporum* f. sp. *cubense*	Banana	Wang et al. (2016a, b)
B. amyloliquefaciens strain BLB369, *B. subtilis* strain BLB277, and *Paenibacillus polymyxa* strain BLB267		*F. graminearum*	Durum wheat	Zalila-Kolsi et al. (2016)
B. amyloliquefaciens Q-426		*F. oxysporum* f. sp. *spinaciae*	Spinach	Zhao et al. (2014)
B. subtilis		*Penicillium digitatum* Sacc.	Citrus fruit	Leelasuphakul et al. (2008)
B. megaterium, *B. subtilis*, *B. subtilis* subsp. *subtilis*		*Aspergillus niger*	peanut	Yuttavanichakul et al. (2012)
B. subtilis CCTCC M 207209		*A. carbonarius* CCTCC AF 2011004	Table grape	Jiang et al. (2014)
B. amyloliquefaciens strain BUZ-14		*Botrytis cinerea*, *Monilinia fructicola*, *M. laxa*, *Penicillium digitatum*, *P. expansum*, and *P. italicum*	Post-harvest fruit disease	Calvo et al. (2017)

in vivo conditions. These fungal species are known as responsible for wheat fungal diseases. *L. plantarum* SLG17 and *B. amyloliquefaciens* FLN13 were used in the experiments. In the field study, durum wheat was treated with a mixture of antagonist strains. Treatments were performed in two ways as weekly from heading until anthesis and at flowering. The results were compared to control groups as untreated and chemical fungicide as treated ones. Both disease incidence and disease severity are combined in FHB index, and the extent of disease is evaluated by using FHB index. A mixture of the bacterial strains, when applied from the heading until anthesis, was found to be efficient to reduce the FHB index. Microbial population pattern of seeds was also determined by molecular tools as PCR-DGGE, and the results showed the presence of *L. plantarum* SLG17 in wheat seeds, and it highlights an endophytic behavior. A mixture of *Lactobacillus* and *Bacillus* strains as BCA starting from the heading period until anthesis of wheat plants was found to be promising for the FHB index reduction.

Durairaj et al. (2018) studied on biological control of fungal phytopathogens that cause root disease and infections in ginseng (*Panax ginseng*). In this work, seven fungal genera (*Ilyonectria* sp., *Neurospora* sp., *Cladosporium* sp., *Eutypella* sp., *Aschersonia* sp., and *Fusarium* sp.) were isolated from infected ginseng root rot samples. Two antagonistic strains, *Pseudomonas aeruginosa* (D4) and *Bacillus stratospherieus* (FW3), were identified by molecular characterization tools. Both bacterial strains produced various bioactive metabolites such as amylase, chitinase, hydrogen cyanide, indole acetic acid (IAA), lipase, protease, and phosphatase. However, they did not show cellulose production activity.

In another study (Wang et al. 2016a, b), the *Bacillus amyloliquefaciens* strain W19 was found to be effective for inhibition of *Fusarium oxysporum* f. sp. *cubense*. Metabolites as iturin and bacillomycin D were responsible for this property. These fungi can treat banana fruit disease and cause *Fusarium* wilt disease. The potential of bioorganic fertilizer containing strain W19 to enhance plant growth and control the *Fusarium* wilt was determined in both pot and field applications. In the results, banana growth and fruit yield were improved. In addition, banana *Fusarium* wilt disease was suppressed. Confocal laser scanning microscopy and scanning electron microscopy were used, and colonization of green fluorescent protein-tagged strain W19 on banana roots was evaluated. Banana root exudates effected the biofilm formation, and it also showed that the banana root exudates may improve colonization. Moreover, production of indole-3-acetic acid (IAA) was determined for strain W19. This compound is known as one of the plant growth promoter hormones. Fungal mycelia and spores resulted in morphological changes. In the absence of the bacteria, fungi mycelia structure was well organized, but in the presence of bacteria, swelling and chlamydospore productions were observed.

Zalila-Kolsi et al. (2016) investigated the potential of *Bacillus* spp. to control plant pathogenic fungi. For this, *Bacillus amyloliquefaciens* BLB369, *Bacillus subtilis* BLB277, and *Paenibacillus polymyxa* BLB267 were isolated, and these strains were identified using biochemical and genetic tools (16S rDNA, gyrA, and rpoB). Production of broad-spectrum antifungal compounds was determined as, respectively, iturin and surfactin, surfactin and fengycin, and fusaricidin and polymyxin.

In vivo studies are performed for wheat seed germination. The efficiency of the tested bacteria was determined by analyzing plant height, phenolic compounds, chlorophyll, carotenoid contents, and antagonist activities. The main aim was to protect Tunisian durum wheat (*Triticum turgidum L.* subsp. *durum*) cultivar Om Rabiia against *F. graminearum* fungus. The microscopic examinations showed morphological variations for fungal mycelia and spores. In the presence of bacterial cells, swelling was observed in the structure of the *F. graminearum* mycelia with chlamydospore productions.

In order to control FHB, Comby et al. (2017) isolated 86 microorganisms from wheat plants, and they were screened for inhibitions against *Fusarium graminearum* and *Fusarium culmorum* growth. Among the tested strains, 22 isolates seemed very effective against both *F. graminearum* (inhibition of 30–51%) and *F. culmorum* (inhibition of 15–53%). Fungal (*Chaetomium globosum, P. glomerata, Clonostachys rosea, M. bolleyi, A. proteae*, and *Sarocladium kiliense)* and bacterial species (*Bacillus amyloliquefaciens, B. pumilus*, and *B. subtilis)* were identified for those 22 strains. In this work, all bacteria belong to *Bacillus* genus. The most effective strains were determined as *B. subtilis* BaSu1 and BaSu3, *B. amyloliquefaciens* BaAm, and *C. globosum* CG1 and CG2, and all of them displayed inhibition by approximately half the mycelial growth.

Fusarium cf. *incarnatum*, also known by the synonyms *F. pallidoroseum* and *F. semitectum*, is often regarded as a secondary colonizer of plant tissues and causes several plant diseases (Song et al. 2014). It has also been isolated from rotten ginseng roots (Lee 2004; Song et al. 2014). For this, it can be considered as a potential cause of ginseng root rot of its strong pathogenicity for ginseng root rots. Song et al. (2014) studied 392 bacteria isolated from ginseng roots and soil environments. Antifungal behavior was characterized for different conditions (inoculum cell concentration, incubation temperature, cell-free filtration, and cell suspension application). Results revealed that bacterial isolate (B2-5) exhibited strong growth inhibition in both bacterial cell suspension and culture filtrates. Improved inhibitory activity was observed at 25 °C, both high (1×10^8 CFU/mL) and low concentrations (1×10^6 CFU/mL) without critical rot symptoms. Additionally, this strain did not produce pectinase enzyme for those conditions; therefore soft rot is not expected, and it is one of the desirable properties for further applications. In pot experiments, higher control efficacy is observed at lower inoculum concentration (1×10^6 CFU/mL). When both the bacteria and the pathogen were inoculated together, establishment and colonization behavior of the *Bacillus* isolates on the rhizosphere are found to be higher than when the bacteria were inoculated alone. Scanning electron microscopy was used to examine morphological deformations. Fungus hyphae were twisted and shriveled by the bacteria, and these changes can be considered as direct damage due to the antifungal metabolites (Fig. 13.2).

In the recent study, the antagonistic property of *Bacillus* spp. to control *Fusarium oxysporum* f. sp. *Lycopersici* was determined (Jangir et al. 2018). Forty-nine *Bacillus* strains were isolated from the rhizosphere environments of tomato plants. These tomato plants were obtained from infested field. Antifungal ability was screened by dual culture antagonism assay. Fungal inhibition and plant promotion

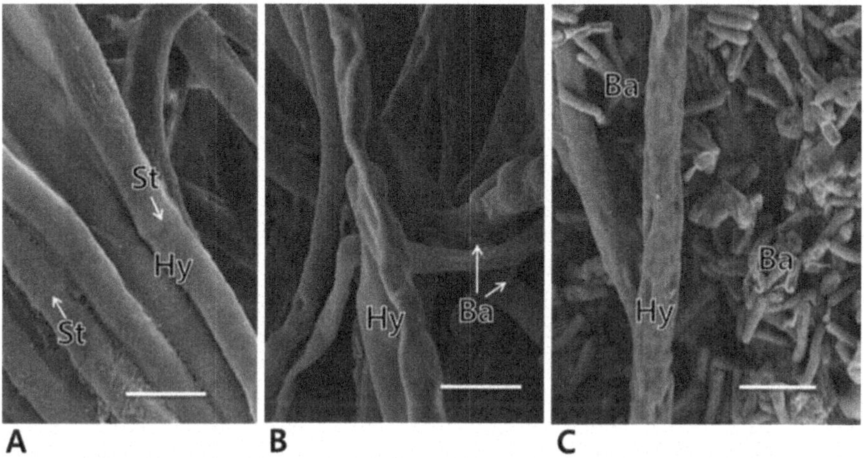

Fig. 13.2 Scanning electron micrographs of *Fusarium* cf. *incarnatum* without bacterial treatment (**a**), and treated with the bacterial isolate B2-5 at inoculum concentrations of (**b**) 10^6 CFU/mL and (**c**) 10^8 CFU/mL, wrinkled, distorted, and shrunken hyphae (Hy) adhered with bacterial cells (Ba), compared to intact hyphae (Hy) with smooth surface showing a contour of septum (St) in (**a**) the untreated control. Bars, 20 mm (Song et al. 2014)

parameters such as β-1,3-glucanase, protease, chitinase, ammonia, siderophore, hydrogen cyanide, IAA, and biofilm formation were evaluated. Strain B44 exhibited best antifungal properties. Production of some enzymes such as β-1,3-glucanase (15.61 U ml^{-1} min^{-1}), protease (1608.15 U ml^{-1} min^{-1}), and chitinase (129 U ml^{-1} min^{-1}) was determined with production of other compounds as volatile and nonvolatile metabolites. The main compounds identified by GC-MS analysis were 1,2-benzenedicarboxylic acid (23.99%), 6-undecylamine (6.61%), 2-methyloctacosane (5.91%), 9-octadecenoic acid (5.13%), and 1-tetradecanamine, N,N-dimethyl (5.05%). Other substances identified by UPLC-MS were tert-butyldifluorophosphine (31.45%), 5-(4,4,5,5-tetramethyl-1,3,2-dioxaborolane-2-yl)-3,6-dihydro-2H-pyran (14.76%), 6-bromo-4-tert-butyl-1,1-dimethyl-1,2-dihydronaphthalene (8.35%), 2-amino-N-butyl-1,8-naphthyridine-3-carboxamide (6%), and N,N-diethyl-2-[(ethylamino) methyl]-3-phenyloxirane-2-sulfonamide (5.12%). Under greenhouse conditions, disease incidence is reduced in 36% for tomato plants.

Black point disease is one of the fungal diseases observed in wheat grains. In the work of El-Gremi et al. (2017), pathogenic fungi were isolated from symptomatic grains and identified as *Cochliobolus sativus*, *Alternaria alternata, and Fusarium graminearum*. Twenty-five microbial strains were isolated from the surface of healthy grains, and they were screened for antifungal properties. The most effective antagonistic strains were *Bacillus megaterium* (B5), *B. amyloliquefaciens* (B28), *Trichoderma harzianum* (T37), and *Epicoccum* sp. (E52). In pot experiments, *B. amyloliquefaciens* (B28) has the best BCA properties in the case of germination and length of roots and shoots. Moreover, in field experiments, the strains enhanced germination and tillering of black-pointed wheat grains. Wheat plants were sprayed with *B. amyloliquefaciens* (B28) which provided dramatic decrease in the incidence of leaf spots and kernel black point diseases.

Bacillus spp. have the potential not only to control fungal plant diseases but also to control mycotoxin production (Nguyen et al. 2017; Khan et al. 2017; Palumbo et al. 2008). Until this point, their antifungal potential was studied without specific attention for their mycotoxins. In this review, special focus is on preventing mycotoxin-producing fungi with possible pre-harvest or post-harvest applications. In the following selected studies, preventing of mold strains with their mycotoxins was evaluated and discussed.

In addition to studies to control fungal plant diseases, one of the important issues is to control and prevent mycotoxigenic fungi (Nguyen et al. 2017; Khan et al. 2017; Palumbo et al. 2008); in this review the potential of *Bacillus* species to control pathogenic and mycotoxigenic fungi is discussed for possible pre-harvest or post-harvest applications. In the following selected studies, special focus is on mycotoxin-producing fungi.

In peanut (*Arachis hypogaea* L.), one of the fungal diseases is root rot caused by *A. niger.*

The inhibition of *A. niger* was investigated by using 765 bradyrhizobial and 350 soil-isolated plant growth-promoting rhizobacteria (PGPR) strains (Yuttavanichakul et al. 2012). *A. niger* and *A. flavus* can produce dangerous mycotoxins such as ochratoxin and aflatoxin. Among the tested strains, only 11 PGPR isolates were found to be able to prevent *A. niger* growth. The best isolates were determined by considering growth inhibition ability and root colonization properties. They were A20, A45, A62, and A106, and they were identified by 16S rRNA sequences as *Bacillus megaterium*, *B. subtilis*, *B. subtilis* subsp. *subtilis*, and *Pseudomonas* sp., respectively. A20, A45, and A62 isolates produced lytic protease, but A106 did not produce this enzyme. Cell-free supernatants of A20 and A62 also showed antifungal activity. Differently, the antifungal activity of isolates A45 and A62 showed resistance to proteinase K enzyme. All PGPR strains produced an IAA hormone and biofilms. This hormone is especially important for promotion of peanut root growth. Inoculation of either isolate A20 or A45 (10^8 cells per ml) together with *Bradyrhizobium* sp. TAL 173 (10^8 cells per ml) provided prevention of the peanut root rot disease caused by *A. niger* (10^5 and 10^6 spores per seed). Within this application, nitrogen fixation is increased and utilization in peanut is reduced. In another study, the antagonist behavior of *Bacillus* species isolated from peanut against *A. parasiticus* was also investigated (Xiao et al. 2014)

In a recent study by Siahmoshteh et al. (2018), soilborne *Bacillus* strains were evaluated for their antifungal properties. *B. subtilis* and *B. amyloliquefaciens* were studied against common field-pathogenic fungi (including *Aspergillus parasiticus* NRRL2999). The fungal growth and aflatoxin production were also determined. In order to understand mode of action, variations in ergosterol content in the membrane, mitochondrial dehydrogenase activity, and fungal cell wall chitin and beta-1,3-glucan content were also evaluated. Both *Bacillus* strains have significant antifungal activity against the wide range of filamentous fungi. High inhibitions for *A. para-*

siticus growth (up to 92%) and aflatoxin production (up to 100%) were obtained and displayed good proteolytic activities. Both *Bacillus* strains decreased ergosterol content of the fungal cell membrane of *A. parasiticus* (9.0–80%). Changes in dehydrogenase activity of fungal cell mitochondria were determined. The structure and shape of fungal hyphae were changed clearly. In the scanning and transmission electron microscopy examinations, folding, wrinkling, cell depletion, and vacuolization were observed.

Toxigenic fungal strains such as *A. flavus* and *A. parasiticus* can be present and grow in a broad range of food crops and result in following production of aflatoxins. Pereyra et al. (2018) showed the biocontrol efficiency of soilborne *Bacillus* strains to control aflatoxigenic *A. parasiticus* growth and aflatoxin productions. Presence of antifungal lipopeptides was determined for the most effective strain. Six *Bacillus* spp. strains reduced *A. parasiticus* growth rate significantly. Among them, *Bacillus* spp. RC1A displayed almost complete fungal growth inhibition, and the growth rate is reduced to 0.16 mm/h with longest lag phase (31.72 h). This strain also reduced AFB1 concentration. Instrumental analysis of supernatant showed the presence of main groups of lipopeptides (surfactin, iturin A, and fengycin), and they were also confirmed by MALDIMS analysis. When they were tested separately, these fractions inhibited *A. parasiticus* growth and AFB1 production. These results highlight effective antifungal compounds produced by *Bacillus* and possible new strategies for AF contamination in crops and feeds.

Bacillus subtilis CICC 10034, *Rhodobacter sphaeroides* CGMCC 1.2182, and *Agrobacterium tumefaciens* CGMCC 1.2554 were tested for their control efficiency against patulin (PAT)-producer *Penicillium expansum* (Wang et al. 2016a, b). Additionally, PAT-removing abilities were evaluated. Among tested strains, *B. subtilis* best inhibited *P. expansum* both on apples and in vivo and in vitro studies. The rot diameter was reduced by 38% compared with the control. Other strains also display inhibition in various degrees as *A. tumefaciens* (27.6%) and *R. sphaeroides* (23.7%). None of the cell-free supernatant prevented pathogen growth. Treatments with *B. subtilis*, *R. sphaeroides*, and *A. tumefaciens* provided suppression of PAT production by 98.5, 93.7, and 94.99%, respectively. Moreover, the three antagonistic strains led to decrease in colony number of *P. expansum*. Another finding is that both live- and heat-treated cells have potential to remove PAT in liquid medium: in the result, the selected strains have potential before harvesting to control PAT-producing fungi and also during processing to remove PAT as a detoxifying agent. Among those tested strains, *Bacillus* strain has the best traits for biocontrol applications.

Wang et al. (2015) showed biocontrol potential of *Bacillus* strain for blue rot decay. This disease is mainly caused by *Penicillium expansum* in harvested sweet cherry fruits. Disease incidence and development are reduced by treatment with *B. cereus* AR156. Moreover, this practice significantly improved some enzyme activities such as chitinase and beta-1,3-glucanase in the fruit. *P. expansum* spores plasma membrane integrity was damaged, and protein and sugar leakage in the pathogen mycelia was observed. The control mechanism is found to be associated with direct fungitoxic characteristic and defense-related enzymes. Controlling this fungus can also provide prevention of PAT contamination in this fruit and its by-products.

13.5 Mechanisms of Biological Control for *Bacillus* Against Plant Fungal Pathogens

In order to maintain efficient biocontrol applications, understanding biocontrol mechanisms is a very important subject. There exist various microbial interactions (Whipps 2001). Biocontrol mechanisms can be explained by mainly three mechanisms: competition for space and nutrients, production of inhibitory substances (lipopeptides, antibiotics, volatile compounds, lytic enzymes, etc.), and induced systemic resistance. Similarly, plant growth can be improved by beneficial organisms that can act as plant probiotics (Rahman et al. 2018; Berlec 2012; Hu et al. 2017). These issues were evaluated below to understand how *Bacillus* inhibit pathogenic and mycotoxigenic fungi.

13.5.1 Competition for Space and Nutrients

Biological control exists in different ways, and one of them can be explained by competition of antagonist directly against pathogens for limited space and nutrients. Competition leads to decrease in growth, productiveness, and other activities of the competing organisms. The main sources of nutrient supply for plants are plant exudates, senesced tissue, or leachates. Generally, the soilborne plant diseases (e.g., *Fusarium* and *Pythium* species) are more exposed to competition because they only infect through mycelial contact. Some nonpathogenic microorganisms protect plants by rapid colonization and consume the main sources; therefore they prevent pathogen microbes. One of the inhibition properties for *Bacillus* species against fungal pathogens is based on this phenomenon.

There is a relationship between competition and colonization behavior. One of the modes of action of *Bacillus* genus is associated with efficient colonization, and good colonization properties provide control of fungal pathogens. Song et al. (2014) reported colonization ability of the *Bacillus* isolates on the ginseng rhizosphere. They found higher colonization when both the bacterial isolate and the pathogen (*Fusarium* cf. *incarnatum*) were inoculated together than when the bacterial isolate was inoculated alone. Cao et al. (2012) showed the role of *Bacillus subtilis* SQR 9 to control *Fusarium* wilt in cucumber by colonizing plant roots. Fusarium wilt is one of the major fungal diseases on cucumber production worldwide. The localization of bacterial cells, pattern of colonization, and survival of *B. subtilis* SQR 9 in the rhizosphere of cucumber were examined by fluorescent microscopy and explored following recovery of the green fluorescent protein labeled.

Nutrient availability is one of the important points for competition, and in the rhizosphere environment, relatively low abundance of substrates can act as a key parameter for competition. For instance, one of the essential elements is known as iron, and its limitation leads in a furious competition. Iron is mostly in ferric form in the presence of oxygen in the soil environment and known as water-insoluble at pH 7.4 (Shafi et al. 2017). Due to its low concentration at 1×10^{-8} mol/L, the growth and microorganism development cannot be supported. For survival, some microorganisms produce low molecular weight compounds called siderophores which help them to maintain their iron requirements from the environment. For biocontrol fungal plant disease, the siderophore-producing ability of *Bacillus* spp. plays a role (Yu et al. 2011).

Rajaofera et al. (2017) report that one of the fungal inhibition abilities of *Bacillus atrophaeus* HAB-5 may be associated with siderophore production and iron bioavailability and phosphate solubilization; therefore this organism could suppress fungal growth by chinase and protease activity.

13.5.2 *Direct Inhibition*

13.5.2.1 Antibiosis (Lipopeptides and Antibiotic Based)

Bacillus species synthesize a broad range of biologically active antimicrobial metabolites such as peptide compounds (Abdel-Mohsein et al. 2011; Ayed et al. 2014), enzymes, lipopeptides (Chen et al. 2008; Cawoy et al. 2015), different antibiotic compounds as a secondary metabolites, and other compounds such as volatiles (Chaves-López et al. 2015); these compounds are very effective for inhibition of different microorganisms. They have various degrees of antifungal properties (Table 13.2).

Peptide molecules represent the predominant class of *Bacillus* antibiotics. They have various sizes and consisted mainly of amino acid. However, some varieties can also contain other residues. Cyclic or linear oligopeptides, basic peptides, and aminoglycoside types are the common ones. Peptide antibiotics have low molecular weight, hydrophobic, or cyclic structures, with unusual molecules like D-amino acids. They have resistance to peptidases and proteases actions with animal and plant origin (Cawoy et al. 2011). *Bacillus brevis* and *Bacillus polymyxa* produce gramicidin S and polymyxin B peptide antibiotics, and it inhibited *Botrytis cinerea* germination in vitro and infield applications. This fungus is associated with *Botrytis* gray mold disease.

Lipopeptides are the main antimicrobial compounds produced by *Bacillus* species, and they play a role in inhibition of fungal growth (Hamley 2015; Ongena and Jacques 2008). Basically, they are amphiphilic compounds and composed of a common structure including a lipid tail linked to a short cyclic oligopeptide (Fig. 13.3). They have a broad range of inhibition activities, and this trait can be explained by physicochemical properties such as surface tension reduction, surface

Table 13.2 Antibiotic compounds by *Bacillus* against mycotoxin-producing fungi

Antibiotic	Producing bacteria	Target fungus	Mode of action	References
Antibiotic plipastatins A	*B. subtilis* NCIB 8872	*Fusarium oxysporum*, *Aspergillus flavus*	İnhibit phospholipase A2	Volpon et al. (2000)
Bacillomycin D	*B. subtilis* fmbj	*A. flavus*	Leads ınjury to cell wall and membrane	Gong et al. (2014)
Fengycin and bacillomycin	*B. subtilis* SQR9	*F. oxysporum* f. sp. *cucumerinum*	İnhibit mycelial growth and conidial germination	Cao et al. (2012)
YrvN protein-based subunit of protease	*B. subtilis* EU07	*F. oxysporum* f sp. *radicis-lycopersici*	İnactivation of AHL	Baysal et al. (2008)
Acyl-homoserine lactonase				
Fengycin, mycosubtilin, subtulune	*B. subtilis* JA	*F. graminearum*	Effective against filamentous fungi but not against bacteria and yeast	Chen et al. (2008)
		R. solani		
		P. irregulare		
		Cladosporium fulvum		
Fengycin, surfactin, and pumilacidin	*B. mojavensis*	Gram (+ve), gram (−ve), and many fungal pathogens	Antibiotic and fungicidal	Ayed et al. (2014)
Bacillomycin D	*B. subtilis* fmbJ	*Aspergillus ochraceus*	Distortion of mycelia and disruption of spores, apoptosis of *A. ochraceus* through cell damage and DNA damage, and induction of more ROS	Qian et al. (2016)

properties modification, and lipid bilayers perturbation. *Bacillus* lipopeptides are synthesized non-ribosomally, and large multienzymes play a role in the synthesis; therefore a significant heterogeneity exists according to the type and sequence of amino acid residues, the nature of the peptide cyclization, and the nature, length, and branching of the fatty acid chain. Based on amino acid sequence, lipopeptides are divided into three main groups: iturins, fengycins, and surfactins (Fig. 13.3).

The surfactins are very strong biosurfactants, and they show antibacterial activity, but limited fungitoxic properties were observed and reported (Pérez-García et al. 2011). On the contrary, iturins display strong antifungal property against a broad range of yeasts and fungi (Klich et al. 1991), but only weak antibacterial activity is reported. Fengycins also display a strong antifungal activity, specifically against filamentous fungi (Pérez-García et al. 2011).

Biocontrol properties of various *Bacillus* strains to inhibit fungal diseases that include soilborne, foliar, and post-harvest have been associated mostly with iturins and fengycins (Pérez-García et al. 2011). Due to the amphiphilic structure of lipopeptides, interaction with biological membranes exists; therefore pore formation occurs. In the recent study, lipopeptide-associated antifungal activity of selected

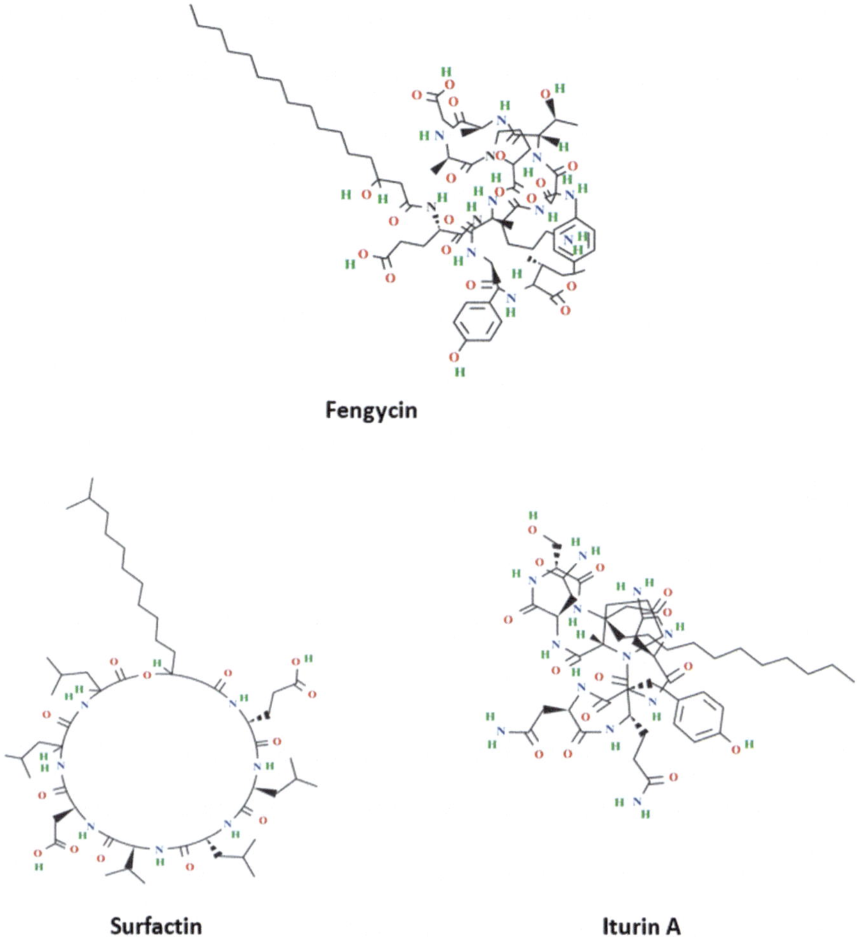

Fig. 13.3 Chemical structure of main lipopeptides

Bacillus strains against *C. acutatum* growth was determined (Yánez-Mendizábal and Falconí 2018). In this study, lipopeptide production and their antifungal-specific activity against *C. acutatum* and the presence of lipopeptide genes were confirmed (Yánez-Mendizábal and Falconí 2018). In another finding, Yuan et al. (2017) reported that lipopeptides produced from *Bacillus* strain have effects on both pathogenic and beneficial soil fungi. Interestingly, they concluded that application of these antifungal agents may cause disruption of the native microbial community diversity, and this community can be responsible for lipopeptide production.

Moreover, two antifungal antibiotics, zwittermicin A and kanosamine, were produced by *B. cereus* UW85, and these compounds are suggested to suppression of damping-off disease of alfalfa. The main responsible organism for this disease was *Phytophthora medicaginis* (Silo-Suh et al. 1994). Zwittermicin A has control properties for cucumber fruit rot (Smith et al. 1993) and prevents other plant diseases (Silo-Suh et al. 1998).

Some enzymes produced by *Bacillus* species can possess antifungal activity. Lytic enzymes can result in the degradation of the polymeric compounds like chitin, cellulose, glucans, protein, and DNA of cell wall of target organisms. Releasing substances can be used as carbon and energy sources in the following period. Especially, cell wall-degrading enzymes such as chitinase and B-1,3-gluconase are effective for fungal plant pathogens (Senol et al. 2014). Another enzyme group, oxidative enzymes such as polyphenol oxidase (PPO), peroxidase (PO), and POL, can play role in fungal inhibition due to defense-related activities.

Agarwal et al. (2017) showed that *Bacillus pumilus* MSUA3 strongly inhibited the growth of *R. solani* and *F. oxysporum*. In these inhibitions, chitinolytic enzymes and surfactin are involved. These fungi cause wilting and root rot disease in *Fagopyrum esculentum*, respectively. The action of cell-free culture supernatant was found effective to inhibit *F. oxysporum* and *R. solani* even in the heat-treated (boiled/autoclaved) cases. The possible involvement of surfactin in disease control was confirmed by colony PCR amplification of *SrfA*. Surfactin activity was found to be heat stable (Fig. 13.3).

13.5.2.2 Plant Resistance Triggering by Bacilli

Plant resistance can be induced by BCAs mainly in two ways. One of them occurs when a pathogen attacks a plant; the non-infected plant tissues become resistant following the attack; this provides plants' long-term ability to resist a wide spectrum of pathogens known as systemic acquired resistance (SAR). Additionally, nonpathogenic or beneficial microorganisms can also trigger plant defense mechanisms, and it is known as induced systemic resistance (ISR). It is used to describe reaction induced in plant by nonpathogenic microorganisms. ISR is activated by colonization of nonpathogenic plant growth-promoting bacteria. This type of ISR was first observed on carnation *(Dianthus caryophyllus)*, and susceptibility to wilt which comes from *Fusarium* sp. is reduced (Van Peer et al. 1991). In this work, the pathogen and antagonist were separated spatially, and the role of ISR was detected experimentally. Some fungi and rhizobacteria are known as inducers of resistance. In this study, accumulation of phytoalexins in stems treated with bacteria and inoculated plants with non-treated bacteria, fungus-infected plants were compared. Then, increased accumulation of phytoalexins was observed. Synthesis of this molecule can be used as indicator of improved defense mechanism in bacteria-treated plants. ISR includes mainly three steps in sequential: (i) the perception by plant cells of elicitors produced by the inducing agents, and it initiates the phenomenon; (ii) signal transduction to propagate the induced state systemically through the plant; and (iii) expression of defense mechanisms limit or inhibit pathogen penetration into the host tissues (Cawoy et al. 2011).

Defense molecules including phytoalexins, pathogenesis-related proteins such as chitinases, β-1,3-glucanases, proteinase inhibitors, and lignin for reinforcement of cell walls were associated with resistance (Cawoy et al. 2011). In the several studies, *Bacillus* genus with ISR property was described. Akram and Anjum (2011) studied the potential of *Bacillus*, viz., *B. fortis* 162 and *B. subtilus* 174, by considering ISR against *Fusarium* wilt disease in tomato. They primed roots of tomato seedlings with *Bacillus* strains. The quantification of total phenolic compounds (PPO), phenyl ammonia lyase (PAL), and peroxidase (PO) was performed to evaluate their role in ISR. Higher levels of antimicrobial phenolic compounds are accumulated, and enzyme activities (PPO, PAL, and PO) were observed for *B. subtilus* 174, and this strain is proposed for biological control of *Fusarium* wilt of tomato.

13.6 Conclusion

Among plant pathogenic fungi, mycotoxin-producing fungi can also cause different plant diseases. *Bacillus*-based biological control products have the potential for managing phytopathogenic fungi and preventing mycotoxin production. This seems an attractive, efficient, and environmentally friendly approach compared with synthetic chemicals. Application of *Bacillus*-based antagonists for managing fungi is closely related to host species and strain diversity, host specificity, mode of action, and active ingredient formulation. Knowledge on in vitro antagonist property, identification and characterization of antifungal compounds, mechanism of action and in vivo suppressions of diseases, and efficient formulations will provide better results for managing fungal diseases of plants and therefore mycotoxin control in the food chain. To date different *Bacillus* species and strains have been reported, and they have remarkable properties to develop environmentally friendly, sustainable strategies for management of fungal plant diseases. However, under field conditions, stable growth seems still a problem due to adverse environmental conditions. Advance knowledge and understanding of active compounds is one of the important issues for developing efficient formulations. Combining different BCA species with different modes of action can improve effectiveness under field conditions. Extensive studies are necessary due to the complexity of field conditions. Besides fundamental research, technical improvements are also required for large-scale productions of *Bacillus*-based BCA. Good coordination between researchers, producers, and practitioners will also promote these agro-biotechnological applications.

References

Ab Rahman SFS, Singh E, Pieterse CM, Schenk PM (2017) Emerging microbial biocontrol strategies for plant pathogens. Plant Sci 267:102–111

Abd-Allah EF, Ezzat SM, Tohamy MR (2007) *Bacillus subtilis* as an alternative biologically based strategy for controlling *Fusarium* wilt disease in tomato: a histological study. Phytoparasitica 35(5):474–478

Abdel-Mohsein HS, Sasaki T, Tada C, Nakai Y (2011) Characterization and partial purification of a bacteriocin-like substance produced by thermophilic *Bacillus licheniformis* H1 isolated from cow manure compost. Anim Sci J 82(2):340–351

Agarwal M, Dheeman S, Dubey RC, Kumar P, Maheshwari DK, Bajpai VK (2017) Differential antagonistic responses of *Bacillus pumilus* MSUA3 against *Rhizoctonia solani* and *Fusarium oxysporum* causing fungal diseases in Fagopyrum esculentum Moench. Microbiol Res 205:40–47

Akram W, Anjum T (2011) Quantitative changes in defense system of tomato induced by two strains of *Bacillus* against *Fusarium* wilt. Indian J Fundam Appl Life Sci 1:7–13

Anderson AJ, Kim YC (2018) Biopesticides produced by plant-probiotic *Pseudomonas chlororaphis* isolates. Crop Prot 105:62–69

Ashworth LJ, McMeans JL (1966) Association of *Aspergillus flavus* and aflatoxins with a greenish yellow fluorescence of cotton seed. Phytopathology 56(9):1104–1105

Ayed HB, Hmidet N, Bechet M, Chollet M, Chataigné G, Leclère V, Jacques P, Nasri M (2014) Identification and biochemical characteristics of lipopeptides from *Bacillus mojavensis* A21. Process Biochem 49(10):1699–1707

Baffoni L, Gaggia F, Dalanaj N, Prodi A, Nipoti P, Pisi A, Di Gioia D (2015) Microbial inoculants for the biocontrol of *Fusarium* spp. in durum wheat. BMC Microbiol 15(1):242

Baquião AC, Zorzete P, Reis TA, Assunção E, Vergueiro S, Correa B (2012) Mycoflora and mycotoxins in field samples of Brazil nuts. Food Control 28(2):224–229

Baysal O, Çalışkan M, Yesilova O (2008) An inhibitory effect of a new *Bacillus subtilis* strain (EU07) against *Fusarium oxysporum* f. sp. radicis-lycopersici. Physiol Mol Plant Pathol 73(1):25–32

Berlec A (2012) Novel techniques and findings in the study of plant microbiota: search for plant probiotics. Plant Sci 193:96–102

Bircan C (2005) The determination of aflatoxins in spices by immunoaffinity column extraction using HPLC. Int J Food Sci Technol 40(9):929–934

Bircan C, Koç M (2012) Aflatoxins in dried figs in turkey: a comparative survey on the exported and locally consumed dried figs for assessment of exposure. J Agric Sci Technol 14:1265–1274

Bisutti IL, Pelz J, Büttner C, Stephan D (2017) Field assessment on the influence of RhizoVital® 42 fl. and Trichostar® on strawberries in the presence of soil-borne diseases. Crop Prot 96:195–203

Bradburn N, Coker RD, Blunden G (1994) The aetiology of turkey 'X' disease. Phytochemistry 35(3):817

Calvo H, Marco P, Blanco D, Oria R, Venturini ME (2017) Potential of a new strain of *Bacillus amyloliquefaciens* BUZ-14 as a biocontrol agent of postharvest fruit diseases. Food Microbiol 63:101–110

Cao Y, Xu Z, Ling N (2012) Isolation and identification of lipopeptides produced by *B. subtilis* SQR 9 for suppressing *Fusarium* wilt of cucumber. Sci Hortic 135:32–39

Cawoy H, Bettiol W, Fickers P, Ongena M (2011) *Bacillus*-based biological control of plant diseases. In: Pesticides in the modern world-pesticides use and management. InTech, Rijkea

Cawoy H, Debois D, Franzil L, De Pauw E, Thonart P, Ongena M (2015) Lipopeptides as main ingredients for inhibition of fungal phytopathogens by *Bacillus subtilis/amyloliquefaciens*. Microb Biotechnol 8(2):281–295

Chaves-López C, Serio A, Gianotti A, Sacchetti G, Ndagijimana M, Ciccarone C, Stellarini A, Corsetti A, Paparella A (2015) Diversity of food-borne *Bacillus* volatile compounds and influence on fungal growth. J Appl Microbiol 119(2):487–499

Chen H, Wang L, Su C (2008) Isolation and characterization of lipopeptide antibiotics produced by *Bacillus subtilis*. Lett Appl Microbiol 47(3):180–186

Chulze SN, Palazzini JM, Torres AM, Barros G, Ponsone ML, Geisen R, Schmidt-Heydt M, Köhl J (2015) Biological control as a strategy to reduce the impact of mycotoxins in peanuts, grapes and cereals in Argentina. Food Addit Contam Part A 32(4):471–479

Comby M, Gacoin M, Robineau M, Rabenoelina F, Ptas S, Dupont J, Profizi C, Baillieul F (2017) Screening of wheat endophytes as biological control agents against *Fusarium* head blight using two different in vitro tests. Microbiol Res 202:11–20

Crane JM, Bergstrom GC (2014) Spatial distribution and antifungal interactions of a *Bacillus* biological control agent on wheat surfaces. Biol Control 78:23–32

da Rocha MEB, Freire FDCO, Maia FEF, Guedes MIF, Rondina D (2014) Mycotoxins and their effects on human and animal health. Food Control 36(1):159–165

Denner WHB, Gillanders TGE (1996) In: Godfrey T, West S (eds) The legislative aspects of user of industrial enzymes in the manufacture of food ingredients, 2nd edn. The Macmillan Press Ltd, Basingstoke, UK, pp 397–411

Dunlap CA, Bowman MJ, Schisler DA (2013) Genomic analysis and secondary metabolite production in *Bacillus amyloliquefaciens* AS 43.3: a biocontrol antagonist of *Fusarium* head blight. Biol Control 64(2):166–175

Durairaj K, Velmurugan P, Park JH, Chang WS, Park YJ, Senthilkumar P, Choi KM, Lee JH, Oh BT (2018) An investigation of biocontrol activity *Pseudomonas* and *Bacillus* strains against *Panax ginseng* root rot fungal phytopathogens. Biol Control 125:138–146

El-Bendary MA, Hamed HA, Moharam ME (2016) Potential of *Bacillus* isolates as bio-control agents against some fungal phytopathogens. Biocatal Agric Biotechnol 5:173–178

El-Gremi SM, Draz IS, Youssef WAE (2017) Biological control of pathogens associated with kernel black point disease of wheat. Crop Prot 91:13–19

Gong Q, Zhang C, Lu F, Zhao H, Bie X, Lu Z (2014) Identification of bacillomycin D from *Bacillus subtilis* fmbJ and its inhibition effects against *Aspergillus flavus*. Food Control 36(1):8–14

Guardado-Valdivia L, Tovar-Pérez E, Chacón-López A, López-García U, Gutiérrez-Martínez P, Stoll A, Aguilera S (2018) Identification and characterization of a new *Bacillus atrophaeus* strain B5 as biocontrol agent of postharvest anthracnose disease in soursop (*Annona muricata*) and avocado (*Persea americana*). Microbiol Res 210:26–32

Gutierrez-Monsalve JA, Mosquera S, González-Jaramillo LM, Mira JJ, Villegas-Escobar V (2015) Effective control of black Sigatoka disease using a microbial fungicide based on *Bacillus subtilis* EA-CB0015 culture. Biol Control 87:39–46

Hamley IW (2015) Lipopeptides: from self-assembly to bioactivity. Chem Commun 51(41):8574–8583

Heidarzadeh N, Baghaee-Ravari S (2015) Application of *Bacillus pumilus* as a potential biocontrol agent of *Fusarium* wilt of tomato. Arch Phytopathol Plant Protect 48(13–16):841–849

Heydari A, Pessarakli M (2010) A review on biological control of fungal plant pathogens using microbial antagonists. J Biol Sci 10(4):273–290

Hu J, Wei Z, Weidner S, Friman VP, Xu YC, Shen QR, Jousset A (2017) Probiotic *Pseudomonas* communities enhance plant growth and nutrient assimilation via diversity-mediated ecosystem functioning. Soil Biol Biochem 113:122–129

Hussein HS, Brasel JM (2001) Toxicity, metabolism, and impact of mycotoxins on humans and animals. Toxicology 167(2):101–134

IARC Working Group on the Evaluation of the Carcinogenic Risk of Chemicals to Humans (1993) IARC monographs on the evaluation of the carcinogenic risk of chemicals to humans, vol 35. IARC, Lyon

Islam MR, Jeong YT, Lee YS, Song CH (2012) Isolation and identification of antifungal compounds from *Bacillus subtilis* C9 inhibiting the growth of plant pathogenic fungi. Mycobiology 40(1):59–65

Jangir M, Pathak R, Sharma S, Sharma S (2018) Biocontrol mechanisms of Bacillus sp., isolated from tomato rhizosphere, against *Fusarium oxysporum* f. sp. *lycopersici*. Biol Control 123:60–70

Jiang C, Shi J, Liu Y, Zhu C (2014) Inhibition of *Aspergillus carbonarius* and fungal contamination in table grapes using *Bacillus subtilis*. Food Control 35(1):41–48

Juan C, Zinedine A, Molto JC, Idrissi L, Manes J (2008) Aflatoxins levels in dried fruits and nuts from Rabat-Salé area, Morocco. Food Control 19(9):849–853

Kamika I, Takoy LL (2011) Natural occurrence of Aflatoxin B1 in peanut collected from Kinshasa, Democratic Republic of Congo. Food Control 22(11):1760–1176

Khan N, Maymon M, Hirsch AM (2017) Combating *Fusarium* infection using *Bacillus*-based antimicrobials. Microorganisms 5(4):75

Kildea S, Ransbotyn V, Khan MR, Fagan B, Leonard G, Mullins E, Doohan FM (2008) *Bacillus megaterium* shows potential for the biocontrol of *septoria tritici* blotch of wheat. Biol Control 47(1):37–45

Klich MA (2007) *Aspergillus flavus:* the major producer of aflatoxin. Mol Plant Pathol 8(6):713–722

Klich MA, Lax AR, Bland JM (1991) Inhibition of some mycotoxigenic fungi by iturin A, a peptidolipid produced by *Bacillus subtilis*. Mycopathologia 116(2):77–80

Lee S-G (2004) Fusarium species associated with ginseng (*Panax ginseng*) and their role in the root-rot of ginseng plant. Res Plant Dis 10(4):248–259

Lee T, Park D, Kim K, Lim SM, Yu NH, Kim S, Kim HY, Jung KS, Jang JY, Park JC, Ham H, Lee S, Hong SK, Kim JC (2017) Characterization of *Bacillus amyloliquefaciens* DA12 showing potent antifungal activity against mycotoxigenic *Fusarium* species. Plant Pathol J 33(5):499–507

Leelasuphakul W, Hemmanee P, Chuenchitt S (2008) Growth inhibitory properties of *Bacillus subtilis* strains and their metabolites against the green mold pathogen (*Penicillium digitatum Sacc.)* of citrus fruit. Postharvest Biol Technol 48(1):113–121

Leslie JF, Summerell BA (2008) The *Fusarium* laboratory manual. Wiley, New York

Logrieco A, Moretti A, Perrone G, Mulè G (2007) Biodiversity of complexes of mycotoxigenic fungal species associated with *Fusarium* ear rot of maize and *Aspergillus* rot of grape. Int J Food Microbiol 119(1–2):11–16

Lucon CMM, Guzzo SD, De Jesus CO, Pascholati SF, De Goes A (2010) Postharvest harpin or *Bacillus thuringiensis* treatments suppress citrus black spot in 'Valencia' oranges. Crop Prot 29(7):766–772

Luo Y, Yoshizawa T, Katayama T (1990) Comparative study on the natural occurrence of *Fusarium* mycotoxins (trichothecenes and zearalenone) in corn and wheat from high-and low-risk areas for human esophageal cancer in China. Appl Environ Microbiol 56(12):3723–3726

Majeed S, Iqbal M, Asi MR, Iqbal SZ (2013) Aflatoxins and ochratoxin A contamination in rice, corn and corn products from Punjab, Pakistan. J Cereal Sci 58(3):446–450

Marin S, Ramos AJ, Cano-Sancho G, Sanchis V (2013) Mycotoxins: occurrence, toxicology, and exposure assessment. Food Chem Toxicol 60:218–237

Marroquín-Cardona AG, Johnson NM, Phillips TD, Hayes AW (2014) Mycotoxins in a changing global environment – a review. Food Chem Toxicol 69:220–230

Mishra J, Tewari S, Singh S, Arora NK (2015) Biopesticides: where we stand? In: Plant microbes symbiosis: applied facets. Springer, New Delhi, pp 37–75

Mnif I, Ghribi D (2015) Potential of bacterial derived biopesticides in pest management. Crop Prot 77:52–64

Moreira RR, Nesi CN, De Mio LLM (2014) *Bacillus* spp. and *Pseudomonas putida* as inhibitors of the *Colletotrichum acutatum* group and potential to control Glomerella leaf spot. Biol Control 72:30–37

Moretti A, Ferracane L, Somma S, Ricci V, Mule G, Susca A, Ritieni A, Logrieco AF (2010) Identification, mycotoxin risk and pathogenicity of *Fusarium* species associated with fig endosepsis in Apulia, Italy. Food Addit Contam 27(5):718–728

Muthomi JW, Ndung'u JK, Gathumbi JK, Mutitu EW, Wagacha JM (2008) The occurrence of *Fusarium* species and mycotoxins in Kenyan wheat. Crop Prot 27(8):1215–1219

Nguyen PA, Strub C, Fontana A, Schorr-Galindo S (2017) Crop molds and mycotoxins: alternative management using biocontrol. Biol Control 104:10–27

Ongena M, Jacques P (2008) *Bacillus* lipopeptides: versatile weapons for plant disease biocontrol. Trends Microbiol 16(3):115–125

Palazzini JM, Dunlap CA, Bowman MJ, Chulze SN (2016) *Bacillus velezensis* RC 218 as a biocontrol agent to reduce *Fusarium* head blight and deoxynivalenol accumulation: genome sequencing and secondary metabolite cluster profiles. Microbiol Res 192:30–36

Palumbo JD, O'Keeffe TL, Abbas HK (2008) Microbial interactions with mycotoxigenic fungi and mycotoxins. Toxin Rev 27(3–4):261–285

Pan HQ, Li QL, Hu JC (2017) The complete genome sequence of *Bacillus velezensis* 9912D reveals its biocontrol mechanism as a novel commercial biological fungicide agent. J Biotechnol 247:25–28

Patriarca A, Pinto VF (2017) Prevalence of mycotoxins in foods and decontamination. Curr Opin Food Sci 14:50–60

Pereyra MG, Martínez MP, Petroselli G, Balsells RE, Cavaglieri LR (2018) Antifungal and aflatoxin-reducing activity of extracellular compounds produced by soil *Bacillus* strains with potential application in agriculture. Food Control 85:392–399

Pérez-García A, Romero D, De Vicente A (2011) Plant protection and growth stimulation by microorganisms: biotechnological applications of Bacilli in agriculture. Curr Opin Biotechnol 22(2):187–193

Pitt J (2000) Toxigenic fungi and mycotoxins. Br Med Bull 56:184–192

Qian S, Lu H, Sun J, Zhang C, Zhao H, Lu F, Bie X, Lu Z (2016) Antifungal activity mode of *Aspergillus ochraceus* by bacillomycin D and its inhibition of ochratoxin A (OTA) production in food samples. Food Control 60:281–288

Rahman M, Sabir AA, Mukta JA, Khan MMA, Mohi-Ud-Din M, Miah MG, Rahman M, Islam MT (2018) Plant probiotic bacteria *Bacillus* and *Paraburkholderia* improve growth, yield and content of antioxidants in strawberry fruit. Sci Rep 8(1):2504

Rahmani A, Jinap S, Soleimany F (2010) Validation of the procedure for the simultaneous determination of aflatoxins ochratoxin A and zearalenone in cereals using HPLC-FLD. Food Addit Contam Part A 27(12):1683–1693

Rajaofera MJN, Jin PF, Fan YM, Sun QQ, Huang WK, Wang WB, Shen HY, Zhang CH, Lin WB, Liu FC, Zheng FC, Miao WG (2017) Antifungal activity of the bioactive substance from *Bacillus atrophaeus* strain HAB-5 and its toxicity assessment on *Danio rerio*. Pestic Biochem Physiol 147:153–161

Reddy KRN, Reddy CS, Muralidharan K (2009) Detection of *Aspergillus* spp. and aflatoxin B1 in rice in India. Food Microbiol 26(1):27–31

Rocha FYO, de Oliveira CM, da Silva PRA, de Melo LHV, do Carmo MGF, Baldani JI (2017) Taxonomical and functional characterization of *Bacillus* strains isolated from tomato plants and their biocontrol activity against races 1, 2 and 3 of *Fusarium oxysporum* f. sp. *Lycopersici*. Appl Soil Ecol 120:8–19

Schöneberg T, Jenny E, Wettstein FE, Bucheli TD, Mascher F, Bertossa M, Musa T, Seifert K, Gräfenhan T, Keller B, Vogelgsang S (2018) Occurrence of *Fusarium* species and mycotoxins in Swiss oats—impact of cropping factors. Eur J Agron 92:123–132

Senol M, Nadaroglu H, Dikbas N, Kotan R (2014) Purification of Chitinase enzymes from *Bacillus subtilis* bacteria TV-125, investigation of kinetic properties and antifungal activity against *Fusarium culmorum*. Ann Clin Microbiol Antimicrob 13(1):35

Senthil R, Prabakar K, Rajendran L, Karthikeyan G (2011) Efficacy of different biological control agents against major postharvest pathogens of grapes under room temperature storage conditions. Phytopathol Mediterr 50(1):55–64

Shafi J, Tian H, Ji M (2017) *Bacillus* species as versatile weapons for plant pathogens: a review. Biotechnol Biotechnol Equip 31(3):446–459

Shoda M (2000) Bacterial control of plant diseases. J Biosci Bioeng 89(6):515–521

Siahmoshteh F, Hamidi-Esfahani Z, Spadaro D, Shams-Ghahfarokhi M, Razzaghi-Abyaneh M (2018) Unraveling the mode of antifungal action of *Bacillus subtilis* and *Bacillus amyloliquefaciens* as potential biocontrol agents against aflatoxigenic *Aspergillus parasiticus*. Food Control 89:300–307

Silo-Suh LA, Lethbridge BJ, Raffel SJ, He HY, Clardy J, Handelsman J (1994) Biological activities of two fungistatic antibiotics produced by *Bacillus cereus* Uw85. Appl Environ Microbiol 60(6):2023–2030

Silo-Suh LA, Stabb EV, Raffel SJ, Handelsman J (1998) Target range of Zwittermicin A, an aminopolyol antibiotic from *Bacillus cereus*. Curr Microbiol 37(1):6–11

Smith KP, Havey MJ, Handelsman J (1993) Suppression of cottony leak of cucumber with *Bacillus cereus* strain Uw85. Plant Dis 77:139–142

Song M, Yun HY, Kim YH (2014) Antagonistic *Bacillus* species as a biological control of ginseng root rot caused by *Fusarium cf. incarnatum*. J Ginseng Res 38(2):136–145

Suárez-Estrella F, Arcos-Nievas MA, López MJ, Vargas-García MC, Moreno J (2013) Biological control of plant pathogens by microorganisms isolated from agro-industrial composts. Biol Control 67(3):509–515

Tekinşen KK, Eken HS (2008) Aflatoxin M1 levels in UHT milk and kashar cheese consumed in Turkey. Food Chem Toxicol 46(10):3287–3289

Tola M, Kebede B (2016) Occurrence, importance and control of mycotoxins: a review. Cogent Food Agric 2(1):1191103

Van der Merwe KJ, Steyn PS, Fourie L, Scott DB, Theron JJ (1965) Ochratoxin A, a toxic metabolite produced by *Aspergillus ochraceus*. Nature 205(4976):1112

Van Peer R, Niemann GJ, Schippers B (1991) Induced resistance and phytoalexin accumulation in biological control of *Fusarium* wilt of carnation by *Pseudomonas* sp. strain WCS 417 r. Phytopathology 81(7):728–734

Volpon L, Besson F, Lancelin JM (2000) NMR structure of antibiotics plipastatins A and B from *Bacillus subtilis* inhibitors of phospholipase A2. FEBS Lett 485(1):76–80

Waewthongrak W, Pisuchpen S, Leelasuphakul W (2015) Effect of *Bacillus subtilis* and chitosan applications on green mold (*Penicillium digitatum Sacc.*) decay in citrus fruit. Postharvest Biol Technol 99:44–49

Wang L, Jin P, Wang J, Jiang L, Zhang S, Gong H, Liu H, Zheng Y (2015) In vitro inhibition and in vivo induction of defense response against *Penicillium expansum* in sweet cherry fruit by postharvest applications of *Bacillus cereus* AR156. Postharvest Biol Technol 101:15–17

Wang B, Shen Z, Zhang F, Raza W, Yuan J, Huang R, Ruan Y, Li R, Shen Q (2016a) *Bacillus amyloliquefaciens* strain W19 can promote growth and yield and suppress *Fusarium* wilt in banana under greenhouse and field conditions. Pedosphere 26(5):733–744

Wang Y, Yuan Y, Liu B, Zhang Z, Yue T (2016b) Biocontrol activity and patulin removal effects of *Bacillus subtilis*, *Rhodobacter sphaeroides* and *Agrobacterium tumefaciens* against *Penicillium expansum*. J Appl Microbiol 121(5):1384–1393

Whipps JM (2001) Microbial interactions and biocontrol in the rhizosphere. J Exp Bot 52:487–511

Winks BL, Williams YN (1965) A wilt of strawberry caused by a new form of *Fusarium oxysporum*. Qld J Agric Anim Sci 22:475–479

Xiao W, Yan PS, Wu HQ, Lin F (2014) Antagonizing *Aspergillus parasiticus* and promoting peanut growth of *Bacillus* isolated from peanut geocarposphere soil. J Integr Agric 13(11):2445–2451

Yánez-Mendizábal V, Falconí CE (2018) Efficacy of *Bacillus* spp. to biocontrol of anthracnose and enhance plant growth on Andean lupin seeds by lipopeptide production. Biol Control 122:67–75

Yu X, Ai C, Xin L, Zhou G (2011) The siderophore-producing bacterium, *Bacillus subtilis* CAS15 has a biocontrol effect on *Fusarium* wilt and promotes the growth of pepper. Eur J Soil Biol 47(2):138–145

Yuan J, Zhao M, Li R, Huang Q, Rensing C, Shen Q (2017) Lipopeptides produced by *B. amyloliquefaciens* NJN-6 altered the soil fungal community and non-ribosomal peptides genes harboring microbial community. Appl Soil Ecol 117:96–105

Yuttavanichakul W, Lawongsa P, Wongkaew S, Teaumroong N, Boonkerd N, Nomura N, Tittabutr P (2012) Improvement of peanut rhizobial inoculant by incorporation of plant growth promoting rhizobacteria (PGPR) as biocontrol against the seed borne fungus, *Aspergillus niger*. Biol Control 63(2):87–97

Zain ME (2011) Impact of mycotoxins on humans and animals. J Saudi Chem Soc 15(2):129–144

Zalila-Kolsi I, Mahmoud AB, Ali H, Sellami S, Nasfi Z, Tounsi S, Jamoussi K (2016) Antagonist effects of *Bacillus* spp. strains against *Fusarium graminearum* for protection of durum wheat (*Triticum turgidum* L. subsp. *durum*). Microbiol Res 192:148–158

Zhao P, Quan C, Wang Y, Wang J, Fan S (2014) *Bacillus amyloliquefaciens* Q-426 as a potential biocontrol agent against *Fusarium oxysporum* f. sp. spinaciae. J Basic Microbiol 54(5):448–456

Zinedine A, Soriano JM, Molto JC, Manes J (2007) Review on the toxicity, occurrence, metabolism, detoxification, regulations and intake of zearalenone: an oestrogenic mycotoxin. Food Chem Toxicol 45(1):1–18

Chapter 14
Application Method and Efficacy of *Bacillus* spp. in Mitigating Abiotic and Biotic Stresses and Enhancing Plant Performance

M. Mahfuz Rahman and Lee Ann Moccaldi

14.1 Introduction

Higher demand of food supply for increasing world population led to synthetic inputs (fertilizer, growth regulator, pesticide) based crop production system to maximize yield by boosting growth and controlling major pests and plant diseases (Godfray et al. 2010). This situation made consumers more and more concerned about the environment, water supply, their health, and food safety. Now it is felt by everyone that this world needs effective, environmentally smart agricultural technologies that are safe for people, use less synthetic inputs, and protect natural resources. To overcome this challenge of increasing food production with a significant reduction of agrochemical use and environmental pollution (especially from synthetic chemicals), a great deal of interest and research have been devoted to beneficial microorganisms/biostimulants in recent days. Many researchers are working to develop these beneficial biological agents as key tools in grower toolboxes and help growers use them effectively. One part of the effort is focused on identifying microbes that enhance crop growth under a range of adverse environmental conditions (e.g., low soil moisture and/or fertility, high soil salinity, temperature extremes) that may ordinarily reduce yield and quality of produce. The other part focuses on identifying methods that allow plant-microbe interactions to be most useful to growers. The science pertaining beneficial plant-microbe interactions is promising. Years of experiments in controlled environments suggest that the potential upside in production will be real, i.e., inoculated crops will outperform non-inoculated ones. There are numerous examples of positive effects on plant growth and fruit quality enhancement as well as disease resistance by biostimulants. As a result, the use of biostimulants is becoming more common in plant production

M. M. Rahman (✉) · L. A. Moccaldi
WVU Extension Service, Morgantown, WV, USA
e-mail: mm.rahman@mail.wvu.edu

M. T. Islam et al. (eds.), *Bacilli and Agrobiotechnology: Phytostimulation and Biocontrol*, Bacilli in Climate Resilient Agriculture and Bioprospecting,
https://doi.org/10.1007/978-3-030-15175-1_14

methods, with the intent of modifying physiological processes in plants to optimize productivity, and due to the fact that crop production challenges such as growing in fluctuating temperatures, drought, excessive moisture, saline soils, acidic or basic soils, or other abiotic stresses are often difficult for the grower to overcome otherwise. Out of many potential remedial measures, soil microorganisms that increase the nutrient- and water-use efficiency and uptake capacity are considered most cost-effective and benign to environment (Armada et al. 2014). If applied early, before environmental stressors are present, bacterial biostimulants such as various strains belonging to *Bacillus* spp. can provide beneficial effects by inducing production of stress tolerance proteins in plant system. Biostimulant organisms act as a buffer in the area close to the root zone where the plant is most often affected by stress. Additionally, these products can aid in resistance to disease causing organisms. A good body of current literatures point to the success of *Bacillus* species and their strains in greenhouse and some cases in field conditions (Kumar et al. 2012b). Scientists have also begun working to unravel the mechanisms involved with positive influence of biostimulants in the production of many different crops. An appreciable number of scientists have been investigating native strains of bacteria associated with target crops including different methods of application of the bacteria to produce best results in greenhouse and field conditions. Name of the major groups of microbes involved with biostimulation may vary depending on their precise role. One major group known as plant probiotic bacteria are naturally occurring plant-associated microorganisms that enhance the growth of the host plants including yield and may suppress diseases when applied in adequate amounts (Borriss 2016). Major genera of plant growth-promoting probiotic bacteria include *Bacillus*, *Paraburkholderia*, *Pseudomonas*, *Acinetobacter*, *Alcaligenes*, *Arthrobacter*, *and Serratia*. Other than microbial biostimulants, numerous commercially available biostimulants belonging to three major non-microbial categories have been described. Humic substances originating predominantly from lignified cell walls that may act as cytokinin or auxin form the first group (Sanderson and Jameson 1986). Second group includes marine algae extracts from seaweed *Ascophyllum nodosum* L. This is considered the most widely used seaweed in agriculture that influences plant growth as cytokinins and auxins due to the presence of special chemical components (Crouch and Van Staden 1993). Animal protein hydrolysates form the third group containing a range of compounds of organic nitrogenous nature known as amino acids (monopeptides) with various molecular weights (Polo et al. 2006). Simultaneous application of both microbial and non-microbial biostimulants for obtaining additive/synergistic effect has been made by many investigators in recent years. Multiple reports mentioned synergistic effect by utilizing both types of plant biostimulants with regard to plant growth and yield enhancements (Nikbakht et al. 2014; Rouphael et al. 2017). However, reports on synergistic use of *Bacillus*-based microbial biostimulant with non-microbial biostimulants are still a scarce. This review updates our knowledge on application methods of *Bacillus*-based microbial biostimulant either as a stand-alone growth-promoting agent or synergistically with microbial and non-microbial biostimulants early in the plant growing stage.

14.2 Influence of Application Methods on Efficacy of *Bacillus* spp.

Like any other biofungicides, *Bacillus* and *Bacillus*-based products are live organisms that interact with the plant growing environment to provide beneficial effects. Plant growth promotion can be facilitated by these organisms through a wide array of modes of action such as enhanced nutrient availability, abiotic and biotic stress tolerance, induction of growth hormone, and systemic resistance. Thus, application method should be well aligned and adjusted based on the situation and targeted crop.

14.2.1 Direct Application on Nutrient Source or Growing Medium

This approach is specifically used in a situation where nutrients are not in the form for easy plant uptake. High demand of food and shrinking arable land in the future will compel growers using less than ideal growing condition. In areas where nutrients are locked in parent materials or growers will have to use raw nutrient source such as rock phosphate, *Bacillus* suspension or *Bacillus*-based products will have to be applied directly on those materials or in the crop growing areas or medium prior to planting. The following examples directly relate to soil fertility or improve fertilizer efficiency by *Bacillus* spp.

Microorganisms in the soil play an important role in the availability of nutrients to plants. Soil microbes decompose or oxidize the chemical compounds in parent materials or organic matter and release nutrients from these compounds in soluble forms for easy uptake by plants. Among the nutrients, P and K are two major ones that *Bacillus* can make available to plants in the absence or lower rate of applications from synthetic sources (Raj 2004; Archana et al. 2012, 2013; Han and Lee 2006). Regardless of the source of P applied to soil, 90% is precipitated by forming complexes with iron, aluminum, and calcium depending on the soil pH (Gyaneshwar et al. 2002). Several *Bacillus* spp. have been shown to solubilize insoluble soil organic and inorganic P through production of organic acids and enzymes (Khan et al. 2007; Panhwar et al. 2009, 2012). Among different possible mechanisms by which bacteria solubilize P, organic and inorganic acid production, release of H+ accompanying NH^{+4}, and production of exo-polysaccharides, acid phosphatases, phytases, and H_2CO_3 are important (Awasthi et al. 2011; Zaidi et al. 2009). Effective microorganisms include the bacteria *Bacillus megaterium*, *B. circulans*, *B. subtilis*, and *B. polymyxa* that can produce organic acids (Jana 2003). Similar situation does exist with K, as 90–98% reserves in soil system are non-exchangeable mineral sources. Efficient rhizospheric microbes (ERMs) are needed to effectively dissolve this mineral and make it available to plants. A diverse group of ERMs such as rhizobacteria (*Bacillus edaphicus*, *B. mucilaginosus*, *Acidithiobacillus ferrooxidans*, *B. circulans*, *Paenibacillus* sp.) is involved in K mineral (orthoclase, muscovite, feldspar, biotite, mica, illite) solubilization. These bacteria can solubilize K into plant available form to some extent, but only few bacterial strains such as *B. mucilaginosus* and *B. edaphicus* are highly efficient in solubilizing K minerals (Meena et al. 2015; Sheng and He 2006; Rajawat et al. 2012). Several studies evaluated the impact of KSMs on plant growth and K mineral solubilization (Parmar and Sindhu 2013; Prajapati et al. 2013; Sugumaran and Janarthanam 2007). Inoculation of *Bacillus mucilaginosus* in nutrient-limited soil showed positive response on growth and yield of eggplant. In addition, maximum K release and uptake were observed in soil amended with K mineral and *B. mucilaginosus* (Basak and Biswas 2009; Han and Lee 2005a, b). Products containing P solubilizer (*Bacillus megaterium*) and K solubilizer (*Bacillus mucilaginous*) (supplied by V-mark Resources, Hong Kong) are now available in the market. These products can be used with other agrochemicals to enhance efficacy and obtain higher-quality produce (Dasan 2012).

14.2.1.1 Inoculation of Transplant Production Mix

Vegetables and other specialty crops are increasingly dependent on transplants that are taken to the production field. This is more common in the areas where growing season is relatively short and transplants are produced in the controlled environment such as greenhouses. *Bacillus*-based products can be mixed directly by agitating the mix in a drum agitator. For a more homogeneous application of the organism, solution or suspension can be sprayed on planting mix laid on a bench. However, survival and colonization of planting mix can be augmented by making a less competitive environment through sterilizing the mix. Many investigators found that the most durable and highest rates of colonization of the rhizosphere by *Bacillus subtilis* were attained in artificial substrates or if the substrate had been sterilized before the application of the bacteria (Batinic et al. 1998; Krebs et al. 1998; Grosch et al. 1996). In these investigations in a competitive environment without sterilization of the medium, *Bacillus subtilis* FZB24® bacteria population showed a distinct decrease in the course of time as a percentage of the total microorganism. Rahman and Jett compared tomato transplant vigor enhancement by growing tomato transplants in sterilized and non-sterilized inoculated planting mix. Transplants grown on pasteurized plus *Bacillus* inoculated medium showed significantly higher vigor compared with non-pasteurized 35 days after seed germination (Fig. 14.1).

14.2.2 Seed Treatment/Seed Dressing

Seed treatment has been the most widely used application method for *Bacillus* strains recovered from all around the world as it ensures early root colonization by this PGPB. Early colonization of roots can competitively preclude pathogenic microorganism invasion through making a barrier to infection site and space (Bais et al. 2002). *Bacillus* strains that have shown rhizosphere competence should

Fig. 14.1 Tomato seedling vigor enhancement due to a combination treatment of seed dressing and planting mix inoculation by Serenade as affected by rate and pasteurization of medium; top panel, pasteurized medium; bottom panel, non-pasteurized; (**a**) 2.1 mL Serenade/L medium; (**b**) 0.21 mL/L; (**c**) 0.0021 mL/L

specifically be applied as seed treatment. Rhizosphere competent Bacilli can also effectively compete with plant pathogens for nutrients from seed and root exudates that normally attract and stimulate pathogen multiplication. Bacilli used as seed treatment are normally efficient in suppressing soilborne pathogens through antibiosis (production of molecules that directly affect other organisms negatively by toxicity or growth inhibition). Additional protection may be provided through development of induced systemic resistance in the host plants as one of the possible mechanisms (Mia and Shamsuddin 2013; Islam and Hossain 2013). As seed treatment facilitates root and rhizosphere colonization by Bacilli, these organisms can play a significant role in preventing root and crown diseases of plants caused by various soilborne pathogens. Examples of major disease suppression by Bacillus are shown below.

14.2.2.1 Seed Treatment by Bacilli Prevents Soilborne Diseases

Various *Bacillus* species can control plant disease caused by fungal pathogens by producing antibiotic metabolites, suppressing plant pathogens, and competing for nutrients and niches (Egamberdieva et al. 2011). They are often used to combat fungal disease in economically important crops (Arrebola et al. 2010). While numerous positive results of disease suppression by *Bacillus* are available in the literature, a few examples include antagonistic activities against soilborne pathogens *Aspergillus flavus* (Moyne et al. 2001), *Penicillium* spp., and *Fusarium oxysporum* (Karuppiah and Rajaram 2011). Araujo et al. (2012) found antagonistic activity of some *Bacillus* species isolated from Brazil to *Fusarium oxysporum* and *Colletotrichum truncatum*. Singh et al. (2008) reported that *B. subtilis* showed promising results against several pathogens causing important crop diseases such as charcoal rot caused by *Macrophomina phaseolina* (Romero et al. 2007), tomato root rot caused by *Fusarium oxysporum* f. sp. *lycopersici* (Baysal et al. 2008; Nihorimbere et al. 2010), root rot of cauliflower caused by *Pythium ultimum* (Abdelzaher 2003), and avocado root rot disease caused by *Rosellinia necatrix* (Cazorla et al. 2007). Several other *Bacillus* species showed disease control potential of tomato Fusarium wilt in in vivo studies, e.g., *B. cereus* (18.75% and 81.2%), *B. amyloliquefaciens* (25% and 75%), *B. pumilus* (37.5% and 62.5%), and *B. subtilis* (37.5% and 62.5%) compared to the negative and positive control with 100% and 0%, respectively (Ajilogba et al. 2013). Strain B44 of *Bacillus* sp., isolated from the tomato rhizosphere, showed a 36% reduction in *Fusarium* disease incidence in tomato under greenhouse conditions (Jangir et al. 2018). *Bacillus subtilis* strain QST 713 decreased *Fusarium* wilt of banana caused by *F. oxysporum cubense* due to treatment (Figueiredo et al. 2010). Zhang et al. (2012) observed that co-inoculation of cotton with *Bacillus vallismortis* HJ-5 and *Glomus versiforme* (AM) decreased disease symptoms caused by *V. dahliae*. The plants inoculated with *Bacillus* sp. culture showed 65% reduction in disease incidence compared to the seed treated with pathogen alone. In another study, *Bacillus amyloliquefaciens* inhibited the growth of *Fusarium oxysporum* and *Ralstonia solanacearum* to some extent and effectively inhibited *R. solanacearum* in the rhizosphere soil of eggplant (Chen et al. 2014). The seed treatment of sunflower seeds with *B. licheniformis* resulted in significant reduction in percentage of infection of *R. solani* damping-off (from 60% to 25%) compared with the pathogen alone (Kamil et al. 2007). Salt-tolerant *B. amyloliquefaciens* BcA12 significantly reduced damping-off of cotton caused by *R. solani* (Egamberdieva and Jabborova 2013). In other studies, Safiyazov et al. (1995) observed that *B. subtilis* 23 and *B. megatherium* 26 were able to control cotton diseases caused by *Xanthomonas malvacearum*, *Rhizoctonia solani*, *Fusarium vasinfectum*, and *Verticillium dahlia* under saline soil conditions. Kumar et al. (2015) isolated and screened soil bacterial isolates for high temperature (50 °C), salinity (7% NaCl), and drought (−1.2 MPa), and observed that *Bacillus* strains were more stress tolerant than other microorganisms. *Bacillus* strains possessed promising antagonistic activity against *Sclerotium rolfsii*, *Macrophomina phaseolina*, *Rhizoctonia solani*, and *Fusarium oxysporum* f. sp. *ricini*. These reports imply that stress-tolerant *Bacillus* species could be a useful approach for protecting plants from various soilborne pathogens under abiotic stress conditions. Bharathi et al. (2004) reported that combined inoculation of *P. fluorescens* (Pf-1) with *B. subtilis* reduced fruit rot infection in chillies caused by *Colletotrichum truncatum* by 51% over control. The strain also increased shoot length (56.14%), number of flowers (270.59%), and fruits (133.3%) compared to uninoculated control plants. In a similar study, chilli seed treatment with *Bacillus amyloliquefaciens* resulted in maximum enhancement of seed germination, seedling vigor, and an increase in growth parameters (Gowtham et al. 2018). Additionally, significant disease protection of 71% against anthracnose (*Colletotrichum truncatum*) was observed in chilli plants pretreated with *Bacillus amyloliquefaciens* under greenhouse conditions (Gowtham et al. 2018). A seed, soil mix, and irrigation treatment with *Bacillus subtilis*

(Serenade) significantly reduced the number of leaves showing symptoms of *Verticillium* in tomato and increased yield (Rahman and Jett 2016). Danielsson et al. (2007) reported enhanced resistance in oilseed rape to fungal pathogens due to seed priming with *Bacillus amyloliquefaciens* 5113. Out of four strains tested, two provided protection against multiple fungal pathogens and improved number of seedling survival, whereas two other strains did not provide any positive effect indicating the need for selecting effective strain. In addition with preventing soilborne diseases, early colonization of root system or endophytic growth of *Bacillus* can stimulate plant growth by improving vigor or providing tolerance against abiotic stresses such as moisture or temperature extremes (Zhao et al. 2011).

14.2.2.2 Biostimulation of Plant Due to Seed Treatment

A biostimulant can be defined as "a formulated product of biological origin that improves plant productivity as a consequence of the novel or emergent properties of the complex of constituents, and not as a sole consequence of the presence of known essential plant nutrients, plant growth regulators, or plant protective compounds" (Yakhin et al. 2017). It is also suggested that these products function at low doses, be ecologically benign, and have reproducible benefits in agriculture (Herve 1994). Many microorganisms from the rhizosphere can be of benefit to plant growth and plant health and are often referred to as plant growth-promoting rhizobacteria (PGPR) (Schippers 1992; 1995; Kilian et al. 2000). Different bacteria species and their strains have been studied and used in production methods as biostimulants (Calvo et al. 2010; Krebs et al. 1998; Idris et al. 2004; Chen et al. 2009). Among PGPRs, the genus *Bacillus* contains gram-positive, endospore-forming bacteria which can be thermophilic, halophilic, acidophilic, and alkalophilic that can live in extreme climatic conditions. Their effects are a complex, cumulative result of various interactions between plant, pathogen, antagonists, and environmental factors (Schippers 1992). Figure 14.2 shows the biostimulation effect of *Bacillus amyloliquefaciens* ssp. *plantarum* resulting from complex interactions among *Bacillus*, plant, and pathogen.

Improved seed and tuber germination, rooting, growth, biomass production, earliness of vegetative growth, generative development, and yield have all been reported to be improved with the use of *Bacillus amyloliquefaciens* ssp. *plantarum*. It can also enhance nutrient availability and nutrient uptake based on plant growth hormone and enzyme production. Increased plant strength, reduced disease pressure, and induced resistance in plants have also been attributed to application of *Bacillus amyloliquefaciens* ssp. *plantarum* in crops. For example, a recent greenhouse and field study on tomato indicated that inoculation with *B. subtilis* QST 713 (Serenade) showed increased tomato yield, seedling vigor (in autoclaved soil), and fruit yield (Rahman and Jett 2016). Multiple examples of increased growth and yield are listed in Table 14.1.

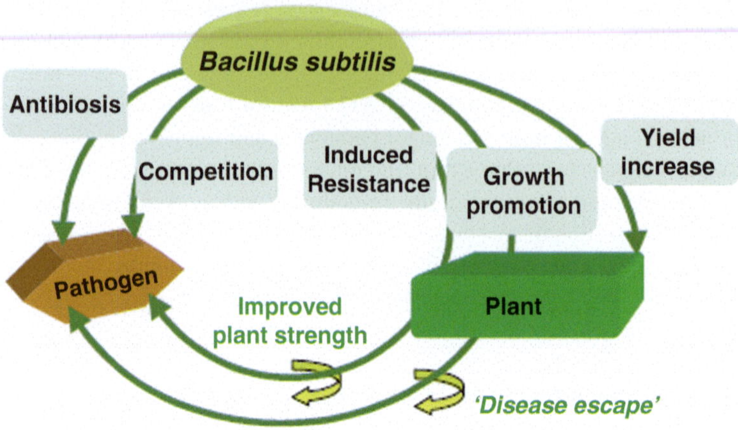

Fig. 14.2 Overview of the modes of action of *Bacillus subtilis* (now *B. amyloliquefaciens* ssp. *plantarum*) and the interaction between *Bacillus subtilis*, the plant, and the pathogen. (Kilian et al. 2000)

Table 14.1 Application rate, active ingredient, method of Serenade, and other treatment application in controlled environment and in field to lower *Verticillium* wilt severity and enhance tomato yield

Treatment and rate	How applied	Active ingredient	Total yield (lb/3 plants)	%Yield increase over non-treated
Non-treated check	N/A	N/A	32.1 c	–
Mustard cover crop "Caliente" 10 lb/A	Incorporated	Isothiocyanate (ITC)	36.9 bc	15
Mustard seed meal 4000 lb/A	Incorporated	Isothiocyanate (ITC)	42.9 b	36
Serenade Soil 1.34ASO 0.25 fl oz/oz	On seeds	*Bacillus subtilis* strain QST 713	42.0 b	36
1 fl oz/gal	Potting mix			
1.5% (v/v)	Transplant water			
Prestop 32WPO 0.25 oz/oz	On seeds	*Gliocladium catenulatum*	39.9 b	26
1 oz/gal	Potting mix			
0.5% (w/v)	Transplant water			
Grafted plant, "Maxifort" rootstock	N/A	Resistant "Maxifort" rootstock	59.1 a	84

14.2.2.3 Enhanced Stress Tolerance of Plants Due to Seed Treatment by *Bacillus*

Abiotic stress in plants is caused by factors such as temperature fluctuation, drought, excess moisture, salinity, and lack of nutrients in soils. These stresses affect plant growth by affecting the hormonal balance, alteration of protein metabolism, inhibition of the activity of enzymes involved in nucleic acid metabolism, the loss of control on nutrient uptake, and susceptibility to diseases (Arbona et al. 2005; Egamberdieva et al. 2011). Plants are colonized by microbes, including endophytes, nitrogen-fixing bacteria, and mycorrhizal fungi that closely cooperate with each other and can help mediate important physiological processes, especially nutrient acquisition and plant resistance to abiotic stresses (Egamberdieva et al. 2013, 2015; Berg et al. 2013; Ahanger et al. 2014). PGPRs can play an important role in crop production as they naturally stimulate root growth, make soil nutrients available to the plant root, fix nitrogen from air, improve soil fertility, and enhance plant tolerance to various environmental stresses including drought, salinity, high temperature, and heavy metals (Adesemoye and Kloepper 2009; Egamberdiyeva and Höflich 2003, 2004). Additionally, they may stimulate physiological parameters of plants and increase photosynthetic pigments, total free amino acids, and proteins compared to uninoculated control plants under saline condition (Han and Lee 2005a, b; Berg et al. 2013; Hashem et al. 2015). There are numerous PGPR from diverse groups of microbes that are naturally abundant in the soil, members of *Bacillus* genus being the most dominating. *Bacillus* is considered an ecologically significant group of bacteria, which are well adapted to the arid and salt-affected environment (Egamberdiyeva 2005; Malfanova et al. 2011). Because they form endospores, *Bacillus* can tolerate high temperature, pH, and osmotic conditions (Ashwini and Srividya 2013). This enables the bacteria to act as effective plant biostimulants in many harsh environmental conditions. Mechanisms in improvement of plant growth include production of growth-stimulating phytohormones, osmoprotectants, siderophores, and antibiotics, solubilization and mobilization of phosphate, inhibition of plant ethylene synthesis, antibiosis against soilborne fungal plant pathogens, and competition for nutrient and niches (Sessitsch et al. 2012; Prashanth and Mathivanan 2010). Notably, ecological factors may affect different mechanisms and limit the interactions between plant and beneficial bacteria, resulting in less than acceptable performance in plant growth promotion and management of diseases (Adesemoye and Egamberdieva 2013).

Excessive Heat and Temperature Fluctuation

Temperature extremes due to global warming and climate change may pose a huge challenge in future crop production efforts. Lobell and Gourdji (2012) reported that climatic uncertainty with elevated temperature would be a major limiting factor of agricultural productivity in many areas in the world. Heat stress effects on plants are very complex, which may alter many essential biochemical, molecular, and developmental processes (Kotak et al. 2007). Severe cellular injury and cell death can also occur in plants exposed to high temperature due to the collapse of cellular organelles leading to decrease in water and eventual plant death. Abd El-Daim et al. (2014) tested two wheat (*Triticum aestivum*) cultivars Olivin and Sids1 for the management of short-term heat stress tolerance by treating with two probiotic bacteria *Bacillus amyloliquefaciens* UCMB5113 and *Azospirillum brasilense* NO40. Seeds were treated by soaking grains in water containing 10^7 bacteria ml^{-1} for 2 h prior to seeding them in pots containing sterile peat mix in a greenhouse. Seedling grown from bacteria-treated and non-treated seeds were either exposed to 45 °C heat stress for 24 h or kept at normal growing condition. Plants of both cultivars were seriously affected by heat stress that caused >95% and 80% plant mortality in cv Olivin and Sids1, respectively. However, plants treated with bacteria had more than 40% survival rate, and those plants recovered early compared with non-treated plants. Mechanism of *Bacillus*-based heat stress tolerance by plants is not clearly understood. However, there are indications that bacterial treatment minimizes the production of reactive oxygen species (ROS) that can be very destructive to chloroplasts under heat stress. To neutralize ROS, higher accumulation of relevant antioxidant enzyme ascorbate peroxidase 1 (APX1) is common under heat stress. It was suggested by Koussevitzky et al. (2008) that protection of chloroplasts from ROS was provided by enhanced accumulation of APX1 after exposing Arabidopsis thaliana to heat and drought. Tiwari et al. (2017) also found that inoculation of rice seedlings with *Bacillus amyloliquefaciens* NBRI-SN13 positively modulated stress-responsive gene expressions under various abiotic stresses including heat that may provide enhanced tolerance to stresses.

Drought or Excessive Moisture

Drought can reduce growth and yield of crop plants by more than 50% by affecting morphological, physiological, and metabolic changes (oxidative stress) (Mundree et al. 2002; Wang et al. 2003). Increasing amounts of amino acids is considered to be an indication of drought tolerance (Greenway and Munns 1980). Plant defense systems protect against oxidative injury by producing enzymes such as superoxide dismutase, peroxidase, glutathione reductase, and catalase (Agarwal and Pandey 2004). In addition, plant-associated drought-tolerant microbial community can provide tolerance to hosts and maintain proper growth and survival under drought condition. More specifically, results from several studies point to success in using *Bacillus* species to temper the effects of drought on crops. Vivas et al. (2003) observed reduced root colonization of lettuce by AM fungus under drought stress. Similar results were reported by Al-Karaki et al. (2001) where mycorrhizal colonization was higher in well-watered plants colonized with AM fungi isolates than water-stressed plants. The combined inoculation of lettuce with fungus and *Bacillus* sp. caused a significant stimulatory effect on AM fungi *G. intraradices* development by enhancing the mycelium growth (Vivas et al. 2003). These results indicate that *Bacillus* not only exert its beneficial effect individually on plant but also can provide additive effect when combined with other beneficial microbes. For example, Ortiz et al. (2015) found that the association of *Pseudomonas putida* and *Bacillus thuringiensis* can decrease stromal conductance to prevent electrolyte leakage in plants. Potential mechanism is accumulation of proline in shoot and root cells. Kasim et al. (2013) reported improved drought tolerance by wheat due to priming seeds with *Bacillus amyloliquefaciens* 5113.

Gusain et al. (2014) studied plant growth-promoting *Bacillus cereus* for their role in enhancing plant growth and induction of stress-related enzymes in rice. They found that plant inoculated with *B. cereus* showed enhanced growth parameters and higher activity of superoxide dismutase, catalase, and peroxidase as compared to uninoculated plants under drought stress. Marulanda et al. (2009) found that the strain *B. megaterium* produced higher levels of proline and indole acetic acid in vitro under drought conditions. *Bacillus* sp. increased root, shoot growth, and nutrient uptake of lettuce (*Lactuca sativa*) under drought stress conditions compared to the control plants (Vivas et al. 2003). Inoculated maize with drought-tolerant strains of *B. amyloliquefaciens*, *B. licheniformis*, *B. thuringiensis*, and *B. subtilis* showed significantly higher root (33–45.3%), shoot length (32.2–42.5%), and dry biomass (46.6–59.06%) as compared to control under drought stress (Vardharajula et al. 2011). Inoculation of *Ocimum basilicum* with PGPR *Bacillus lentus* showed increased chlorophyll synthesis and photosynthetic electron transport and also mitigated the negative impact of water stress (Heidari and Golpayegani 2012). Kavamura et al. (2013) observed increased plant growth and yield of maize under drought stress (30% of field capacity) by two *Bacillus* spp. Increased germination, root and shoot length, fresh weight, and proline content were observed by Sharma et al. (2013) by inoculating chickpea seedlings with *Bacillus* sp. under osmotic potential of up to 0.4 MPa over uninoculated control.

14.2.3 Drench or Drip Application in the Rhizosphere

Bacillus-based products are formulated in the natural carrier materials that are easily dissolved in water and can be applied by drenching the rhizosphere of seedlings or transplants. Although most of the investigations with strain of Bacilli were conducted in vitro, commercially available products such as Serenade, Rhizoplus, Ballad Plus, Taegro Eco, Integral, or Kodiak can be directly applied by drenching rhizosphere or applying through drip irrigation system. Drenching of soil immediately after sowing of kohlrabi and again 4 weeks later with 0.2 g FZB24®WG/L water led to a 5% increase in the dry root weight with concurrent increase in yield up to 12% in a greenhouse experiment. Lahlali et al. (2013) drenched Serenade at 5% concentration (vol/vol) to a *Plasmodium brassicae* infested planting mix at seeding of canola 7 or 14 days after seeding and found a reduction of clubroot severity by 62–83%.

14.2.4 *Foliar Spray of* **Bacillus** *and* **Bacillus**-*Based Products*

Preventative spray of *Bacillus*-based product or strains of *Bacillus* were proven effective against foliar diseases in many crops. *Bacillus subtilis* re-isolated from the biological control products FZB24® and Phytovit® showed efficacy against several important foliar tomato diseases causing pathogens (late blight, early blight, powdery mildew, and leaf mold) with higher activity when applied before the pathogen infection (Bochow et al. 2001; Borriss 2011). Kilian et al. (2000) also reported 25% reduction of powdery mildew severity on wheat that also provided evidence of induced resistance of plants due to treatment with *Bacillus* spp.

14.2.5 *Combination of Methods for Synergistic Effect*

Compatible strains or products especially fungicides are selected to provide broad-spectrum pathogen control by combining different modes of actions or organisms with different ecological competencies and additive effects (Roberts et al. 2008). Combination of application methods such as soil drench and foliar spray was also found effective in *Camelia sinensis* for plant growth promotion (Chakraborty et al. 2012). Kiewnick et al. (2001) reported that *Rhizoctonia* root and crown rot of sugar beet caused by the soilborne fungus *Rhizoctonia solani* AG 2-2 was best controlled with a combination treatment of azoxystrobin applied at 76 g a.i./ha and the *Bacillus* isolate MSU-127 than that of separate application of fungicide and BCA. This treatment also provided greatest root and sucrose yield increase. Roberts et al. (2008) also tested diverse combination of products that included *B. subtilis* to manage bacterial spot on tomato. In spray programs containing acibenzolar-S-methyl (ASM) alternated with *B. subtilis* or copper hydroxide, treated plants showed significantly reduced disease compared to the untreated control plants. These treatments were as effective as copper-mancozeb standard. To control wheat sharp eyespot caused by *Rhizoctonia cerealis*, combined effect of *B. subtilis* NJ-18 with the fungicides flutolanil and difenoconazole was determined by Peng et al. (2014). Authors found that the growth of NJ-18 was not affected by flutolanil in a broth medium and the survival of NJ-18 endospores on wheat seed remained unaffected by difenoconazole. Disease control obtained with a combination of NJ-18 and either of the two fungicides was better than that of the bacterium or fungicides alone. Combination of *Bacillus* spp. strains with chemicals proved compatible and contributed to significant disease reduction with remarkably lower chemical use except for a few study results. Reduced use of chemical should also contribute to reduced selection pressure on the organism and resistance development against synthetic products. Co-application of *B. subtilis* ABS-S14 and chitosan to manage postharvest green mold on mandarin fruit showed that bacterial endospores, the crude extract, and chitosan resulted in significant reduction of fruit decay compared with the individual effect (Waewthongrak et al. 2014). In addition, same *Bacillus* sp. can be applied as a combination of seed treatment, planting mix amendment, transplant drench, and foliar spray to maximize the beneficial effect. Rahman and Jett (2016) obtained significant *Verticillium* wilt suppression on heirloom tomato "Mortgage Lifter" by combining different application methods of Serenade (Tables 14.1 and 14.2).

14.3 Conclusion and Future Perspectives

In the quest of biological control agents for sustainable agriculture, discovery of *Bacillus* spp. is very meaningful. Findings from many important and relevant studies indicated that *Bacillus* bacteria especially the ones isolated from the native environment could be used as natural agents for sustainable production of various crops with very little additional use of expensive synthetic inputs. However, researchers found that the effects of *Bacillus* are more pronounced in nutrient poor growing conditions when its use was considered for nutrient availability. Additive effect of utilization of *Bacillus* spp. in growing environment with balanced nutrition has not been sufficiently researched. In addition, with the disappearance of soil fumigant methyl bromide that played an essential role in managing soilborne diseases in high-value crops and limitation of synthetic products due to resistance development in pathogen population, more research should be directed to finding natural alternatives for sustainable management of both foliar and soilborne diseases. Optimization of the efficacy of these beneficial organisms should be confirmed by conducting research on formulation and various application methods. Numerous *Bacillus* strains have been screened for antimicrobial properties with positive outcomes enabling plants to resist important phytopathogens and provide plant growth-promoting effects. However, only a few of these products are commercially available to growers. Large-scale use of these products did not occur due to the variability and inconsistency of results in field conditions. Lowering the variability and increasing the consistency of results from these products are among a few challenges that will have to be addressed. Scientific community must determine the factors that interfere with the reproducibility of results from one location to another or controlled condition to field condition over time. Integration of these products with other management options should help in reducing the variability of results and produce additive effects. The continuing need for microbial-based products supporting sustainable crop production will make discovery and commercialization of *Bacillus*-based products as an attractive and profitable pursuit in the coming days.

Table 14.2 Different application methods of *Bacillus*-based products or *Bacillus* strains that enhanced growth and yield of different crop species through enhanced nutrient uptake, stress tolerance, or disease suppression in vitro or in vivo (field/greenhouse condition)

Bacillus strain	Application method	Crop species	Observed growth/yield enhancement effect	Reference
B. pumilus strain S2	Seed inoculation	*Triticum aestivum*	Enhance growth, yield, nutrient uptake	Abbasi et al. (2011)
B. pumilus S6-05	Pot soil inoculation	*Triticum aestivum*	Enhance growth, yield, nutrient uptake	Upadhyay et al. (2012)
B. mojavensis	Rock phosphate inoculation	*Zea Mays*	Enhanced vegetative growth and P availability	Manzoor et al. (2017)
B. subtilis SJ-101	Seed inoculation	*Brassica juncea*	Increased shoot length, fresh and dry weights	Zaidi et al. (2006)
B. amyloliquefaciens UCMB-5113, 5036	Seed priming/ inoculation	*Brassica napus*	Higher plant survival rate, vegetative and reproductive growth	Danielsson et al. (2007)
B. subtilis *B. pumulus*	Pot soil inoculation	*Sorghum bicolor*	Increase root shoot biomass	Abou-Shanab et al. (2008)
B. cereus	Pot soil inoculation	*Pinus thunbergii*	Growth, nutrient uptake	Wu et al. (2011)
B. amyloliquefaciens W19	Pot soil inoculation	*Musa paradisiaca*	Plant growth, increased biomass	Baset Mia et al. (2010)
B. subtilis, B. megaterium	Seed inoculation	*Raphanus sativus*	Yield	Kaymak et al. (2009)
B. pumilus, B. lichenformis	Pot soil inoculation	*Medicago sativa*	Growth, yield, Nutrient uptake	Medina et al. (2003)
Bacillus subtilis	spraying on foliage at 4–5 leaf stage	*Cucumis sativus*	Higher peroxidase and b-1,3-glucanase enzyme activities, growth enhancement	El-Borollosy and Oraby (2012)
B. megaterium RC07	Seed inoculation	*Beta vulgaris*	Root, shoot weight	Çakmakçi et al. (2006)
B. polymyxa	Inoculation of rock phosphates that were used as P-source	*Sorghum bicolor*	Increase grain and dry matter yields	Alagawadi and Gaur (1992)

(continued)

Table 14.2 (continued)

Bacillus strain	Application method	Crop species	Observed growth/yield enhancement effect	Reference
B. megaterium	Seed inoculation	*Cicer arietinum*	Increase dry matter and grain yield	Verma at al. (2012)
B. megaterium M-13	Seed inoculation	*Helianthus annuus*	Increased yield, oil, protein content	Ekin (2010)
B. polymyxa	Tuber inoculation with culture suspension	*Solanum tuberosum*	Increased yield	Kundu and Gaur (1980)
B. megaterium	Root inoculation	*Rubus idaeus*	Increased crop yield, cane length, cluster per cane, berries per cane	Orhan et al. (2006)
B. simplex, B. cereus	Inoculation of rock phosphates that were used as P-source	*Ammi visnaga*	Increased shoot, root length, dry weight	Hassan et al. (2010)
B. megaterium OSU-142 and M-3	Root inoculation and floral/foliar spray inoculation	*Fragaria* x *ananassa*	Increased fruit yield, plant growth, and leaf P and Zn contents	Esitken et al. (2010)
B. megaterium	Preplant rhizome inoculation	*Curcuma longa*	Plant growth and yield	Sumathi et al. (2011)
B. subtilis	Seed priming	*Momordica charantia*	Enhanced yield, quality, root length, and dry root weight	Kumar et al. (2012a)
B. coagulans	Biocoating of seeds	*Begonia malabarica*	Biomass yield/ more flowering capacity	Selvaraj et al. (2008)
B. megaterium M3	Cutting inoculation	*Mentha piperita*	Root length, dry matter	Kaymak et al. (2008)
B. subtilis with *Trichoderma* and *Glomus*	Seed treatment	*Sphaeranthus amaranthoides*	Enhanced growth, biomass	Sumithra and Selvaraj (2011)
B. megaterium	Seedling dip inoculation	*Withania somnifera*	Increased plant height, root length, alkaloid content	Rajasekar and Elango (2011)
B. megaterium *B. circulanse*	Soil drench	*Rosmarinus officinalis*	Increased oil content, yield in fresh herb, and total CHO	Abdullah et al. (2012)

(continued)

Table 14.2 (continued)

Bacillus strain	Application method	Crop species	Observed growth/yield enhancement effect	Reference
B. amyloliquefaciens BChi1	Root and foliar inoculation	*Fragaria* x *ananassa*	Vegetative growth, fruit yield, antioxidant activity	Rahman et al. (2018)
B. amyloliquefaciens CMB5113	Seed priming	*Oryza sativa*	Improved survival rate, recovery from stress, and vegetative growth	Abd El-Daim (2014)
B. subtilis SM21 and *B. cereus* AR156	Seed soaking+rhizosphere drenching	*Capsicum annum*	Enhanced growth, yield and fruit nutrient content	Zhou et al. (2014)

References

Abbasi MK, Sharif S, Kazmi M, Sultan T, Aslam M (2011) Isolation of plant growth promoting Rhizobacteria from wheat rhizosphere and their effect on improving growth, yield and nutrient uptake of plants. Plant Biosyst 145:159–168

Abd El-Daim IA, Bejai S, Meijer J (2014) Improved heat stress tolerance of wheat seedlings by bacterial seed treatment. Plant Soil. https://doi.org/10.1007/s11104-014-2063-3

Abdelzaher HMA (2003) Biological control of using selected antagonistic rhizospheric strains of *Bacillus subtilis*. N Z J Crop Hortic Sci 31:209–220

Abdullah AT, Hanafy MS, El-Ghawwas EO, Ali ZH (2012) Effect of compost and some biofertilizers on growth, yield, essential oil productivity and chemical composition of *Rosmarinus officinalis* L. plants. J Hortic Sci Ornamental Plants 4(2):201–214

Abou-Shanab RA, Ghanem K, Ghanem N, Al-Kolaibe A (2008) The role of bacteria on heavy-metal extraction and uptake by plants growing on multi-metal-contaminated soils. World J Microbiol Biotechnol 24:253–262

Adesemoye AO, Egamberdieva D (2013) Beneficial effects of plant growth promoting rhizobacteria on improved crop production: the prospects for developing economies. In: Maheshwari DK, Saraf M, Aeron A (eds) Bacteria in agrobiology: crop productivity. Springer, Berlin/Heidelberg, pp 45–63

Adesemoye AO, Kloepper JW (2009) Plant-microbes interactions in enhanced fertilizer use efficiency. Appl Microbiol Biotechnol 85:1–12

Agarwal S, Pandey V (2004) Antioxidant enzyme responses to NaCl stress in *Cassia angustifolia*. Biol Plant 48:555–560

Ahanger MA, Hashem A, Abd Allah EF, Ahmad P (2014) Arbuscular mycorrhiza in crop improvement under environmental stress. In: Ahmad P, Rasool S (eds) Emerging technologies and management of crop stress tolerance, vol 2, pp 69–95

Ajilogba CF, Babalola OO, Ahmad F (2013) Antagonistic effects of *Bacillus* species in biocontrol of tomato *Fusarium* wilt. Ethno Med 7(3):205–216

Alagawadi AR, Gaur AC (1992) Inoculation of Azospirillum brasilense and phosphate solubilizing bacteria on yield of sorghum [Sorghum bicolor (L.) Moench] in dry land. Trop Agric 69:347–350

Al-Karaki GN, Hammad R, Rusan M (2001) Response of two cultivars differing in salt tolerance to inoculation with mycorrhizal fungi under salt stress. Mycorrhiza 11:43–44

Araujo FFD, Souza EC, Guerreiro RT, Guaberto LM, Araugo ASF (2012) Diversity and growth-promoting activities of *Bacillus* sp. in maize. Revista Caatinga Mossoró 25(1):1–7

Arbona V, Iglesias DJ, Jacas J, Primo-Millo E, Talon M, Gomez-Cadenas A (2005) Hydrogel substrate amendment alleviates drought effects on young citrus plants. Plant Soil 270:73–82

Archana DS, Nandish MS, Savalagi VP, Alagawadi AR (2012) Screening of K-solubilizing bacteria (KSB) for plant growth promotional activity. Bioinfolet 9(4):627–630

Archana DS, Nandish MS, Savalagi VP, Alagawadi AR (2013) Characterization of potassium solubilizing bacteria (KSB) from rhizosphere soil. Bioinfolet 10:248–257

Armada A, Roldan A, Azcon R (2014) Differential activity of autochthonous bacteria in controlling drought stress in native *Lavandula* and Salvia plants species under drought conditions in natural arid soil. Microb Ecol 67:410–420

Arrebola E, Sivakumar D, Bacigalupo R, Korsten L (2010) Combined application of antagonist *Bacillus amyloliquefaciens* and essential oils for the control of peach postharvest diseases. Crop Prot 29:369–377

Ashwini N, Srividya S (2013) Potentiality of *Bacillus subtilis* as biocontrol agent for management of anthracnose disease of chilli caused by *Colletotrichum gloeosporioides*OGC1. 3. Biotech 4(2):127–136. https://doi.org/10.1007/s13205-013-0134-4

Awasthi R, Tewari R, Nayyar H (2011) Synergy between plants and P-solubilizing microbes in soils: effects on growth and physiology of crops. Int Res J Microbiol 2:484–503

Bais HP, Walker TS, Schweizer HP, Vivanco JM (2002) Root specific elicitation and antimicrobial activity of rosmarinic acid in hairy root cultures of Ocimum basilicum. Plant Physiol Biochem 40:983–995

Basak BB, Biswas DR (2009) Influence of potassium solubilizing microorganism (Bacillus mucilaginosus) and waste mica on potassium uptake dynamics by sudan grass (Sorghum vulgare Pers.) grown under two Alfisols. Plant Soil 317:235–255

Baset Mia MA, Shamsuddin ZH, Wahab Z, Marziah M (2010) Effect of plant growth promoting rhizobacterial (PGPR) inoculation on growth and nitrogen incorporation of tissue-cultured musa plantlets under nitrogen-free hydroponics condition. Aust J Crop Sci 4(2):85–90

Batinic T, Schmitt J, Schulz UM, Werner D (1998) Konstruktion von RAPD-Sonden für die Quantifizierung von Bazillus subtilis FZB C und dessen antagonistische Wirksamkeit im System Cucumis sativus Pythium ultimum. Z. Pflanzenkrankh. Pflanzenschutz 105:168–180

Baysal O, Caliskan M, Yesilova O (2008) An inhibitory effect of a new *Bacillus subtilis* strain (EU07) against *Fusarium oxysporum* f. sp. *radicis-lycopersici*. Physiol Molecular Plant Pathol 73:25–32

Berg G, Zachow C, Müller H, Phillips J, Tilcher R (2013) Next-generation bio-products sowing the seeds of success for sustainable agriculture. Agronomy 3:648–656

Bharathi R, Vivekananthan R, Harish S, Ramanathan A, Samiyappan R (2004) Rhizobacteria-based bioformulations for the management of fruit rot infection in chilies. Crop Prot 23:835–843

Bochow H, El-Sayed SF, Junge H, Stavropoulou A, Schmiedeknecht G (2001) Use of *Bacillus subtilis* as biocontrol agent. IV. Salt-stress tolerance induction by *Bacillus subtilis*FZB24 seed treatment in tropical vegetable field crops, and ist mode of action. Z. PflKrankh. Pflanzenschutz 108:21–30

Borriss R (2011) Use of plant-associated Bacillus strains as biofertilizers and biocontrol agents. In: Maheshwari DK (ed) Bacteria in agrobiology: plant growth responses. Springer, Heidelberg/Dordrecht/New York, pp 41–76

Borriss R (2016) Phytostimulation and biocontrol by the plant-associated *Bacillus amyloliquefaciens* FZB42: an update. In: Islam M, Rahman M, Pandey P, Jha C, Aeron A (eds) Bacilli and Agrobiotechnology. Springer, Switzerland, pp 163–184

Brunetti G, Farrag K, Soler-Rovira P, Nigro F, Senesi N (2011) Greenhouse and field studies on Cr, Cu, Pb and Zn phytoextraction by Brassica napus from contaminated soils in the Apulia region, Southern Italy. Geoderma 160:517–523

Calvo P, Ormeno-Orrillo E, Martinez-Romero E, Zuniga D (2010) Characterization of Bacillus isolates of potato rhizosphere from Andean soils of Peru and their potential PGPR characteristics. Braz J Microbiol 41:899–906

Çakmakçi R, Dönmez F, Aydın A, Şahin F (2006) Growth promotion of plants by plant growth-promoting rhizobacteria under greenhouse and two different field soil conditions. Soil Biol Biochem 38(6):1482–1487

Cazorla FM, Romero D, Pérez-García A, Lugtenber BJJ, de Vicente A, Bloemberg G (2007) Isolation and characterization of antagonistic *Bacillus subtilis* strains from the avocado rhizoplane displaying biocontrol activity. J Appl Microbiol 103:1950–1959

Chakraborty U, Chakraborty BN, Chakraborty AP (2012) Induction of plant growth promotion in Camellia sinensis by Bacillus megaterium and its bioformulations. World J Agric Sci 8(1):104–112

Chen X-H, Scholz R, Borriss M, Junge H, Mögel G, Kunz S (2009) Borriss R:Difficidin and bacilysin produced by plant-associated Bacillus amyloliquefaciens are efficient in controlling fire blight disease. J Biotechnol 140:38–44

Chen Y, Fan JB, Du L, Xu H, Zhang QH, He YQ (2014) The application of phosphate solubilizing endophyte Pantoea dispersa triggers the microbial community in red acidic soil. Appl Soil Ecol 84:235–244

Crouch IJ, van Staden J (1993) Evidence for the presence of plant growth regulators in commercial seaweed products. Plant Growth Regul 13:21–29

Danielsson J, Reva O, Meijer J (2007) Protection of oilseed rape (Brassica napus) towards fungal pathogens by strains of plant-associated Bacillus amyloliquefaciens. Microb Ecol 54:134–140

Dasan AS (2012) Compatibility of agrochemicals on the growth of phosphorous mobilizing bacteria Bacillus megaterium var. phosphaticum and potassium mobilizing bacteria Frateuria aurantia. Appl Res Dev Inst J 6:118–134

Egamberdieva D, Jabborova D (2013) Biocontrol of cotton damping-off caused by rhizoctonia solani in salinated soil with rhizosphere bacteria. Asian Australas J Plant Science Biotechnol 7(2):31–38

Egamberdieva D, Kucharova Z, Davranov K, Berg G, Makarova N, Azarova T, Chebotar V, Tikhonovich I, Kamilova F, Validov S, Lugtenberg B (2011) Bacteria able to control foot and root rot and to promote growth of cucumber in salinated soils. Biol Fertil Soils 47:197–205

Egamberdieva D, Berg G, Lindström K, Räsänen LA (2013) Alleviation of salt stress of symbiotic Galega officinalis L. (goat's rue) by co-inoculation of rhizobium with root colonising Pseudomonas. Plant Soil 369(1):453–465

Egamberdieva D, Jabborova D, Berg G (2015) Synergistic interactions between Bradyrhizobium japonicum and the endophyte Stenotrophomonas rhizophila and their effects on growth, nodulation and nutrition of soybean under salt stress. Plant Soil:1–11

Egamberdiyeva D (2005) Plant growth promoting rhizobacteria isolated from calcisol soil in a semiarid region of Uzbekistan: biochemical characterisation and effectiveness. Plant Nutr Soil Sci 168:94–99

Egamberdiyeva D, Hoflich G (2003) Influence of growth promoting bacteria on the growth of wheat at different soils and temperatures. Soil Biol Biochem 35:973–978

Egamberdiyeva D, Höflich G (2004) Importance of plant growth promoting bacteria on growth and nutrient uptake of cotton and pea in semi-arid region Uzbekistan. J Arid Environ 56:293–301

Ekin Z (2010) Performance of phosphate solubilizing bacteria for improving growth and yield of sunflower (Helianthus annuus L.) in the presence of phosphorus fertilizer. Afr J Biotechnol 9:3794–3800

El-Borollosy AM, Oraby MM (2012) Induced systemic resistance against cucumber mosaic cucumovirus and promotion of cucumber growth by some plant growth-promoting rhizobacteria. Ann Agric Sci 57:91–97

Esitken A, Yildiz HE, Ercisli S, Donmez MF, Turan M, Gunes A (2010) Effects of plant growth promoting bacteria (PGPB) on yield, growth and nutrient contents of organically grown strawberry. Sci Hortic 124:62–66

Figueiredo MVB, Seldin L, de Araujo FF, Mariano RLR (2010) Plant growth promoting rhizobacteria: fundamentals and applications. In: Maheshwari DK (ed) Plant growth and health promoting bacteria microbiology monographs. Springer, Berlin/Heidelberg

Godfray HCJ, Beddington JR, Crute IR, Haddad L, Lawrence D, Muir JF et al (2010) Food security: the challenge of feeding 9 billion people. Science 327:812–818

Gowtham HG, Murali M, Brijesh Singh S, Lakshmeesha TR, Narasimha Murthy K, Amruthesh KN, Nirangana SR (2018) Plant growth promoting rhizobacteria- Bacillus amyloliquefaciens improves plant growth and induces resistance in chili against anthracnose disease. Biol Control 126:209–221

Greenway H, Munns R (1980) Mechanism of salt tolerance in non halophytes. Annu Rev Plant Physiol 31:149–190

Grosch R, Malies U, Bochow H (1996) Population dynamics of biocontrol agent Bacillus subtilis in closed hydroponic plant cultivation systems after application of different cell numbers. Bull Oilb Scrop 19:134–144

Gusain YS, Singh US, Sharma A (2014) Enzymatic amelioration of drought stress in rice through the application of plant growth promoting rhizobacteria (PGPR). Int J Curr Res 6(1):4487–4491

Gyaneshwar P, Kumar GN, Parekh LJ, Poole PS (2002) Role of soil microorganisms in improving P nutrition of plants. Plant Soil 245:83–93

Han HS, Lee KD (2005a) Plant growth promoting rhizobacteria effect on antioxidant status, photosynthesis, mineral uptake and growth of lettuce under soil salinity. Res J Agric Biol Sci 1(3):210–215

Han HS, Lee KD (2005b) Phosphate and potassium solubilizing bacteria effect on mi- neral uptake, soil availability and growth of eggplant. Res J Agric Biol Sci 1:176–180

Han HS, Lee KD (2006) Effect of co-inoculation with phosphate and potassium solubilizing bacteria on mineral uptake and growth of pepper and cucumber. Plant Soil Environ 52:130–136

Hashem A, Abd-Allah EF, Alqarawi AA, Al-Huqail AA, Alshalawi SRM, Wirth S, Egamberdieva D (2015) Salt tolerant PGPR Bacillus subtilis improved growth and physiological parameters of Indian bassia (Bassia indica) under salt stress condition. Pak J Bot 47(5):1735–1741

Hassan EA, Hassan EA, Hamad EH (2010) Microbial solubilization of phosphate-potassium rocks and their effect on khella (Ammi visnaga) growth. Ann Agric Sci 55(1):37–53

Heidari M, Golpayegani A (2012) Effects of water stress and inoculation with plant growth promoting rhizobacteria (PGPR) on antioxidant status and photosynthetic pigments in basil (*Ocimum basilicum* L.). J Saudi Soc Agric Sci 11:57–61

Herve JJ (1994) Biostimulants, a new concept for the future; prospects offered by the chemistry of synthesis and biotechnology. Comptes Redus Acad Agric Fr 80:91–102

Idris EE, Bochow H, Ross H, Borriss R (2004) Use of Bacillus subtilis as biocontrol agent. VI. Phytohormonelike action of culture filtrates prepared from plant growth promoting Bacillus amyloliquefaciens FZB24, FZB42, FZB45 and Bacillus subtilis FZB37. J Plant Dis Protect 111(6):583–597. issn: 0340-8159

Jana BB (2003) Distribution pattern and role of phosphate solubilizing bacteria in the enhancement of fertilizer value of rock phosphate in aquaculture ponds: state-of-the-art. In: Velázquez E, Rodríguez-Barrueco Salamanca C (eds) First International meeting on microbial phosphate solubilization, 16–19 July, 2002, Spain, pp 229–238

Jangir M, Pathk R, Sharma S, Sharma S (2018) Biocontrol mechanisms of Bacillus sp., isolated from tomato rhizoshere, against Fusarium oxysporum f. sp. lycopersici. Biol Control 123:60–70

Kamil Z, Rizk M, Saleh M, Moustafa S (2007) Isolation and identification of rhizosphere soil chitinolytic bacteria and their potential in antifungal biocontrol. Global J Mol Sci 2:57–66

Karuppiah P, Rajaram S (2011) Exploring the potential of chromium reducing Bacillus sp. and there plant growth promoting activities. J Microbiol Res 1:17–23

Kasim WA, Osman ME, Omar MN, Abd El-Daim IA, Bejai S, Meijer J (2013) Control of drought stress in wheat using plantgrowth-promoting bacteria. J Plant Growth Regul 32:122–130
Kavamura VN, Santos SN, da Silva JL, Parma MM, Ávila LA, Visconti A, Zucchi TD, Taketani RG, Andreote FD, de Melo IS (2013) Screening of Brazilian cacti rhizobacteria for plant growth promotion under drought. Microb Res 168:183–191
Kaymak HC, Yarali F, Guvenc I, Figen DM (2008) The effect of inoculation with plant growth rhizobacteria (PGPR) on root formation of mint (Mentha piperita L.) cuttings. Afr J Biotechnol 7(24):4479–4483
Kaymak HC, Guvenc I, Yarali F, Donmez MF (2009) The effects of biopriming with PGPR on germination of radish (Raphanus sativus L.) seeds under saline conditions. Turk J Agric For 33(2):173–179
Khan MS, Zaidi A, Wani PA (2007) Role of phosphate-solubilizing microorganisms in sustainable agriculture – a review. Agron Sustain Dev 27:29–43
Kiewnick S, Jacobsen BJ, Braun-Kiewnick A, Eckhoff JLA, Bergman JW (2001) Integrated control of rhizoctonia crown and root rot of sugar beet with fungicides and antagonistic bacteria. Plant Dis 85:718–722
Kilian M, Steiner U, Krebs B, Junge H, Schmiedeknecht G, Hain R (2000) FZB24® Bacillus subtilis – mode of action of a microbial agent enhancing plant vitality. Pflanzenschutz-Nachrichten Bayer 1:72–93
Kotak S, Larkindale J, Lee U, von Koskull-Döring P, Vierling E, Scharf KD (2007) Complexity of the heat stress response in plants. Curr Opin Plant Biol 10:310–316
Koussevitzky S, Suzuki N, Huntington S, Armijo L, Sha W, Cortes D, Shulaev V, Mittler R (2008) Ascorbate peroxidase 1 plays a key role in the response of Arabidopsis thaliana to stress combination. J Biol Chem 283:34197–34203
Krebs B, Höding B, Kübart S, Alemayehu MW, Junge H, Schmiedeknecht G, Grosch R, Bochow H, Hevesi M (1998) Use of Bacillus subtilis as biocontrol agent. I. Activities and characterization of Bacillus subtilis strains. Z Pflanzenkrankh Pflanzenschutz 105:181–197
Kumar A, Kumar A, Devi S, Patil S, Payal C, Negi S (2012a) Isolation, screening and characterization of bacteria from rhizospheric soils for different plant growth promotion (PGP) activities an in-vitro study. Recent Res Sci Technol 4:1–5
Kumar P, Dubey RC, Maheshwari DK (2012b) Bacillus strains isolated from rhizosphere showed plant growth promoting and antagonistic activity against phytopathogens. Microbiol Res 67:493–499
Kumar A, Bahadur I, Maurya BR, Raghuwanshi R, Meena VS, Singh DK, Dixit J (2015) Does a plant growth-promoting rhizobacteria enhance agricultural sustainability? J Pure Appl Microbiol 9(1):715–724
Kundu BS, Gaur AC (1980) Effect of phosphobacteria on the yield and phosphate uptake of potato crop. Curr Sci 49:159–160
Lahlali R, Peng G, Gossen BD, McGregor L, Yu FQ, Hynes RK, Hwang SF, McDonald MR, Boyetchko SM (2013) Evidence that the biofungicide Serenade (Bacillus subtilis) suppresses clubroot on canola via antibiosis and induced host resistance. Phytopathology 103:245–254
Lobell DB, Gourdji SM (2012) The influence of climate change on global crop productivity. Plant Physiol 160:1686–1697
Malfanova N, Kamilova F, Validov S, Shcherbakov A, Chebotar V, Tikhonovich I, Lugtenberg B (2011) Characterization of Bacillus subtilis HC8, a novel plant-beneficial endophytic strain from giant hogweed. Microbial Biotechnol 4(4):523–553
Manzoor M, Kaleem MA, Sultan T (2017) Isolation of phosphate solubilizing Bacteria from maize rhizosphere and their potential for rock phosphate Solubilization–mineralization and plant growth promotion. Geomicrobiol J 34:81–95
Marulanda A, Barea JM, Azcón R (2009) Stimulation of plant growth and drought tolerance by native microorganisms (AM fungi and bacteria) from dry environments: mechanisms related to bacterial effectiveness. J Plant Growth Regul 28:115–124

Medina A, Probanza A, Gutierrez Mañero FJ, Azcón R (2003) Interactions of arbuscular-mycorrhizal fungi and Bacillus strains and their effects on plant growth, microbial rhizosphere activity (thymidine and leucine incorporation) and fungal biomass (ergosterol and chitin). Appl Soil Ecol 22:15–28
Meena OP, Maurya BR, Meena VS (2013) Influence of K- solubilizing bacteria on release of potassium from waste mica. Agric Sustain Dev 1(1):53–56
Meena VS, Maurya BR, Verma JP (2014) Does a rhizospheric microorganism enhance K+ availability in agricultural soils? Microbiol Res 169:337–347
Meena VS, Maurya BR, Verma JP, Aeron A, Kumar A, Kim K, Bajpai VK (2015) Potassium solubilizing rhizobacteria (KSR): Isolation, identification, and K-release dynamics from waste mica. Ecol Eng 81:340–347
Mia MAB, Hossain MM, Shamsuddin ZH, Islam MT (2013) Plant associated bacteria in nitrogen nutrition in crops, with special reference to rice and banana. In: Maheshwari DK, Saraf M, Aeron A (eds) Bacteria in agrobiology: crop productivity. Springer, Berlin, pp 97–118
Moyne AL, Shelby R, Cleveland TE, Tuzun S (2001) Bacillomycin D: an iturin with antifungal activity against Aspergillus flavus. J Appl Microbiol 90:622–662
Mundree SG, Baker B, Mowla S, Peters S, Marais S, Willigen CV, Govender K, Maredza A, Muyanga S, Farrant JM, Thomson JA (2002) Physiological and molecular insights into drought tolerance. Afr J Biotechnol 1:28–38
Nihorimbere V, Ongena M, Cawoy H, Brostaux Y, Kakana P, Jourdan E, Thonart P (2010) Beneficial effects of Bacillus subtilis on field-grown tomato in Burundi: reduction of local Fusarium disease and growth promotion. Afr J Microbiol Res 4(11):1135–1142
Nikbakht A, Pessarakli M, Daneshvar-Hakimi-Maibodi N, Kafi M (2014) Perennial ryegrass growth responses to mycorrhizal infection and humic acid treatments. Agron J 106:585–595. https://doi.org/10.2134/agronj2013.0275
Orhan E, Esitken A, Ercisli S, Turan M, Sahin F (2006) Effects of plant growth promoting rhizobacteria (PGPR) on yield, growth and nutrient contents in organically growing raspberry. Sci Hortic 111(1):38–43
Ortiz N et al (2015) The contribution of arbuscular mycorrhizal fungi and/or bacteria to enhancing plant drought tolerance under natural soil conditions: effectiveness of autochthonous or allochthonous strains. J Plant Physiol 174:87–96
Panhwar QA, Radziah O, Sariah M, MohdRazi I (2009) Solubilization of phosphate forms by phosphate solubilizing bacteria isolated from aerobic rice. Int J Agric Biol 11:667–673
Panhwar QA, Othman R, Rahman ZA, Meon S, Ismail MR (2012) Isolation and characterization of phosphate-solubilizing bacteria from aerobic rice. Afr J Biotechnol 11(11):2711–2719
Parmar P, Sindhu SS (2013) Potassium solubilization by rhizosphere bacteria: influence of nutritional and environmental conditions. J Microbio Res 3(1):25–31
Peng D, Li S, Chen C, Zhou M (2014) Combined application of Bacillus subtilis NJ-18 with fungicides for control of sharp eyespot of wheat. Biol Control 70:28–34
Prajapati K, Sharma MC, Modi HA (2013) Growth promoting effect of potassium solubilizing microorganisms on Abelmoschus esculentus. Int J Agric Sci 3(1):181–188
Prashanth S, Mathivanan N (2010) Growth promotion of groundnut by IAA producing rhizobacteria Bacillus licheniformis MML2501. Arch Phytopathol Plant Prot 43:191–208
Polo J, Barroso R, Ródenas J, Azcón-Biet J, Cáceres R, Marfà O (2006) Porcine haemoglobin hydrolysate as a biostimulant for lettuce plants subject to conditions of thermal stress. HortTechnology 16(3):483–487
Rahman M, Jett L (2016) Evaluation of bio-fumigants and biological control agents for managing Verticillium wilt on tomato, 2015. Plant Disease Management Reports 10 V056. https://www.plantmanagementnetwork.org/pub/trial/PDMR/reports/2016/V056.pdf
Rahman M, Sabir AA, Mukta JA, Khan MA, Mohi-Ud-Din M, Miah MG, Rahman M, Islam MT (2018) Plant probiotic bacteria Bacillus and Paraburkholderia improve growth, yield and content of antioxidants in strawberry fruit. Sci Rep 8:2504

Raj SA (2004) Solubilization of silicate and concurrent release of phosphorus and potassium in rice ecosystem. In: Book chapter conference paper Biofertilizers technology, Coimbatore, India, pp 372–378

Rajasekar S, Elango R (2011) Effect of microbial consortium on plant growth and improvement of alkaloid content in Withania somnifera (Ashwagandha). Curr Bot 2(8):27–30

Rajawat MVS, Singh S, Singh G, Saxena AK (2012) Isolation and characterization of K-solubilizing bacteria isolated from different rhizospheric soil. In: Proceeding of 53rd Annual conference of Association of Microbiologists of India, p 124

Roberts PD, Momol MT, Ritchie L, Olson SM, Jones JB, Balogh B (2008) Evaluation of spray programs containing famoxadone plus cymoxanil, acibenzolar-S-methyl, and Bacillus subtilis compared to copper sprays for management of bacterial spot on tomato. Crop Prot 27:1519–1526

Romero D, de Vicente A, Olmos JL, Dávila JC, Pérez-García A (2007) Effect of lipopeptides of antagonistic strains of Bacillus subtilis on the morphology and ultrastructure of the cucurbit fungal pathogen Podosphaera fusca. J Appl Microbiol 103:969–976

Rouphael Y, Cardarelli M, Bonini P, Colla G (2017) Synergistic action of a microbial-based biostimulant and a plant derived-protein hydrolysate enhances lettuce tolerance to alkalinity and salinity. Front Plant Sci 8:131. https://doi.org/10.3389/fpls.2017.00131

Safiyazov JS, Mannanov RN, Sattarova RK (1995) The use of bacterial antagonists for the control of cotton diseases. Field Crops Res 43:51–54

Sanderson KJ, Jameson PE (1986) The cytokinins in a liquid seaweed extract: could they be active ingredients? Acta Hortic 179:113–116

Sessitsch A, Weilharter A, Gerzabek MH, Kirchmann H, Kandeler E (2001) Microbial population structures in soil particle size fractions of a long-term fertilizer field experiment. Appl Environ Microbiol 67:4215–4224

Sessitsch A, Hardoim P, Doring J, Weilharter A, Krause A, Woyke T, Mitter B, Hauberg-Lotte L, Friedrich F, Rahalkar M (2012) Functional characteristics of an endophyte community colonizing rice roots as revealed by metagenomic analysis. Mol Plant-Microbe Interact 25:28–36

Singh B, Ravneet K, Kashmir C (2008) Characterization of rhizobium strain isolated from the roots of *Trigonella foenumgraecum* (fenugreek) Afri. J Biotech 7(20):3671–3676

Schippers B (1992) Prospects for management of natural suppressiveness to control soilborne pathogenes. In: Tjamos EC, Papavizas GC, Cook RJ (eds) Biological control of plant diseases, Progress and challenges for the future, NATO ASI Series A: Life Sciences, vol 230. Plenum Press, New York, pp 21–34

Schippers B, Scheffer RJ, Lugtenberg JJ, Weisbek PJ (1995) Biocoating of seed with plant growth promoting rhizobacteria to improve plant establishment. Outlook Agric 24:179–185

Selvaraj T, Rajeshkumar S, Nisha MC, Wondimu L, Tesso M (2008) Effect of Glomus mosseae and plant growth promoting rhizomicroorganisms (PGPR's) on growth, nutrients and content of secondary metabolites in Begonia malabarica Lam. Mj Int J Sci Technol 2(3):516–525

Sharma P, Khanna V, Kumari P (2013) Efficacy of aminocyclopropane-1-carboxylic acid (ACC)-deaminase-producing rhizobacteria in ameliorating water stress in chickpea under axenic conditions. Afr J Microb Res 7:5749–5757

Sheng XF, He LY (2006) Solubilization of potassium bearing minerals by a wild type strain of Bacillus edaphicus and its mutants and increased potassium uptake by wheat. Can J Microbiol 52(1):66–72

Sugumaran P, Janarthanam B (2007) Solubilization of potassium containing minerals by bacteria and their effect on plant growth. World J Agic Sci 3(3):350–355

Sumathi CS, Ramesh N, Balasubramanian V, Rajesh KV (2011) Microbial bioinoculant potential on turmeric (Curcuma Longa L.) growth improvement under tropical nursery conditions. Asian J Exp Biol Sci 2(4):612–623

Sumithra P, Selvaraj T (2011) Influence of Glomus walkeri and plant growth promoting rhizomicroorganisms on growth, nutrition and content of secondary metabolites in Sphaeranthes amaranthoides (L.) Burm. J Agril Technol 7(6):1685–1692

Tiwari S, Prasad V, Chauhan PS, Lata C (2017) bacillus amyloliquefaciens confers tolerance to various abiotic stresses and modulates plant response to phytohormones through osmoprotection and gene expression regulation in rice. Front Plant Sci 8:1510. https://doi.org/10.3389/fpls.2017.0151

Upadhyay SK, Singh JS, Saxena AK, Singh DP (2012) Impact of PGPR inoculation on growth and antioxidants status of wheat plant under saline condition. Plant Biol 4:605–611

Vardharajula S, Ali SZ, Grover M, Reddy G, Bandi V (2011) Drought-tolerant plant growth promoting Bacillus spp. effect on growth, osmolytes, and antioxidant status of maize under drought stress. J Plant Interact 6(1):1–14

Verma JP, Yadav J, Tiwari KN, Kumar A (2012) Effect of indigenous Mesorhizobium spp. and plant growth promoting rhizobacteria on yields and nutrients uptake of chickpea (Cicer arietinum L.) under sustainable agriculture. Ecol Eng 51:282–286

Vivas A, Marulanda A, Ruiz Lozano JM, Barea JM, Azcon R (2003) Influence of Bacillus sp. on physiological activities of two arbuscular mycorrhizal fungi and on plant responses to PEG induced drought stress. Mycorrhiza 13:249–256

Waewthongrak W, Leelasuphakul W, McCollum G (2014) Cyclic Lipopeptides from Bacillus subtilis ABS–S14 elicit defense-related gene expression in citrus fruit. PLos One. http://journals.plos.org/plosone/article?id=10.1371/journal.pone.0109386

Wang W, Vinocur B, Altman A (2003) Plant responses to drought, salinity and extreme temperatures: towards genetic engineering for stress tolerance. Planta 218:1–14

Wu XQ, Hou LL, Sheng JM, Ren JH, Zheng L, Chen D, Ye JR (2011) Effects of ectomycorrhizal fungus Boletus edulis and mycorrhiza helper Bacillus cereus on the growth and nutrient uptake by Pinus thunbergii. Biol Fertil Soils 48:385–391

Yakhin O, Lubyanov A, Yakhin I, Brown P (2017) biostimulants in plant science: a global perspective. Front Plant Sci 7:2049

Zaidi S, Usmani S, Singh BR, Musarrat J (2006) Significance of Bacillus subtilis strain SJ-101 as a bioinoculant for concurrent plant growth promotion and nickel accumulation in Brassica juncea. Chemosphere 64:991–997

Zaidi A, Khan MS, Ahemad M, Oves M, Wani PA (2009) Recent advances in plant growth promotion by phosphate-solubilizing microbes. In: Khan MS et al (eds) Microbial strategies for crop improvement. Springer, Berlin/Heidelberg

Zhang G, Raza W, Wang X, Ran W, Ahen Q (2012) Systemic modification of cotton root exudates induced by arbuscular mycorrhizal fungi and Bacillus vallismortis HJ-5 and their effects on Verticillium wilt disease. Appl Soil Ecol 61:85–91

Zhao QY, Shen QR, Ran W, Xiao TJ, Xu DB, Xu YC (2011) Inoculation of soil by Bacillus subtilis Y-IVI improves plant growth and colonization of the rhizosphere and interior tissues of muskmelon (Cucumis melo L.). Biol Fertil Soils 47:507–514

Zhou DM, Wang KP, Liu HX, Guo C (2014) Field evaluation of different application methods of the mixture of Bacillus cereus strain AR156 and Bacillus subtilis strain SM21 on pepper growth and disease resistance. Biocontrol Sci Technol 24:1451–1468

Chapter 15
Bacillus thuringiensis-Based Gene Pyramiding: a Way Forward for a Combined Horizontal and Vertical Resistance in Plant

Jane Choene Segolela, Obiro Cuthbert Wokadala, and Naser Aliye Feto

15.1 Introduction

Crop loss due to pests is a major concern worldwide which could reach as high as 70% if preventive measures with either pesticide, natural enemies, host plant resistance or other controls are not utilized. About 67,000 pest species damage crops, of which 9000 contribute to insect species and mites (Ibrahim and Shawer 2014). Furthermore, insects are the primary direct cause of crop losses, whereas the indirect object is by the impaired quality of the products and their roles as vectors of various plant pathogens (Kumar et al. 2006). Apart from that, crops contribute a significant part of the world food supply to maintain the growing human population (Osman et al. 2015; Oerke 2006). Most developing countries still rely on agriculture as their primary source of food. Hence, the development and protection of agriculture are very critical in sustaining the growing human population worldwide.

Crop protection could be achieved, though not sustainably, through conventional methods like application of chemical pesticides and other cultural approaches. Yet, bioprotection of crop plants from insect pests via application of natural enemies is relatively sustainable and environmentally friendly. *Bacillus thuringiensis* (Bt)-based technologies like Bt crops have widely been used with relative success. However, recent reports indicated that Bt crops are losing sustainability as insects pests are learning to somehow develop resistance to withstand pressure exerted by Bt crops, thereby compromising Bt crop's resistance to pests [for further reading

J. C. Segolela · N. A. Feto (✉)
OMICS Research Group & Facility, Department of Biotechnology, Vaal University of Technology, Vanderbijlpark, South Africa
e-mail: naserf@vut.ac.za

O. C. Wokadala
Agricultural Research Council, Institute of Tropical and Subtropical Crops, Nelspruit, Mpumalanga, Republic of South Africa

M. T. Islam et al. (eds.), *Bacilli and Agrobiotechnology: Phytostimulation and Biocontrol*, Bacilli in Climate Resilient Agriculture and Bioprospecting,
https://doi.org/10.1007/978-3-030-15175-1_15

you may refer to a review by Feto (2016)]. Hence, the decades old 'single-gene-Bt crop' technology started to fail. Thus, there is a need to come up with an alternative sustainable tool to fill the void.

On the other hand, the novel Bt-based gene pyramiding could serve as an alternative to 'single-gene-Bt crop'. Though the tool has yet to go mainstream, success has already been reported (Jain et al. 2017).

Therefore, this review will explore further on Bt as the most suitable natural source of genes that could be used to provide durable resistance in crops, since the multiple genes that make up the pyramid will render the transgene resistant to multiple pests.

Therefore, in this chapter different crop protection strategies are compared, and efforts have been made to underline that the gene pyramiding could be a better alternative if not the only one for sustainable pest management.

15.2 Prevalence of Crop Loss Worldwide

The incidence of crop loss worldwide is mainly due to insect attacks which can be as high as 70% if preventive measures are not used (Maxmen 2013). Previous reports have summarized the loss of crops to various insects and bacterial and fungal pests (Table 15.1). Most affected crops are wheat (Bahri et al. 2011), rice (Niu

Table 15.1 Percentage yield loss of crops mainly attributed to insect, bacterial, viral or fungal pests

Crop	Yield loss (%)	Pest	Origin	References
Wheat	70	Yellow rust (*Puccinia striiformis*)	Pakistan	Bahri et al. (2011)
Rice	5–10 and can reach 60% if the conditions are favourable	Rice stem borer (*Helicoverpa zea*), Brown planthopper (*Nilaparvata lugens*)	Pakistan	Liu et al. (2016)
Cowpea	40–68 if heavily infected	*Xanthomonas axonopodis pv. Vignicola*	Nigeria	Okechukwu et al. (2010) and Neya et al (2015)
Soybean	61	*Helicoverpa zea* (Corn earworm)	USA	Abudulai et al. (2012)
	25.8–42.8	*Aspavia amigera* (Stinkbug)	Ghana	Musser et al. (2016)
		Nezara viridula (Southern green stinkbug)		
Maize	Total of 57	Insects and pathogens	Kenya	Grisley (1997) and Anderson et al. (2016)
	47	Stalk borers (*Busseola fusca*) and weevils (*Sitophilus oryzae*)		
	10	Head smut (*Sporisorium reilianum*), *MSV* (Maize streak virus)		

et al. 2017; Liu et al. 2016), cowpea (Okechukwu et al. 2010), soybean (Musser et al. 2016; Abudulai et al. 2012) and maize (Anderson et al. 2016; Grisley 1997). These crops are usually affected by pests such *Xanthomonas axonopodis pv. Vignola* (Xav), *Nezara viridula* and many more which require immediate attention.

15.3 Crop Protection Methods Against Pests

In the past, humans have searched for crops that can survive and produce under different biotic and abiotic stresses. Furthermore, farmers avoided yield loss through searching pest-resistant crops by collecting the seeds from only the highest yielding crops in their fields (Ibrahim and Shawer 2014) or through the application of chemical pesticides. Although chemical pesticides do result in reduced crop yield loss, more money is spent each year globally for inadequate control measures. Hence there is still a need to search for adequate protection of crops against pests.

15.3.1 Conventional or Traditional Methods and Its Drawbacks

Traditional crop protection method such as chemical control of pests was used as the most effective and attractive strategy in the previous century, during the 1940s and 1950s (Malav et al. 2016; Oerke 2006; Graves et al. 1999). Moreover, conventionally grown crops use more pesticides, and that represents the worst effect of chemically dependent agriculture. Although chemical pesticides are effective and had guaranteed a production increase in agriculture during the last 40 years, their continuous use is a primary cause of resistance and environmental concern (Mekonnen et al. 2017; Oerke 2006; Graves et al. 1999). These led to contamination of water and food sources, as well as the poisoning of nontarget beneficial pests and development of pests that are resistant to the chemical pesticides (Kumar et al. 2008; Scheyer et al. 2005).

Thus, the global public concern to seek alternative methods to control pests such as insects and fungal pathogens has increased due to the adverse effect of the application of chemical pesticides (Ibrahim and Shawer 2014). Furthermore, one approach could be the use of biological control methods such as biopesticides and entomopathogenic microorganisms like bacteria, fungi and viruses that include the development of technologies that would allow the insertion and functional expression of foreign genes in plant cells (Malav et al. 2016; Danny et al. 1992). Moreover, biological control reduces expenses and health hazards associated with pesticide formulations (Kouser and Qaim 2011). Hence, Bt has been used for several years as an alternative crop protection method to conventional methods.

15.3.2 Bt Biopesticide Methods

Currently, the application of Bt as a biological control method has increased crop production. Furthermore, Bt biopesticides are more effective as compared to the use of chemical pesticides which attribute more adverse effect due to contamination of the environment and food products and lead to human health problems (Tu et al. 2000). The insecticidal activity of Bt bacterium is due to the presence of the parasporal crystals (*cry*) which are formed during the sporulation phase of the bacterium and are assembled by the *cry* proteins that are expressed by the *cry* genes (Crickmore et al. 2017; Schnepf et al. 1998).

The Bt species has considerable variability due to the number of strains isolated around the world (Palma et al. 2014; Roh et al. 2007), the number of serotypes known to date (Roh et al. 2009) and the high number of crystal (*cry*) gene sequences accumulated so far. Despite the variability observed within this species, there is some uniformity in at least part, which shows some reflection on the five conserved blocks in the gene structure that is present in almost all the *cry* genes (De Maagd et al. 2001). Bt toxins create a heterozygous family of 74 different types of proteins (*cry*1–*cry*74) that are toxic to numerous insect pests such as *lepidopteran*, *coleopteran*, *dipteran*, *hemipteran*, some nematodes and snails' species that cause a severe damage to economically important crops (Crickmore et al. 2017; Palma et al. 2014).

The *cry* toxins belonging to three domain families share similar and conserved three domain structures (Fig. 15.1) which display their differences in amino acid sequences (Pardo-Lopez et al. 2013). Also, domain I constitutes of seven α-helix clusters that are subjected to proteolytic cleavage in all three-domain *cry* proteins during toxin activation (Fig. 15.1). It is usually referred to as perforating domain and is located towards the N-terminus which may be responsible for toxin membrane insertion and pore formation (Ben-Dov 2014; Xu et al. 2014). Moreover, domain II or middle domain is responsible for toxin-receptor interactions, and it consists of three antiparallel β-sheets (Xu et al. 2014; Jenkins and Dean 2000). Besides, domain III which is usually referred to as the galactose-binding domain has two antiparallel β-sheet sandwiches (Fig. 15.1), which are also involved in receptor binding and pore formation (Xu et al. 2014). In addition, domain IV

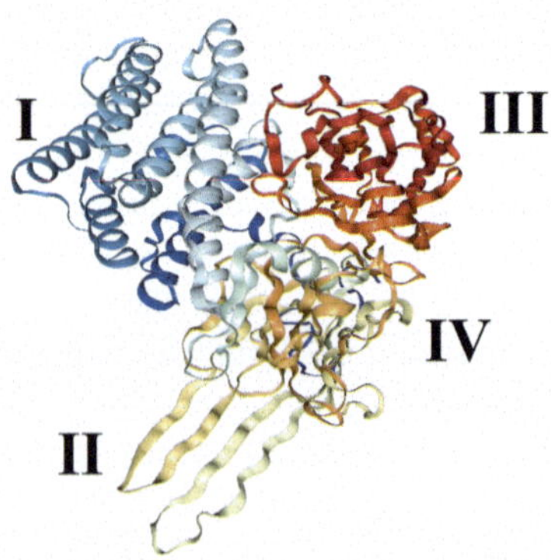

Fig. 15.1 The 3-D structure of Cry2Aa toxin (PDB accession number 1I5P) showing the four domain toxins (I–IV) produced by Bt (Soberon et al. 2016)

(Fig. 15.1) is mainly composed of alpha helices which resemble structural domains such as spectrin- or fibrinogen-binding complement inhibitor (Soberon et al. 2016).

15.3.2.1 Mechanism of Bt Biopesticides

The processing of the crystals relies on the solubilization of the toxins in the alkaline midgut of the insect pest and then activated by proteolytic digestion of the specific serine proteases (Palma et al. 2014). Interestingly, consumption of Bt toxins is found to be safe to humans because the intestinal walls of mammals do not have endotoxin receptor necessary for the toxic effect mainly due to the acidic conditions, and thus, the proteins tend to get degraded quickly in the stomach (Mekonnen et al. 2017). Some reports showed that *cry* gene specificity and activity could be influenced by other factors such as associated with toxin processing or stability in the insect midgut apart from the receptor binding (Jurat-Fuentes and Crickmore 2017). Moreover, *cry* genes are co-localized with other genes such as vegetative insecticidal proteins (*vip*) forming the insecticidal pathogenicity island (PAI) (Zhu et al. 2015).

15.4 Possible Challenges of Bt Toxins and Resistant Breakdown of Bt Crops

Bt has been studied for decades and is a used bacterial control agent to date. Nevertheless, several pest species have acquired field resistance to the most used Bt toxins and more severely to those included in transgenic crops (Peralta and Palma 2017). The currently used Bt toxins have not provided durable resistance due to the observed Bt resistance breakdown in the current Bt crops (Table 15.2) (Peralta and Palma 2017). An explanation for this may be due to the insects and pathogenic

Table 15.2 Resistance breakdown in the Bt crops

Bt crop	Insect pest	Resistance type	Origin	References
Cotton	Corn earworm (*Helicoverpa zea*)	A mild resistance that led to reduced efficacy of 2nd generation of crops	Tucson (USA)	Brévault et al. (2013)
	Pink bollworm (*Pectino gossypiella*)	Field resistance to Cry1Ac Bollgard® I	India	Tabashnik and Carrière (2010) and Dhurua and Gujar (2011)
Corn	Fall armyworm (*Spodoptera frugiperda*)	A mild resistance that led to reduced efficacy of 2nd generation of crops	Brazil	Santos-Amaya et al. (2015)
Maize	African caterpillar (*Busseola fusca*)	Field resistance and is dominant	South Africa	Campagne et al. (2013)

diversity displayed by most pests which lead to a rapid breakdown of specific resistance genes (Peralta and Palma 2017; Geffroy et al. 1999). Even though numerous reports involve the insect resistance over Bt-based formulations, field-evolved resistance has occurred which is promoted by the selective pressure applied over some insect populations (e.g. the *lepidopteran Trichoplusiani*) (Song et al. 2015). Furthermore, this occurs frequently in the most used Bt crops in agriculture, especially those from the first generation that express only one protein and plants that have multiple genes which have similar toxins.

Previously, some reports indicated the reduced efficacy of second-generation Bt cotton and corn harbouring *cry*1Ac + *cry*2Ab and *cry*1A.105 + *cry*2Ab against *Helicoverpa zea* and *Spodoptera frugiperda*, respectively (Table 15.2) (Santos-Amaya et al. 2015; Brévault et al. 2013). Furthermore, some authors reported on field resistance of Bt spray (Table 15.2) containing *cry*1C or *cry*1Ac observed in maize and cotton (Brévault et al. 2013; Campagne et al. 2013). Hence, there is a need in continuous search for novel Bt strains that have a broad spectrum range and could potentially circumvent the resistant issue, thus requiring the novel strategies that could anticipate the evolutionary responses of insects pests (Peralta and Palma 2017). Therefore, the best possible crop protection strategy could be pyramiding genes in such a way that could address both vertical and horizontal resistance in plants.

15.5 Transgenic Bt Crops

Genes from Bt have currently received increased attention due to their broad range of biotechnological applications, especially in agriculture for biocontrol of harmful insects and fungal pathogens (Kuddus and Ahmad 2013). These single or multiple *cry*-based genes could be inserted into crops, resulting in transgenic crops that are resistant to insects and fungal pathogens.

15.5.1 Single Cry-Based Bt Crops

Single *cry*-based Bt crops are crops incorporated with only a single *cry* toxin. However, single *cry*-based Bt crops are most likely prone to resistant breakdown than multiple *cry*-based Bt crops (Keshavareddy and Kumar 2018). These might be due to the pest developing resistance towards the crop mainly because the pests tend to adapt to the treatment conditions very quickly than in multiple *cry*-based crops. Moreover, commercialization of Bt crops such as maize, cotton and soybean worldwide has significantly reduced the application of synthetic pesticides (Keshavareddy and Kumar 2018; Ferré and Van Rie 2002). In addition, some reports showed the effective control of Bt rice such as KMD (*cry*1Ab), T1c-9 (*cry*1C) and T2A-1 (*cry*2A) to target *lepidopteran* insects (Table 15.3) including stem borers and leaf folders (Wang et al. 2016; Zheng et al. 2011; Chen et al. 2005).

15.5.2 Multiple Cry-Based Bt Crops

The multiple *cry*-based Bt crops are crops incorporated with two or more *cry* toxins. Even though multiple *cry*-based genes were used in transgenic crops before such as chickpea and brassica (Table 15.3), the broad-spectrum range has not been considered, or the incorporated multiple genes shared similar toxins. Hence, there is a resistant development as well (Meenakshi et al. 2011; Cao et al. 2008). Therefore, a wide range of sequences known to date is attributed to intense interest in finding novel *cry* proteins with alternative toxins that has a broad spectrum to manage the resistant breakdown observed in the current Bt crops (Ibrahim and Shawer 2014). These could be done with the application of Bt-based gene pyramiding as a genetic tool for inserting multiple genes that do not share the same toxins. As studies show Bt crops consisting of a single *cry*-based gene or multiple *cry*-based genes sharing the same toxins tends to be more prone to pest resistance (Keshavareddy and Kumar 2018).

Table 15.3 Successful genetically engineered crops with Bt genes with broad spectrum that provide durable resistance

Crop type	Trait	Engineered genes	References
Corn	Asian corn borer	*Cry1Ie* and *Cry1Ac*	Jiang et al. (2016)
Rice	Yellow stem borer resistance	*Cry1Ab/Cry1Ac*	Datta et al. (2002) and Cheng et al. (1998)
	Striped stem borers resistance	*Cry1Ab, Cry1Ac,* and *Cry2A*	Wang et al. (2016)
	Stem borers and bacterial blight disease resistance	*Cry1Ab/Cry1Ac* and *Xa21*	Jiang et al. (2004)
	Leaf folder, yellow stem borer, and brown planthopper resistance and increasing lectin content	*Cry1Ac, Cry2A,* and *gna*	Maqbool et al. (2001)
	Lepidopteran resistance and increasing lysine content	*Cry1Ac* and *LRP*	Liu et al. (2016)
Cotton	Insect pest resistance	*Cry1Ac*, and *Cry2Ac*	Gahan et al. (2005)
	Bollworm resistance	*Cry1Ac*, and *Cry2Ab*	Jackson et al. (2004)
Broccoli	Diamondback moth resistance	*Cry1Ac* + *Cry1C*	Cao et al. (2002)
Brassica	Diamondback moth larvae and lepidopteran insect resistance	*Cry1Ac* + *Cry1C*	Cao et al. (2008)
Chickpea	Lepidopteran resistance	*Cry1Ac* + *Cry1Ab*	Meenakshi et al. (2011) and Ahmed et al. (2017)
Soya bean	Lepidopteran resistance	*Cry1Ac* + *corn earworm QTL*	Walker et al. (2002) and Malav et al. (2016)

15.6 Gene Pyramiding Method

Gene pyramiding is a method of assembling or stacking multiple genes to improve durable resistance in crops against insects or diseases which is crucial for stable food production. Moreover, breeding resistance crops with either single or multiple Bt-based *cry* genes is the most cost-effective and environment-friendly strategy for resistance management. The advantage of gene pyramiding is that it uses the same strategy as that of the pesticidal mixture to broaden the resistance spectrum in crops. In addition, if two or more resistant genes are incorporated in a crop, it is less likely for the crop to be attacked by a pathogen race resistant to both genes or for the plant to lose both genes at the same time (Meenakshi et al. 2011). Furthermore, due to biotic factors, gene pyramiding is a cost-effective and environmentally friendly method used to manage crop production. Hence it has become the most used method for developing durable resistance in crops against pests (Meziadi et al. 2016; Fukuoka et al. 2015).

Previous reports on Bt-based gene pyramiding have shown an outstanding performance against insects where Bt toxins were incorporated in rice (Ye et al. 2009; Chen et al. 2008). Moreover, the integrated genes into elite cultivars with different genetic background were introduced by sexual crossing. Hence, in field evaluations, the improved lines also showed excellent efficacy against the target insects (Liu et al. 2016; Yang et al. 2011). So recently, different types of gene pyramiding such as conventional gene pyramiding and molecular gene pyramiding are widely used to obtain durable resistance in crops (Meenakshi et al. 2011).

15.6.1 Conventional Gene Pyramiding

Conventional gene pyramiding also known as serial gene pyramiding is a method where genes are arranged in the same plant one after another. These include pedigree crossing, backcross breeding and recurrent selection (Table 15.4). The identification of sources of useful genes is very slow using traditional methods. Hence, breeders' capability to trace the presence or absence of the target genes is limited, thus resulting in the limited number of genes incorporated into selected cultivars (Malav et al. 2016).

15.6.2 Molecular Gene Pyramiding

Molecular gene pyramiding also referred to as simultaneous gene pyramiding is a method where genes are arranged at the same time in a plant (Srivastava et al. 2017). These include marker-assisted selection and transgenic methods (Table 15.4). The differences among the two gene pyramiding methods are summarized below

Table 15.4 Differences between conventional and molecular gene pyramiding

Conventional gene pyramiding	Molecular gene pyramiding
Pedigree: *Is suitable when resistance is administered by the significant genes (Malav et al.* 2016*).*	Marker-assisted: *It involves the use of molecular markers for the selection of desired traits and identification of genomic regions associated with different major diseases (e.g., blast resistance). These markers are highly precise and reduce the selection time which makes this approach outstand conventional approach (Mekonnen et al.* 2017*; Srivastava et al.* 2017*).*
Backcrossing: *It involves the substitution of the desired gene from the donor parent to the recipient parent. It is mainly used to decline the donor genome content into the progenies (Allard and Allard* 1999*; Mekonnen et al.* 2017*).*	Transgenic: *It involves methods such as Agrobacterium transformation which is used to transfer a gene/s of interest into plant cells. This method ensures the stable integration of DNA of the desired gene into the genome (Srivastava et al.* 2017*).*
Re-current selection: *It allows for shorter breeding cycles. Besides, more specific follow-up of genetic gains is involved and provides an opportunity to develop a broad range of genetic diversity in breeding lines (Srivastava et al.* 2017*).*	

(Table 15.4). In addition, molecular gene pyramiding such as transgenic method is more advantageous over other pyramiding methods (Keshavareddy and Kumar 2018). Although there has been a success in Bt crop production, there are some drawbacks concerning Bt-based gene pyramiding.

15.7 Potential Challenges of Gene Pyramiding

Although gene pyramiding is a widely adopted strategy for improvement of crops against resistant effects, there are certain drawbacks associated with this strategy. In addition, the reliability of phenotyping at an individual level is minimal since the presence of target traits must first be confirmed. Phenotyping influences the inheritance model of genes for the target traits, linkage and pleiotropism between the target traits at an individual level (Malav et al. 2016; Riaz et al. 2006).

Another drawback involves the limitation of successfully pyramided transgenic crops for enhanced fungal and bacterial resistance (Summers and Brown 2013; Punja 2006; Schnepf et al. 1998). These may be due to two primary life strategies of pathogens, namely, biotrophy and necrotrophy. Thus, biotrophic pathogens essentially act as a sink for the hosts' anabolic assimilates which keep it alive, while necrotrophic pathogens consume the hosts' tissues as invaded. As a result, plants developed different approaches to deal with these two strategies (Summers and Brown 2013; Punja 2006) which are not obtained through genetic engineering.

Lastly, to avoid recognition by host (R) genes, the pathogen avirulence (Avr) gene undergoes strong diversifying selection or mutation (Ferry et al. 2004). The low level of pathogenic resistance by some transgenic crops coupled with a negative perception of genetic engineering-modified crops has resulted in few transgenic crops (Palma et al. 2014) being brought to the market due to the relatively small number of transgenic crops available (Mekonnen et al. 2017).

Apart from those resistant to fungal and bacterial pathogens, virus-resistant crops are not commercially available (Collinge et al. 2007). Hence, many transformation strategies have been used to increase fungal, bacterial and viral resistance in crops (Mekonnen et al. 2017). In addition, this includes introgressing R genes and introducing genes coding for antimicrobial compounds such as chitinase and glucanase enzymes that break down the fungal cell walls (chitin or glucan) and also upregulating defence pathways through promoter transfer, disarming host susceptibility genes, detoxifying pathogen virulence factors (toxins), increasing structural barriers and silencing essential pathogen genes (RNA silencing, RNA interference or RNAi) (Vincelli 2016; Collinge et al. 2007; Schnepf et al. 1998). Hence, two R genes were introgressed to develop rice cultivars resistant to bacterial blight and bacterial streak diseases in a study conducted by Zhou et al. (2008).

15.8 Conclusion and Future Prospects

The addressed reports presented insights into the fundamental basis of Bt isolates with broad spectrum, subjected to screening programmes to evaluate their insecticidal activity. The current review shows that the production and continuing development of Bt crops has been a major scientific success up to date which is deployed by the expression of Bt toxins. However, several studies documented that pests are developing resistance to Bt crops or Bt biopesticides. This situation is mostly observed in the current Bt crops that are incorporated with *cry*2 genes and lower. Hence, there is still a need to explore other effective strategies that could stand on its own or could be integrated with other control measures to diversify the resistance management tools.

Therefore, in this chapter, we tried to compare different crop protection strategies and make a point that gene pyramiding could be a better alternative if not the only one. These could include the involvement of pyramiding Bt toxins with other genes such as phytase, *vip*3 and other genes to broaden the spectrum. Another management tool could consist of the crop rotation method of cultivating Bt crops with other non-Bt crops to try and confuse the pests. However, there must be an assurance that the development of pest resistance genes does not compromise the protection of produced Bt crops. Furthermore, there are commercially available Bt crops with single or multiple toxins which reduced the application of chemical pesticides.

This review stipulated possible challenges that can inhibit the efficiency of gene pyramiding. Hence, extensive and precise phenotyping is required to counteract the difficulties in gene pyramiding. These involve the dissection of phenotypes into components that can improve the heritability, thus aiding the understanding of biological systems causing the phenotype (Varshney et al. 2005). Another strategy is phenotyping characterization of large mutagenized populations and tilling populations which could link a gene with phenotype. Apart from the challenges of gene pyramiding and current resistant breakdown in Bt crops, Bt will continue to play a significant role as a candidate bacterium for pyramiding multiple toxin genes into crops for resistant management due to its broad spectrum of resistance from the natural origin. Very recently some studies have been carried out to pyramid different Bt-sourced *cry* genes. Such kind of strategy is a relatively recent advancement that should be explored further.

References

Abudulai M, Salifu AB, Opare-Atakora D, Haruna M, Denwar NN, Baba II (2012) Yield loss at the different growth stages in soybean due to insect pests in Ghana. Arch Phytopathol Plant Protect 45:1796–1809

Ahmed N, Saini J, Sharma R, Seth MJHJoAR (2017) Performance of chickpea under organic and inorganic sources of nutrients at different soil moisture regimes in chickpea-okra cropping system. Himachal J Agric Res 43:23–28

Allard RW, Allard RW (1999) Principles of plant breeding. Wiley, New York

Anderson SJ, Simmons HE, French-Monar RD, Munkvold GP (2016) Susceptibility of maize inbreds and incidence of symptomless infection by the head smut pathogen, sphacelotheca reilana. Plant Health Prog 17:1–5

Bahri B, Shah S, Hussain S, Leconte M, Enjalbert J, De Vallavieille-Pope C (2011) Genetic diversity of the wheat yellow rust population in Pakistan and its relationship with host resistance. Plant Pathol 60:649–660

Ben-Dov E (2014) *Bacillus thuringiensis subsp. Israelensis* and its dipteran-specific toxins. *Toxins* 6:1222–1243

Brévault T, Heuberger S, Zhang M, Ellers-Kirk C, Ni X, Masson L et al (2013) Potential shortfall of pyramided transgenic cotton for insect resistance management. Proc Natl Acad Sci 110:5806–5811

Campagne P, Kruger M, Pasquet R, Le Ru B, Van Den Berg J (2013) Dominant inheritance of field-evolved resistance to bt corn in busseola fusca. PLoS One 8:e69675

Cao J, Zhao J-Z, Tang J, Shelton A, Earle E (2002) Broccoli plants with pyramided cry1ac and cry1c bt genes control diamondback moths resistant to cry1a and cry1c proteins. Theor Appl Genet 105:258–264

Cao J, Shelton AM, Earle ED (2008) Sequential transformation to pyramid two bt genes in vegetable indian mustard (*brassica juncea* l.) and its potential for control of diamondback moth larvae. Plant Cell Rep 27:479

Chen H, Tang W, Xu C, Li X, Lin Y, Zhang Q (2005) Transgenic indica rice plants harboring a synthetic cry2a* gene of *bacillus thuringiensis* exhibit enhanced resistance against lepidopteran rice pests. Theor Appl Genet 111:1330

Chen H, Zhang G, Zhang Q, Lin Y (2008) Effect of transgenic *bacillus thuringiensis* rice lines on mortality and feeding behavior of rice stem borers (lepidoptera: Crambidae). J Econ Entomol 101:182–189

Cheng X, Sardana R, Kaplan H, Altosaar I (1998) Agrobacterium-transformed rice plants expressing synthetic cryia (b) and cryia (c) genes are highly toxic to striped stem borer and yellow stem borer. Proc Natl Acad Sci 95:2767–2772

Collinge DB, Lund OS, Thordal-Christensen H (2007) What are the prospects for genetically engineered, disease resistant plants? In: Sustainable disease management in a european context. Springer, Dordrecht

Crickmore N. (2017) *Bacillus thuringiensis* Toxin Classification. In: Fiuza L., Polanczyk R., Crickmore N. (eds) *Bacillus thuringiensis* and *Lysinibacillus sphaer*icus. Springer, Cham

Danny J, Llewellyn M, Brown Y, Cousins Y, Hartweck D, Last A, et al (1992) The science behind transgenic cotton plants. In: Proceedings of 6th Australian cotton conference, Broadbeach and Queensland, Australia

Datta K, Baisakh N, Thet KM, Tu J, Datta S (2002) Pyramiding transgenes for multiple resistance in rice against bacterial blight, yellow stem borer and sheath blight. Theor Appl Genet 106:1–8

De Maagd RA, Bravo A, Crickmore N (2001) How *bacillus thuringiensis* has evolved specific toxins to colonize the insect world. Trends Genet 17:193–199

Dhurua S, Gujar GT (2011) Field-evolved resistance to bt toxin cry1ac in the pink bollworm, pectinophora gossypiella (saunders)(lepidoptera: Gelechiidae), from India. Pest Manag Sci 67:898–903

Ferré J, Van Rie J (2002) Biochemistry and genetics of insect resistance to b acillus thuringiensis. Annu Rev Entomol 47:501–533

Ferry N, Edwards MG, Gatehouse JA, Gatehouse AM (2004) Plant–insect interactions: molecular approaches to insect resistance. Curr Opin Biotechnol 15:155–161

Feto NA (2016) Bacillus spp. and their biotechnological roles in green industry. Bacilli and Agrobiotechnology. In: Islam MT, Rahman MM, Pandey P, Jha CK, Aeron A (eds) Bacilli and Agrobiotechnology. Springer Verlag. ISBN: 978-3-319-44408-6, pp 143–162. https://doi.org/10.1007/978-3-319-44409-3

Fukuoka S, Saka N, Mizukami Y, Koga H, Yamanouchi U, Yoshioka Y et al (2015) Gene pyramiding enhances durable blast disease resistance in rice. Sci Rep 5:7773

Gahan LJ, Ma Y-T, Macgregorcoble ML, Gould F, Moar WJ, Heckel DG (2005) Genetic basis of resistance to cry1ac and cry2aa in heliothis virescens (lepidoptera: Noctuidae). J Econ Entomol 98:1357–1368

Geffroy V, Sicard D, De Oliveira JC, Sévignac M, Cohen S, Gepts P et al (1999) Identification of an ancestral resistance gene cluster involved in the coevolution process between phaseolus vulgaris and its fungal pathogen colletotrichum lindemuthianum. Mol Plant-Microbe Interact 12:774–784

Graves JB, Leonard BR, Ottea J (1999) Chemical approaches to managing arthropod pests. In: Handbook of Pest Management. Marcel Dekker, Inc, New York, pp 449–486

Grisley W (1997) Crop-pest yield loss: a diagnostic study in the Kenya highlands. Int J Pest Manag 43:137–142

Ibrahim RA, Shawer DM (2014) Transgenic bt-plants and the future of crop protection (an overview). Int J Agric Food Res 3(1):14–40

Jackson R, Bradley J Jr, Van Duyn J (2004) Performance of feral and cry1ac-selected helicoverpa zea (lepidoptera: Noctuidae) strains on transgenic cottons expressing one or two *bacillus thuringiensis ssp. Kurstaki* proteins under greenhouse conditions. J Entomol Sci 39:46–55

Jain D, Sunda SD, Sanadhya S, Nath DJ, Khandelwal SK (2017) Molecular characterization and pcr-based screening of cry genes from *bacillus thuringiensis* strains. 3 Biotech 7:4

Jenkins JL, Dean DH (2000) Exploring the mechanism of action of insecticidal proteins by genetic engineering methods. In: Genetic engineering. Springer, Boston

Jiang G, Xu C, Tu J, Li X, He Y, Zhang Q (2004) Pyramiding of insect-and disease-resistance genes into an elite indica, cytoplasm male sterile restorer line of rice, 'minghui 63'. Plant Breed 123:112–116

Jiang F, Zhang T, Bai S, Wang Z, He K (2016) Evaluation of Bt corn with pyramided genes on efficacy and insect resistance management for the Asian corn borer in China. PLoS One 11:e0168442

Jurat-Fuentes JL, Crickmore N (2017) Specificity determinants for cry insecticidal proteins: insights from their mode of action. J Invertebr Pathol 142:5–10

Keshavareddy G, Kumar A (2018) Characterization of bt transgenic plants: a review. Int J Curr Microbiol App Sci 7:3035–3051

Kouser S, Qaim M (2011) Impact of bt cotton on pesticide poisoning in smallholder agriculture: a panel data analysis. Ecol Econ 70:2105–2113

Kuddus M, Ahmad I (2013) Isolation of novel chitinolytic bacteria and production optimization of extracellular chitinase. J Genet Eng Biotechnol 11:39–46

Kumar S, Chandra A, Pandey K (2006) Genetic transformation of lucerne (*medicago sativa* l.) for weevil (*hypera postica*) resistance. In: Extended Summaries, national seminar transfiguration crops Indian agriculture: status, risks and acceptance, Hisar, India, pp 35–37

Kumar S, Chandra A, Pandey K (2008) *Bacillus thuringiensis* (bt) transgenic crop: an environment friendly insect-pest management strategy. J Environ Biol 29:641–653

Liu Y, Chen L, Liu Y, Dai H, He J, Kang H et al (2016) Marker assisted pyramiding of two brown planthopper resistance genes, bph3 and bph27 (t), into elite rice cultivars. Rice 9:27

Malav AK, Chandrawat I, Chandrawat KS (2016) Gene pyramiding: an overview. Int J Curr Res Biosci Plant Biol 3:22–28

Maqbool SB, Riazuddin S, Loc NT, Gatehouse AM, Gatehouse JA, Christou P (2001) Expression of multiple insecticidal genes confers broad resistance against a range of different rice pests. Mol Breed 7:85–93

Maxmen A (2013) Crop pests: under attack. Nature 501:S15–S17

Meenakshi MA, Singh AK, Sanyal I, Altosaar I, Amla DV (2011) Pyramiding of modified cry1ab and cry1ac genes of *bacillus thuringiensis* in transgenic chickpea (*cicer arietinum* l.) for improved resistance to pod borer insect helicoverpa armigera. Euphytica 182:87

Mekonnen T, Haileselassie T, Tesfaye K (2017) Identification, mapping and pyramiding of genes/ quantitative trait loci (qtls) for durable resistance of crops to biotic stresses. J Plant Pathol Microbiol 8:142

Meziadi C, Richard MM, Derquennes A, Thareau V, Blanchet S, Gratias A et al (2016) Development of molecular markers linked to disease resistance genes in common bean based on whole genome sequence. Plant Sci 242:351–357

Musser F, Catchot A Jr, Davis J, Herbert D Jr, Lorenz G, Reed T et al (2016) 2015 soybean insect losses in the southern us. Midsouth Entomologist 9:5–17

Neya BJ, Zida PE, Sereme, D, Lund OS, Traore O (2015) Evaluation of yield losses caused by cowpea aphid-borne mosaic virus (CABMV) in 21 cowpea (*Vigna unguiculata* (L.) Walp.) varieties in Burkina Faso. Pak J Biol Sci 18:304–313

Niu L, Mannakkara A, Qiu L, Wang X, Hua H, Lei C et al (2017) Transgenic bt rice lines producing cry1ac, cry2aa or cry1ca have no detrimental effects on brown planthopper and pond wolf spider. Sci Rep 7:1940

Oerke E-C (2006) Crop losses to pests. J Agric Sci 144:31–43

Okechukwu RU, Ekpo EJA, Okechukwu OC (2010) Seed to plant transmission of *Xanthomonas campestris* pv. *vignicola* isolates in cowpea. Afr J Agric Res 5(6):431–435

Osman G, Already R, Assaeedi A, Organji S, El-Ghareeb D, Abulreesh H, Althubiani A (2015) Bioinsecticide *bacillus thuringiensis* a comprehensive review. Egypt J Biol Pest Control 25:271

Palma L, Muñoz D, Berry C, Murillo J, Caballero P (2014) *Bacillus thuringiensis* toxins: an overview of their biocidal activity. Toxins 6:3296–3325

Pardo-Lopez L, Soberon M, Bravo A (2013) *Bacillus thuringiensis* insecticidal three-domain cry toxins: mode of action, insect resistance and consequences for crop protection. FEMS Microbiol Rev 37:3–22

Peralta C, Palma L (2017) Is the insect world overcoming the efficacy of *bacillus thuringiensis*? Toxins 9:39

Punja ZK (2006) Recent developments toward achieving fungal disease resistance in transgenic plants. Can J Plant Pathol 28:S298–S308

Riaz N, Husnain T, Fatima T, Makhdoom R, Bashir K, Masson L et al (2006) Development of indica basmati rice harboring two insecticidal genes for sustainable resistance against lepidopteran insects. S Afr J Bot 72:217–223
Roh JY, Choi JY, Li MS, Jin BR, Je YH (2007) *Bacillus thuringiensis* as a specific, safe, and effective tool for insect pest control. J Microbiol Biotechnol 17:547
Roh JY, Liu Q, Lee DW, Tao X, Wang Y, Shim HJ et al (2009) *Bacillus thuringiensis* serovar mogi (flagellar serotype 3a3b3d), a novel serogroup with a mosquitocidal activity. J Invertebr Pathol 102:266–268
Santos-Amaya OF, Rodrigues JV, Souza TC, Tavares CS, Campos SO, Guedes RN, Pereira EJ (2015) Resistance to dual-gene bt maize in spodoptera frugiperda: selection, inheritance, and cross-resistance to other transgenic events. Sci Rep 5:srep18243
Scheyer A, Graeff C, Morville S, Mirabel P, Millet M (2005) Analysis of some organochlorine pesticides in an urban atmosphere (Strasbourg, east of France). Chemosphere 58:1517–1524
Schnepf E, Crickmore N, Van Rie J, Lereclus D, Baum J, Feitelson J et al (1998) Bacillus thuringiensis and its pesticidal crystal proteins. Microbiol Mol Biol Rev 62:775–806
Soberon M, Monnerat R, Alejandra B (2016) Mode of action of cry toxins from bacillus thuringiensis and resistance mechanisms. In: Microbial toxins. Springer, Dordrecht, pp 1–13
Song X, Kain W, Cassidy D, Wang P (2015) Resistance to *bacillus thuringiensis* toxin cry2ab in *trichoplusia ni* is conferred by a novel genetic mechanism. Appl Environ Microbiol 81:5184–5195
Srivastava D, Shamim M, Kumar M, Mishra A, Pandey P, Kumar D et al (2017) Current status of conventional and molecular interventions for blast resistance in rice. Rice Sci 24:299–321
Summers R, Brown J (2013) Constraints on breeding for disease resistance in commercially competitive wheat cultivars. Plant Pathol 62:115–121
Tabashnik BE, Carrière Y (2010) Field-evolved resistance to bt cotton: bollworm in the us and pink bollworm in India. Southwest Entomol 35:417–424
Tu J, Zhang G, Datta K, Xu C, He Y, Zhang Q et al (2000) Field performance of transgenic elite commercial hybrid rice expressing *bacillus thuringiensis* δ-endotoxin. Nat Biotechnol 18:1101
Varshney RK, Graner A, Sorrells ME (2005) Genomics-assisted breeding for crop improvement. Trends Plant Sci 10:621–630
Vincelli P (2016) Genetic engineering and sustainable crop disease management: opportunities for case-by-case decision-making. Sustainability 8:495
Walker D, Boerma HR, All J, Parrott W (2002) Combining cry1Ac with QTL alleles from PI 229358 to improve soybean resistance to lepidopteran pests. Mol Breed 9:43–51
Wang YN, Ke KQ, Li YH, Han LZ, Liu YM, Hua HX, Peng YF (2016) Comparison of three transgenic bt rice lines for insecticidal protein expression and resistance against a target pest, chilo suppressalis (lepidoptera: Crambidae). Insect Sci 23:78–87
Xu C, Wang B-C, Yu Z, Sun M (2014) Structural insights into *bacillus thuringiensis* cry, cyt and parasporin toxins. Toxins 6:2732–2770
Yang Z, Chen H, Tang W, Hua H, Lin Y (2011) Development and characterisation of transgenic rice expressing two *bacillus thuringiensis* genes. Pest Manag Sci 67:414–422
Ye R, Huang H, Yang Z, Chen T, Liu L, Li X et al (2009) Development of insect-resistant transgenic rice with cry1c∗-free endosperm. Pest Manag Sci 65:1015–1020
Zheng X, Yang Y, Xu H, Chen H, Wang B, Lin Y, Lu Z (2011) Resistance performances of transgenic bt rice lines t2a-1 and t1c-19 against cnaphalocrocis medinalis (lepidoptera: Pyralidae). J Econ Entomol 104:1730–1735
Zhou Y, Choi Y-L, Sun M, Yu Z (2008) Novel roles of *bacillus thuringiensis* to control plant diseases. Appl Microbiol Biotechnol 80:563–572
Zhu X, Lei Y, Yang Y, Baxter SW, Li J, Wu Q et al (2015) Construction and characterisation of near-isogenic plutella xylostella (lepidoptera: Plutellidae) strains resistant to cry1ac toxin. Pest Manag Sci 71:225–233

Chapter 16
Probiotic Bacilli in Sustainable Aquaculture

Md. Mahbubur Rahman, A. Q. M. Robiul Kawser, and Md Tofazzal Islam

16.1 Introduction

Aquaculture is one of the fastest-growing food-producing industries in the world which became an important economic activity in many countries (Perez-Sanchez et al. 2014). It provides 47% of the total food fish supply to the global population. Fish and shrimp are important sources of animal protein. The global fish production from the capture fishery remained almost static for the last two decades, whereas, aquaculture production has been rapidly increasing and doubled within the last decade due to expansion, diversification, and intensification of aquaculture activities (FAO 2018).

The rapid expansion of aquaculture also increased the occurrence of infectious diseases resulting from high stocking densities and stress conditions that favored the spread of pathogens (Al-Faragi and Alsaphar 2012). The emergence of many different infectious diseases is now considered as one of the major limiting factors in aquaculture (Perez-Sanchez et al. 2014; Rahman et al. 2017). For the prevention and control of diseases, synthetic chemicals and antibiotics have been used in aquaculture (Cabello 2006; Taylor et al. 2011). However, the indiscriminate use of antibiotics in aquaculture has posed a threat to the environment and human health, and various pathogenic organisms including fish and human pathogens can grow resistance against these chemicals (Sharifuzzaman and Austin 2009; Rahman et al. 2017). To overcome this situation, several alternative strategies such as herbal

M. M. Rahman (✉) · M. T. Islam
Institute of Biotechnology and Genetic Engineering (IBGE), Bangabandhu Sheikh Mujibur Rahman Agricultural University, Gazipur, Bangladesh

A. Q. M. Robiul Kawser
Department of Aquaculture, Bangabandhu Sheikh Mujibur Rahman Agricultural University, Gazipur, Bangladesh

M. T. Islam et al. (eds.), *Bacilli and Agrobiotechnology: Phytostimulation and Biocontrol*, Bacilli in Climate Resilient Agriculture and Bioprospecting,
https://doi.org/10.1007/978-3-030-15175-1_16

supplements, immunostimulants, and vaccines have been proposed (Gatesoupe 1999; Balcázar et al. 2006; Rahman et al. 2017).

The most attractive alternative strategy to the user to reduce the use of antibiotics is to use natural substances with the broad-spectrum inhibitory activities, including use of probiotic bacteria antagonistic to fish pathogens (Chen et al. 2016). Probiotic microbes are friendly to the host and environment and can reduce the use of antibiotics and synthetic drugs in aquaculture. As probiotics secrets a wide and diverse range of antimicrobial substances, the chance of development of resistance in pathogens against probiotics is very low. Probiotics are also reported to boost the immune system of the host fishes. Besides the antagonistic and immune-inducing effects, probiotics are also reported to improve the nutritional quality of food; enhance digestion; secrete different hydrolytic enzymes, vitamins, and growth factors; accelerate growth performance; reduce larvae and fry mortality; and improve the water quality of the aquaculture system.

The genus *Bacillus* is ubiquitous and one of the most important groups of microorganism that produce a wide range of antimicrobial substances. In recent years, a varied range of antagonistic activities of *Bacillus* spp. against the most important and harmful aquatic pathogens have been reported (Ran et al. 2012; Paul 2018; Yi et al. 2018). *Bacillus* has also been reported to induce immunity and promote growth and survival of fish and shrimp. The fish probiotic bacilli produce a high number and diverse kind of enzymes and chemicals that can degrade organic wastes in aquaculture facilities supporting bioremediation and consequently preventing viral and bacterial diseases (Guo et al. 2009; Olmos et al. 2011; Olmos and Paniagua-Michel 2014). Considering the antagonistic, immunity-inducing, health-promoting, bioremediation and other beneficial properties, this review updates current knowledge on bacilli and discusses their application for promoting the sustainable aquaculture.

16.2 Aquaculture: Present Status and Challenges

Aquaculture is the production of aquatic organisms in the aquatic medium. It includes farming of fish, crustaceans, mollusks, aquatic plants, algae, and other organisms in freshwater and/or marine water under controlled condition. According to the Food and Agriculture Organization (FAO 2016), "aquaculture is defined as the farming of aquatic organisms including fish, mollusks, crustaceans, and aquatic plants. Farming implies some form of intervention in the rearing process to enhance production such as regular stocking, feeding, protection from predators, etc. Farming also implies individual or corporate ownership of the stock being cultivated."

Aquaculture is an important part of the global food and seafood supplies. Fish is one of the major animal protein sources for humans and other mammals. It comes from both capture fishery and aquaculture. During the year 2016, global fish production reached to a pick at 171 million tons with aquaculture representing 47% of the total and 53% if non-food uses are excluded (FAO 2018). Although the total

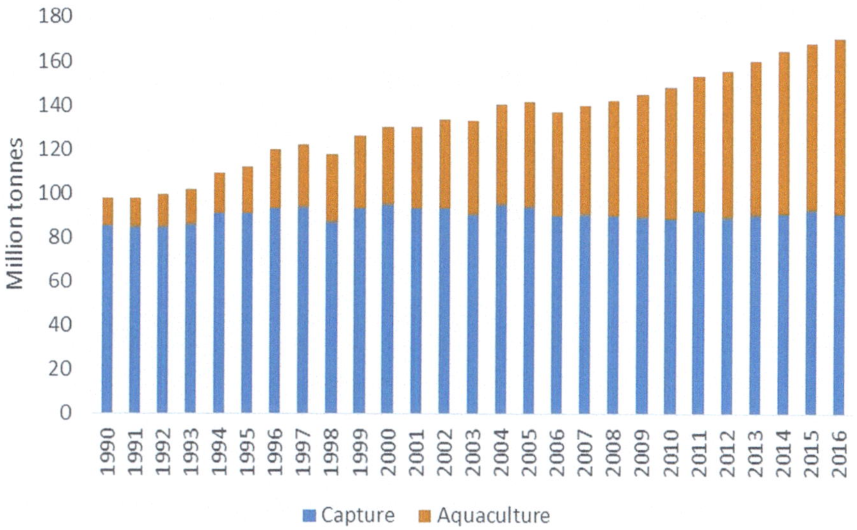

Fig. 16.1 World capture fisheries and aquaculture production (data source FAO 2018)

global production of capture fish remains relatively static since the late 1980s, aquaculture continues its impressive growth in the supply of fish for human consumption (Fig. 16.1). Due to the stagnation of captured fish from wild sources and overexploitation of popular marine species, combined with the growing demand for high-quality protein, farmers and commercial aquaculturist were encouraged to expand aquaculture with the domestication of new species.

Aquaculture sector significantly contributes to the income and livelihood of millions of peoples in the world. Approximately, 19.3 million people are engaged in aquaculture, while another 40.3 million people are engaged in fisheries activities (FAO 2018). Aquaculture is practiced in 194 out of 201 countries of which China is the highest producer of farmed fish in the year 2018. Other major aquaculture producers are India, Bangladesh, Myanmar, Cambodia, Indonesia, Vietnam, Thailand, and Egypt (Figs. 16.2, 16.3 and 16.4).

Although global aquaculture production is gradually increasing, the sector is facing several problems that hamper the expected production. The outbreak of different diseases is one of the major constraints in sustainable aquaculture. Due to the intensification of aquaculture, fish are stocked at high densities, and excess use of feed and fertilizer cause stress in fish. The stressed fish become susceptible to pathogens and easily infected by various pathogens. Huge numbers of these maladies or pathogens have no specific prescribed medications and consequently remain a critical obstacle for the financial practicality of aquaculture (Rodger 2016). Diseases cause substantial production and economic losses every year. Another important problem is the increased mortality of fish fry. The fish fry is highly vulnerable to biotic and abiotic stresses. Most of the fry mortality occurs during the weaning period when larvae go through the transition period from endogenous to exogenous food supply.

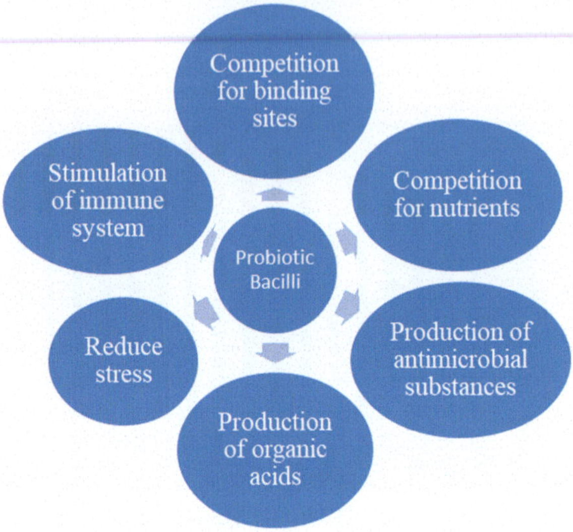

Fig. 16.2 Schematic diagram of the major modes of action of probiotic *Bacillus* spp. in fish and shrimp

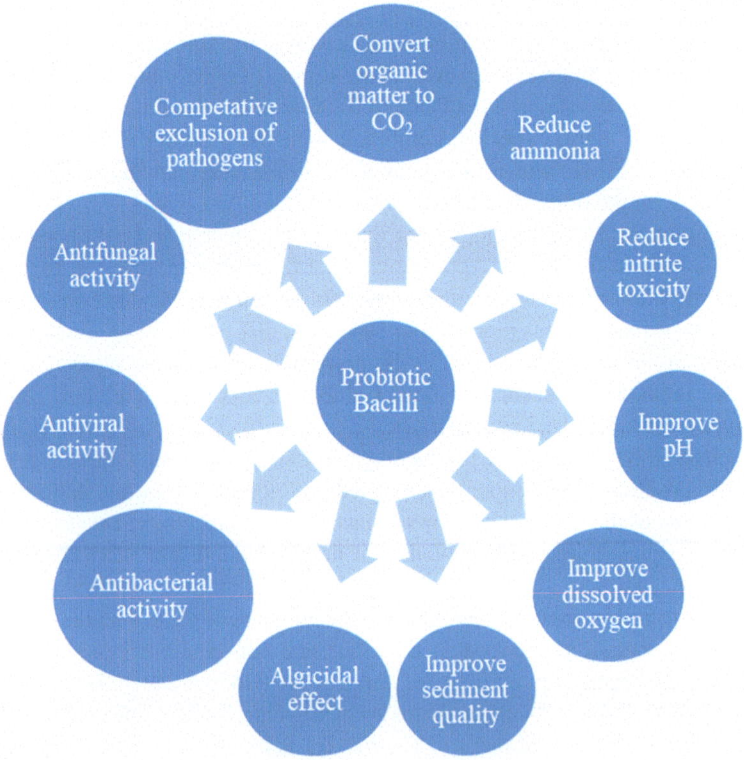

Fig. 16.3 Schematic diagram of the mode of action of probiotic *Bacillus* in the aquatic environment

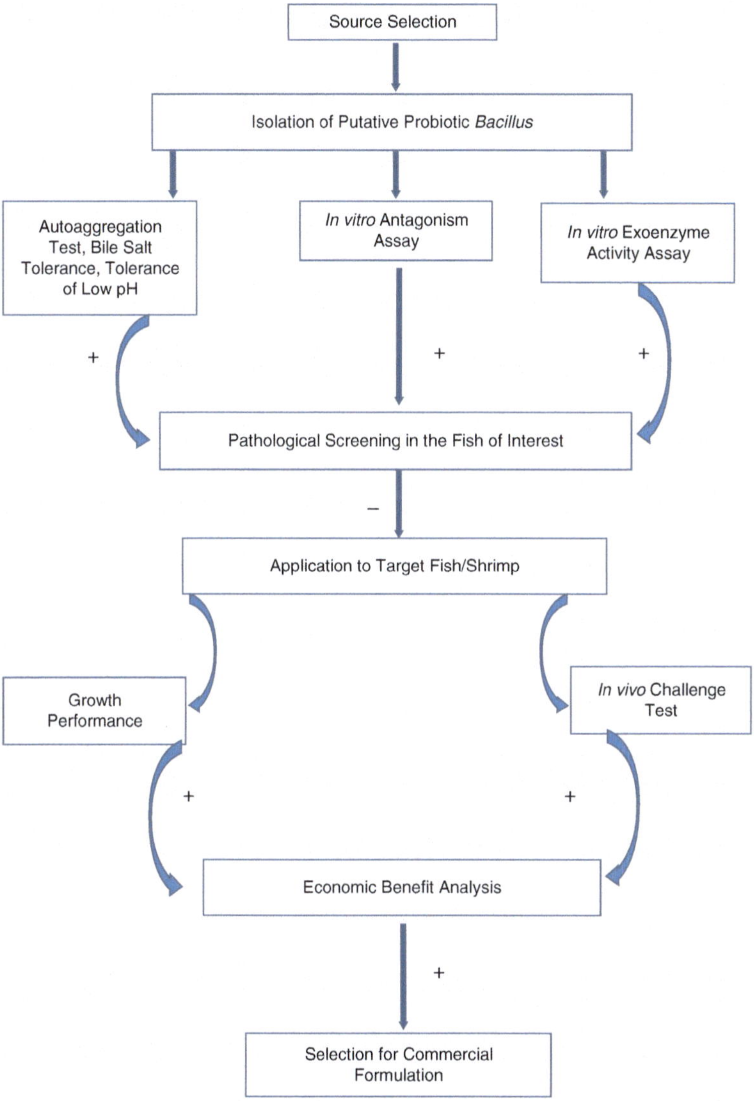

Fig. 16.4 Schematic diagram of the selection process of probiotic *Bacillus*

One of the major challenges in aquaculture is the pollution in the aquatic environment through discharge of agricultural pesticides, domestic wastes, industrial effluent, and oil spills. Pollution also occurs due to decomposition of excess food used in the culture facilities, metabolic wastes of fish, and drugs applied to fish for prevention and control of diseases. Pollution causes oxidative stress on the aquatic organism (Livingstone 2003) resulting in mortality. Availability of quality fish feed and increasing the price of fish feed is another obstacle in the development of aquaculture. Fish growth is mostly depended on the nutritional quality of fish feed. Low-quality feed is one of the major causes of low productivity in aquaculture.

16.3 Prospects of Probiotics in Aquaculture

The word "probiotic" derived from two Greek words "pro" and "bios." Pro means for and bios mean life; thus probiotic means for life (Reid et al. 2003). In 1905, Dr. Elie Metachnikoff first spotlighted the beneficial role played by some bacteria among farmers who consumed microbes-containing milk. However, the term probiotic was first introduced by Lilly and Stillwell (1965) to describe "substances secreted by one microorganism that stimulate the growth of another." According to the currently adopted definition of the Food and Agricultural Organization and World Health Organization, "probiotics are live microorganisms which when administered in adequate amounts confer a health benefit on the host" (FAO 2001; WHO 2001).

The concepts of probiotics have evolved over the years, integrating new ideas that are related to new investigations regarding its application (Sayes et al. 2018). Merrifield et al. (2010) propose a modified definition for probiotic used in aquaculture as "alive, dead or components of a microbial cell, which when administered via the feed or to the rearing water, benefiting the host by improving disease resistance, health status, growth performance, feed utilization, stress response or general vigor, which is achieved at least in part via improving the host microbial balance or the microbial balance of the ambient environment." Banerjee et al. (2017) reported that probiotic microorganisms improve immunity, help digestion, protect against pathogens, improve water quality, and promote growth and reproduction of fish that can be used as an alternative to antibiotics. Probiotics are reported to enrich nutritional value of feed, enhance enzymatic digestion of feed, provide growth-promoting factors, contribute to better feed conversion ratio, decrease fry mortality, promote growth, inhibit pathogenic microorganisms and antimutagenic and carcinogenic activity, improve the host immune system, and also improve the water quality (Irianto and Austin 2002a, b; Yanbo and Zirong 2006; Wang 2007; Kuhlwein et al. 2013; Ambas et al. 2013; Akhter et al. 2015).

Probiotics are suggested as a new tool in disease control and water quality improvement (Das et al. 2017) and considered as new eco-friendly measures of health management in aquaculture (Irianto and Austin 2002a, b; Dahiya et al. 2012; Sihag and Sharma 2012). The beneficial effects of probiotic for disease resistance and production of antimicrobial substances to promote innate immunity and increase survivability are well studied (Sugita et al. 1996, 1998; Ringø et al. 2007). The role of microbes in digestion as well as in production of enzymes, amino acids, and vitamins is also documented (Moriarty, 1990; Sugita et al. 1992, 1997; Thompson et al. 1999; Verschuere et al. 2000; Bairagi et al. 2002; Ramirez and Dixon 2003). Several reports suggested that ecological factors regulate the growth, multiplication, and other activity of pathogenic microbes such as *Aeromonas hydrophila* (Majumdar et al. 2006), *Edwardsiella tarda* (Swain and Nayak, 2003), *Vibrio parahaemolyticus* (Cheng et al. 2004), and some other *Vibrio* spp. (Jakšić et al. 2002) in the aquatic environment. In all types of aquaculture, microbes are associated with the food web and productivity of the water body (Moriarty 1997). The aquaculture production largely depends on the environment where the fish lives in. A better environment can reduce the stress and thus reduce the chance of pathogenic infection in fish, and there is no better way known to date that can act as efficiently as probiotics. Based on the above discussion, it is evident that probiotic has a potential role in culture ecosystem and in internal physiology of fish toward better and sustainable aquaculture production.

At the early stage, research works on fish probiotic microbes were mainly focused on Gram-positive bacteria particularly of the genus *Bifidobacterium*, *Lactobacillus*, and *Streptococcus*. Bacteria representing the genus *Bacillus* have also been widely studied. Gram-positive bacterial genera *Carnobacterium*, *Pediococcus*, *Enterococcus*, *Lactococcus*, *Micrococcus*, and *Weissella* and Gram-negative bacteria *Alteromonas*, *Photobacterium*, *Pseudomonas*, and *Vibrio* and some strains of *Aeromonas* are also reported as good probiotic candidates. Certain fungi belong to the genus *Aspergillus* and yeast *Saccharomyces* are also reported as potential probiotics for aquaculture. Among the bacteria, a good number of bacilli have been commercialized for aquaculture industry. The beneficial roles of probiotic *Bacillus* spp. in aquaculture, especially in fish and shrimp culture, are discussed in the following sections.

16.4 How Probiotic Bacilli Exert their Beneficial Roles in Aquaculture?

Probiotic bacilli play a very important role in the host and in the aquatic environment. They have a very precise mechanism of action in the host. The major mechanisms of action of probiotic *Bacillus* spp. in fish and shrimp are (1) competition for binding sites; (2) probiotic bacilli bind with the binding sites in the intestinal mucosa of fish and shrimp, forming a physical barrier and preventing the connection by pathogenic bacteria; (3) production of antimicrobial substances; (4) synthesis of hydrogen peroxide, bacteriocins, and bacteriocins-like inhibitory substances (e.g., subtilin and coagulin) that have antimicrobial action against pathogenic bacteria; (5) production of a wide range of antimicrobial substances such as surfacing, bacilysin, etc. that inhibit or directly kill pathogenic microbes; (7) production of organic acids that lower the pH of the gastrointestinal tract, preventing the growth of pathogens, and enhance growth of *Lactobacillus*; (6) competition for essential nutrient elements; and (7) stimulation of the host immune system by increasing the production of antibodies, activation of macrophages, T-cell proliferation, and production of interferon.

Probiotic bacilli also significantly improve the water quality by maintaining the balance of different physicochemical parameters, kill pathogens through antagonism, decompose different organic wastes, and keep the environment healthy for aquaculture. The mechanisms of action of probiotic *Bacillus* spp. in the improvement of water quality have been briefly discussed in the following subsections.

16.4.1 Convert Organic Matter to CO_2

As Gram-positive aerobic bacteria, *Bacillus* spp. are more capable of mineralization of organic matter into CO_2 compared to Gram-negative bacteria (Balcázar et al. 2006; Mohapatra et al. 2012). A large accumulation of organic matter in pond increases the biological oxygen demand and favors the development of anaerobic patches. Under aerobic decomposition, 50% of the organic matter converted into bacterial cells (Henze et al. 2001).

16.4.2 Reduce Ammonia and Nitrite Toxicity

Ammonia and nitrite concentration increases as the culture move toward intensive operation. Ammonia and nitrite toxicity can be eliminated by the application of nitrifying cultures into the fish environment (Mohapatra et al. 2012). They can be treated with *Nitrosomonas* spp. to keep water quality in an optimal range, which might be possible because of various roles played by the probiotic bacteria.

16.4.3 Improvement of pH

A buffering system to circumvent wide pH fluctuation is vital for aquaculture (Swann 1990). Brauner (2008) reported that exchange of HCO^{3-} as a waste product of respiration reduced the pH, which led to physiological stress. *Bacillus* spp. have known to utilize multiple nitrogen sources, including both NH_3 and NH_4^+ for catabolism of proteins and thus utilize H^+ ion.

16.4.4 Improve Dissolved Oxygen

Probiotic *Bacillus* produce substances which inhibit harmful algae by algicidal compound and prevent blooms. Algal bloom can potentially cover the surface of pond water and prevent the sunlight to penetrate into the water to provide the energy to phytoplankton to produce oxygen.

16.4.5 Improve Sediment Quality

The probiotic *Bacillus* precipitated at the bottom when applied in the culture system and unleash their beneficial effects. They facilitate the decomposition and prevent the formation of toxic H_2S and thus keep the sediment nontoxic and healthy for the bottom dweller fishes.

16.4.6 Competitive Exclusion of Pathogens

Bacillus in water and in sediment competes for space and nutrition with pathogens as well as produces antibacterial, antiviral, and antifungal substances which help to keep these pathogens in check.

16.5 Isolation and Development of Probiotic *Bacillus*

In order to develop a probiotic candidate strain of *Bacillus*, the first step is to isolate the microbe from different sources. Although *Bacillus* is a ubiquitous bacterium, the digestive tract of fish or shrimp is the best source for isolation of a putative probiotic strain. *Bacillus* isolated from soil, marine sediment, marine sponge, and pickle of fermented food may also be screened since several probiotic *Bacilli* have been isolated from these sources. After isolation, in vitro screening of antagonistic effects of the putative probiotic strain against different fish pathogens is to be carried out. In vitro screening may be done by exposing different pathogens to putative probiotic bacteria or their extracellular products in liquid and/or solid medium. In vitro antiviral activity of a candidate strain may also be screened, since some *Bacilli* possess antiviral properties. In vitro screening of autoaggregation, bile salt tolerance, tolerance at low pH, etc. should be performed to select a good probiotic candidate. In vitro hydrolytic enzyme activity (such as protease, lipase, amylase, etc.) of the strain may also be screened.

Pathological effects of the candidate strain should be screened in fish and/or shrimp of interest for which the probiotic is going to be developed. Then, the putative probiotic candidate is to be administered to the target fish or shrimp to evaluate the effects of the bacteria on growth and survival of fish or shrimp. After that, in vivo artificial infection challenge test in the putative probiotic-treated fish is to be done with different virulent pathogens. Multiplication field trial should be conducted to evaluate the performance of the strain that exhibits significantly satisfactory performance in laboratory condition. If the performance of any strain seems satisfactory both in laboratory and field condition, the economic benefit of its application is needed to be evaluated. Finally, the strain can be selected for commercial formulation.

16.6 Selection Criteria of a Good Probiotic Bacilli

Selection of probiotic bacteria in aquaculture facilities is an important process based on proven scientific evidence. In most cases, probiotics provide expected beneficial effects on some specific host species in appropriate environmental condition. It is essential to know the mechanism of probiotic action before selecting a potential probiotic. A *Bacillus* strain should fulfill the following criteria to be selected as a good probiotic bacterium: (a) resistant to acidic environment of the stomach, bile, and digestive enzymes; (b) ability to access to the intestinal mucosa; (c) capacity of colonization to the intestinal mucosa; (d) ability to produce different hydrolytic enzymes; (e) production of antimicrobial substances against a wide range of pathogens; (f) absence of translocation; and (g) staying viable for a long time during transportation and storage. In addition, a good probiotic *Bacillus* sp. should also possess the capacity to (a) improve food digestibility of host species, (b) enhance the growth and survival of the host, (c) boost up the immunity of the host, (d) decompose organic wastes, (e) detoxify waste materials, etc.

16.7 Mode of Application of Probiotics

Probiotics are normally applied to the targeted aquatic host either via feed or bioencapsulation through live food like artemia and rotifer. Another important way of probiotic administration is directly applied in culture water which is not only to improve the health status of the fish but also to improve the ecosystem (Verschuere et al. 2000). No matter the way of application of probiotics, the most important factor is the viability of probiotics which is often impaired due to poor processing and storage condition (Havenaar and Huis 1992).

(a) *Dietary Administration*

Dietary administration of probiotics can improve the intestinal microbial community balance of fish toward the better health and growth performance as reported by many scientists (Fuller 1989; Purwandari and Chen 2013; Villamil et al. 2002; Al-Dohail et al. 2009; Saenz de Rodriguez et al. 2009; Chabrillón et al. 2005; Vine et al. 2004). However, it has to administer at a suitable temperature for the probiotic bacteria in feed; otherwise probiotic may not be available to fish at needed dose or with full potentiality (Gatesoupe 1999; El-Ezabi et al. 2011; Korkea-Aho et al. 2012; Díaz-Rosales et al. 2009). Probiotics are typically supplemented to feed as freeze-dried cultures, blended with lipids as top dressings in the pellet (Robertson et al. 2000; Nikoskelainen et al. 2001) or added with suitable binders.

(b) *Bioencapsulation*

Bioencapsulation is a process of enrichment of live foods (like artemia and rotifer) with nutrients which are often deficient with a vital compound like eicosapentaenoic acid (EPA) or docosahexaenoic acid (DHA) but are essential to fish especially for the larval survival. These live foods are pretty attractive to larvae; therefore, encapsulating them with probiotics can exert immense benefit in larval diet. Bioencapsulation with probiotic is a widely used method of probiotic administration that can improve growth, increase immunity, reduce feed conversion ratio, and increase the survivability of fish (Venkat et al. 2004; Vaseeharan and Ramasamy 2003; De et al. 2014; Austin et al. 1995). Probiotic bioencapsulation has the potential to protect microbes and to transport them into the gut (Gbassi and Vandamme 2012).

(c) *Application in Water*

Probiotic can be used directly into the water to improve the culture condition by converting organic matter into CO_2 and minimize the buildup of organic carbon. By improving water quality, probiotic helps to improve the health status and increase growth and survival of fish (Dalmin et al. 2001).

16.8 *Bacillus* in Sustainable Aquaculture

The beneficial roles of *Bacillus* as probiotic bacteria have been extensively studied during the last decade. Although, in earlier sections, the functions of probiotic *Bacillus* have briefly been outlined, the following subsections elaborately discuss the major roles of probiotic bacilli in aquaculture industry.

16.8.1 Bacillus *Promotes Growth of Fishes and Shrimp*

Several probiotic *Bacillus* strains have been reported to increase the growth of aquaculture species (Table 16.1). However, the exact mechanism of probiotics as growth promoter has not yet been well understood. Probiotic microbes are able to colonize in the gastrointestinal tract and adhere to the intestinal mucosa of fish and exert their multiple benefits (Balcázar et al. 2006). Ghosh et al. (2003) reported that *B. circulans* significantly improve growth, reduce feed conversion ratio, and increase protein efficiency ratio in *Labeo rohita* when fed at a rate of 1.5×10^5 CFU 100 g^{-1} feed at 3% of the body weight of fish. Diet supplemented with bacilli and lactic acid bacteria (LAB) had good survival and the best growth performances suggesting that bacteria are appropriate growth-stimulating additives in tilapia cultivation (Apún-Molina et al. 2009). Faramarzi et al. (2011) assessed the effects of various concentrations of probiotic *Bacillus* spp. on the growth performance and survival rate of Persian ganoid fish (*Acipenser persicus*) larvae. They reported significant conversion efficiency ratio (CER), specific growth rate (SGR), food conversion ratio (FCR), condition factor (CF), and daily growth coefficient (DGC) ($p > 0.05$) when ganoid fish larvae were fed at 30% of their body weight for five times a day. Mohapatra et al. (2012) studied the effects of different probiotic in the diet of Rohu

Table 16.1 Bacillus used in aquaculture to promote growth and/or increase survival

Probiotics	Host species	Special effects	Reference (s)
Bacillus spp.	Common carp (*Cyprinus carpio*)	Better digestive enzyme activities; better growth performance and feed efficiency	Yanbo and Zirong (2006)
Bacillus spp.	Rainbow trout (*Oncorhynchus mykiss*)	Better growth performance and survival	Bagheri et al. (2008)
B. pumilus and *B. Clausii*	Grouper (*Epinephelus coioides*)	Improved growth performance and immune responses of *E. coioides*	Sun et al. (2010)
B. circulans	Rohu (*L. rohita*)	Improved growth, reduced feed conversion ratio, and increased protein efficiency ratio	Ghosh et al. 2003
Bacillus sp.	Tilapia (*Oreochromis niloticus*)	Good survival and impressive growth performances	Apún-Molina et al. 2009
Bacillus sp.	Persian ganoid fish (*Acipenser persicus*)	Increased specific growth rate (SGR) and decreased food conversion ratio (FCR)	Faramarzi et al. 2011
B. subtilis, *L. lactis*, and *S. cerevisiae*	Rohu fingerling (*L. rohita*)	Higher growth, protein efficiency ratio, nutrient digestibility	Mohapatra et al. (2012)
Bacillus sp. strain B51f and *B. siamensis* strain B44v	hybrid catfish (*Clarias microcephalus* × *C. gariepinus*)	Significantly increased weight gain and lower FCR	Meidong et al. (2017)
Bacillus sp.	Tilapia (*Oreochromis niloticus*)	Increased growth performance, decreased feed conversion ratio	Elsabagh et al. (2018)
B. subtilis and Mannanoligosaccharide	Japanese eel (*Anguilla japonica*)	Synergistically improved growth performance	Lee et al. (2018)
B. velezensis strain AP193	Catfish	Stimulated growth of fish	Thurlow et al. (2019)
Bacillus sp.	Juvenile *Penaeus monodon*	Improved the growth and survival rate	Dalmin et al. (2001)
Enterococcus casseliflavus, *Citrobacter koseri*, *B. subtilis*, and *Staphylococcus* sp.	*Litopenaeus vannamei*	Increased specific growth rate (SGR)	Sanchez-Ortiz et al. (2015)
B. subtilis strain E20	*Litopenaeus vannamei*	Increased growth performance and survivability	Liu et al. (2009)
B. subtilis	*Macrobrachium malcolmsonii*	Increased weight gain	John et al. (2018)

(continued)

Table 16.1 (continued)

Probiotics	Host species	Special effects	Reference (s)
Bacillus sp.	Channel catfish (*Ictalurus punctatus*)	Increased survival rate	Queiroz and Boyd (1998)
B. subtilis	*L. vannamei*	Significantly increased survival rate	Xue et al. (2016)
B. licheniformis and *B. subtilis*	Pacific white shrimp (*L. vannamei*)	Improved survival rate	Jamali et al. (2015)
Bacillus strain OJ	*L. vannamei*	Significantly increased survival rate	Li et al. (2009)
B. subtilis	*L. vannamei*	Increased survivability against *Vibrio harveyi* infection	Balcázar and Rojas-Luna (2007)
B. pumilus strain RI06-95	Eastern oyster (*Crassostrea virginica*)	Increased survivability against *V. tubiashii*	Karim et al. (2013)

fingerling (*L. rohita*) on growth, nutrient digestibility, digestive enzyme, and intestinal microbiota. They claimed that the fish fed with a combination of three probiotics (*B. subtilis*, *Lactococcus lactis*, and *Saccharomyces cerevisiae*) showed a higher growth, protein efficiency ratio, nutrient digestibility, and lower feed conversion ratio compared to other groups.

Meidong et al. (2017) evaluated the growth performance of hybrid catfish (*Clarias microcephalus* × *C. gariepinus*) in 30 days feeding trial experiment with *Bacillus* sp. strain B51f and *B. siamensis* strain B44v derived from Thai fermented food. They reported significant growth performance (SGR and weight gain) of hybrid catfish fed diets containing 10^7 $CFUg^{-1}$ of either bacteria than the control group. The feed conversion ratio (FCR) of the fish fed diets containing the *Bacillus* strains was also found significantly lower than the FCR of control fish. Elsabagh et al. (2018) observed a notable increase in growth performance, feed conversion ratio, and blood profiles in tilapia fed on *Bacillus* treated diets. Supplementation of *B. subtilis* at 0.5×10^7 $CFUg^{-1}$ diet and mannanoligosaccharide at 5 g kg^{-1} diet have a synergistic effect on growth performance, non-specific immune responses, intestinal morphology, and disease resistance in Japanese eel, *Anguilla japonica* (Lee et al. 2018). *B. velezensis* strain AP193 can stimulate catfish growth and improve the water quality of rearing pond when used as a feed amendment (Thurlow et al. 2019).

Bacillus sp. also improved the growth and survival rate and the health status of juvenile *Penaeus monodon* (Dalmin et al. 2001). Sanchez-Ortiz and co-workers (2015) isolated several bacterial isolates from homogenates of the gastrointestinal tract of adult mangrove cockle, *Anadara tuberculosa*. They selected four bacteria, viz., *Enterococcus casseliflavus*, *Citrobacter koseri*, *B. subtilis*, and *Staphylococcus* sp., to evaluate the specific growth rate (SGR) and cellular immune response of shrimp *Litopenaeus vannamei* and reported the best SGR for two treatments related to *B. subtilis* and highest cellular immune response in the treatment with *B. subtilis*. *B. subtilis* strain E20 isolated from natto is reported as an extraordinary protease producer and is able to increase the growth performance of *L. vannamei* through increasing the digestibility of food (Liu et al. 2009). *B. subtilis* coated diet has been reported to increase the weight of *Macrobrachium malcolmsonii* as 3.5 times compared to control diet and also increase the amount of polyunsaturated fatty acid (38%) (John et al. 2018).

16.8.2 **Bacillus** ***Improves Fish Nutrition***

Probiotic bacteria play a significant role in fish nutrition. They detoxify the potentially harmful compounds in the diet by hydrolytic enzymes. The hydrolytic enzymes including amylases, proteases, and lipases break down the complex food materials and make simple molecules, hence improving the digestibility of food. Balcázar et al. (2006) suggested that probiotics have a beneficial role on aquatic animals as probiotic strains synthesize enzymes such as proteases, amylases, and lipases as well as provide growth factors like vitamins, fatty acids, and amino acids. Some *Bacillus* strains produce vitamin B12 (Warren et al. 2002) and riboflavin (Bacher et al. 2000; Perkins and Pero 2002). The extracellular enzymes secreted by probiotic bacteria have been reported to play an important role in the digestion and assimilation process of nutrients in the gut of the host by modifying the gut flora (Farzanfar 2006).

Bacillus spp. have been reported to improve the digestive enzyme activities, better growth performance, and feed efficiency in common carp, *C. carpio* (Yanbo and Zirong 2006) (Table 16.2). Dietary administration of *B. pumilus* and *B. clausii*, in grouper (*Epinephelus coioides*), has reported to improve feed conversion ratio, phagocytic activity, and phagocytic index but did not increase the weight gain or specific growth rate after 60 days of rearing (Sun et al. 2010). *Bacillus* spp. also increased the specific activities of alkaline and acid protease and improved the husbandry parameters and nutritional condition in larvae of gilthead sea bream (*Sparus aurata*) (Arığ et al. 2013). Diet supplemented with live *Bacillus* sp. DDKRC1 at 2.94×10^7 CFU 100 g^{-1} feed and diet fermented (48 h) with the same bacteria showed the better effect on growth, digestibility, FCR, survival, and immune response of *P. monodon* compared to control diet (De et al. 2018). *B. subtilis* ANSB060 could also improve digestive enzyme activities of hepatopancreas and intestine, as well as decrease aflatoxin B1 residues in the hepatopancreas and gonad of yellow river carp (*C. carpio haematopterus*) (Fan et al. 2018). By modulating intestinal microflora, *B. licheniformis* fb11 at the concentration of 5×10^6 $CFUg^{-1}$ significantly improved digestion in juvenile chitala (*Chitala chitala*) (Mitra et al. 2018).

Table 16.2 *Bacillus* used for fish nutrition

Probiotics	Species	Special effects	Reference (s)
Bacillus spp.	Common carp (*C. carpio)*	Provided better digestive enzyme activities; better growth performance and feed efficiency	Yanbo and Zirong (2006)
B. pumilus and *B. clausii*	Grouper (*Epinephelus coioides*)	Improved feed conversion ratio, phagocytic activity, and phagocytic index	Sun et al. (2010)
Bacillus sp.	Gilthead sea bream (*S. aurata*, L.)	Increased the specific activities of alkaline and acid protease; improved nutritional condition in larvae	Arığ et al. (2013)
Bacillus sp. strain DDKRC1	Giant tiger shrimp, *P. monodon*	Diet fermented with bacteria provided better growth, digestibility, FCR, survival, and immune response	De et al. (2018)
B. subtilis strain ANSB060	Yellow river carp, *C. carpio haematopterus*	Improved digestive enzyme activities of hepatopancreas and intestine and decreased aflatoxin B1 residues in hepatopancreas and gonad	Fan et al. (2018)
B. licheniformis strain fb11	Chitala (*C. chitala*)	Modulated intestinal microflora and significantly improved digestion	Mitra et al. (2018)

16.8.3 Bacillus *Improves Survival Rate of Fish*

Probiotics maintain a healthy environment by improving the water quality, degrading organic waste materials, inhibiting the growth of pathogens, and reducing the load of harmful microbes in aquaculture systems. Such a healthy environment provides better survival of larvae and growing fish. Several studies demonstrated that probiotic bacteria can reduce mortalities during bacterial infection in fish. The commercially prepared mixture of *Bacillus* spp. has been reported to increase the survival and production of channel catfish (*Ictalurus punctatus*) when applied in rearing water (Queiroz and Boyd 1998). Administration of either individual or combined probiotics *Bacillus* sp. or LAB increased the survival and growth performance in Nile tilapia (Apún-Molina et al. 2009). Similarly, Liu et al. (2012) reported that dietary administration of *B. subtilis* strain E20 at a dose of 10^4, 10^6, and 10^8 CFU g^{-1} for 28 days to grouper (*Epinephelus coioides*) enhanced the survival against *Streptococcus* sp. and an *Iridovirus* infection with relative survival percentage of 22.8, 40.9, and 45.5 and 21.7, 30.4, and 52.2, respectively.

B. subtilis delivered to larvae singly at concentrations of 1.0×10^8 CFU L^{-1} exhibited a significantly higher survival rate of the Pacific white shrimp *(L. vannamei*) larvae from nauplii 3–4 to zoea three stages (Xue et al. 2016). Artemia and rotifer enriched with *B. licheniformis* and *B. subtilis* at 1×10^6 CFU mL^{-1} reported to improve growth and survival rate of *L. vannamei* larvae (Jamali et al. 2015). Liu et al. (2009) observed that postlarvae of *L. vannamei* had significantly higher survival in the treatments with *B. subtilis* strain E20 at 10^8 and 10^9 CFU L^{-1} compared to the control during their exposure to freshwater. *L. vannamei* fed with *Bacillus* strain OJ at a dose of 10^{10} CFU g^{-1} of feed produced significantly higher survivals and immune parameters than the control groups (Li et al. 2009). Balcázar and Rojas-Luna (2007) suggested that *B. subtilis* UTM 126 at a dose of 10^5 CFU g^{-1} can be used to increase the survivability of *L. vannamei* against *V. harveyi* infection. Karim et al. (2013) reported *B. pumilus* strain RI06–95 as a potential agent to stop larval and juvenile mortality by the bacteria *V. tubiashii* and *Roseovarius crassostreae* in eastern oyster (*Crassostrea virginica*).

16.8.4 Antimicrobial Activity of **Bacillus**

Oral administration of antibiotics with fish feed is the most commonly practiced treatment method for bacterial fish diseases. But a prolonged and indiscriminate application of antibiotics may result in the development antibiotic resistance in pathogenic bacteria (Rahman et al. 2017) and spread of the resistant bacteria in aquaculture system and in the environment. The antibiotic resistance gene of these bacteria may spread to other pathogenic, nonpathogenic, and environmental bacteria through horizontal gene transfer. Antibiotics are also hazardous for the environment and a threat for safe food (Ran et al. 2012).

Probiotic *Bacillus* are reported to exhibit antimicrobial activities against fish and shrimp pathogens and can be used as an eco-friendly and safe alternative to antibiotics (Mondol et al. 2013). Most bacilli produce antibiotics such as difficidin, oxydifficidin, bacitracin, polymyxin, subtilin, mycobacillin, bacillin, gramicidin, or bacillomycin B that are antagonistic to pathogenic bacteria both in vitro and in vivo conditions (Korzybski et al. 1978; Zimmerman et al. 1987; Al-Faragi and Alsaphar 2012). Sugita et al. (1998) isolated *Bacillus* sp. strain NM 12 from the intestine of a dragonet (*Callionymus* sp.) that produced an antibacterial substance and suggested the strain as a suitable biocontrol agent for fish. Ran et al. (2012) isolated *Bacillus* strains from soil and intestine of channel catfish and screened antagonism of the strains against fish pathogens. All of the 50 *Bacillus* strains showed inhibitory activity against 2 strains of *Edwardsiella ictaluri*, and all but one *Bacillus* strain showed antagonism against another *E. ictaluri* strain. They also reported that 21 of these strains expressed antagonism against multiple catfish pathogens including *A. hydrophila*, *E. tarda*, *S. iniae*, *Yersinia ruckeri*, *Flavobacterium columnare*, and the oomycete *Saprolegnia ferax*. Ambas et al. (2015) evaluated the antagonistic activity of *Bacillus* strains isolated from the intestine of healthy marron (*Cherax cainii*) and commercial probiotic products. They reported that *B. subtilis* and *Bacillus* sp. (PM4) from commercial product exhibited strong inhibition activity against *V. mimicus* and *V. cholera* non-01, whereas *B. mycoides* isolated from marron inhibited only *V. mimicus*. Yi et al. (2018) reported antimicrobial activity of *B. velezensis* strain JW isolated from healthy grass carp against a broad range of fish pathogenic bacteria including *A. hydrophila*, *A. salmonicida*, *L. garvieae*, *S. agalactiae*, and *V. parahaemolyticus*. In a recent study, Paul (2018) reported that *B. subtilis* strain WS1A isolated from a sponge of the Bay of Bengal exhibited antibacterial activity against fish pathogenic *A. veronii* strains (Fig. 16.5). Rahman (2018) iso-

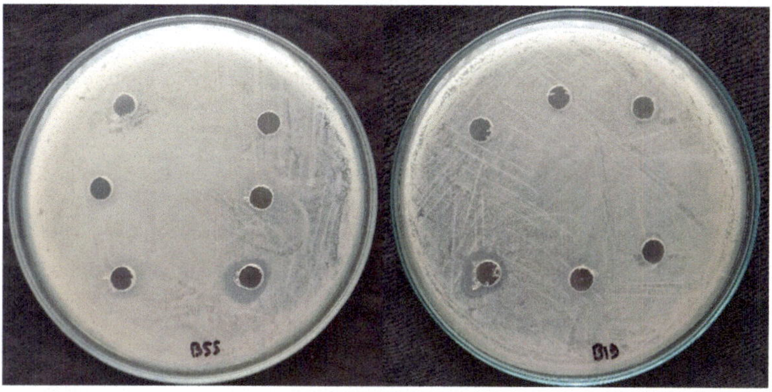

Fig. 16.5 Inhibition of fish pathogenic *A. veronii* strain B55 and B19 by the cell-free extracts of *Bacillus subtilis* strain WS1A isolated from a sponge

lated some bacterial strains from the sediment of the Bay of Bengal adjacent to the Chera deep and reported that *B. haynesii* strain CD223 exhibited antibacterial activity against different strains of fish pathogen: *A. veronii*, *Enterococcus faecalis*, and *Stenotrophomonas* sp.

16.8.5 Bacillus *Prevents Diseases*

The probiotic *Bacillus* spp. are efficient in preventing fish diseases. Ran et al. (2012) reported that *B. subtilis* strains conferred significant benefit in reducing mortality of channel catfish challenged by *E. ictaluri* after 14 days feeding with *B. subtilis* spores. They also reported protective effects of four out of five strains of *Bacillus* against *E. ictaluri* infection in striped catfish. Bhatnagar and Lamba (2017) conducted 90 days feeding experiment where mrigal (*Cirrhinus mrigala*) were fed on different levels of *B. cereus* (strain SL-1) incorporated feed. The *B. cereus* (strain SL-1) was isolated from the intestine of *C. mrigala*. After 90 days of feeding, the fish were challenged with pathogenic *A. hydrophila* by immersion method. They reported that all of the probiotic-fed fish group showed high survival in comparison to control group fish. Thy et al. (2017) also obtained higher survival of striped catfish (*Pangasianodon hypophthalmus*) fed with a mixture of *B. amyloliquefaciens* strain 54A and *B. pumilus* strain 47B at 5×10^7 CFU 100 g^{-1} against *E. ictaluri* in a 90-day feeding trial. Meidong et al. (2018) obtained higher survival (77.3%) of *Pangasius bocourti* fed with *B. aerius* strain B81e supplemented diet than the control fish when challenged with virulent strain of *A. hydrophila* (FW52). In a recent experiment, Rahman (Unpublished) evaluated the effects of dietary *B. subtilis* strain WS1A on the disease resistance of *L. rohita* against *Aeromonas* septicaemia infection. Fish were fed bacteria at a dose of 5×10^7 CFU 100 g^{-1} for 90 days, and another group of fish was fed only the basal diet. Then, both groups of fish were

Fig. 16.6 Probiotic *B. subtilis* strain WS1A prevents motile *Aeromonas* septicaemia in *Labeo rohita* after feeding for 90 days. Left, normal fish (no disease symptom) which was fed with *B. subtilis* strain WS1A for 90 days and challenged with *A. veronii* strain B55; right, hemorrhages as external infection symptom in control fish (arrow indicates hemorrhages)

Table 16.3 Use of *Bacillus* to enhance immunity and antimicrobial activity

Probiotics	Species	Special effects	Reference (s)
B. subtilis and *B. licheniformis*	Rainbow trout (*O. mykiss*)	Increased resistance to *Yersinia ruckeri*	Raida et al. (2003)
B. subtilis	Gilthead seabream	Stimulated cellular innate immune response	Salinas et al. (2005)
B. subtilis (ATCC 6633)	Nile tilapia (*O. niloticus*)	Stimulated the gut immune system; enhanced the immune and health status	Aly et al. (2008)
Bacillus	Channel catfish (*I. punctatus*)	Good potential to mitigate the enteric septicemia of catfish (ESC)	Ran et al. (2012)
B. coagulans and *B. subtilis*	Artemia nauplii	Produced antimicrobial activity against the pathogenic *Vibrio* species, including *V. alginolyticus*	Mahdhi et al. (2011)
B. subtilis AB1	Rainbow trout (*O. mykiss*)	Controlled *Aeromonas* infection	Newaj-Fyzul et al. (2007)
B. subtilis	Orange spotted grouper (*E. coioides*)	Increased the innate immunity and intestinal microbial population	Purwandari and Chen (2013)

intramuscularly challenged with a previously identified virulent *A. veronii* strain (B55) where, all of the probiotic treated fish survived at the end of the experiment, but 100% mortality occurred in the control group fish with distinct external symptoms of hemorrhages in the body surface (Fig. 16.6). In shrimp, several studies focused on probiotics such as *B. cereus*, *Paenibacillus polymyxa*, and *Pseudomonas* sp. as biocontrol agents against pathogens of various *Eubacterium* spp. (Ravi et al. 2007; Vijayan et al. 2006).

16.8.6 Bacillus *Enhances Immunity*

Enhancement of host immunity by the probiotic bacilli has been reported by many researchers (Table 16.3). Zhao et al. (2011) found that *Bacillus* strain TC22 and prebiotic fructo-oligosaccharide (FOS), when fed to ocean cucumbers as TC22 at 10^9 CFU g^{-1} feed and 0.5% FOS alone or in combination, resulted in enhanced activity, respiratory burst, and phenoloxidase activity of ocean cucumber coelomocytes and disease resistance against *V. splendidus* infection.

Yi et al. (2018) reported a significant increase of humoral immune responses such as acid phosphatase (ACP), alkaline phosphatase (AKP), and glutathione peroxidase (GSP-PXP) activities in *Carassius auratus* fed with the probiotic *B. velezensis* strain JW supplemented feed. They also measured the mRNA expression of immune-related genes in the head kidney of the probiotic-treated fish. Among them, the interferon gamma gene (IFN-γ) and tumor necrosis factor-α (TNF-α) showed higher expression after 3 and 4 weeks of feeding. They also reported significant upregulation of the expression of interleukin-1 (IL-1) in fish fed with 10^9 CFU g^{-1} of the probiotic bacteria. *B. amyloliquefaciens* strain R8 can improve nutrient metabolism and hepatic stress tolerance, immunity, and disease resistance against *A. hydrophila* and *S. agalactiae* in zebra fish (*Danio rerio*) (Lin et al. 2019). Zheng et al. (2019) demonstrated that *B. amyloliquefaciens* (strain GB-9) enhance non-specific immune defense system of *Anguilla japonica*, providing them with higher resistance to pathogens. *B. amyloliquefaciens* can be used in aquaculture to improve health status and disease resistance in catla (*Catla catla*) with an optimal dietary supplementation of 10^9 CFU g^{-1} (Das et al. 2013). Bandyopadhyay and Mohapatra (2009) found that *B. circulans* strain PB7, isolated from the intestine of *C. catla*, improved phagocytic ratio, phagocytic index, and leucocrit value when fed at 2×10^5 cells 100 g^{-1} feed. Total serum protein and globulin content increased significantly if the fingerling of *L. rohita* fed *B. subtilis* at a rate of 10^8 CFU g^{-1} of feed for a period of 60 days (Nayak et al. 2007). Highest respiratory burst activity and serum bactericidal activity were induced in *L. rohita* when the fish fed *B. subtilis* at 1.5×10^7 CFU g^{-1} feed (Kumar et al. 2008). Tilapia (*O. niloticus*) treated with *B. coagulans* strain B16 at a rate of 1.9×10^7 CFU ml^{-1} every 2 days in the rearing tank water showed ominously higher final weight, daily weight gain, specific growth rate, average myeloperoxidase, and increased respiratory burst activity compared with those treated with *B. subtilis* strain B10 (Zhou et al. 2010). Probiotic *B. subtilis* added diet at 5×10^6 CFU g^{-1} promoted the innate immune system of Nile tilapia (Telli et al. 2014). Cobia (*Rachycentron canadum*) fed the diet containing 2.0 g kg^{-1} *B. subtilis* and 6.0 g kg^{-1} chitosan had significantly higher post-challenge survival against *V. harveyi* infection (Geng et al. 2011). Ran et al. (2012) isolated *Bacillus* strain from soil and/or intestine of channel catfish (*I. punctatus*) and reported the strain as very potential to mitigate the enteric septicemia of catfish. Combination of dietary fructo-oligosaccharide (FOS) and *B. licheniformis* at the rate of 0.3% FOS and 1×10^7 CFU g^{-1} diet can significantly enhance the innate immunity and antioxidant capability of triangular bream (*Megalobrama terminalis*), as well as improve its disease resistance (Zhang et al. 2013). Leukocyte counts, alternative complement activity, and total serum protein and globulin contents all increased significantly as dietary *B. licheniformis* levels increased from 0 to 1×10^7 CFU g^{-1} diet (Zhang et al. 2013). In contrast, addition of FOS with *B. subtilis* had no effects on the growth performance, immune response, and disease resistance of large yellow croaker (*Larimichthys crocea*) (Ai et al. 2011). Thy et al. (2017) fed a mixture of probiotic bacteria (*B. amyloliquefaciens* strain 54A and

B. pumilus strain 47B) to striped catfish (*Pangasius hypophthalmus*) at different concentrations for 90 days and obtained higher immune parameters such as phagocytic activity, respiratory burst, and lysozyme activity in fish fed diet containing probiotics at 5×10^7 CFU g^{-1} than the control. Meidong et al. (2018) reported that dietary administration of *B. aerius* strain B81e significantly increased the immune parameters such as serum lysozyme and bactericidal activities, alternative complement, and phagocytic and respiratory burst activities in catfish (*P. bocourti*).

Moriarty (1998) suggested a commercialized *Bacillus* product that prevented infection by detrimental vibrios in shrimp. *Bacillus* provided disease protection by activating both cellular and humoral immune defenses, as well as competitive exclusion of *V. harveyi* in the gut of tiger shrimp (*P. monodon*) (Rengpipat et al. 2000). NavinChandran et al. (2014) suggested that *B. cereus* screened from the wild shrimp *P. monodon*, in turn, can be added as an effective growth and immune enhancer in *P. monodon*. Zokaeifar et al. (2012) reported that the expression of all immune-related genes was suggestively upregulated ($P < 0.05$) in the white shrimp (*L. vannamei*) when fed with *B. subtilis* strain L10 and G1 at 10^5 and 10^8 CFU g^{-1} diets compared to the control group.

16.8.7 Bacillus *Improves Water Quality*

Bacillus spp. has beneficial roles in the improvement of water quality in aquaculture (Table 16.4). They have a better conversion efficiency of organic matter into CO_2 than Gram-negative bacteria, which converts a greater proportion of organic matters into bacterial biomass (Balcázar et al. 2006; Mohaparta et al. 2012). Application of probiotics in lentic water bodies could improve fish health by boosting up many water quality parameters since they modify the bacterial composition of the water and sediments (Ali 2000; Rao 2007). The temperature, pH, dissolved oxygen, NH_3, and H_2S were reported to be of higher quality when *Bacilli* were added in the shrimp larvae culture in the green water system (Banerjee and Ray 2017; Aguirre-Guzman et al. 2012). Probiotic bacilli enhance decomposition of organic matters, reduce nitrogen and phosphorus concentrations, and control ammonia, nitrate, and hydrogen sulfide (Cha et al. 2013).

Bacillus spp. was found to have a wonderful water purification capability as Gram-positive bacterium are better converters of organic matter back to carbon-dioxide than Gram-negative bacterium (Moriarty 1997; Anik et al. 2011). *Bacillus* spp. improved water quality, survival, and growth rates, improved the health status of juvenile *P. monodon*, and reduced the number of infective vibrios (Dalmin et al. 2001; Devaraja et al. 2013). For edible fish, *Bacillus* spp. could reduce the chemical load of high concentrations of nitrogen in trout production (Maillard et al. 2005) and reduce the total ammonia in tilapia production in recirculating systems (Rafiee and Saad 2005). Moreover, commercial probiotics created from *E. licheniformis* and *B. subtilis* were used in Nile tilapia (*O. niloticus*) farming to optimize the dissolved oxygen level (El-Haroun et al. 2006). *Bacillus* isolated from fish (*C. carpio*)

Table 16.4 *Bacillus* used as probiotics in water in aquaculture

Probiotics	Origin	Impacts on water quality and fish/shrimp	Doses	References
Bacillus sp.	Commercial product	Total nitrogen and ammonia concentrations reduced up to a lower extent	10^8 and 10^5 CFU mL^{-1}	Matias et al. (2002)
Bacillus sp.	Commercial product	Reduced concentrations of nitrogen and phosphorus and increased yields of shrimp	10^4–10^9 CFU mL^{-1}	Wang et al. (2005)
Bacillus sp.	–	Probiotics maintained optimum transparency and low organic load	–	Dalmin et al. (2001)
Mixed bacillus	Commercial product	Levels of pH, ammonia, and nitrite significantly decreased	–	Nimrat et al. (2012)
B. subtilis and *B. megaterium* with soybean meal	Commercial product	Enhanced stress tolerance and hemolymph metabolites	1.2×10^4 CFU g^{-1}	Olmos et al. (2011)
B. circulans and *B. licheniformis*	Commercial product	Improved growth and health of cultured fish	1×10^6 CFU mL^{-1}	Sahandi et al. (2012)
B. pumilus, *B. licheniformis*, and *B. subtilis*	Marine water and soil samples	Reduced total ammonia nitrogen (TAN) and improved growth and survival in shrimp postlarvae	–	Devaraja et al. (2013)
B. subtilis strain LR1	Gut of rohu (*L. rohita*) fingerling	pH, dissolved oxygen, NH_3, and H_2S were in higher quality	–	Banerjee and Ray (2017)
B. subtilis, *B. pumilus* or *B. licheniformis*	–	Significantly decreased ammonia concentration	1×10^4 CFU mL^{-1}	Cha et al. (2013)
Bacillus spp.	–	Converters of organic matter to CO_2	–	Moriarty (1997)
Pseudomonas sp., *Nitrobacter* sp., *Nitrosomonas* sp., *Bacillus* sp., and *Aerobacter* sp.	Commercial product	Decreased accumulation of total organic matter, TAN, and nitrite concentration	1×10^{10} CFU mL^{-1}	Anik et al. (2011)
Bacillus sp.	Commercial product	Reduced total organic carbon	1×10^9 CFU mL^{-1}	Dalmin et al. (2001)

(continued)

Table 16.4 (continued)

Probiotics	Origin	Impacts on water quality and fish/shrimp	Doses	References
B. pumilus, *B. licheniformis*, and *B. subtilis*	Isolated from marine water and soil samples	Reduced TAN	–	Devaraja et al. (2013)
Eubacterium licheniformis and *B. subtilis*	Commercial product	Increased dissolved oxygen level	–	El-Haroun et al. (2006)
Bacillus sp.	–	Bioremediation of organic wastes	–	Thomas et al. (1992)
Bacillus, *Eubacterium*, *Pseudomonas*	–	Decomposition of undesirable organic substances	–	Xiang-Hong et al. (2003)

was used to minimize the pathogenicity of *A. hydrophila* in decorative fish culture (Lalloo et al. 2007), but in catfish fish *I. punctatus*, no significant distinction was found between the treated and management cluster of fishes for the examined water quality variables (Queiroz and Boyd 1998). This may due to the intrinsic capability of catfish, allowing the fish to withstand themselves even at the threshold limit.

Bacterial species under the genera *Bacillus*, *Eubacterium*, *Pseudomonas*, *Acinetobacter*, *Cellulomonas*, *Rhodopseudomonas*, and *Nitrosomonas* are reported to be potent bioremediators for organic wastes (Thomas et al. 1992). These probiotic bacteria regulate the microflora of cultivation water and control infective microorganisms to enhance the decomposition of undesirable organic substances within the water and sediment due to the improved ecological atmosphere of cultivation (Xiang-Hong et al. 2003; Rao 2007). Devaraja et al. (2013) isolated indigenous *E. pumilus*, *B. licheniformis*, and *B. subtilis* from marine water and soil samples and investigated these bacteria for their bioremediation ability in *P. monodon* culture, and recommended *Bacillus* spp. as potential candidates for bioremediation for *P. monodon* culture systems (Devaraja et al. 2013). In situ bioremediation has also been widely applied in cultivation through bioaugmentation mistreatment endemic or exogenous probiotics, which amend water quality (Wang et al. 2005).

16.8.8 Bacillus-*Based Commercial Probiotics*

To flourish the growth of probiotics, the environment in which the probiotic will inhabit such as the gut of fish or the water body where the fish live in must be congenial for that specific probiotic. *Bacillus* isolated from one environmental condition may not respond equally in another environment because the growth of *Bacillus* not only depends on the environmental parameters but also varies in accordance to

Table 16.5 List of some commercially available *Bacillus*-based probiotics

Trade Name	Country	Pack size	Composition/information
Promax-Aqua	India	200 g	*Bacillus subtilis*, *B. licheniformis*, *B. megaterium*, *B. polymyxa*, *B. firmus*, *B. mesentericus*, *Aspergillus oryzae*, *A. niger*, *Nitrosomonas*, *Nitrobacter*, *Cellulomonas*, *Pediococcus*, *Thiobacillus denitrificans*, *Paracoccus denitrificans*, herbals, and needful stabilizers.
Sanolife pro-F	Thailand	500 g	Mixture of *Bacillus* strain
PRO-2	Thailand	500 g	Mixture of *Bacillus* strain
Pro-W	Thailand	500 g	Mixture of *Bacillus* strain
Eco Toxnil	India	1 kg	*Bacillus subtilis*, *B. natto*, *B. mesentericus*
Ariake 3	Indonesia	1 kg	*B. amyloliquefaciens* strain D203
			B. licheniformis strain D3270
			B. pumilus strain D1729
pH fixer	India	5 kg	Concentrated strain of beneficial *Bacillus* bacteria

its association with other microbes. For this reason probiotics may not be equally effective in all cases. A list of commercially available probiotic *Bacillus* is given in Table 16.5.

16.9 Concluding Remarks

The Gram-positive and spore-forming probiotic *Bacillus* spp. have profound beneficial effects on aquaculture industry and are thus considered an important tool for promoting sustainable aquaculture. They exert their beneficial effects through multiple mechanisms including promotion of growth of host fishes and shrimps, suppress pest and disease infestation by secreting antimicrobial compounds and induction of immunity in the hosts, promote nutrition, and improve aquatic ecosystems for growth and reproduction of fishes and shrimps. Isolation, identification, and selection of the right strain of *Bacillus* and its proper application to the aquaculture largely regulate the beneficial effects in fish farming. Current genomics and post-genomics approaches provide opportunity to precisely understand the underlying molecular mechanisms of the beneficial effects of the probiotic bacilli and their genetic engineering for the development of highly efficient strains for boosting aquaculture in a sustainable manner. Larger application of probiotic bacilli instead of the hazardous synthetic chemicals would promote eco-friendly low-input sustainable aquaculture for food and nutritional security of the increasing world population.

References

Aguirre-Guzmán G, Lara-Flores M, Sánchez-Martínez JG, Campa-Córdova AI, Luna-González A (2012) The use of probiotics in aquatic organisms: a review. Afr J Microbiol Res 6(23):4845–4857

Ai Q, Xu H, Mai K, Xu W, Wang J (2011) Effects of dietary supplementation of *Bacillus subtilis* and fructo-oligosaccharide on growth performance, survival, non-specific immune response and disease resistance of juvenile large yellow croaker, *Larimichthys crocea*. Aquaculture 317:155–161

Akhter N, Wu B, Memon AM, Mohsin M (2015) Probiotics and prebiotics associated with aquaculture: a review. Fish Shellfish Immunol 45(2):733–741

Al-Dohail MA, Hashim R, Aliyu-Paiko M (2009) Effects of the probiotic, *Lactobacillus acidophilus*, on the growth performance, hematology parameters and immunoglobulin concentration in African Catfish (*Clarias gariepinus*, Burchell 1822) fingerling. Aquac Res 40(14):1642–1652

Al-Faragi JK, Alsaphar SA (2012) Isolation and identification of *Bacillus subtilis* as probiotic from intestinal microflora of common carp *Cyprinus carpio* L. Iraqi J Vet Med 36(2):355–361

Ali A (2000) Probiotics in fish farming. Evaluation of a bacterial mixture. Rapport-Sveriges Lantbruksuniversitet, Vattenbruksinstitutionen (Sweden)

Aly SM, Ahmed YAG, Ghareeb AAA, Mohamed MF (2008) Studies on Bacillus subtilis and Lactobacillus acidophilus, as potential probiotics, on the immune response and resistance of *Tilapia nilotica* (*Oreochromis niloticus*) to challenge infections. Fish Shellfish Immunol 25(1–2):128–136

Ambas I, Suriawan A, Fotedar R (2013) Immunological responses of customised probiotics-fed marron, *Cherax tenuimanus*,(Smith 1912) when challenged with *Vibrio mimicus*. Fish Shellfish Immunol 35(2):262–270

Ambas I, Buller N, Fotedar R (2015) Isolation and screening of probiotic candidates from marron, *Cherax cainii* (Austin, 2002) gastrointestinal tract (GIT) and commercial probiotic products for the use in marron culture. J Fish Dis 38(5):467–476

Anik MH, Sudirdjo A, Soemarno M (2011) The effect of consortia bacteria on accumulation rate of organic matter in Tiger shrimp, *Penaeus monodon* culture. J Biodivers Environ Sci 1:592–596

Apún-Molina JP, Santamaría-Miranda A, Luna-González A, Martínez-Díaz SF, Rojas-Contreras M (2009) Effect of potential probiotic bacteria on growth and survival of tilapia, *Oreochromis niloticus* L., cultured in the laboratory under high density and suboptimum temperature. Aquac Res 40(8):887–894

Arığ N, Süzer C, Gökvardar A, Başaran F, Çoban D, Yıldırım Ş et al (2013) Effects of probiotic (*Bacillus* sp.) supplementation during larval development of Gilthead Sea bream (*Sparus aurata*, L.). Turk J Fish Aquat Sci 13(3):407–414

Austin B, Stuckey LF, Robertson PAW, Effendi I, Griffith DRW (1995) A probiotic strain of *Vibrio alginolyticus* effective in reducing diseases caused by *Aeromonas salmonicida*, *Vibrio anguillarum* and *Vibrio ordalii*. J Fish Dis 18(1):93–96

Bacher A, Eberhardt S, Fischer M, Kis K, Richter G (2000) Biosynthesis of vitamin B2 (riboflavin). Annu Rev Nutr 20(1):153–167

Bagheri T, Hedayati SA, Yavari V, Alizade M, Farzanfar A (2008) Growth, survival and gut microbial load of rainbow trout (*Oncorhynchus mykiss*) fry given diet supplemented with probiotic during the two months of first feeding. Turk J Fish Aquat Sci 8(1):43–48

Bairagi A, Ghosh KS, Sen SK, Ray AK (2002) Enzyme producing bacterial flora isolated from fish digestive tracts. Aquac Int 10:109–121

Balcázar JL, Rojas-Luna T (2007) Inhibitory activity of probiotic *Bacillus subtilis* UTM 126 against *Vibrio* species confers protection against vibrios is in juvenile shrimp (*Litopenaeus vannamei*). Curr Microbiol 55(5):409–412

Balcázar JL, De Blas I, Ruiz-Zarzuela I, Cunningham D, Vendrell D, Muzquiz JL (2006) The role of probiotics in aquaculture. Vet Microbiol 114(3–4):173–186

Bandyopadhyay P, Mohapatra PKD (2009) Effect of a probiotic bacterium *Bacillus circulans* PB7 in the formulated diets: on growth, nutritional quality and immunity of *Catla catla* (Ham.). Fish Physiol Biochem 35(3):467–478

Banerjee G, Ray AK (2017) The advancement of probiotics research and its application in fish farming industries. Res Vet Sci 115:66–77

Banerjee G, Nandi A, Ray AK (2017) Assessment of hemolytic activity, enzyme production and bacteriocin characterization of *Bacillus subtilis* LR1 isolated from the gastrointestinal tract of fish. Arch Microbiol 199(1):115–124

Bhatnagar A, Lamba R (2017) Molecular characterization and dosage application of autochthonous potential probiotic bacteria in *Cirrhinus mrigala*. J Fish Sci com 11(2):46

Brauner CJ (2008) Acid-base balance. In: Fish larval physiology. Science Publisher, Enfield, pp 185–198

Cabello FC (2006) Heavy use of prophylactic antibiotics in aquaculture: a growing problem for human and animal health and for the environment. Environ Microbiol 8(7):1137–1144

Cha JH, Rahimnejad S, Yang SY, Kim KW, Lee KJ (2013) Evaluations of *Bacillus* spp. as dietary additives on growth performance, innate immunity and disease resistance of olive flounder (*Paralichthys olivaceus*) against *Streptococcus iniae* and as water additives. Aquaculture 402:50–57

Chabrillón M, Rico RM, Balebona MC, Moriñigo MA (2005) Adhesion to sole, *Solea senegalensis* Kaup, mucus of microorganisms isolated from farmed fish, and their interaction with *Photobacterium damselae* subsp. *piscicida*. J Fish Dis 28(4):229–237

Chen Y, Li J, Xiao P, Zhu W, Mo Z (2016) The ability of marine *Bacillus* spp. isolated from fish gastrointestinal tract and culture pond sediment to inhibit growth of aquatic pathogenic bacteria. Iran J Fish Sci 15(2):701–714

Cheng W, Hsiao IS, Chen JC (2004) Effect of ammonia on the immune response of Taiwan abalone *Haliotis diversicolor supertexta* and its susceptibility to *Vibrio parahaemolyticus*. Fish Shellfish Immunol 17(3):193–202

Dahiya T, Gahlawat SK, Sihag RC (2012) Elimination of pathogenic bacterium (Micrococcus sp.) by the use of probiotics. Turkish J Fish Aquat Sci 12(1):185–187

Dalmin G, Kathiresan K, Purushothaman A (2001) Effect of probiotics on bacterial population and health status of shrimp in culture pond ecosystem. Indian J Exp Biol 39:939–942

Das A, Nakhro K, Chowdhury S, Kamilya D (2013) Effects of potential probiotic *Bacillus amyloliquefaciens* FPTB16 on systemic and cutaneous mucosal immune responses and disease resistance of catla (*Catla catla*). Fish Shellfish Immunol 35(5):1547–1553

Das S, Mondal K, Haque S (2017) A review on application of probiotic, prebiotic and synbiotic for sustainable development of aquaculture. J Entomol Zool Stud 5(2):422–429

De BC, Meena DK, Behera BK, Das P, Mohapatra PD, Sharma AP (2014) Probiotics in fish and shellfish culture: immunomodulatory and ecophysiological responses. Fish Physiol Biochem 40(3):921–971

De D, Ananda Raja R, Ghoshal TK, Mukherjee S, Vijayan KK (2018) Evaluation of growth, feed utilization efficiency and immune parameters in tiger shrimp (*Penaeus monodon*) fed diets supplemented with or diet fermented with gut bacterium *Bacillus* sp. DDKRC1 isolated from gut of Asian seabass (*Lates calcarifer*). Aquac Res 49(6):2147–2155

Devaraja T, Banerjee S, Yusoff F, Shariff M, Khatoon H (2013) A holistic approach for selection of *Bacillus* spp. as a bioremediator for shrimp postlarvae culture. Turk J Biol 37(1):92–100

Díaz-Rosales P, Arijo S, Chabrillón M, Alarcón FJ, Tapia-Paniagua ST, Martínez-Manzanares E et al (2009) Effects of two closely related probiotics on respiratory burst activity of Senegalese sole (*Solea senegalensis*, Kaup) phagocytes, and protection against *Photobacterium damselae* subsp. *piscicida*. Aquaculture 293(1–2):16–21

El-Ezabi MM, El-Serafy SS, Essa MA, Lall S, Daboor SM, Esmael NA (2011) The viability of probiotics as a factor influencing the immune response in the Nile tilapia, *Oreochromis niloticus*. Egypt J Aquat Biol Fish 15(1):105–124

El-Haroun ER, Goda AS, Kabir Chowdhury MA (2006) Effect of dietary probiotic Biogen® supplementation as a growth promoter on growth performance and feed utilization of Nile tilapia, *Oreochromis niloticus* (L.). Aquac Res 37(14):1473–1480

Elsabagh M, Mohamed R, Moustafa EM, Hamza A, Farrag F, Decamp O et al (2018) Assessing the impact of *Bacillus* strains mixture probiotic on water quality, growth performance, blood profile and intestinal morphology of Nile tilapia, *Oreochromis niloticus*. Aquac Nutr 24(6):1613–1622

Fan Y, Liu L, Zhao L, Wang X, Wang D, Huang C et al (2018) Influence of *Bacillus* subtilis ANSB060 on growth, digestive enzyme and aflatoxin residue in Yellow River carp fed diets contaminated with aflatoxin B1. Food Chem Toxicol 113:108–114

FAO (2001) Health and nutritional properties of probiotics in food including powder milk with live lactic acid bacteria, Rome

FAO (2016) The state of the world fisheries and aquaculture. Food and Agriculture Organization. Rome, Italy: United Nations. 2016. p. 77. ISBN 978-92-5-109185-2. Retrieved 2016-10-30

FAO (2018) The state of world fisheries and aquaculture 2018 – meeting the sustainable development goals. Rome. 227 pp

Faramarzi M, Kiaalvandi S, Iranshahi F (2011) The effect of probiotics on growth performance and body composition of common carp (*Cyprinus carpio*). J Anim Vet Adv 10(18):2408–2413

Farzanfar A (2006) The use of probiotics in shrimp aquaculture. FEMS Immunol Med Microbiol 48(2):149–158

Fuller R (1989) Probiotic in man and animals. J Appl Bacteriol 66:131–139

Gatesoupe FJ (1999) The use of probiotic in aquaculture. Aquaculture 180:147–165

Gbassi GK, Vandamme T (2012) Probiotic encapsulation technology: from microencapsulation to release into the gut. Pharmaceutics 4(1):149–163

Geng X, Dong XH, Tan BP, Yang QH, Chi SY, Liu HY, Liu XQ (2011) Effects of dietary chitosan and *Bacillus subtilis* on the growth performance, non-specific immunity and disease resistance of cobia, *Rachycentron canadum*. Fish Shellfish Immunol 31(3):400–406

Ghosh K, Sen SK, Ray AK (2003) Supplementation of an isolated fish gut bacterium, *Bacillus circulans*, in formulated diets for rohu, *Labeo rohita* fingerlings. Isr J Aquacult Bamidgeh 55(1):13–21

Guo JJ, Liu KF, Cheng SH, Chang CI, Lay JJ, Hsu YO et al (2009) Selection of probiotic bacteria for use in shrimp larviculture. Aquac Res 40(5):609–618

Havenaar R, Huis JH (1992) Probiotics: a general view. In: The lactic acid bacteria volume 1. Springer, Boston, pp 151–170

Henze M, Harremoes P, la Cour Jansen J, Arvin E (2001) Wastewater treatment: biological and chemical processes. Springer Science & Business Media, Berlin

Irianto A, Austin B (2002a) Probiotics in aquaculture. J Fish Dis 25(11):633–642

Irianto A, Austin B (2002b) Use of probiotics to control furunculosis in rainbow trout, *Oncorhynchus mykiss* (Walbaum). J Fish Dis 25(6):333–342

Jakšić S, Uhitil S, Petrak T, Bažulić D, Karolyi LG (2002) Occurrence of Vibrio spp. in sea fish, shrimps and bivalve molluscs harvested from Adriatic sea. Food Control 13(8):491–493

Jamali H, Imani A, Abdollahi D, Roozbehfar R, Isari A (2015) Use of probiotic *Bacillus* spp. in rotifer (*Brachionus plicatilis*) and Artemia (*Artemia urmiana*) enrichment: effects on growth and survival of Pacific white shrimp, *Litopenaeus vannamei*, Larvae. Prob Antimicrob Prot 7(2):118–125

John G, Mohamed RR, Kolanchinathan P, Balasundaram A (2018) Nutritional value of two bacterial strains *Bacillus subtilis* RCMB21 and *Pseudomonas fluorescens* RCMB39 as feed supplement for freshwater prawn *Macrobrachium malcolmsonii*. Asian J Anim Vet Adv 13:282–294

Karim M, Zhao W, Rowley D, Nelson D, Gomez-Chiarri M (2013) Probiotic strains for shellfish aquaculture: protection of eastern oyster, *Crassostrea virginica*, larvae and juveniles against bacterial challenge. J Shellfish Res 32(2):401–408

Korkea-Aho TL, Papadopoulou A, Heikkinen J, Von Wright A, Adams A, Austin B, Thompson KD (2012) *Pseudomonas* M162 confers protection against rainbow trout fry syndrome by stimulating immunity. J Appl Microbiol 113(1):24–35

Korzybski T, Kowszyk-Gindifer Z, Kurylowicz W (1978) Antibiotics isolated from the genus *Bacillus* (Bacillaceae). Antibiot-Ori Nat Prop 3:1529–1661

Kühlwein H, Emery MJ, Rawling MD, Harper GM, Merrifield DL, Davies SJ (2013) Effects of a dietary β-(1, 3)(1, 6)-D-glucan supplementation on intestinal microbial communities and intestinal ultrastructure of mirror carp (*Cyprinus carpio* L.). J Appl Microbiol 115(5):1091–1106

Kumar R, Mukherjee SC, Ranjan R, Nayak SK (2008) Enhanced innate immune parameters in *Labeo rohita* (Ham.) following oral administration of *Bacillus subtilis*. Fish Shellfish Immunol 24(2):168–172

Lalloo R, Ramchuran S, Ramduth D, Görgens J, Gardiner N (2007) Isolation and selection of *Bacillus* spp. as potential biological agents for enhancement of water quality in culture of ornamental fish. J Appl Microbiol 103(5):1471–1479

Lee S, Katya K, Hamidoghli A, Hong J, Kim DJ, Bai SC (2018) Synergistic effects of dietary supplementation of *Bacillus subtilis* WB60 and mannanoligosaccharide (MOS) on growth performance, immunity and disease resistance in Japanese eel, Anguilla japonica. Fish Shellfish Immunol 83:283–291

Li J, Tan B, Mai K (2009) Dietary probiotic *Bacillus* OJ and isomaltooligosaccharides influence the intestine microbial populations, immune responses and resistance to white spot syndrome virus in shrimp (*Litopenaeus vannamei)*. Aquaculture 291(1–2):35–40

Lilly DM, Stillwell RH (1965) Probiotics: growth-promoting factors produced by microorganisms. Science 147(3659):747–748

Lin YS, Saputra F, Chen YC, Hu SY (2019) Dietary administration of *Bacillus amyloliquefaciens* R8 reduces hepatic oxidative stress and enhances nutrient metabolism and immunity against *Aeromonas hydrophila* and *Streptococcus agalactiae* in zebrafish (*Danio rerio*). Fish Shellfish Immunol 86:410–419

Liu CH, Chiu CS, Ho PL, Wang SW (2009) Improvement in the growth performance of white shrimp, *Litopenaeus vannamei*, by a protease-producing probiotic, *Bacillus subtilis* E20, from natto. J Appl Microbiol 107(3):1031–1041

Liu CH, Chiu CH, Wang SW, Cheng W (2012) Dietary administration of the probiotic, *Bacillus subtilis* E20, enhances the growth, innate immune responses, and disease resistance of the grouper, *Epinephelus coioides*. Fish Shellfish Immunol 33(4):699–706

Livingstone DR (2003) Oxidative stress in aquatic organisms in relation to pollution and aquaculture. Revue de Medecine Veterinaire 154(6):427–430

Mahdhi A, Hmila Z, Chaieb K, Kamoun F, Bakhrouf A (2011) Probiotic properties of halophilic Bacillus strains enhance protection of Artemia culture against pathogenic vibrio. Aquat Biol 13(3):225–231

Maillard VM, Boardman GD, Nyland JE, Kuhn DD (2005) Water quality and sludge characterization at raceway-system trout farms. Aquac Eng 33(4):271–284

Majumdar T, Ghosh S, Pal J, Mazumder S (2006) Possible role of a plasmid in the pathogenesis of a fish disease caused by *Aeromonas hydrophila*. Aquaculture 256(1–4):95–104

Matias HB, Yusoff FM, Shariff M, Azhar O (2002) Effects of commercial microbial products on water quality in tropical shrimp culture ponds. Asian Fish Sci 15(3):239–248

Meidong R, Doolgindachbaporn S, Jamjan W, Sakai K, Tashiro Y, Okugawa Y, Tongpim S (2017) A novel probiotic *Bacillus siamensis* B44v isolated from Thai pickled vegetables (Phak-dong) for potential use as a feed supplement in aquaculture. J Gen Appl Microbiol 63(4):246–253

Meidong R, Khotchanalekha K, Doolgindachbaporn S, Nagasawa T, Nakao M, Sakai K, Tongpim S (2018) Evaluation of probiotic *Bacillus aerius* B81e isolated from healthy hybrid catfish on growth, disease resistance and innate immunity of Pla-mong *Pangasius bocourti*. Fish Shellfish Immunol 73:1–10

Merrifield DL, Harper GM, Dimitroglou A, Ringø E, Davies SJ (2010) Possible influence of probiotic adhesion to intestinal mucosa on the activity and morphology of rainbow trout (*Oncorhynchus mykiss*) enterocytes. Aquac Res 41(8):1268–1272

Mitra A, Mukhopadhyay PK, Homechaudhuri S (2018) Probiotic effect of *Bacillus licheniformis* fb11 on the digestive efficiency and growth performance in juvenile *Chitala chitala* (Hamilton, 1822). Proc Zool Soc 71(4):403–414. Springer India

Mohapatra S, Chakraborty T, Prusty AK, Das P, Paniprasad K, Mohanta KN (2012) Use of different microbial probiotics in the diet of rohu, *Labeo rohita* fingerlings: effects on growth, nutrient digestibility and retention, digestive enzyme activities and intestinal microflora. Aquac Nutr 18(1):1–11

Mondol MAM, Shin HJ, Islam MT (2013) Diversity of secondary metabolites from marine *Bacillus* species: chemistry, biosynthesis and biological activity. Mar Drugs 11:2846–2872

Moriarty DJW (1990) Interactions of microorganisms and aquatic animals, particularly the nutritional role of the gut flora. In: Lesel R (ed) Microbiology of poecilotherms. Elsevier, Amsterdam/New York, pp 217–223

Moriarty DJW (1997) The role of microorganisms in aquaculture ponds. Aquaculture 151(1–4):333–349

Moriarty DJW (1998) Control of luminous *Vibrio* sp. in penaeid aquaculture ponds. Aquaculture 164:351–358

NavinChandran M, Iyapparaj P, Moovendhan S, Ramasubburayan R, Prakash S, Immanuel G, Palavesam A (2014) Influence of probiotic bacterium *Bacillus cereus* isolated from the gut of wild shrimp *Penaeus monodon* in turn as a potent growth promoter and immune enhancer in P. *monodon*. Fish Shellfish Immunol 36(1):38–45

Nayak SK, Swain P, Mukherjee SC (2007) Effect of dietary supplementation of probiotic and vitamin C on the immune response of Indian major carp, *Labeo rohita* (Ham.). Fish Shellfish Immunol 23(4):892–896

Newaj-Fyzul A, Adesiyun AA, Mutani A, Ramsubhag A, Brunt J, Austin B (2007) Bacillus subtilis AB1 controls Aeromonas infection in rainbow trout (*Oncorhynchus mykiss*, Walbaum). J Appl Microbiol 103(5):1699–1706

Nikoskelainen S, Salminen S, Bylund G, Ouwehand AC (2001) Characterization of the properties of human-and dairy-derived probiotics for prevention of infectious diseases in fish. Appl Environ Microbiol 67(6):2430–2435

Nimrat S, Suksawat S, Boonthai T, Vuthiphandchai V (2012) Potential *Bacillus* probiotics enhance bacterial numbers, water quality and growth during early development of white shrimp (*Litopenaeus vannamei*). Vet Microbiol 159(3–4):443–450

Olmos J, Paniagua-Michel J (2014) *Bacillus subtilis* a potential probiotic bacterium to formulate functional feeds for aquaculture. Microb Biochem Technol 6(7):361–365

Olmos J, Ochoa L, Paniagua-Michel J, Contreras R (2011) Functional feed assessment on *Litopenaeus vannamei* using 100% fish meal replacement by soybean meal, high levels of complex carbohydrates and *Bacillus* probiotic strains. Mar Drugs 9(6):1119–1132

Paul SI (2018) Marine sponge and sponge associated bacteria inhibiting fish pathogen. *Master of Science Thesis*. Department of Biotechnology. Bangabandhu Sheikh Mujibur Rahman Agricultural University, Bangladesh, p 94

Pérez-Sánchez T, Ruiz-Zarzuela I, de Blas I, Balcázar JL (2014) Probiotics in aquaculture: a current assessment. Rev Aquac 6(3):133–146

Perkins JB, Pero J (2002) Vitamin biosynthesis. In *Bacillus subtilis* and its closest relatives *American Society of Microbiology*, 271–286

Purwandari AR, Chen HY (2013) Effects of probiotic *Bacillus subtilis* on intestinal microbial diversity and immunity of orange spotted grouper *Epinephelus coioides*. J Appl Biotechnol 1(1):25

Queiroz JF, Boyd CE (1998) Effects of a bacterial inoculum in channel catfish ponds. J World Aquacult Soc 29(1):67–73

Rafiee G, Saad CR (2005) Nutrient cycle and sludge production during different stages of red tilapia (*Oreochromis* sp.) growth in a recirculating aquaculture system. Aquaculture 244(1–4):109–118

Rahman MM (2018) Identification of marine and terrestrial soil microbes inhibiting fish and shrimp pathogen. *Master of Science Thesis*. Department of Biotechnology. Bangabandhu Sheikh Mujibur Rahman Agricultural University, Bangladesh, p 85

Rahman M, Rahman MM, Deb SC, Alam MS, Alam MJ, Islam MT (2017) Molecular identification of multiple antibiotic resistant fish pathogenic *Enterococcus faecalis* and their control by medicinal herbs. Sci Rep 7(1):3747

Raida MK, Larsen JL, Nielsen ME, Buchmann K (2003) Enhanced resistance of rainbow trout, *Oncorhynchus mykiss* (Walbaum), against Yersinia ruckeri challenge following oral administration of Bacillus subtilis and B. licheniformis (BioPlus2B). J Fish Dis 26(8):495–498

Ramirez RF, Dixon BA (2003) Enzyme production by obligate intestinal anaerobic bacteria isolated from oscars (*Astronotus ocellatus*), angelfish (*Pterophyllum scalare*) and southern flounder (*Paralichthys lethostigma*). Aquaculture 227(1–4):417–426

Ran C, Carrias A, Williams MA, Capps N, Dan BC, Newton JC et al (2012) Identification of *Bacillus* strains for biological control of catfish pathogens. PLoS One 7(9):e45793

Rao V (2007) Bioremediation to restore the health of aquaculture pond ecosystem. Adv Aquat Ecol 1:85–94

Ravi AV, Musthafa KS, Jegathammbal G, Kathiresan K, Pandian SK (2007) Screening and evaluation of probiotics as a biocontrol agent against pathogenic Vibrios in marine aquaculture. Lett Appl Microbiol 45(2):219–223

Reid G, Jass J, Sebulsky MT, McCormick JK (2003) Potential uses of probiotics in clinical practice. Clin Microbiol Rev 16(4):658–672

Rengpipat S, Rukpratanporn S, Piyatiratitivorakul S, Menasaveta P (2000) Immunity enhancement in black tiger shrimp (*Penaeus monodon*) by a probiont bacterium (*Bacillus* S11). Aquaculture 191(4):271–288

Ringø E, Myklebust R, Mayhew TM, Olsen RE (2007) Bacterial translocation and pathogenesis in the digestive tract of larvae and fry. Aquaculture 268(1–4):251–264

Robertson PAW, O'Dowd C, Burrells C, Williams P, Austin B (2000) Use of Carnobacterium sp. as a probiotic for Atlantic salmon (*Salmo salar* L.) and rainbow trout (*Oncorhynchus mykiss*, Walbaum). Aquaculture 185(3–4):235–243

Rodger HD (2016) Fish disease causing economic impact in Global Aquaculture. In: Fish vaccines. Springer, Basel, pp 1–34

Saenz de Rodriguez MA, Diaz-Rosales P, Chabrillon M, Smidt H, Arijo S, Leon-Rubio JM (2009) Effect of dietary administration of probiotics on growth and intestine functionally of juvenile Senegalese sole (*Solea senegalensis*, Kaup1858). Aquac Nutr 15:177–185

Sahandi J, Jafariyan H, Dehghan M, Adineh H, Shohreh P (2012) Direct inoculation of *Bacillus* to rearing fish tanks effect on growth performance of two carp species fed with Artemia sp. World Appl Sci J 20(5):687–690

Salinas I, Cuesta A, Esteban MÁ, Meseguer J (2005) Dietary administration of Lactobacillus delbrüeckii and Bacillus subtilis, single or combined, on gilthead seabream cellular innate immune responses. Fish Shellfish Immunol 19(1):67–77

Sanchez-Ortiz AC, Luna-Gonzalez A, Campa-Cordova AI, Escamilla-Montes R, Flores-Miranda MC, Mazon-Suastegui JM (2015) Isolation and characterization of potential probiotic bacteria from pustulose ark (*Anadara tuberculosa*) suitable for shrimp farming. Lat Am J Aquat Res 43(1):123–136. https://doi.org/10.3856/vol43-issue1-fulltex-11

Sayes C, Leyton Y, Riquelme C (2018) Probiotic bacteria as a healthy alternative for fish aquaculture. In Antibiotic Use in Animals. InTech. https://doi.org/10.5772/intechopen.71206

Sharifuzzaman SM, Austin B (2009) Influence of probiotic feeding duration on disease resistance and immune parameters in rainbow trout. Fish Shellfish Immunol 27(3):440–445

Sihag RC, Sharma P (2012) Probiotics: the new ecofriendly alternative measures of disease control for sustainable aquaculture. J Fish Aquat Sci 7(2):72–103

Sugita H, Takahashi J, Deguchi Y (1992) Production and consumption of biotin by the intestinal microflora of cultured freshwater fishes. Biosci Biotechnol Biochem 56(10):1678–1679

Sugita H, Shibuya K, Shimooka H, Deguchi Y (1996) Antibacterial abilities of intestinal bacteria in freshwater cultured fish. Aquaculture 145(1–4):195–203

Sugita H, Kawasaki J, Deguchi Y (1997) Production of amylase by the intestinal microflora in cultured freshwater fish. Lett Appl Microbiol 24(2):105–108

Sugita H, Hirose Y, Matsuo N, Deguchi Y (1998) Production of the antibacterial substance by *Bacillus* sp. strain NM 12, an intestinal bacterium of Japanese coastal fish. Aquaculture 165(3–4):269–280

Sun YZ, Yang HL, Ma RL, Lin WY (2010) Probiotic applications of two dominant gut *Bacillus* strains with antagonistic activity improved the growth performance and immune responses of grouper *Epinephelus coioides*. Fish Shellfish Immunol 29(5):803–809

Swain P, Nayak SK (2003) Comparative sensitivity of different serological tests for seromonitoring and surveillance of *Edwardsiella tarda* infection of Indian major carps. Fish Shellfish Immunol 15(4):333–340

Swann LD (1990) A basic overview of aquaculture: history, water quality, types of aquaculture, and production methods Illinois-Indiana Sea Grant Program Extension Bulletin AS-457 and IL-IN-SG-E-90-2. Purdue University West Lafayette, Indiana, p 10

Taylor NG, Verner-Jeffreys DW, Baker-Austin C (2011) Aquatic systems: maintaining, mixing and mobilising antimicrobial resistance. Trends Ecol Evol 26(6):278–284

Telli GS, Ranzani-Paiva MJT, de Carla Dias D, Sussel FR, Ishikawa CM, Tachibana L (2014) Dietary administration of *Bacillus subtilis* on hematology and non-specific immunity of Nile tilapia *Oreochromis niloticus* raised at different stocking densities. Fish Shellfish Immunol 39(2):305–311

Thomas GM, Ward CH, Raymond RL, Wilson JT, Loehr RC (1992) Bioremediation. In: Leperberg J (ed) Encyclopedia of microbiology. Academic Press, London, pp 369–385

Thompson FL, Abreu PC, Cavalli R (1999) The use of microorganisms as food source for *Penaeus paulensis* larvae. Aquaculture 174:139–153

Thurlow CM, Williams MA, Carrias A, Ran C, Newman M, Tweedie J et al (2019) *Bacillus velezensis* AP193 exerts probiotic effects in channel catfish (*Ictalurus punctatus*) and reduces aquaculture pond eutrophication. Aquaculture 503:347–356

Thy HTT, Tri NN, Quy OM, Fotedar R, Kannika K, Unajak S, Areechon N (2017) Effects of the dietary supplementation of mixed probiotic spores of *Bacillus amyloliquefaciens* 54A, and *Bacillus pumilus* 47B on growth, innate immunity and stress responses of striped catfish (*Pangasianodon hypophthalmus*). Fish Shellfish Immunol 60:391–399

Vaseeharan BARP, Ramasamy P (2003) Control of pathogenic Vibrio spp. by *Bacillus subtilis* BT23, a possible probiotic treatment for black tiger shrimp *Penaeus monodon*. Lett Appl Microbiol 36(2):83–87

Venkat HK, Sahu NP, Jain KK (2004) Effect of feeding Lacto*bacillus*-based probiotics on the gut microflora, growth and survival of postlarvae of *Macrobrachium rosenbergii* (de Man). Aquac Res 35(5):501–507

Verschuere L, Rombaut G, Sorgeloos P, Verstraete W (2000) Probiotic bacteria as biological control agents in aquaculture. Microbiol Mol Biol Rev 64(4):655–671

Vijayan KK, Singh IB, Jayaprakash NS, Alavandi SV, Pai SS, Preetha R et al (2006) A brackish water isolate of Pseudomonas PS-102, a potential antagonistic bacterium against pathogenic vibrios in penaeid and non-penaeid rearing systems. Aquaculture 251(2–4):192–200

Villamil L, Tafalla C, Figueras A, Novoa B (2002) Evaluation of immunomodulatory effects of lactic acid bacteria in turbot (*Scophthalmus maximus*). Clin Diagn Lab Immunol 9(6):1318–1323

Vine NG, Leukes WD, Kaiser H, Daya S, Baxter J, Hecht T (2004) Competition for attachment of aquaculture candidate probiotic and pathogenic bacteria on fish intestinal mucus. J Fish Dis 27(6):319–326

Wang YB (2007) Effect of probiotics on growth performance and digestive enzyme activity of the shrimp *Penaeus vannamei*. Aquaculture 269(1–4):259–264

Wang YB, Xu ZR, Xia MS (2005) The effectiveness of commercial probiotics in northern white shrimp *Penaeus vannamei* ponds. Fish Sci 71(5):1036–1041

Warren MJ, Raux E, Schubert HL, Escalante-Semerena JC (2002) The biosynthesis of adenosylcobalamin (vitamin B12). Nat Prod Rep 19(4):390–412

World Health Organization (2001) Antimicrobial resistance fact sheet 194. http://www.who.int/inf-fs/en/fact194.html

Xiang-Hong W, Jun L, Wei-shang J, Huai-shu X (2003) Application of probiotics in aquaculture. Ocean University of Qungo, Qingdao. *Online* pp 1–10

Xue M, Wen C, Liang H, Ding M, Wu Y, Li X (2016) In vivo evaluation of the effects of commercial *Bacillus* probiotics on survival and development of *Litopenaeus vannamei* larvae during the early hatchery period. Aquac Res 47(5):1661–1669

Yanbo W, Zirong X (2006) Effect of probiotics for common carp (*Cyprinus carpio*) based on growth performance and digestive enzyme activities. Anim Feed Sci Technol 127(3–4):283–292

Yi Y, Zhang Z, Zhao F, Liu H, Yu L, Zha J, Wang G (2018) Probiotic potential of *Bacillus velezensis* JW: antimicrobial activity against fish pathogenic bacteria and immune enhancement effects on *Carassius auratus*. Fish Shellfish Immunol 78:322–330

Zhao Y, Mai K, Xu W, Zhang W, Ai Q, Zhang Y et al (2011) Influence of dietary probiotic *Bacillus* TC22 and Prebiotic fructooligosaccharide on growth, immune responses and disease resistance against *Vibrio splendidus* infection in sea cucumber, *Apostichopus japonicus*. J Ocean Univ China 10(3):293–300

Zhang C-N, Li X-F, Xu W-N, Jiang G-Z, Lu K-L, Wang L-N, Liu W-B (2013) Combined effects of dietary fructooligosaccharide and Bacillus licheniformis on innate immunity, antioxidant capability and disease resistance of triangular bream (*Megalobrama terminalis*). Fish Shellfish Immunol 35(5):1380–1386

Zheng CC, Cai XY, Huang MM, Mkingule I, Sun C, Qian SC et al (2019) Effect of biological additives on Japanese eel (*Anguilla japonica*) growth performance, digestive enzymes activity and immunology. Fish Shellfish Immunol 84:704–710

Zhou X, Tian Z, Wang Y, Li W (2010) Effect of treatment with probiotics as water additives on tilapia (*Oreochromis niloticus*) growth performance and immune response. Fish Physiol Biochem 36(3):501–509

Zimmerman SB, Schwartz CD, Monaghan RL, Pleak BA, Weissberger B, Gilfillan EC, Mochales S, Hermande S, Currie SA, Tejera E, Stapley EO (1987) Difficidin and oxydifficidin: novel broad spectrum antibacterial antibiotics produced by *Bacillus subtilis*. J Antibiot 40(12):1677–1681

Zokaeifar H, Balcázar JL, Saad CR, Kamarudin MS, Sijam K, Arshad A, Nejat N (2012) Effects of *Bacillus subtilis* on the growth performance, digestive enzymes, immune gene expression and disease resistance of white shrimp, *Litopenaeus vannamei*. Fish Shellfish Immunol 33(4):683–689